Physical Science

AN INTRODUCTORY STUDY

Prentice-Hall Intermediate Science Series

SERIES EDITOR: **William A. Andrews**
Professor of Science Education
Faculty of Education, University of Toronto

Understanding Science 7
Understanding Science 8
Physical Science: An Introductory Study
Biological Science: An Introductory Study

Physical Science

AN INTRODUCTORY STUDY

William A. Andrews

T.J. Elgin Wolfe

John F. Eix

PRENTICE-HALL OF CANADA, LTD., SCARBOROUGH, ONTARIO

Canadian Cataloguing in Publication Data

Andrews, William A., 1930-
Physical science

Includes index.
ISBN 0-13-671446-3

1. Science. I. Wolfe, Thomas J. Elgin, 1938-
II. Eix, John F., 1941- III. Title.

Q161.2.A54 500.2 C77-001617-0

Prentice-Hall, Inc., Englewood Cliffs, New Jersey
Prentice-Hall International, Inc., London
Prentice-Hall of Australia, Pty., Ltd., Sydney
Prentice-Hall of India, Pvt., Ltd., New Delhi
Prentice-Hall of Japan, Inc., Tokyo
Prentice-Hall of Southeast Asia (PTE) Ltd., Singapore

ISBN 0-13-671446-3

The Metric Commission Canada has granted use of the National Symbol for Metric Conversion.

Art Direction: Julian Cleva
Design: Artplus Limited
Illustration: James Loates Illustrating
Composition: Webcom Limited
Printed and bound in Canada by Bryant Press

2 3 4 5 82 81 80 79 78

Contents

UNIT 3 Density: A Characteristic Physical Property

UNIT 4 The Three States of Matter: More Characteristic Physical Properties

UNIT 5 Using the Particle Theory

UNIT 6 Water, Solutions, and Mixtures

A. Water

B. Solutions

Acknowledgments

The authors wish to acknowledge the competent professional help received from the staff of Prentice-Hall of Canada, Ltd. in the production of this text. In particular, we extend our thanks to Marta Tomins, Veronica Orocio and Barb Steel for their skillful editorial work and to Steve Lane for his untiring and valuable assistance in the planning and development of this text. This book owes its final shape and form largely to the efforts of Julian Cleva and Rand Paterson.

We wish, further, to thank the many teachers who reviewed the manuscript and offered their constructive criticisms, and we would be remiss if we did not express our appreciation of the imaginative, attractive, and accurate art work of Jim Loates.

The writing of a text is a time-consuming undertaking that leaves little time for one's family. Thus we thank our families for their understanding and support during the many months that we were writing this text. We extend a special word of appreciation to Barbara Wolfe for her assistance in the typing of the manuscript and to Lois Andrews for her careful and dedicated preparation of the final manuscript.

W.A.A. / T.J.E.W. / J.F.E.

Preface

This is an introductory course in physical science. Its main purpose is to provide you with knowledge about the physical world. In this textbook you investigate the nature of solids, liquids, and gases; you learn about the properties of water and air; you perform interesting chemical experiments; you study force, work, and power; and you examine the nature of heat and light. You learn about these and other topics in two ways: through investigations and through narratives. Investigations are experiments that either you or your teacher carry out in the laboratory. We believe that you can best learn the content and methods of science by doing investigations. Narratives are reading sections that contain important information. With the knowledge gained from the narratives and investigations, you often are able to develop a theory to explain what you have observed. You then consider how this knowledge can be applied in your day-to-day experiences.

The second main purpose of this course is to acquaint you with the scientific method of inquiry. After you have done several investigations, you should be able to approach problems much as scientists do. Often in our daily lives we are faced with problems that can be solved only by using the scientific method.

The training and knowledge you acquire in this course should be useful to you for many years to come. They will be even more useful if you follow this course with others in science—biology, physics, or chemistry.

This book has much more in it than your teacher will expect you to study in one year. Your teacher will select those units and sections that will be most useful to you and your classmates.

You should enjoy being able to solve problems in a scientific way and to explain many of the interesting things that surround you. For example, before the end of the course, you will be able to answer questions such as these: Why is the sky blue? What is a flame made of? How does a refrigerator work? What is a molecule? How large is an atom? What is heat? More important, you may start asking questions about the world around you. If that happens, this course will have been successful.

W.A.A. / T.J.E.W. / J.F.E.

Physical Science

AN INTRODUCTORY STUDY

UNIT

1 How Scientists Investigate Problems

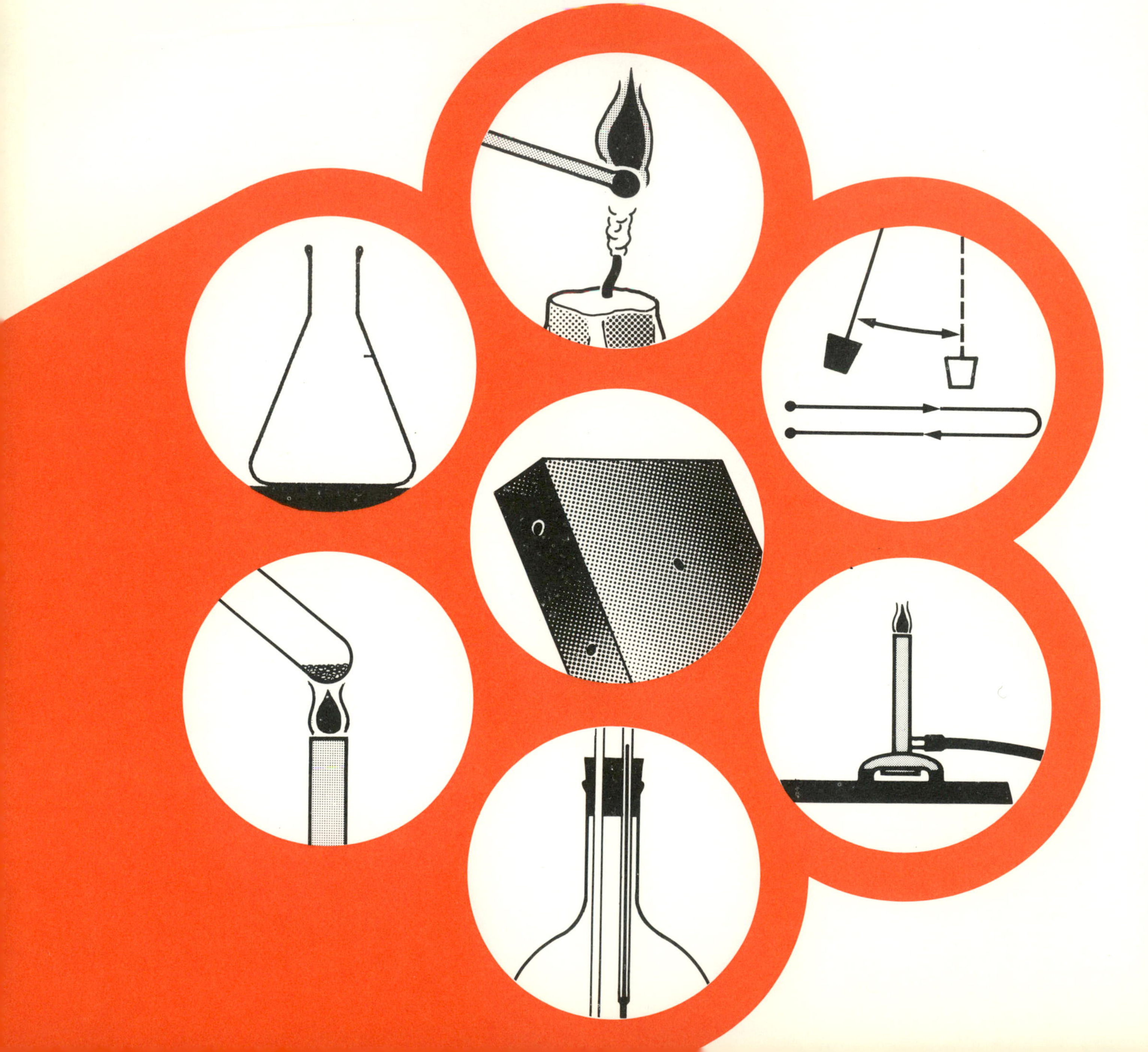

What is science? Is it a collection of facts that scientists have discovered over the ages? Is it a collection of theories that scientists have developed? Is it a method that is used for discovering the facts and for forming the theories? Is science a combination of any or all of the above? What does a scientist do? How does he attack and solve a problem?

As you study this unit and do the investigations that are in it, you will gradually become aware of the answers to these and related questions. In this chapter you explore the nature of scientific problems, facts, and theories. You discover how they interact. You investigate problems much as a scientist does. As a result, you will become familiar with the scientific method of investigation. It is the method that you will use throughout this course and, hopefully, throughout your entire life.

You will find many new terms in this unit. You need to understand them to get the most from this course. However, some of them may be difficult for you to understand when you meet them the first time. Don't worry if that happens. After you have done the investigations in this unit, most of the terms will make sense to you. By the end of this course, all of them should be a part of your vocabulary.

1.1 Why Study Science?

One of the greatest thinkers of all time was a scholar by the name of Aristotle. He lived in Greece over 2 300 years ago. His authority was so great that men throughout the world accepted without question anything that he said. Aristotle noticed that leaves, feathers, and other light objects fell more slowly than did stones and other heavy things. From these observations he concluded that the mass of an object determines how fast it falls. According to Aristotle's reasoning, two bricks that are fastened together should fall faster than a single brick (Fig. 1-1). Aristotle never tested this theory. In fact, Aristotle's conclusion seemed so reasonable that no one took the trouble to test it until about 350 years ago. At that time, so the story goes, an Italian scientist, Galileo Galilei, decided that he would not accept Aristotle's conclusion simply because people had believed it for a long time. To test it, he took several cannon balls of different masses to the top of the Leaning Tower of Pisa and dropped them at the same time. All of them fell at the same rate. Many scholars came to watch Galileo repeat his experiment. But they remained unconvinced because Aristotle's authority was so great.

Figure 1-1 Will two bricks fall faster than one brick? Try it!

Here is another example of the power of Aristotle's words. One of Aristotle's projects was the creation of a scale of perfection for all things. At the top of the scale he placed the human male. Next on the list was the human female. One of Aristotle's reasons for classifying women as being less perfect than men was his belief that they had fewer teeth. Strange as it may seem, Aristotle never bothered to count the teeth in men and women. He probably made his conclusion from the observation that the male horse usually has four more teeth than the female horse. Many years later, however, someone did count the number of teeth in many men and women. Of course, he found it to be the same. Nevertheless, Aristotle's authority was so great that, instead of concluding that Aristotle had erred, the investigator concluded that women had changed during the period of time since Aristotle's statement!

In many ways Aristotle was a good scientist. He made careful observations and then drew conclusions that greatly aided the advance of science. However, like many of the early philosophers, he tended to give explanations that were based on how he thought things ought to be rather than on observation of how they actually were.

How much has mankind changed since the days of Aristotle? Certainly scientists today realize that direct observation is the only basis upon which reliable conclusions can be made. In this scientific age, it is important that all of us be aware of this fact. Are we? Consider carefully the following example.

A television commercial that has been used for many years states with authority that a certain product "consumes 47 times its weight in excess stomach acid". The long life of this commercial indicates that it has been successful in selling the product. People believe it. Perhaps the statement is true; perhaps it is not true. In either case, an inquiring mind would ask some questions: What does "consume" really mean? What is "excess" stomach acid? How does the product tell the difference between "excess" stomach acid and "normal" stomach acid? Do all stomachs have the same normal acid concentration? How was such an exact number as 47 obtained? Consider the claim carefully and think of two or three more questions that you could ask the manufacturer. What one question could you ask that would make most of the other questions unnecessary?

Why do few people bother to challenge statements like the one in this commercial? The answer to this question provides us with the most important reason for studying science. Be sure that you discuss this matter thoroughly with your classmates and your teacher before you continue with this course.

Discussion

1. a) Will two bricks tied together fall faster than one brick?
b) Why do leaves and feathers fall more slowly than stones?

2. a) In what ways was Aristotle a good scientist?
b) In what ways was he a poor scientist?

3. Explain why few people question statements that are made in television commercials.

1.2 INVESTIGATION: The Study of a Burning Candle

Section 1.1 stressed the importance of careful observation and description. This study of a burning candle gives you a chance to test your powers of observation. It also tests your ability to describe accurately the observations that you make.

In order to make your observations in a scientific fashion you must do the experiment under **controlled conditions**. Often the conditions that must be controlled are very difficult to recognize. Nevertheless, the use of controlled conditions must be a part of every experiment. For example, if you do this experiment near an open window, near an open door, or too close to the mouth of an observer, the resulting draft may affect your results. Thus location is one **variable** that must be controlled. Some conditions that need not be controlled in this experiment are the time of day at which it is done and the floor of the school on which it is done.

As you carry out this investigation, keep in mind that a scientist does at least five things as he observes:

1. He makes a written record of each observation right after making the observation.
2. He records all observations, no matter how trivial some may seem to be.
3. He makes both **qualitative** and **quantitative** observations. (Look up the meanings of these words before you continue if you do not already know them.) The colour of the wax is a qualitative observation. The measured length of the candle is a quantitative one.
4. He tries to separate important from unimportant observations. This usually means that he repeats the experiment at least two or three times. (You can examine your neighbours' results instead.) For example, if you observed in one trial that the burning wick of the candle leaned toward the window, you should attach little importance to this observation. However, if this happens in several trials, the observation is important and an explanation should be sought.
5. He is careful not to confuse observations with conclusions. For example, if you state that "the wax melted when the candle was lit", you have made a conclusion. What you really observed was a colourless liquid being formed. This may be melted wax. On the other hand, it may be water or some other colourless liquid formed as a result of the burning. Do not draw conclusions without sufficient evidence.

Materials

candle
matches
wick
wax
carbon paper

Procedure

a. Establish the controlled conditions. For example, control the location by making all observations in the same place.
b. Divide a page in half lengthwise. Place the word "observation" at the top of the left half and the word "conclusion" at the top of the right half.
c. Carefully examine an unlit candle. Record your observations. Be sure to number them.
d. Light the candle and make as many observations as you can during a time interval of 20 min.
e. Mark the observations that you feel are important. Be prepared to defend your choices.
f. Record as many conclusions as you can in the right half of the page. Indicate by number the observations that led to each conclusion.
g. You may wish to continue your observations at home. Add any further discoveries to your list.

Discussion

1. A trained observer can make over 50 observations during this investigation. How close did you come to this number?
2. If you finish early, try to find out whether it is the wick, the wax, or something else that is burning. Hint: Blow out the flame and immediately hold a lit match about 1 cm above the wick (Fig. 1-2). Your teacher has pieces of wax and wick.

1.3 Recognizing a Problem: Curiosity

Most of us share with scientists a basic human characteristic called **curiosity**. We wonder about the how and why of the things that happen around us. Scientists differ from us only in that they have made curiosity their profession. They are always asking questions about the observations that they make. In fact, they usually produce more questions than they do answers. If this were not so, the wheels of science would grind to a halt because no new frontiers would be opened for study. Curiosity, then, is a necessary characteristic of the scientist. It is also a necessary characteristic of the student of science. But the kind of curiosity

Figure 1-2 What is it that burns?

that "killed the cat" will not put a man on Mars. Scientists learn to direct their curiosity. They put controls on their experiments; they think before they act; they try to develop explanations for their results. These careful methods are called **processes of science**. Together they form a pattern of thought and action that is called the **process of scientific inquiry** or the **scientific method**.

Observation, **description**, and **conclusion making** are processes of science that you have already met. **Recognition of a problem** as a result of curiosity is another. If you question the claim by a car manufacturer that free air conditioning is included in his cars, you have recognized a problem. If you question a cigarette advertisement that emphasizes the "clean fresh taste", you have recognized a problem. If you wonder why sewage promotes algal growth in lakes, you have recognized a problem. If you wonder why a candle flame is yellow, you have recognized a problem.

Recognition of a problem is a key step in the process of scientific inquiry. We all need training in this process. Otherwise many interesting things pass us by without getting a second glance. In most of the investigations in this book, you will be expected to use this process. Test your ability to recognize a problem by trying Investigations 1.4 and 1.5.

Discussion

1. **a)** Why is curiosity an important characteristic of a scientist?
   b) How does a scientist's curiosity differ from "ordinary" curiosity?
2. Give an example of a problem that you recognized in a television or newspaper advertisement.
3. Explain the difference between an observation and a conclusion. Use an example to illustrate your answer.

1.4 INVESTIGATION: Can You Recognize a Problem?

The purpose of this investigation is to see if you can recognize a problem. Maybe you can even solve it!

Before you start, turn back to Section 1.2. Read again what a scientist does as he observes. Be sure you do the same here.

Materials

Bunsen burner
clean dry test tube
cobalt chloride crystals
adjustable clamp

CAUTION: Wear safety goggles during this investigation.

Procedure

a. Place a few crystals of cobalt chloride in the bottom of a clean dry test tube. Describe the colour of the crystals carefully.
b. Heat the crystals gently with a low flame. Incline the test tube as shown in Figure 1-3. Heat only the bottom centimetre of the test tube. Record your observations.
c. Now heat the cobalt chloride and the whole test tube strongly. Continue until no more changes occur. Record your observations.

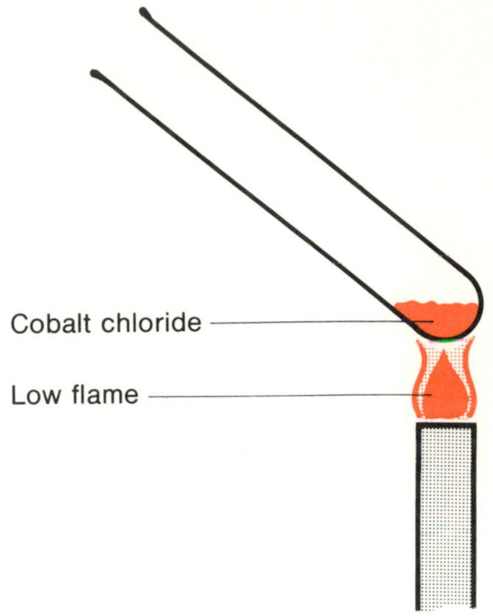

Figure 1-3 Studying the effects of heat on cobalt chloride.

Discussion

1. Can you recognize a problem? That is, did you see something that makes you curious? If so, describe it in your notebook.
2. What can you do with the material in the test tube to solve the problem? Try it.
3. Do you have in your home a humidity gauge that uses cobalt chloride? If so, describe it and explain how it works. Bring it to class, if possible, for others to see.

1.5 INVESTIGATION: Another Problem

The purpose of this investigation is the same as the one in Section 1.4. You are to see if you can recognize and, perhaps, partly solve a problem. This is a more difficult investigation. Proceed carefully.

Materials

Bunsen burner
clean dry test tube
ammonium dichromate crystals
glass wool
adjustable clamp
ruler
CAUTION. Wear safety goggles during this investigation.

Procedure

a. Place ammonium dichromate crystals to a depth of about 5 mm in the bottom of a clean dry test tube. **CAUTION:** Do not touch this substance with your hands!
b. Place a loose plug of glass wool in the open end of the test tube as shown in Figure 1-4.

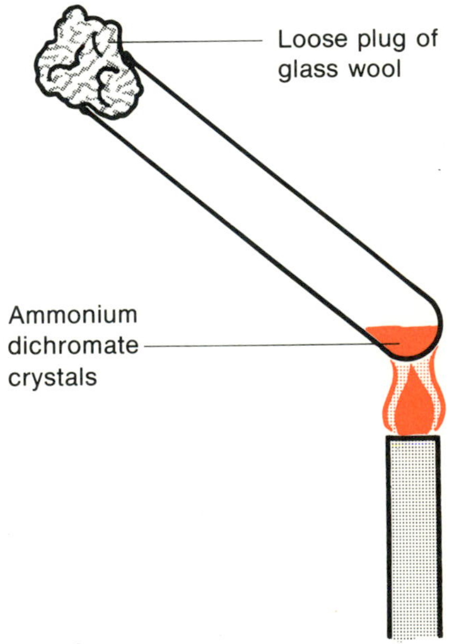

Figure 1-4 Studying the effects of heat on ammonium dichromate crystals.

c. Describe the crystals carefully.
d. Heat the crystals gently with a low flame until a change occurs. Record your observations. **CAUTION:** Do not touch the substance formed here!

Discussion

1. What do you see as an interesting problem? Describe it in your notebook.
2. What quantitative observations might help you solve the problem? If you wish to make such observations, ask your teacher for a new sample of ammonium dichromate.
3. You probably cannot solve this problem completely. Why?

1.6 Prediction: Another Process of Science

The following story is a fable. But it will help to increase your understanding of the process of scientific inquiry and therefore prepare you for the following investigations.

Iron, copper, silver, and oxygen are chemical elements. Years ago when such elements were just being discovered, four scientists were asked by the director of the laboratory in which they worked to do a study of the effects of heat on a new chemical element. Scientist A hurried off to his laboratory and put a piece of the element in a flame that he made by burning wood. The flame of burning wood is relatively cool. After observing his experiment for some time, he sent in his report to the director. It stated that "no changes occurred when the element was heated other than the fact that it gradually became hotter". Scientist B began the project by heating a piece of the element in a bed of burning charcoal. This source of heat has a higher temperature than burning wood. Her report stated that "the element gradually became hotter as it was heated and then turned red". Scientist C began by heating a piece of the element in a bed of burning charcoal through which she passed a stream of compressed air. This produced a temperature even higher than that used by Scientist B. Her report said that "the element gradually became hotter as it was heated, turned red, then yellow, and finally white. Then it melted".

The fourth member of the team, Scientist D, used quite a different approach. The other three scientists had submitted their reports. But he had not yet heated his sample of the element. What had he been doing? He had looked at the element and noticed that it had many of the properties of such metals as iron, nickel, and chromium. For example, it was silvery in appearance, it was very dense, it conducted electricity, and it could be hammered without chipping. He reasoned, therefore, that the new element might be a metal similar to iron, nickel and chromium. On this basis he **predicted**, or **hypothesized**, that the new element might also behave like these metals when it was heated. He knew that iron, nickel, and chromium had been thoroughly studied. So he went to the library and looked up the effects of heat on these and similar metals (Fig. 1-5). In all cases the metals became hotter when they were heated, turned red, yellow, white, and finally, they melted. Scientist D now felt that he was ready to go to the laboratory and heat the metal.

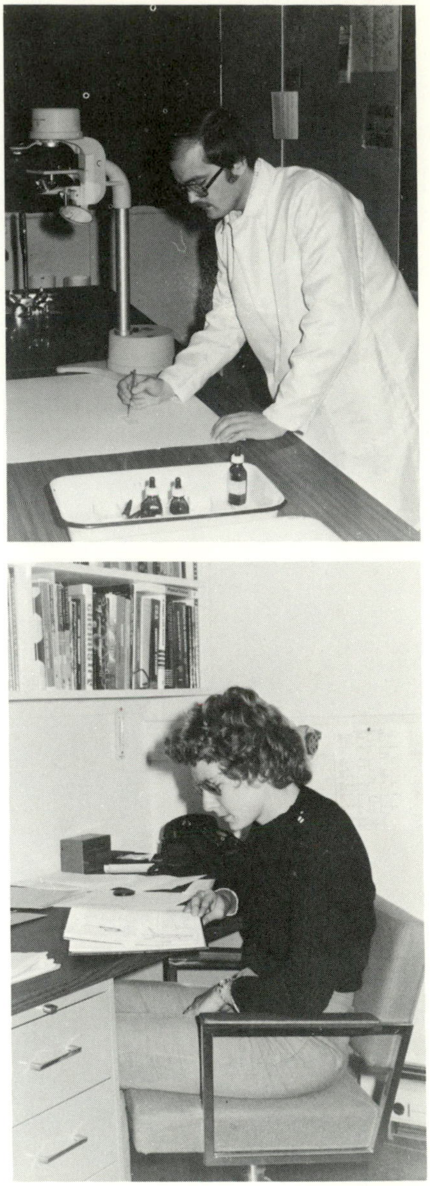

Figure 1-5 What does a scientist look like?

Discussion

1. a) Explain how Scientist D used the scientific processes of problem recognition and prediction.
b) With all of the information that Scientist D collected in the library, is there really any need to do an experiment? Why?

2. a) What is the difference between a prediction (hypothesis) and a guess?
b) What values are there in predicting before you do an experiment?
c) Often you cannot find enough information to enable you to predict with confidence the outcome of an experiment. Yet you may have a "hunch" as to the outcome. Should you ignore or follow up on a "hunch"? Defend your answer.

1.7 INVESTIGATION: The Study of a Simple Pendulum

In this investigation you use all of the scientific processes that you have met so far: problem recognition, prediction, experiment design, experimentation, observation, and description. Also, two new processes are used, the organization of data and the synthesis of data. **Organization** of data means the orderly arrangement of the data that you collect. Usually this is done in a table. If you arrange the data this way, you are more likely to spot regularities or patterns in the data. You then use the patterns to **synthesize** (put together) conclusions. Since this may be the first time you have done such a hard experiment, we will help you with the difficult parts.

Many people could look at a swinging pendulum all day and not recognize a problem. That is, they could not see anything worth investigating. How do you compare?

Set up a pendulum using the materials provided by your teacher. Start it swinging. Then ask yourself this question: What variables affect how fast the pendulum swings back and forth? If you can list some of these, you have recognized some problems. We will help you investigate one of these. You will investigate the others on your own in Section 1.8.

The problem that you are to investigate here is the effect that the displacement has on the period of the pendulum. **Displacement** is the distance that the bob is pulled to one side. The **period** is the time required for the bob to travel through one complete **cycle** (from its starting position to the other extreme and back again).

Points to Remember

1. Begin by stating the problem or the purpose of the experiment.
2. Predict the outcome of the experiment. Write your hypothesis in your notebook. Be sure that you are not merely guessing.
3. Design a method to test your hypothesis. Record it carefully. Remember that your method is only as good as your controls.
4. Do you have a suitable control? You can change only one variable at a time and still get useful results.
5. Prepare for the recording of the data in a systematic way. Data that are recorded in a table are more easily studied.
6. Do not make conclusions on the basis of only one or two trials.
7. Quantitative observations often tell you more than do qualitative observations.

Problem

To study the relationship between the displacement and the period of a simple pendulum.

Hypothesis

Record your prediction of the outcome of this experiment. What effect do you think displacement will have on the period?

Materials

string
rubber stopper (bob)
ring stand
clamp
ruler
watch or clock with second hand

Procedure

a. Set up a pendulum as shown in Figure 1-6. Make the length of the pendulum 70 cm. The length should be measured from where the string is tied to the clamp to the middle of the bob.

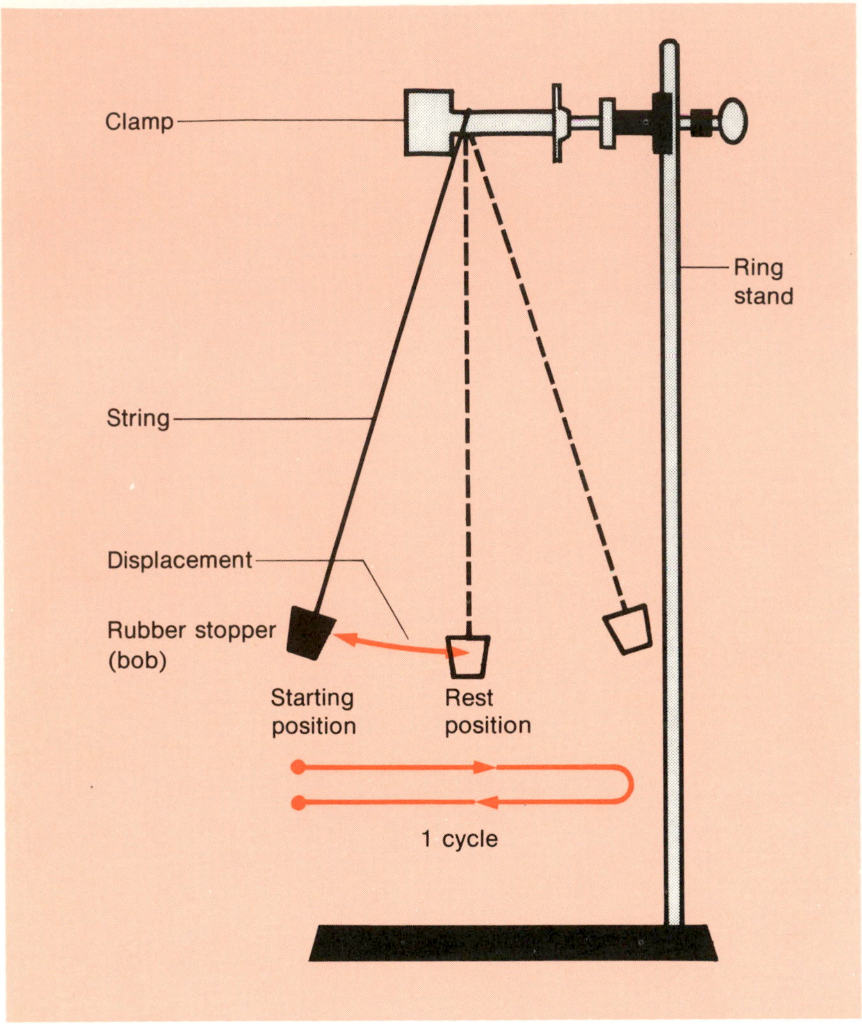

Figure 1-6 A simple pendulum.

b. Pull the bob to one side a distance of 5 cm. This is the displacement. Release the bob and determine the period (the time required to complete 1 cycle). Do this by timing the bob for 5 complete cycles and dividing the total time by 5. Record the time in seconds in Table 1. For example, if 7 s were required for 5 cycles, then the period is $\frac{7\text{ s}}{5} = 1.4\text{ s}$.

c. Repeat step b. 2 more times. Then average your 3 answers to get the average period for a displacement of 5 cm.

d. Repeat steps b. and c. using displacements of 10 cm, 20 cm, and 30 cm.

e. Record your results in a data table like Table 1.

Observations

TABLE 1 The Simple Pendulum

Displacement	Trial	Period (s)	Average period for 3 trials
	1		
5 cm	2		
	3		
	1		
10 cm	2		
	3		
	1		
20 cm	2		
	3		
	1		
30 cm	2		
	3		

Conclusion

Examine the data and make up a conclusion that summarizes your findings. Your conclusion should answer the problem. If your hypothesis was correct, it should be the same as the conclusion. Do not worry if your hypothesis was wrong. If scientists could always make a correct hypothesis, they would never have to do any experiments. Also, we often learn a great deal even when the hypothesis is wrong.

1.8 INVESTIGATION: The Study of a Simple Pendulum —Other Variables

We helped you in your first experiment with a simple pendulum. Now, let's see how well you can do on your own!

In Section 1.7 you studied the effect of one variable, displacement, on the period of the pendulum. List the other variables that you think might affect the period. State each of these as a problem. Then study each problem in a scientific way. Keep looking back to Section 1.7 for ideas on how to do your experiments.

Write a separate report on the effects of each variable. Use the headings: Problem, Hypothesis, Materials, Procedure, Observations, and Conclusions for each report. Follow the format used in Section 1.7.

1.9 Building a Model from Indirect Evidence

In Sections 1.7 and 1.8 you made conclusions based on **direct** observations. But scientists often obtain their information by **indirect** methods because direct observation may be impossible. For example, until Neil Armstrong walked on the moon and picked up samples of lunar material, most of our knowledge about the moon's surface had been obtained by indirect methods. Yet, in spite of the fact that indirect observations had been made, scientists had developed a fairly good model of the moon's surface long before samples were brought back to earth.

Many books have been written on the structure of the atom. Yet no one has observed an atom directly. What we read is just a **model** for the structure of the atom. This model, or mental picture, is a description of the atom as scientists think it must be in order to have the properties that they have indirectly observed.

If someone handed you a sealed metallic can that must not be opened and asked you what was in it, you would probably shake the can before you answered the question (Fig. 1-7). If you noticed a sloshing sound, you would conclude that the can contains a liquid. You might even say it contains water. You have formed a model, or a mental picture, of what is in the can in order to explain your observations. But you have not actually observed the liquid. Your evidence is indirect. You really do not know for sure what is in the can, do you? You do not know the colour, the taste, the density, or the odour of the liquid. In fact, your model is not very complete at all.

Figure 1-7 Can you find out what is in a sealed can without opening it?

How could you make your model more complete? If you follow the scientific method you would hypothesize as to what would happen if you put the can in a refrigerator at 0°C. You would then design and perform an experiment to test your hypothesis. The results would help you to perfect your model. If the liquid freezes at 0°C, you would be more sure that it is water. But you still cannot be certain. Other liquids may freeze at 0°C.

You could do further experiments like calculating the density of the liquid. Each experiment tells you more about the liquid and makes your model more perfect. But, without opening the can, you will never have a complete model. That is, you will not know for sure what is in the can.

Remember this, however. Whenever you must use indirect evidence to build a model, do all of the experiments you can. The more evidence you collect to support your model, the more confidence you can have in it.

Discussion

1. **a)** What is a variable in an experiment?
b) What variables did you consider in the two investigations you did with the simple pendulum?
c) You are shopping in a supermarket for a box of corn flakes. Assuming that you want the best value for your money, what 3 variables should you consider before making your selection from the shelves?

2. **a)** What is a control in an experiment?
b) Explain why it is necessary to use controls in an experiment. Refer to the simple pendulum to illustrate your answer.
c) Aristotle made an incorrect conclusion because the variables were not controlled while he was making observations on falling objects (see Section 1.1). Explain this statement.

3. Explain the difference between qualitative and quantitative observations. Which are usually more useful? Why?

4. **a)** What is the difference between direct and indirect observations?
b) What is a model?
c) Describe an example of a model.
d) How can you increase your confidence in a model?

1.10 INVESTIGATION: Building a Model from Indirect Evidence

You are to find out how many sides of a small wooden cube have a dot on them. You are not allowed to handle the cube directly. This means that you must form a model from indirect evidence.

Materials

wooden cube
marking pen
tall container with non-transparent sides

Procedure

a. Ask your laboratory partner (or a friend if you are trying this at home) to use a marking pen to place a single dot on one or more of the 6 surfaces of the cube. You are not to be told the number of dots.

b. Have your helper place the cube in a tall container that has non-transparent sides. A cardboard milk container or dark brown jar will do. You must be able to see only the top face of the cube when you look into the container.

c. Look down into the container at the cube. Note whether or not a spot is showing. Record your observation.

d. Cover the top of the container and shake it so that the cube tumbles. Look inside and record whether or not a dot is showing. Do this at least 50 times. If time permits, do it 100 times.

Discussion

1. Calculate how many sides have a dot on them using only the information gathered in this experiment. If you have trouble doing this, use the following procedure:
   a) Record the total number of faces that you saw during the experiment.
   b) Record the total number of dots that you saw.
   c) Use a and b to calculate the average number of dots per face.
   d) Calculate the average number of dots per cube (6 faces).
2. Since the cube has 6 sides and each side can have only 1 dot on it, one might expect your answer to part 1 to be 1, 2, 3, 4, 5, or 6. That is, your model states that it should be a whole number. Was it? Why?
3. Repeat part 1 using only the information from the first 5 trials. Is your answer the same? Why? Discuss the importance of the number of trials when an average is to be calculated. How could you change this experiment to get a better model?

1.11 INVESTIGATION: Building a Model for a Black Box

If the can referred to in Section 1.9 could never be opened, it would be called a **black box**. A television set is a black box for you if you have never seen inside one. However, you may know a great deal about the interior of a television set because of "experiments" that you have done on its exterior. For example, you know that turning knobs produces certain results. Therefore, you form a mental picture or model of the interior.

If you have never seen inside an automobile engine, it is a black box. But you know that you must provide it with gasoline; and you know that it gives off heat. Thus you form a mental image of gasoline burning within the engine, even though you have never seen the flames. You have developed a model for the automobile engine.

Here is another model for an automobile engine. Little creatures from a distant planet who eat gasoline were placed in the engine by the car builder. If you feed them well, they pedal rapidly inside the engine and make the car go. When they work hard, they get hot. As a result, the engine gets hot. They even sweat. You probably have seen the resulting water coming out of the exhaust pipes of cars. Can you prove that this is an incorrect model without taking the engine apart? How?

Scientists make predictions based on models and then do experiments to test the predictions. In this investigation you make up a model for the contents of the black box that is given to you by your teacher (Fig. 1-8). Then you predict what will happen if you do certain things to the box. Finally you design and do experiments to test your predictions.

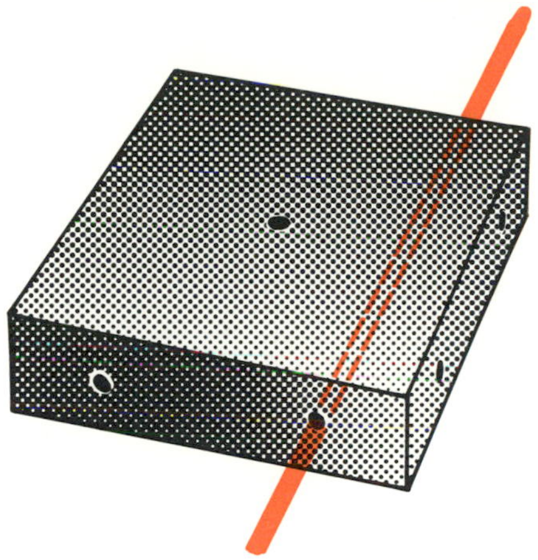

Figure 1-8 A black box.

Materials

black box with probe
identical but empty black box
balance

Procedure

a. Move the box in any manner that you wish, but do not open it. You must not enlarge the holes in the box. Carefully planned movements of the box will tell you more than will erratic, violent shaking. Move the box back and forth, tilt it, and use the probe to "examine" the interior of the box. Perform any other operations that you think will help you discover more information about the object in the box.

b. Once you think you know what is in the box, ask your teacher for an empty box identical to yours. Ask also for objects similar to what you think is in the black box. You should be able to use these, along with a balance, to get a still better idea of what is in your black box.

c. Summarize your experiment as shown in Table 2.

TABLE 2 Studying a Black Box

Procedure	Observation(s)	Conclusion(s)

NOTE: Obviously you cannot be sure that your answer is right without opening the box. Your objective should be to get the most reasonable answer possible. Try to get data that will let you make closer and closer estimates of the shape, size, and other properties of the object. Do not guess. After you have developed a conclusion, test it by developing other experiments. Then use your new observations to modify your first conclusion. Keep doing this until your model is as complete as you can make it.

Discussion

1. A good model should explain experimental observations. It should also help you to predict with some success the results of further experiments. Does yours?
2. Write a detailed description of the object. This is your model for the contents of the black box.
3. What is a black box?
4. List 3 things that are black boxes to you.
5. For each thing in 4, name a group of people for whom it is not a black box.
6. A pop dispensing machine is a black box, provided you have never seen inside it. We know that when money is placed in the slot a can of pop is delivered. Here is a model which explains that observation: A person inside the machine takes the coin that you deposit and checks to see that it is the correct amount. If it is, he releases a can.
   a) Make a list of the experiments that you would perform on the machine to disprove this model. Remember, you cannot open the machine. And don't kill the person inside!
   b) Pair up with another student in your class. Give this student your list. Explain to her why you selected each experiment. Then ask her to criticize and, if possible, "shoot down" each of your experiments. For example, your list may suggest that you test the model by trying to coax a free drink out of the person inside. Your partner should point out that the manufacturer of the machine anticipated sneaky people like you and put ear plugs on the person. As an added precaution he used someone who does not understand your language.

1.12 The Process of Scientific Inquiry: A Summary

You have used most of the processes of scientific inquiry. Figure 1-9 arranges these processes in the order in which they are commonly used. Refer to this figure as you carefully read the following description. Make sure you understand how the pro-

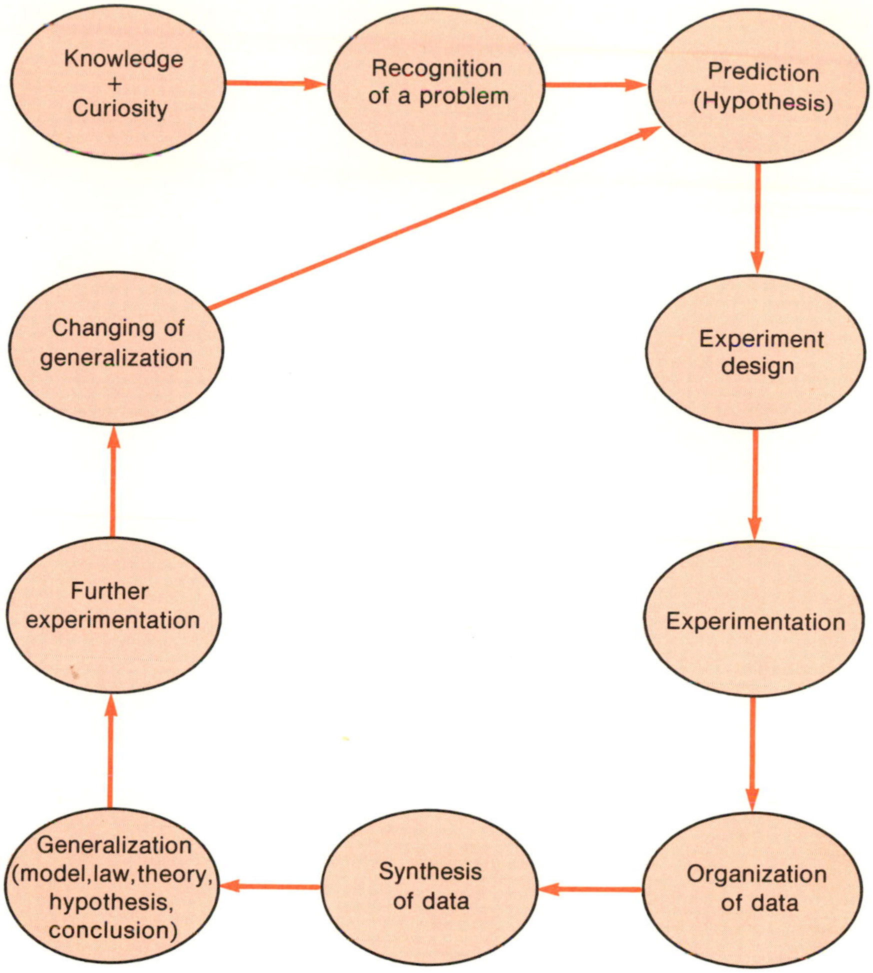

Figure 1-9 The method of scientific inquiry.

cesses fit together. You will use part or all of this scheme many times in the months ahead.

Scientific inquiry usually begins when a person's **curiosity** is aroused. This only happens, however, when that person has a certain fund of knowledge in his brain. Here is an example. Engineers often construct weirs across a river. They do this when the river runs through a city where it becomes polluted and loses some of its oxygen. Most people never wonder why the weirs were built. But the person with some knowledge about air and how it dissolves in water might be curious about the purpose of the weir. In other words, this person **recognizes a problem**. He watches the water tumble over the weir, mixing with air as it does so. He then **predicts** or **hypothesizes** that the weir was placed in the river to aerate (add air to) the water (Fig. 1-10). He might even **design an experiment** to test this prediction. He may use a test kit to measure the oxygen content of the water above and below the weir. This **experiment** is easily done. The experimenter remembers that he must have **controls**. Thus he gets both water

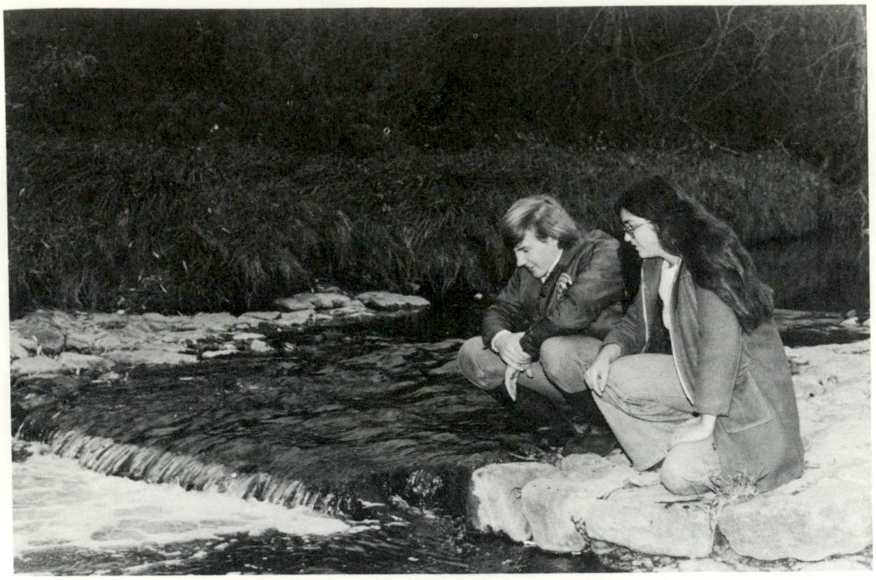

Figure 1-10 Why was a weir constructed in this river?

samples from the same depth, at the same distance from the bank, and at the same time of day. He **repeats the experiment** three times, **organizing the results** in a data table. The table shows that the oxygen level is indeed higher below the weir. The experimenter can now **synthesize** (put together) his data into a **generalization** (conclusion). It states that the weir increases the oxygen content of the river. He now proceeds to do **further experiments** at other weirs in the same river and at weirs in other rivers. He finds that he must **modify his generalization** to read as follows: A weir increases the oxygen content of the water, provided the water above the weir does not already contain all of the oxygen that it can hold.

This example of scientific inquiry was not hard to follow, was it? That's because it is just common sense. In fact, many people use scientific inquiry without even knowing that they are doing so. We want *all* of you to be able to use it whenever the need arises. Therefore, study this section again and discuss with your teacher any problems that you may have.

Discussion

1. A toothpaste commercial states that "regular use of Brand X toothpaste with its special fluoride ingredient reduces cavities by up to 42%".

a) What does "up to" mean? Why do you think the manufacturer included these words in the commercial?

b) Describe an experiment that could be used to test this statement. Pay attention to the control of variables.

c) Describe an experiment that could be used to find out how important it is to have fluoride in your toothpaste.

d) What studies should you conduct before you allow yourself to be impressed by the figure 42%?

2. Nancy noticed that the spokes of her bicycle rusted very quickly when she rode it on streets where salt had been used to melt snow. She knows that the spokes are made of iron. She also knows that air is a mixture of many gases. Nitrogen is the main gas and oxygen is the next most abundant. As a result of her observation and her knowledge, Nancy made up this hypothesis: Rust is caused by a combination of water, air (most likely nitrogen), and salt. She then designed and performed a number of experiments to test the hypothesis.

   a) She placed an iron nail in a test tube containing salt water and air. The nail rusted quickly. She repeated the experiment using fresh water instead of salt water. The nail rusted slowly. What conclusion can be made at this point in her investigation?

   b) She boiled some fresh water in a test tube to drive out the air. She added an iron nail and then stoppered the test tube. No rust formed. What conclusion can be made from this observation?

   c) She placed an iron nail in a test tube that contained only a drying agent to remove all water from the inside of the test tube. She then stoppered the test tube. No rust formed. What conclusion can be made here?

   d) She repeated the experiment outlined in b but, before inserting the stopper, she bubbled nitrogen gas through the water. No rust formed. What conclusion can be made here?

   e) What do you think causes the rusting of iron? What experiment(s) would you perform to test your prediction?

3. Two people, Greg and Brenda, notice that they feel cooler when a fan blows air over them. In order to explain this observation, Greg hypothesizes that the fan actually makes the air colder. Brenda says that this is not so. Her hypothesis is that the fan simply increases the rate of evaporation of water from the skin. This evaporation, in turn, cools the body. Design and try an experiment to determine whose hypothesis is correct. (Hint: Your teacher has a supply of thermometers.)

Highlights

Scientific inquiry is a logical and useful method for solving problems of many types. It involves a number of steps that are often, but not always, in this order: recognition of a problem, formation of a hypothesis, designing of the experiment, performing the experiment, observation, organizing the results, making a generalization. The generalization often suggests further experiments to do. These, in turn, may lead to a new hypothesis and a repeating of the problem-solving sequence called scientific inquiry.

The method of scientific inquiry is used throughout this text. Learn it well and try to apply it to problems you encounter both in and out of school.

UNIT

2 The Measurement of Matter

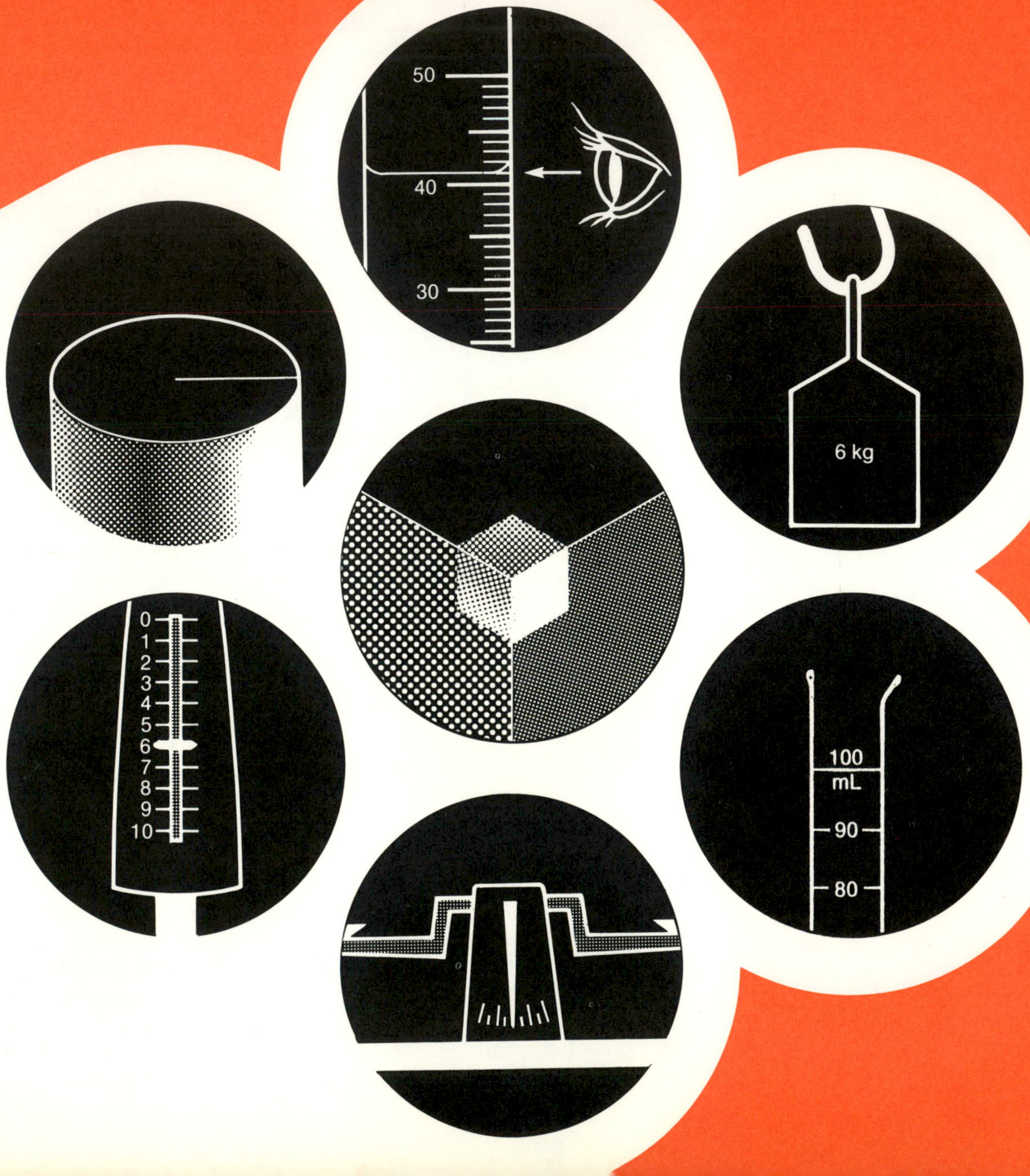

2.1 Introduction

You know that one of the key steps in an experiment is observation. There are two main kinds of observations: qualitative and quantitative. **Qualitative observations** note characteristics such as colour, odour, and taste. **Quantitative observations** deal with measurements. Length, volume, mass, density, and temperature are common measurements.

Measurements are stated using a numeral and a unit. Thus we may say that an object has a length of 6 m, a volume of 15 m^3, a mass of 45 000 kg, a density of 3 000 kg/m^3, and a melting point of 150 °C. You probably already know the meanings of most of these units. You will find out later what the others mean.

Measurements are usually more useful to a scientist than are qualitative observations. Here is an example: Lead orthoarsenate is a deadly poison. Yet it is a white solid that looks like sugar. How could you tell the difference between lead orthoarsenate and sugar? Examine Table 3. The only qualitative observation that would help is taste. Would you want to do that experiment? Scientists have determined quantitative properties of these substances. Note that lead orthoarsenate has a density of 7 800 kg/m^3 and a melting point of 1 042 °C. Sucrose has a density of 1 600 kg/m^3 and a melting point of 185 °C. Clearly, then, you could distinguish between these two substances by finding their densities and melting points. Later in this course you will learn how to do that.

TABLE 3 Comparing Sugar and Lead Orthoarsenate

	Property	Sugar	Lead orthoarsenate
Qualitative	colour odour taste smell	colourless odourless sweet none	colourless odourless somewhat sweet none
Quantitative	density melting point	1 600 kg/m^3 185 °C	7 800 kg/m^3 1 042 °C

Accuracy is important in measuring. Therefore this unit gives you practice in making measurements so you will not make mistakes in the quantitative experiments that follow. Keep in mind during this unit that all measurements are estimates. There is no such thing as an exact measurement. How close you get to the "exact" answer depends on the accuracy of the measuring instruments and on how you use them. Your teacher will explain, at suitable times, the methods and rules used when you make and record measurements.

Our country replaced the imperial system of measurement with the metric system. By the end of this unit you should be a master of the common metric units that make up the modernized metric system. This system, called **SI** (after the French name **Le Système International d'Unités**), is being adopted throughout the world. Make sure you learn this unit well.

Discussion

1. **a)** Make a list of 5 qualitative and 5 quantitative observations about yourself. Mark the 3 observations that you feel would be most useful to a police department if the police were trying to identify you in a line-up.
   b) Explain why the 3 you marked are the most useful ones.
2. Methyl alcohol and ethylene glycol are colourless liquids with pleasant odours. You probably could not tell them apart just by looking at them or smelling them. Ethylene glycol is used as an antifreeze for cars while methyl alcohol is not recommended. Suppose you were given a few gallons of antifreeze but were not told what kind it was. What property could you investigate to find out whether it was ethylene glycol or methyl alcohol?

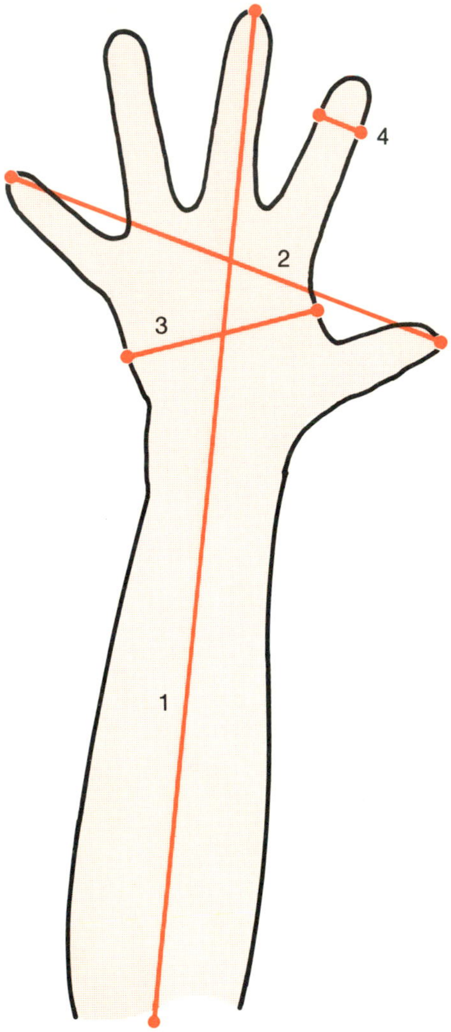

Figure 2-1 Some early units of measurement based on the length of parts of the human body: cubit (1), span (2), palm (3), and digit (4).

2.2 A History of Measurement: The Need for Standards

Early civilizations (Egyptian, Babylonian, and Hebrew) saw the need for standard units of measurement. Without these, people could not communicate measurements to one another in a meaningful way.

Before the development of standard units, parts of the body served as units of length. The **cubit** that was commonly used in Biblical times is the distance from the tip of the middle finger to a point in the centre of the elbow. Other units of length used at the same time were the **span**, the **palm**, and the **digit**. Examine Figure 2-1 and your own arm to get an idea of the sizes of these units.

It was during the building of the pyramids that the Egyptians first saw the need for a standard unit of length. Imagine what would have happened if each worker had used his own arm to measure one cubit! The local leader made up a standard one-cubit measure based on his own arm. All the workers used it (or copies of it) to make their measurements. This was the first

standard unit of linear measurement (length) to be made by man.

Many centuries later, the early Romans introduced the units **mile**, **foot**, and **inch**. These were also based on body measurements. The mile was 1 000 double paces of the size that a Roman legionnaire would take on a long march. The foot was the length of a large human foot. The inch was the width of the thumb at its widest point.

As late as the twelfth century, units for length measurement were still based on the human body. King Henry I of England (1068-1135) defined the **yard** as the distance from the tip of his nose to the end of his thumb when his arm was fully extended to one side. The longest unit based on body measurements is the **fathom**. It is the distance from middle fingertip to middle fingertip of a large man with both arms fully extended (2 yards).

New units were introduced as the need for them arose. The league, furlong, chain, and rod are examples. Each of these was developed for a certain purpose. They were not based on a common standard. As a result, we have confusing relationships between them, as shown in Table 4.

TABLE 4 Some Imperial Units For Length Measurement

1 league	=	3 miles
1 mile	=	8 furlongs
1 furlong	=	10 chains
1 chain	=	4 rods
1 rod	=	5.5 yards
1 yard	=	3 feet
1 foot	=	12 inches

As time passed a **standard yard** was developed. It is a bronze bar that has a gold plug near each end. Each plug has a scratch on it. The yard is defined as the distance between the two scratches when the bar is at 62°F and supported by eight equally-spaced rollers.

This system of measurement which had been introduced into Britain by the Romans became known as the **imperial system**. It was modified by the British. Standards were added but confusing relationships still existed. Try calculating the number of inches in 35 rods. Some garden farmers have to make calculations like that when they are sowing seeds in a field.

Suppose the imperial units for length measurement had been defined as shown in Table 5.

TABLE 5 Possible Definitions for Imperial Units		
1 league	=	10 miles
1 mile	=	10 furlongs
1 furlong	=	10 chains
1 chain	=	10 rods
1 rod	=	10 yards
1 yard	=	10 feet
1 foot	=	10 inches

Now try calculating the number of inches in 35 rods. Easy, isn't it?

By the end of the eighteenth century, scientists saw the need for a new system of measurement. They wanted one that was easy to use. Also, they wanted one that was not based on such arbitrary things as body measurements. Thus the **metric system** was developed. It is based on standard units and uses multiples of 10 as we did in Table 5. Let us see how it works by studying Section 2.3.

Discussion

1. Discuss the need for standard units of measurement.
2. A standard unit is also called a **base unit**. What is the base unit of length that is used in this country?
3. **a)** What is the main disadvantage of the imperial system of measurement?
b) What is the main advantage of the metric system of measurement?
4. Here is a list of units of linear measurement that have interesting origins: chain; rod; league; barleycorn; furlong. Select 2 of these. Then visit your school or public library and find out:
a) The origin of each unit;
b) A definition for each unit.

2.3 The Metric System of Length Measurement

History of the Metre

The base unit of length in the metric system is the **metre**. It was first defined in 1799 as one ten-millionth of the distance from the equator to the North Pole. Two French scientists surveyed the meridian that runs through Paris—from Barcelona to Dunkirk—to determine the length of a metre.

The metre was re-defined in 1889 to be the distance between two scratches on a platinum-iridium bar at 0°C. This bar, called the International Prototype Metre, was kept near Paris. Copies were given to many other countries.

By the 1960s modern scientific work required a better standard. As a result, the metre is now defined as that length equal to 1 650 763.73 wave lengths in a vacuum of the radiation corresponding to the transition between the levels $2p_{10}$ and $5d_5$ of the krypton-86 atom. Of course you are not expected to understand this definition. But you should appreciate its precision. With this new standard and with modern instruments, scientists can now measure lengths of one metre to an accuracy of 1 part in 100 000 000.

Measuring Length

The main advantage of the metric system is that it uses a decimal system. This system is compatible with our base-ten number system. Examine Table 6 to see how this decimal system works.

In this table the following symbols are used:

kilometre = km
hectometre = hm
decametre = dam
metre = m
decimetre = dm
centimetre = cm
millimetre = mm

Note that there is really only one unit of length, the metre. The others are just multiples or sub-multiples of it. Thus the metre is called the base unit of length and the others are called derived units of length.

TABLE 6 Measuring Length

1 km = 10 hm		1 km = 1 000 m
1 hm = 10 dam		1 hm = 100 m
1 dam = 10 m	**OR**	1 dam = 10 m
1 m = 10 dm		1 dm = 0.1 m
1 dm = 10 cm		1 cm = 0.01 m
1 cm = 10 mm		1 mm = 0.001 m

Only the kilometre (km), metre (m), centimetre (cm), and millimetre (mm) are commonly used.

Discussion

1. For decades scientists throughout the world have used the metric system to report their findings. What do you think are the two main reasons for doing this?
2. Most countries in the world are now metric or are converting to metric.
   a) Why was the metric system selected instead of some other system?
   b) What are the advantages in having the entire world using the same system of measurement?
   c) Suppose a country decides not to convert to the metric system. What disadvantages will that country experience?

2.4 INVESTIGATION: Relationships among Common Length Units

This investigation is intended for those who have not had much experience with the metric system. It is a good review for the rest of you.

Materials

metre stick
pencil and notebook

Procedure

a. Copy Table 7 into your notebook.
b. Complete the table. Refer to Table 6 for help when necessary.
c. Use the metre stick to get an idea of the sizes of the various metric measurements of length as you do the calculations.

TABLE 7 Using Common Length Units

10 mm	=	______	cm	10 cm	=	______	m	150 m	=	______	km
100 cm	=	______	m	100 m	=	______	km	5 000 mm	=	______	cm
1 000 m	=	______	km	10 m	=	______	km	5 000 mm	=	______	m
100 mm	=	______	cm	5 km	=	______	m	1.03 m	=	______	mm
1 000 cm	=	______	m	1 km	=	______	cm	0.52 m	=	______	cm
10 000 m	=	______	km	50 cm	=	______	mm	0.52 cm	=	______	m
1 000 mm	=	______	cm	50 cm	=	______	m	358 m	=	______	km
10 000 cm	=	______	m	50 mm	=	______	cm	0.35 km	=	______	m
100 000 m	=	______	km	70 km	=	______	m	1.5 km	=	______	m
1 mm	=	______	cm	1 500 m	=	______	km	5 800 mm	=	______	m

2.5 INVESTIGATION: Using Common Length Measurements

The studies in this section will give you practice in using the metric system for measuring length.

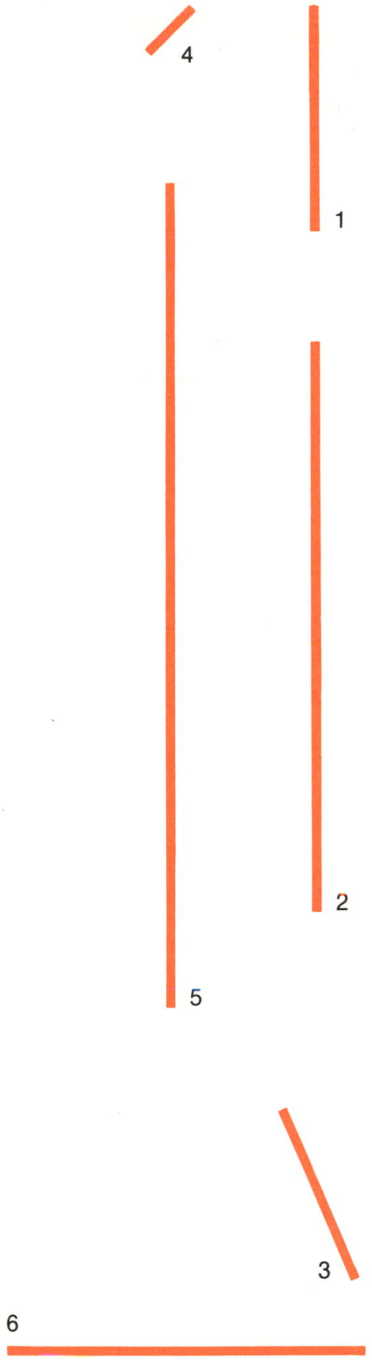

Figure 2-2 How long are these lines?

Materials

pencil and notebook
ruler
cloth or plastic tape measure

Procedure A Estimating and Measuring the Lengths of Lines

a. Estimate in centimetres the length of each line in Figure 2-2. Enter your estimate in a table like Table 8.
b. Now measure each line in centimetres. Enter your measurement in the table.
c. Record the difference (+ or −) between each estimate and the measured length. (Subtract the measured length from the estimate.)

TABLE 8 Lengths of Some Lines

Line	Estimate	Length (cm)	Difference (+ or −)
1			
2			
3			
4			
5			
6			

Procedure B Drawing Lines, by Estimate, of Given Lengths

a. Use a straight edge (but not a ruler) to draw lines that you think are the following lengths: 15 mm, 2.5 cm, 130 mm, 7.0 cm, 25 mm, 0.8 cm.
b. Measure each line with the ruler. Put your results in a table like Table 9.
c. Record the difference (+ or −) between the given and measured lengths.

TABLE 9 Drawing and Measuring Lines		
Given length	**Measured length (mm or cm)**	**Difference (+ or −)**
15 mm		
2.5 cm		
130 mm		
7.0 cm		
25 mm		
0.8 cm		

Procedure C Do You Know Yourself Metrically?

a. Copy Table 10 into your notebook.
b. Estimate each of the measurements in the table.
c. Measure each factor with the ruler or tape. Work with a partner, if necessary (Fig. 2-3).

Figure 2-3 Measuring sleeve length using the metric system.

TABLE 10 My Metric Measurements		
Factor	**Estimate**	**Measurement**
Height		
Hips to floor		
Vertical reach		
Neck size		
Sleeve length		
Wrist size		
Chest size		
Waist size		
Length of foot		
Length of pace		

Procedure D Selecting the Proper Prefix

By now you are probably wondering if it matters whether you give your answer in millimetres, centimetres, metres, or kilometres. Here is the rule that is commonly used: *Choose a prefix so that the numeral lies between 0.1, and 1 000*. Thus 12 km is better than 12 000 m; 0.2 mm is better than 0.02 cm.

a. Copy Table 11 into your notebook.
b. For each length given, state the unit that you think would be the best one to use.

TABLE 11 Selecting Units

Measurement	Unit
Inter-city distances	
Your height	
Length of a pencil	
Room dimensions	
Cloth for a dress	
Sheet of paper	
Circumference of the earth	
Height of a door	
Height of Niagara Falls	
Length of a swimming pool	
Height of a horse	
Length of your arm	
Diameter of a dime	
Thickness of a dime	

2.6 Metric Prefixes

You know that the prefix "**kilo**" means 1 000. Thus a kilometre is 1 000 m. The same prefix can be used with metric units other than length. For example, a kilogram is 1 000 g. The prefix "**centi**" means $\frac{1}{100}$ or 0.01. Thus a centimetre is 0.01 m. The prefix "**milli**" means $\frac{1}{1\,000}$ or 0.001. Thus a millimetre is 0.001 m and a millilitre is 0.001 L.

Many prefixes have been developed for the metric system. This was done so that we would not have to use cumbersome numbers when we are dealing with very large and very small measurements. For example, the distance from the Sun to Pluto is 6 000 000 000 000 m. A number of that size is very awkward to use. Even when we convert it to kilometres (6 000 000 000 km) the number is still very large. But if we convert it to terametres, it becomes 6 Tm. That number looks more reasonable, doesn't it?

Here is another example. When atomic scientists measure the diameter of the nucleus of an atom, they get values like 0.000 000 000 000 1 cm—another awkward number. However, when this is converted to femtometres it becomes 1 f, a more reasonable number.

Table 12 includes the prefixes you already know, as well as tera, pica, and several other new ones. You should know from memory the ones in the white rectangle. You need not memorize the others; but make sure you can use them.

TABLE 12 The Metric Prefixes

Prefix	Symbol	Factor by which Unit is Multiplied	Exponential notation
exa	E	1 000 000 000 000 000 000	10^{18}
peta	P	1 000 000 000 000 000	10^{15}
tera	T	1 000 000 000 000	10^{12}
giga	G	1 000 000 000	10^{9}
mega	M	1 000 000	10^{6}
kilo	k	1 000	10^{3}
hecto	h	100	10^{2}
deca	da	10	10^{1}
THE UNIT (e.g. metre)		1	10^{0}
deci	d	0.1	10^{-1}
centi	c	0.01	10^{-2}
milli	m	0.001	10^{-3}
micro	μ	0.000 001	10^{-6}
nano	n	0.000 000 001	10^{-9}
pico	p	0.000 000 000 001	10^{-12}
femto	f	0.000 000 000 000 001	10^{-15}
atto	a	0.000 000 000 000 000 001	10^{-18}

Discussion

1. **a)** Convert the following measurements to metres: 1 000 mm, 100 mm, 10 mm, 1 000 cm, 100 cm, 10 cm, 10 km, 570 cm, 650 mm, 1.5 km.
b) Convert the following measurements to centimetres: 1 000 mm, 50 mm, 6.5 mm, 720 mm, 8.0 m, 22.5 m.
c) Convert the following measurements to kilometres: 1 000 m, 7 500 m, 12 200 m, 600 m, 72 m, 3 500 cm.
d) Convert the following measurements to millimetres: 20 cm, 35 cm, 1.5 cm, 152 cm, 3.0 m, 21 m.

2. You should enjoy doing this question! Complete the following statements by picking the proper metric prefixes from Table 12. The first one has been done as an example.

10^{-2} pedes = 1 centipede
10^{6} phones = 1 ________
10^{-6} phones = 1 ________

10^{-3} pedes = 1 ________
10^{-1} mate = 1 ________
10^{-2} mental = 1 ________
10^{15} bulls = 1 ________
10^{-9} goats = 1 ________
10^{1} cards = 1 ________
10^{12} bulls = 1 ________
10^{-18} boys = 1 ________
10^{-12} boos = 1 ________
10^{9} los = 1 ________
10^{12} pins = 1 ________

3. Perform the following conversions. Refer to Table 12, when necessary.

10^{3} m = ____ km
10^{6} m = ____ Mm
10^{9} m = ____ Gm
10^{12} m = ____ Tm
10^{3} km = ____ Mm
10^{6} km = ____ Gm
10^{9} km = ____ Tm
10^{-3} m = ____ mm
10^{-6} m = ____ μm
10^{-9} m = ____ nm
10^{-12} m = ____ pm
10^{-15} m = ____ fm

4. If you had little trouble with the conversions in question 3, try these:

8.0×10^{9} km = ____ Tm
2.0×10^{-12} cm = ____ pm
26 Mm = ____ km
3.1×10^{3} μm = ____ mm
2.5 nm = ____ μm
35 Gm = ____ Mm
0.2 Gm = ____ km
10^{12} μm = ____ Mm
200 cm = ____ μm
7.2 Tm = ____ Mm

2.7 INVESTIGATION: Using Metric Prefixes for Large and Small Measurements

The purpose of this exercise is to give you practice in choosing and using the proper metric prefixes.

Materials

pencil and notebook
Table 12

Procedure

a) Copy Table 13 into your notebook.
b) Form a group of 3 people and work together on this investigation.
c) For each given length, select a better prefix and convert the length to that prefix. Record your answers in the table that you prepared. We have done the first one for you. Refer to Table 12 as you do this exercise. Remember the rule on page 30.

TABLE 13 Using Metric Prefixes

Distance measured	Given length	Your conversion
Halifax to Winnipeg	4 000 000 m	4.0 Mm
Thickness of a dime	0.001 m	
Diameter of earth's orbit	322 000 000 000 m	
Diameter of our galaxy	9 600 000 000 000 000 000 km	
Size of a bacterium	0.000 001 m	
Diameter of a proton	0.000 000 000 000 001 m	
Diameter of a human hair	0.000 07 m	
Thickness of a sheet of paper	0.000 1 m	
Height of a table	0.000 8 km	
Common Olympic race	10 000 cm	
Length of loaf of bread	300 000 μm	
Length of a small car	37 000 mm	
Height of a horse	1 600 000 μm	
Height of a basketball player	2 100 mm	
Height of a door	0.002 2 km	

2.8 Area

Area is defined as the amount of surface. It is two-dimensional. That is, it is flat. It does not extend into the space above. A rectangle, triangle, and circle are two-dimensional and all have an area.

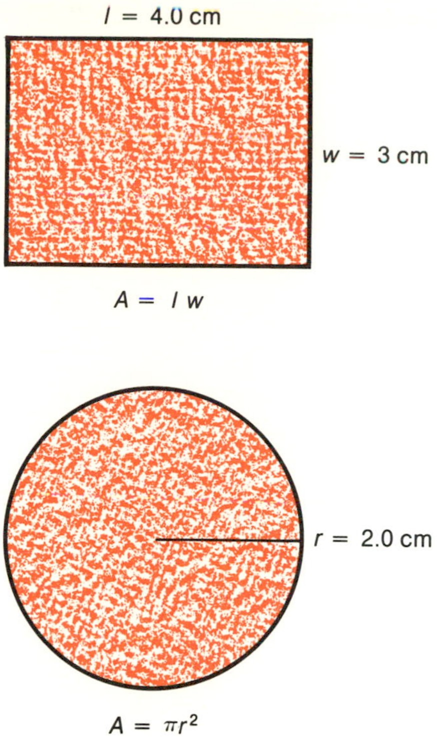

Figure 2-4 Area of a rectangle and a circle.

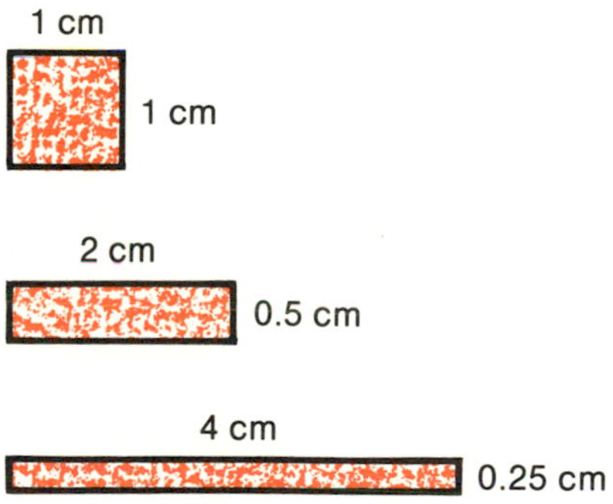

Figure 2-5 Each of these rectangles has an area of 1cm².

Examine Figure 2-4. The area of a rectangle is calculated using $\boldsymbol{A} = \boldsymbol{l} \times \boldsymbol{w}$, where $\boldsymbol{l}$ = length and $\boldsymbol{w}$ = width. Thus the area of this rectangle is:

$$\begin{aligned} A &= l \times w \\ &= 4.0\ \text{cm} \times 3.0\ \text{cm} \\ &= 12.0\ \text{cm}^2 \end{aligned}$$

The symbol cm^2 is read "*square centimetre*".

The area of a circle is calculated using $\boldsymbol{A} = \boldsymbol{\pi r^2}$, where π("pi") = 3.14 and $\boldsymbol{r}$ = radius. Thus the area of this circle is:

$$\begin{aligned} A &= \pi r^2 \\ &= 3.14 \times (2.0\ \text{cm})^2 \\ &= 3.14 \times 2.0\ \text{cm} \times 2.0\ \text{cm} \\ &= 3.14 \times 4.0\ \text{cm}^2 \\ &= 12.6\ \text{cm}^2 \end{aligned}$$

Some commonly used units of area are listed in Table 14. Study this table carefully before you start this investigation. Think about the sizes of the units as you do so. Figure 2-5 should help you. The square metre is the base unit of area. The others are just multiples and sub-multiples of it.

TABLE 14 Common Area Units and Uses

Unit	Symbol	Uses
Square kilometre	km^2	to measure areas of large surfaces such as land masses, oceans, lakes, and forests
Square hectometre or Hectare	hm^2 ha	to measure areas of smaller surfaces such as farms, parks, and playing fields
Square metre	m^2	to measure areas of such things as floors, lawns, small gardens, and offices
Square centimetre	cm^2	to measure areas of small surfaces like the sole of a shoe, a card, a leaf, an animal's footprint

2.9 INVESTIGATION: Measuring Area

In this investigation you measure the areas of some surfaces provided by your teacher.

Materials

pencil and notebook
metre stick
tape measure (preferably 30 m)

Procedure

a. Copy Table 15 into your notebook. Your teacher may add to or delete from the list of items given. He/she will tell you what portion of the lawn and playing field to measure.
b. Estimate the area of each of the surfaces listed in the table. Be sure to use suitable units. (See Table 14.)
c. Measure each of the surfaces listed in the table.
d. Calculate the area of each surface.

Discussion

1. a) Compare your estimated and calculated areas. Are you better at estimating small or large areas? Why?
b) If your estimated and calculated areas differ greatly, you are having trouble ''thinking metrically'' about area. Discuss the problem with your teacher.
2. Wall-to-wall broadloom costs $14.95/m²$ installed. What would it cost to carpet a room that is 6.0 m long and 4.5 m wide?
3. a) What would it cost to cover your classroom floor with carpet that costs $15.50/m²$ installed?
b) What would it cost to cover the same floor with 30 cm × 30 cm tiles that cost 65¢ each installed?

TABLE 15 Measuring Area

Area to be measured	Estimate of area	Linear measurements	Calculated area
Playing card	40 cm^2	8.8 cm × 5.7 cm	50 cm^2
Dollar bill			
Postage stamp			
Floor tile			
Desk top			
Section of chalkboard			
Classroom floor			
Portion of school lawn			
Portion of playing field			
Quarter			
45-r/min record			
L-P record			
Circular area in schoolyard			

2.10 Volume

Volume is defined as the amount of space an object occupies. It is three-dimensional. That is, it covers an area on a surface *and* extends into the space above. A brick, a baseball, and a glass of water all have volume. They take up space. You, too, have volume. You can change your volume quickly by either exhaling or inhaling. You can change it slowly by eating too much or too little. If you are large, you take up more space than you would if you were small.

Volume units are of two types. Some are **cubic units** and some are **capacity units**. In the metric system, the common cubic units of volume are the cubic metre (m^3), cubic decimetre (dm^3), and cubic centimetre (cm^3). The common capacity units of volume are the kilolitre (kL), litre (L), and millilitre (mL). Table 16 shows how these units are related. The cubic metre is the primary unit of volume.

TABLE 16 Volume Units

A. Relationships Among Cubic Units		
1 m^3	=	1 000 dm^3
1 dm^3	=	1 000 cm^3
B. Relationships Among Capacity Units		
1 kL	=	1 000 L
1 L	=	1 000 mL

Don't memorize the relationships in Table 16. You can reason them out. For example, 1 dm = 10 cm. Therefore 1 dm^3 = 10 cm × 10 cm × 10 cm, or 1 000 cm^3. Also, kilo means 1 000. Thus 1 kL = 1 000 L. See if you can reason out the other two relationships in Table 16.

TABLE 17 Comparison of Cubic and Capacity Units

Cubic unit		Capacity unit
1 m^3	=	1 kL
1 dm^3	=	1 L
1 cm^3	=	1 mL

The fact that there are two systems for measuring volume may seem confusing. Table 17 shows that they are actually very similar. Generally cubic units are used for solids. Capacity units are used for things that pour. However, it is not wrong to speak of 50 mL or 50 cm^3 of water. Nor is it wrong to speak of 50 cm^3 or 50 mL of wood.

Study Figure 2-6 carefully to make sure you have an idea of the sizes of and relationships among the volume units. Your teacher may have an actual cubic decimetre to show you.

Now that you have an idea of the sizes of the volume units, look at Table 18 to see when you would use each unit. Think of other uses for each unit.

TABLE 18 Common Volume Units and Uses

Unit	Symbol	Uses
Cubic metre	m^3	to measure large volumes of earth and gravel, the volume of a building, the volume of the hold of a ship
Kilolitre	kL	to measure the volume of water in a reservoir, the volume of gasoline in a tanker truck
Cubic decimetre	dm^3	no common usage
Litre	L	to measure volumes of milk, gasoline, paint, ice cream; to measure capacities of pails, kettles, auto gas tanks, refrigerators, freezers
Cubic centimetre	cm^3	to measure volumes of small boxes and other small objects of regular shapes (rectangular, spherical, cylindrical, etc.)
Millilitre	mL	to measure volume of materials (usually fluid) that come in containers smaller than 1 L; for example, toothpaste, a glass of milk, soft drinks, hair shampoo, shaving lotion

Discussion

1. a) Explain the term ''volume''.
b) Explain the difference between cubic and capacity units of volume.

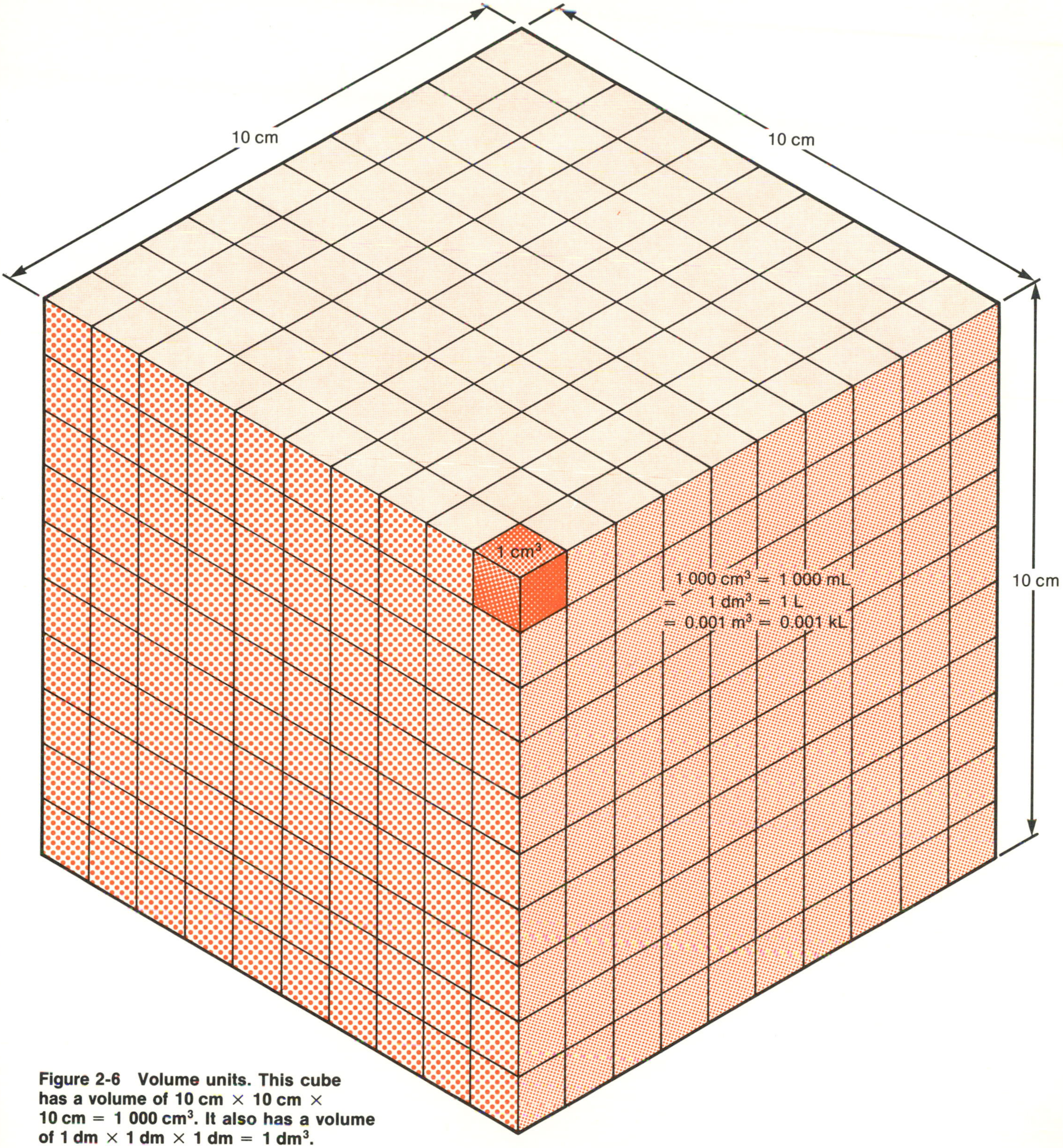

Figure 2-6 Volume units. This cube has a volume of 10 cm × 10 cm × 10 cm = 1 000 cm^3. It also has a volume of 1 dm × 1 dm × 1 dm = 1 dm^3.

2. Complete the following:

1 m^3 = ____ dm^3
1 dm^3 = ____ cm^3
1 m^3 = ____ cm^3
1 kL = ____ L
1 L = ____ mL
1 kL = ____ mL
1 m^3 = ____ kL
1 dm^3 = ____ L
1 cm^3 = ____ mL
1 m^3 = ____ mL
1 dm^3 = ____ mL
1 m^3 = ____ L
573 mL = ____ cm^3
573 mL = ____ L
1 370 cm^3 = ____ mL
1 370 cm^3 = ____ dm^3
27 L = ____ mL
27 L = ____ cm^3

3. What unit of volume would you use to measure each of the following:

a) capacity of a railway boxcar
b) capacity of a refrigerator
c) storage space in a farm silo
d) capacity of a swimming pool
e) capacity of a pail for watering horses
f) size of a 4-drawer filing cabinet
g) size of a loaf of bread
h) capacity of a coffee mug
i) size of a golf ball
j) size of a dice
k) displacement of a car engine

4. Look around your home and find 10 examples of the use of the metric system to describe the volume of household products. Examine grocery items, cosmetic containers, medicine containers, cleansing agent containers, and so on. For each example, give the name and the volume.

5. a) On page 37 you were told that the common cubic units of volume are the cubic metre, cubic decimetre, and cubic centimetre. However, other units are possible, even though they are not widely used. Refer to Table 12 on page 32. Then complete the following to see what these other units are.

1 km^3 = ____ hm^3 (cubic hectometres)
1 hm^3 = ____ dam^3 (____ ________)
1 dam^3 = 1 000 ____ (____ ________)
1 m^3 = ____ dm^3 (____ ________)
1 dm^3 = 1 000 ____ (____ ________)
1 cm^3 = ____ mm^3 (____ ________)

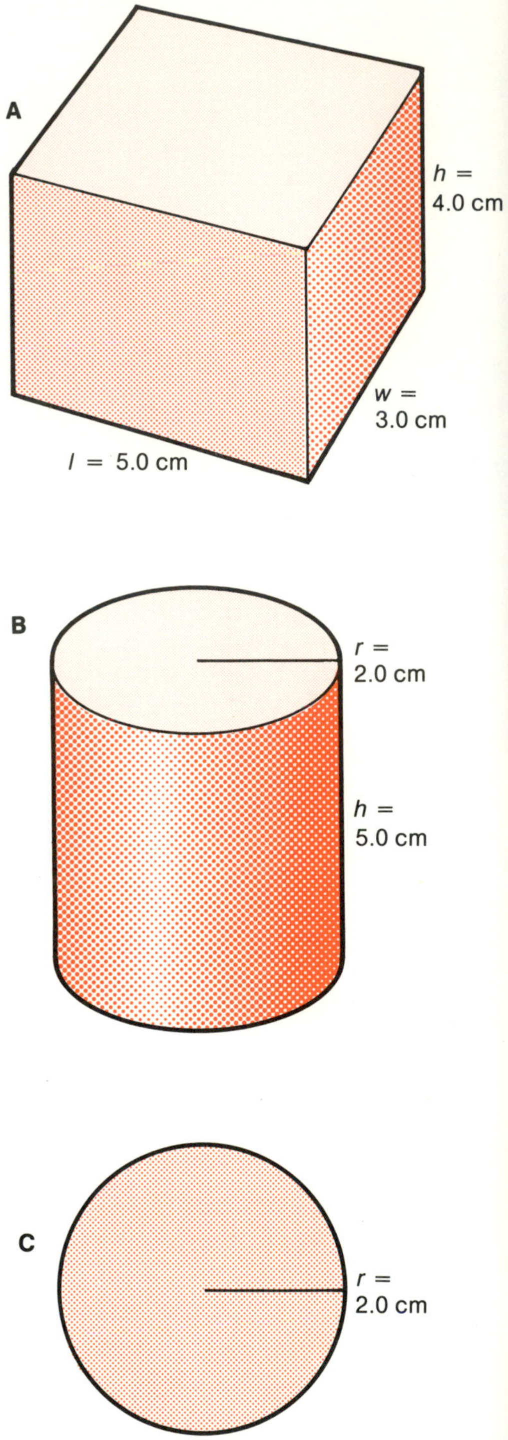

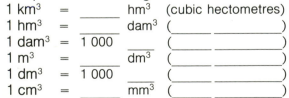

Figure 2-7 Volume of a rectangular solid (A), a cylinder (B), and a sphere (C).

b) Page 37 also said that the common capacity units of volume are the kilolitre, the litre, and the millilitre. Here, too, other units are possible, though rarely used. Complete the following to see what these units are.

1 kL	=	______	hL	(hectolitre)
1 hL	=	______	daL	(__________)
1 daL	=	______	L	(__________)
1 L	=	______	dL	(__________)
1 dL	=	______	cL	(__________)
1 cL	=	______	mL	(__________)

6. What will it cost to fill a 55 L automobile gasoline tank with gasoline that costs 25¢/L?

7. a) The gasoline consumption of a car is 9.2 L/100 km. What does this mean?

b) What will it cost for gasoline to drive this car 250 km, if gasoline costs 25¢/L?

2.11 INVESTIGATION: Determining the Volume of a Rectangular Solid

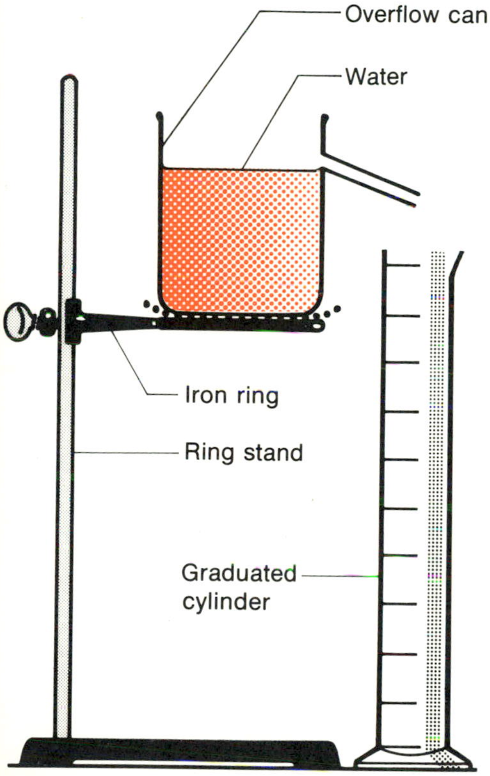

Figure 2-8 Determining the volume of a solid by displacement of water.

A rectangular solid is said to be a **regular** solid. It has a definite shape. You can get its volume by making some measurements and calculations. For example, the rectangular solid, A, in Figure 2-7 is a regular solid. Its volume is calculated using the formula $V = l \times w \times h$, where l = length, w = width, and h = height. Thus $V = 5.0\ \text{cm} \times 3.0\ \text{cm} \times 4.0\ \text{cm} = 60\ \text{cm}^3$.

You can also get the volume of this solid by the **displacement** method. Suppose a rectangular solid with a volume of 25 cm^3 was lowered into the overflow can in Figure 2-8. How much water would be displaced into the graduated cylinder?

In this investigation you calculate the volumes of four rectangular solids both by measurement and by displacement. You then compare the two methods by plotting a graph of your results.

Materials

4 rectangular solids with different volumes
ruler
ring stand and iron ring
overflow can
graduated cylinder (100 mL)
fine thread
pin
graph paper
beaker (150 mL)

Procedure

a. Measure the length, width, and height of each rectangular solid. Record the results in your notebook, using a table like Table 19.

b. Calculate the volume of each solid using the formula $V = l \times w \times h$.

c. Set up the apparatus as shown in Figure 2-8. Place a finger over the spout of the overflow can. Then fill the can above the spout with water. Remove your finger and catch all of the water that runs out the spout in a beaker. Place the graduated cylinder under the spout.

d. Slowly lower Solid A into the water in the overflow can using a piece of thread. If necessary, slowly push the solid under the water with a pin. Keep it there until no more water runs out the spout. Do not drop the solid into the water nor use your fingers to push it under. Why?

e. Measure the volume of water in the graduated cylinder. (See Fig. 2-9.) Record the result in the table.

f. Repeat steps c, d, and e two more times with Solid A. Record the results in the table.

g. Average the three answers for Solid A and record the result in the table.

h. Repeat steps c to g for the other 3 solids. Be sure to record the results in the table.

i. Label the axes of your graph paper as shown in Figure 2-10.

j. Plot a graph of the measured volumes of the solids versus the volumes of displaced water.

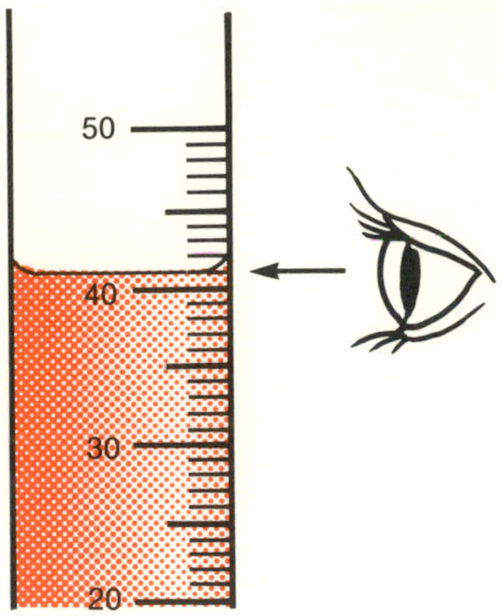

Figure 2-9 Note that the surface of the liquid is curved. The curved surface is called a *meniscus*. Always read the volume at the bottom of the curve. Make sure that your eye is directly opposite the curve. The correct reading here is 41.0 mL.

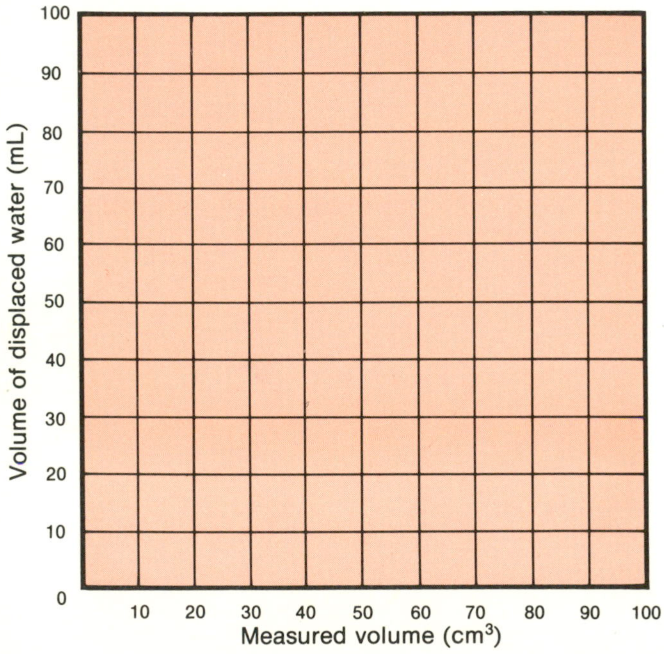

Figure 2-10 Graph of measured volume versus volume of displaced water.

TABLE 19 Volume of a Rectangular Solid

Solid	Length (cm)	Width (cm)	Height (cm)	Volume (cm^3)	Volume of displaced water (mL)			
					Trial 1	Trial 2	Trial 3	Average
A								
B								
C								
D								

Discussion

1. Describe and explain your results and the graph.
2. If your results were not exactly what you expected them to be, include an explanation.

2.12 INVESTIGATION: Determining the Volumes of Other Regular Solids

A cylindrical solid is a regular solid. Its volume can be calculated using $\boldsymbol{V} = \pi \boldsymbol{r}^2\boldsymbol{h}$, where $\pi = 3.14$, $\boldsymbol{r} =$ radius of an end, and $\boldsymbol{h} =$ height. Thus the volume of the cylindrical solid, B, in Figure 2-7 is

$V = 3.14 \times (2.0\text{ cm})^2 \times 5.0\text{ cm} = 3.14 \times 2.0\text{ cm} \times 2.0\text{ cm} \times 5.0\text{ cm} = 62.8\text{ cm}^3$.

A spherical solid is also a regular solid. Its volume can be calculated using $\boldsymbol{V} = \frac{4}{3}\pi \boldsymbol{r}^3$. Thus the volume of the spherical solid, C, in Figure 2-7 is $V = \frac{4}{3} \times 3.14 \times (2.0)^3 = 33.5\text{ cm}^3$.

In this investigation you calculate the volumes of a cylindrical solid and a spherical solid both by measurement and by displacement.

Materials

a cylindrical solid
a spherical solid
calipers
the same materials used in Investigation 2.11

Procedure

a. Use a procedure similar to that of Investigation 2.11. You should use calipers to get the radius of the cylinder and sphere. Your teacher will tell you how to use them.

b. The plotting of graphs is not necessary. Simply prepare a report that shows the results of all of your trials and the final average volumes.

2.13 INVESTIGATION: Determining the Volumes of Irregular Solids

In this investigation you are to determine the volumes of several irregular solids. No procedure is given. Make up your own. When you have finished the experiment, write a full report that includes purpose, procedure, results, and conclusions.

Materials

a stone
metal pellets (shot)
nails
sand
graduated cylinder

Procedure

Your teacher will give you a stone and samples of the metal pellets, nails, and sand. You are to measure the total volume of each sample, not the volume of one pellet or nail.

Consider the following questions before you prepare your procedure:

a. Why can you *not* get the volumes of the pellets, nails, and sand simply by pouring each into the graduated cylinder?
b. How can you get the volumes of the pellets, nails, and sand using the displacement method and *only* the graduated cylinder and water?

Results

Prepare a suitable data table and record your results in it.

Discussion

1. Would the method that you used for sand work for table sugar or table salt? Why?
2. Calculate the volume of one nail. Could you have determined its volume by using only one nail and the displacement method? Why?

2.14 Mass

The **mass** (***m***) of an object is the amount of material in it. Study the following carefully to make sure you know the difference between mass, volume, and weight.

Mass and Volume

An object's volume can change but its mass remains constant. For example, when a thermometer is heated, the liquid in it rises. The volume of the liquid (the space it takes up) becomes larger. But the mass (amount of material in the liquid) stays the same; no liquid was added or removed.

Mass and Weight

The **weight** of an object is the **force of gravity** on that object. Unlike mass, it changes with location. For example, if you "weighed" yourself on a bathroom scale on the earth and on the moon, you would weigh only one sixth as much on the moon. The moon pulls you down on the scales only one sixth as hard as the earth does. However, your mass (the amount of material in you) would be the same in both places.

The force of gravity between a planetary body and an object like yourself depends on the mass of the planet and on how far you are from its centre. Thus you would weigh much more on Jupiter than on Earth, and less on a mountain-top than in a valley. But, no matter where you are, your mass is the same.

Many people use "weight" to mean both "mass" and "force of gravity". To prevent confusion, we will not use the word weight in this text.

The Standard Kilogram

The standard of mass in the metric system is the **kilogram**. It is a platinum-iridium cylinder kept at Sèvres, France. All countries using the metric system standardize their kilogram against this one.

Measuring Mass

The mass of an object can be measured using a double-arm balance. If the double-arm balance is an equal-arm balance (one with both arms the same length), you measure the mass of an object by balancing the object with known masses (Fig. 2-11).

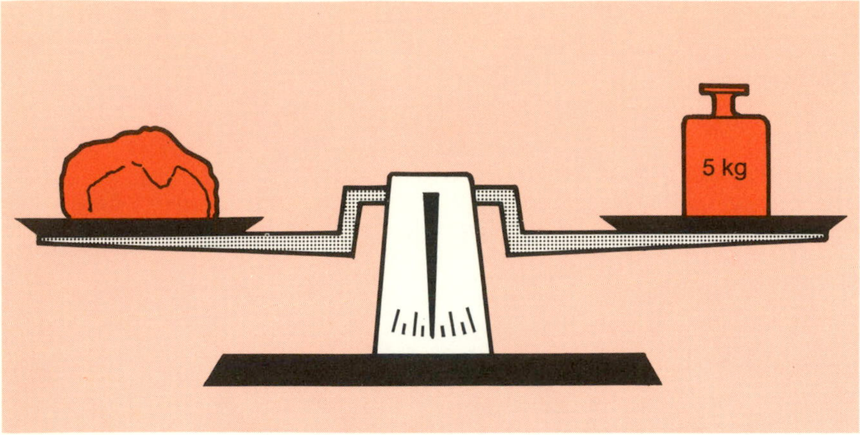

Figure 2-11 An equal-arm balance. The mass of the stone is 5 kg since it balances with a known 5 kg mass.

You may use some other type of balance such as a four-beam balance (Fig. 2-12). If you examine it closely, you will see that it is still a double-arm balance. However, one arm is shorter than the other. Your teacher will explain how your balance works.

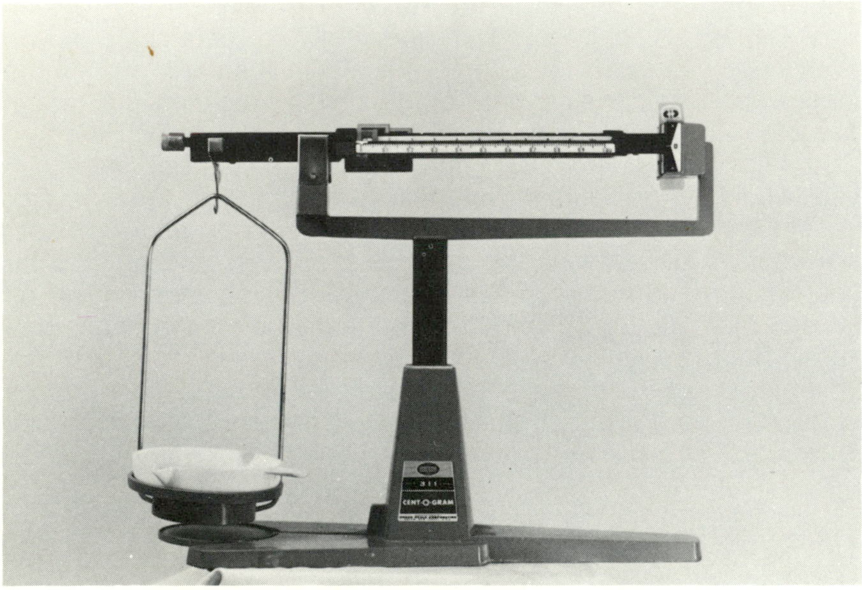

Figure 2-12 A four-beam balance.

Common Mass Units

The common mass units in the metric system are the tonne (t), kilogram (kg), gram (g), and milligram (mg). Table 20 shows how these units are related. The kilogram is the base unit of mass. The others are simply multiples and sub-multiples of it and are called derived units.

Since "tonne" sounds like "ton" (2 000 pounds), in spoken language we commonly call a tonne a "metric ton". You could also call a tonne a megagram (Mg), since it is 1 000 000 g.

TABLE 20 Mass Units

1 t	=	1 000 kg
1 kg	=	1 000 g
1 g	=	1 000 mg

Examine the set of masses given to you by your teacher to get an idea of the sizes of the mass units. Then study Table 21 to see when you would use each unit. Try to think of other uses for each unit.

TABLE 21 Common Mass Units and Uses

Unit	Symbol	Uses
Tonne	t	Masses of large objects like trucks, tractors, airplanes, and ships; masses of loads of earth, grain, ore, etc.
Kilogram	kg	Masses of sugar, flour, meat, and other grocery items; masses of people, horses, and other animals.
Gram	g	Masses of smaller grocery items like butter, powdered milk, yogurt, cheese, and meat slices.
Milligram	mg	Masses of vitamins and minerals in pills, cereal, or bread; masses of ingredients in medical products.

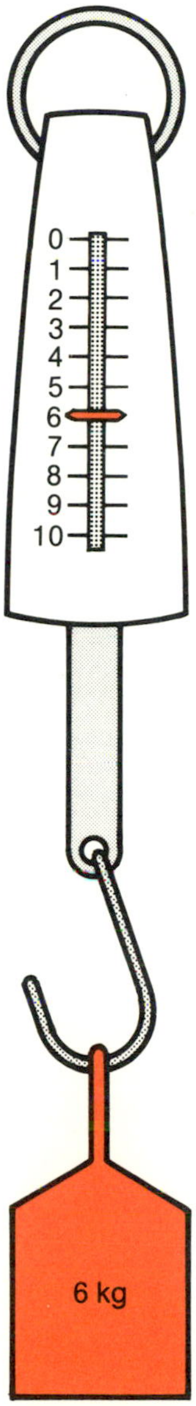

Figure 2-13 A spring balance. Can it measure mass?

Discussion

1. **a)** Define the term "mass".
   b) Explain the difference between mass and volume.
   c) Explain the difference between mass and force of gravity.
2. Could you determine the mass of an object on the moon using an equal arm balance and a set of standard masses? How would your answer compare with the result of a similar measurement on earth? Explain.
3. Figure 2-13 shows a spring balance with a 6 kg mass hanging on it. Notice that no units are shown on the scale of the balance. However, the needle points to 6.
   a) Could this balance be used to measure mass on the earth, provided the result did not have to be too accurate? Explain.
   b) Could this balance be used to measure mass on another planet or on the moon? Explain.
4. Dry cereal such as cornflakes is sold by mass and not by volume. Why?
5. Complete the following:
   1 t = ____ kg
   1 t = ____ g

1 t	=	____ mg
1 kg	=	____ g
1 kg	=	____ mg
1 g	=	____ mg
1 mg	=	____ g
1 g	=	____ kg
1 kg	=	____ t
50 t	=	____ kg
12 kg	=	____ g
26 g	=	____ mg
250 mg	=	____ g
2 500 g	=	____ kg
1 370 kg	=	____ t
530 g	=	____ kg
6 530 mg	=	____ g
535 kg	=	____ t

6. What unit of mass would you use to measure each of the following:

a) a jet plane
b) an elephant
c) a car
d) yourself
e) a chicken
f) a bag of cement
g) a jar of peanut butter
h) a box of cereal
i) an aspirin tablet
j) a handful of peanuts

7. A box of cereal has a label on it which says that "100 g of contents contain 20.2 mg of niacinamide, 4.2 mg of riboflavin, 2.6 mg of thiamine, and 13.0 mg of iron."

a) What is the total mass of vitamins and minerals in 100 g of cereal?
b) How much of the 100 g of cereal is *not* vitamin or mineral?

8. The common mass units are shown in Table 20 on page 46. Other units are possible, though not widely used. Refer to Table 12 on page 32. Then complete the following to see what these other units are.

1 t	=	______	kg	(kilograms)
1 kg	=	______	hg	(hectograms)
1 hg	=	______	dag	(__________)
1 dag	=	10	___	(__________)
1 g	=	______	dg	(__________)
1 dg	=	10	___	(__________)
1 cg	=	______	mg	(__________)

9. Look around your home and find 10 examples of the use of the metric system to measure the mass of household products. Examine grocery items, hardware items, and so on. For each example, list the product (or item) and the mass.

2.15 INVESTIGATION: Measuring Mass

In this investigation you measure the masses of several objects. The purpose is to make sure you can use the balance properly before you do investigations where you must not make mistakes.

Materials

balance
rubber stopper
cork
piece of wood
stone
piece of metal

Procedure

a. Follow your teacher's instructions carefully regarding how to use the balance. A balance is a delicate, expensive instrument. It can be easily damaged.
b. Estimate the mass of each object and place your answer in a table like Table 22.
c. Use the balance to find the mass of each of the five objects. Record your answers in the table. Compare your results with those of your classmates and with your estimates. Repeat any measurements that seem wrong.

TABLE 22 Measuring Mass

Object	Estimate (g)	Mass (g)

Discussion

1. Will objects with equal masses have equal volumes? Why?
2. Will objects with equal volumes have equal masses? Why?
3. How will temperature affect the volume and the mass of a liquid? (Hint: Think about the liquid in a thermometer.)

Highlights

The modernized metric system (SI) uses the decimal system and a set of basic definitions that are designed to make calculations easy.

The base unit of length is the metre. The kilometre (1 000 m) is the most commonly used multiple of the metre. The decimetre (0.1 m), centimetre (0.01 m), and millimetre (0.001 m) are widely used sub-multiples of the metre.

Area is the amount of surface. Units of area are derived from units of length. The square kilometre (km^2), square metre (m^2), and square centimetre (cm^2) are some common units of area. The hectare (square hectometre) is commonly used to give the area of farms and parks.

Volume is the amount of space an object occupies. Units of volume are also derived from units of length. The cubic metre (m^3), cubic decimetre (dm^3), and cubic centimetre (cm^3) are cubic units of volume. The kilolitre (kL), litre (L), millilitre (mL) are capacity units of volume.

$$1\ dm^3 = 1\ L \text{ and } 1\ cm^3 = 1\ mL.$$

The kilogram (kg) is the base unit of mass. Other common mass units are the tonne (t), gram (g), and milligram (mg).

UNIT

3 Density

A Characteristic Physical Property

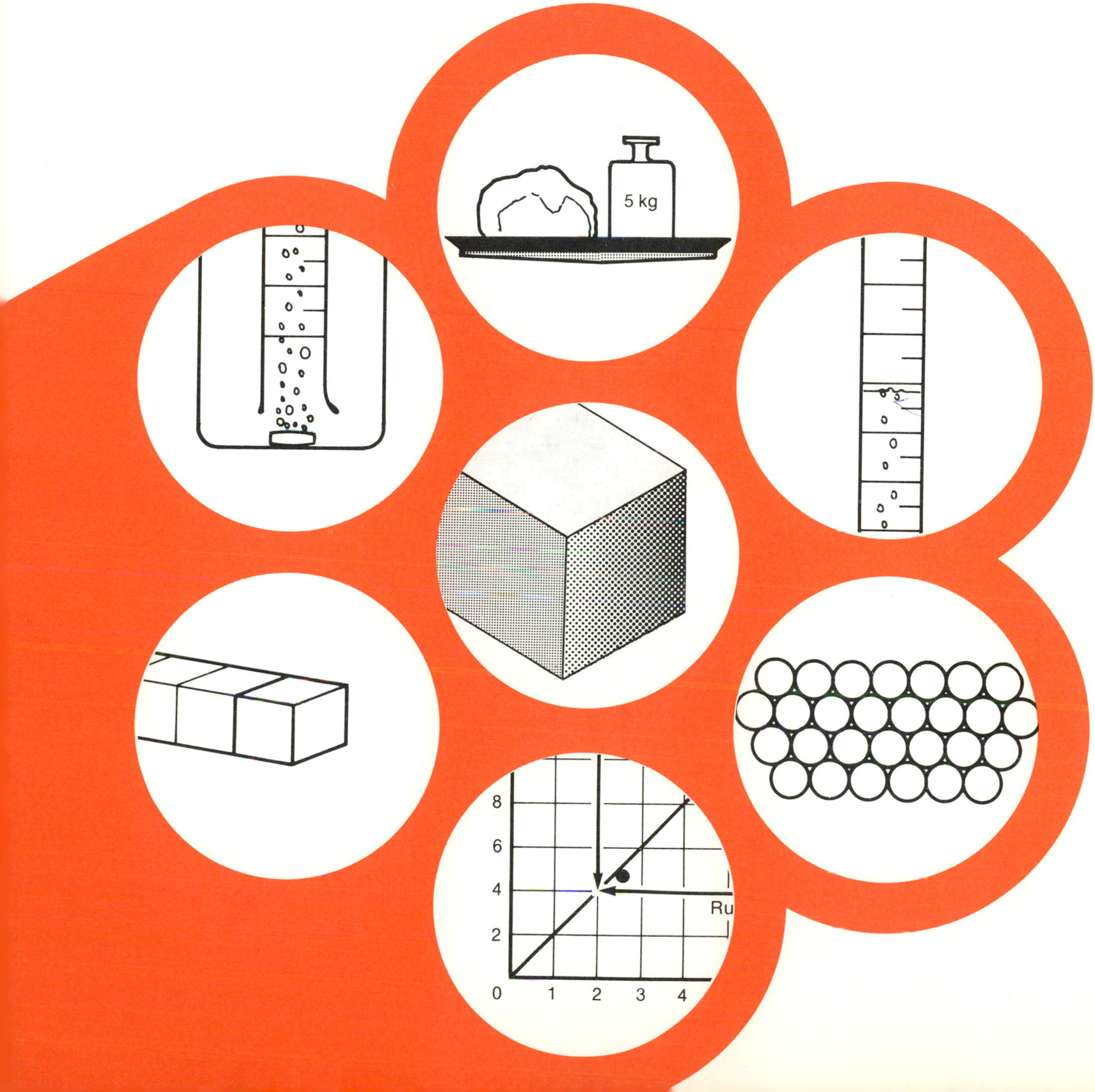

3.1 Physical and Chemical Properties

Matter is anything that has mass. All types of matter (substances) have **properties**. These are of two types—chemical and physical.

Chemical properties are those that involve the formation of a new substance. The fact that iron rusts is a chemical property. A new substance, rust, is formed. Look back to Investigation 1.5 on page 8. A chemical property of ammonium dichromate is the fact that it forms a new substance (the green powder) when heated.

Physical properties do not involve the formation of a new substance. Colour and taste are physical properties. The temperatures at which a substance melts (its melting point) and boils (its boiling point) are also physical properties. Read Section 2.1, page 23, again. There we learned that some physical properties like colour are not much help in identifying a substance. Yet others, like the melting point, can be very helpful. Very few substances have exactly the same melting point. But thousands may have the same colour.

Physical properties of a substance that are different from those of most other substances are called **characteristic physical properties**. For example, ice melts at 0°C; few other solids do. Thus melting point is a characteristic physical property. Why does a substance have characteristic physical properties? We can guess that it has something to do with what is "inside" the substance. No doubt you have heard that substances are made of atoms and molecules. Perhaps these determine in some way the characteristic physical properties of a substance.

In this unit you investigate density, one characteristic physical property of matter. By the time you have studied this unit and the next, you will know a great deal about matter and its characteristic properties. You will then be able to develop a theory about the structure of matter. This theory will help you explain the characteristic physical properties and many other interesting properties of matter.

Discussion

1. What is the difference between a chemical and a physical property?
2. Which of the following are chemical changes and which are physical changes: the freezing of water, the burning of a piece of paper, the crushing of a rock, the crumpling of a car fender in an accident, the rusting of a car fender, the melting of snow, the formation of rain, the dissolving of sugar in tea, the explosion of a firecracker?

3.2 Density: A Characteristic Physical Property

Chromium and nickel look very much alike. They are both hard, shiny metals that are used to make some of the trim on cars. Suppose you are given a piece of hard shiny metal and are told that it is either chromium or nickel (Fig. 3-1). Its mass is 250 g and its volume is 28 cm^3. How can you find out which metal it is? The answer, of course, is to study its characteristic physical properties. For example, the melting point of chromium is 1 615°C and that of nickel is 1 455°C. Therefore, if you heat some of the substance and it melts at 1 615°C, you will know that it is chromium. However, your school does not have the equipment to do this experiment. So let's look for another characteristic physical property.

You know how to measure mass and volume. And your school has the equipment for this. Are mass and volume characteristic physical properties? Volume is not since any volume of both metals is possible. You could buy 28 cm^3 of either nickel or chromium. So, knowing that the volume is 28 cm^3 will not tell you which metal it is. Nor is mass a characteristic physical property. Any mass of both metals is possible. You could buy 250 g of either metal. So, knowing that the mass is 250 g will not tell you which metal it is.

Figure 3-1 Is this piece of metal nickel or chromium? Its mass is 250 g and its volume is 28 cm^3.

Figure 3-2 Are volume and mass characteristic properties of nickel and chromium?

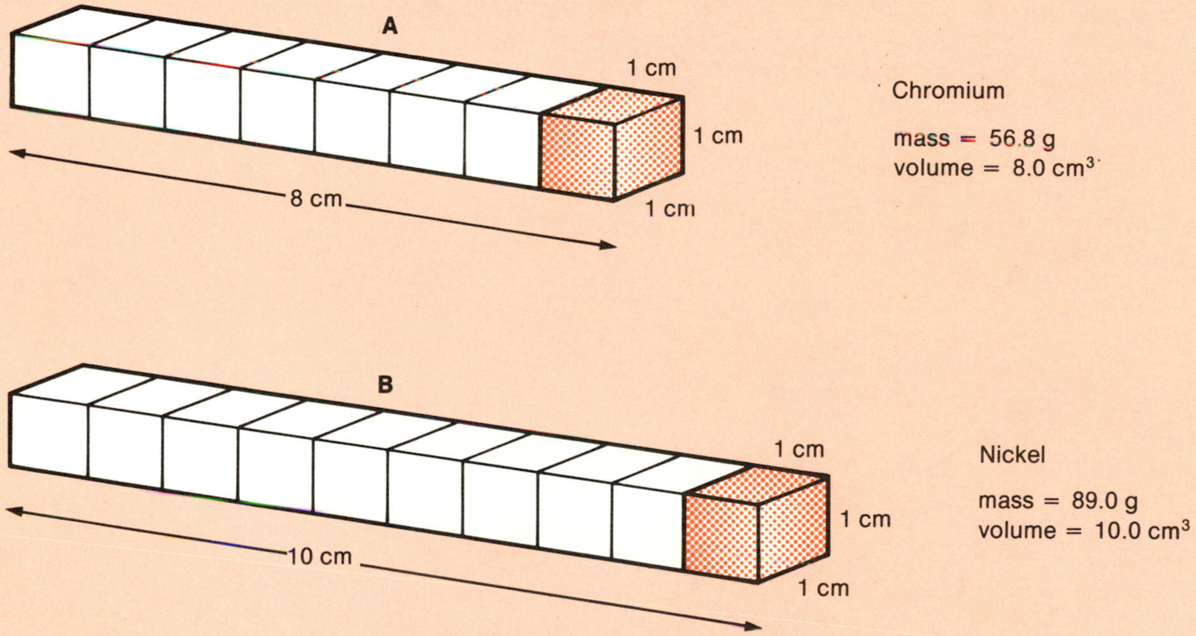

Look at Figure 3-2. By considering mass and volume together, we may be able to distinguish the metals. Bar A is made of chromium and Bar B is made of nickel. Let us compare the masses of 1 cm^3 of each metal.

Chromium: 8.0 cm^3 have a mass of 56.8 g

therefore 1.0 cm^3 has a mass of $\frac{56.8\ g}{8.0} = 7.1\ g$

Nickel: 10.0 cm^3 have a mass of 89.0 g

therefore 1.0 cm^3 has a mass of $\frac{89.0\ g}{10.0} = 8.9\ g$

Now, go back to the unknown metal in Figure 3-1 and find the mass of 1 cm^3 of it.

28 cm^3 have a mass of 250 g

therefore 1.0 cm^3 has a mass of $\frac{250\ g}{28} = 8.9\ g$

Since 1 cm^3 of the unknown metal has the same mass as 1 cm^3 of nickel, the unknown metal must be nickel.

Definition of Density

The mass of 1 cm^3 of nickel differs from the mass of 1 cm^3 of chromium. In fact, experiments have shown that it is different from the mass of 1 cm^3 of most substances. Thus the mass of 1 cm^3 ("a unit volume") of a substance is a characteristic physical property. It is called the density of the substance. **Density** *is the mass per unit volume* of a substance. If you are told that the density of nickel is 8.9 g/cm^3, you know that 1 cm^3 has a mass of 8.9 g. If you are told that the density of nickel is 8 900 kg/m^3, you know that 1 m^3 has a mass of 8 900 kg.

Units of Density

Since density is mass per unit volume, some units of density are:

grams per cubic centimetre (g/cm^3)
grams per millilitre (g/mL)
grams per cubic decimetre (g/dm^3)
grams per litre (g/L)
kilograms per cubic metre (kg/m^3)

Our example introduced density in grams per cubic centimetre (g/cm^3) because you should find it easy to visualize this unit. However, this unit is not widely used today. Instead, the preferred unit is kilograms per cubic metre (kg/m^3). Become familiar with it by using it wherever you can. It will make your work easier later in this course. The following should help you understand the meaning of kilograms per cubic metre.

Since $1\ dm^3 = 1\ L$
then $1\ g/dm^3 = 1\ g/L$

Also, since $1\ kg = 1\ 000\ g$ and $1\ m^3 = 1\ 000\ dm^3$,
then $1\ kg/m^3 = 1\ g/dm^3$

Remember that $\mathbf{1\ kg/m^3 = 1\ g/dm^3 = 1\ g/L}$

Since $1\ 000\ g = 1\ kg$ and $1\ 000\ 000\ cm^3 = 1\ m^3$,
then $\mathbf{1\ g/cm^3 = 1\ 000\ kg/m^3}$

Discussion

1. **a)** Define density.
b) What evidence shows that density is a characteristic physical property?

2. Calculate the density of people in your classroom. Express your answer in people per cubic metre.

3. What would you have to know in order to calculate the density of a rock?

4. **a)** What is the preferred unit of density?
b) State two other units that are equivalent to the preferred unit.

5. **a)** $1\ m^3$ of chromium has a mass of 5 680 kg. What is the density of chromium?
b) 2 700 kg of aluminum occupy a volume of $1\ m^3$. What is the density of aluminum?
c) How much more dense is chromium than aluminum?

6. **a)** The density of iron is $7\ 900\ kg/m^3$. What is the mass of $1\ m^3$ of iron?
b) The density of copper is $8\ 900\ kg/m^3$. How much space will 8 900 kg occupy?

3.3 INVESTIGATION: The Density of a Regular Solid

In Investigation 2.11 you learned how to find the volume of a regular solid. In Investigation 2.15 you learned how to measure mass. Here you will combine the two methods to determine the density of a regular solid.

Materials

4 rectangular solids (same material but different volumes)
ruler
balance
graph paper

Procedure

a. Determine the volume in cubic centimetres of each rectangular solid as outlined in Investigation 2.11 on page 42. Use the measurement method. Record your measurements and calculations in a table like Table 23.

b. Determine the mass in grams of each rectangular solid as outlined in Investigation 2.15 on page 49. Record your measurements in the table.

c. Calculate the density in grams per cubic centimetre of each rectangular solid as outlined in Section 3.2 on page 54. Record your results in the table.

d. Convert the density in grams per cubic centimetre to kilograms per cubic metre. Record your answers in the table.

e. Average your values for density. Record your answers in the table.

f. Plot a graph with mass on the vertical axis and volume on the horizontal axis. In addition to the 4 points your data give, you can also use zero as a point (when mass = 0, volume = 0). Draw a straight line beginning at zero and coming as close as possible to each of the other 4 points.

TABLE 23 Density of a Regular Solid

Solid	Length (cm)	Width (cm)	Height (cm)	Volume (cm^3)	Mass (g)	Density (g/cm^3)	Density (kg/m^3)
A							
B							
C							
D							
					Average density		

TABLE 24 Results of a Density Experiment

Solid	Volume (m^3)	Mass (kg)	Density (kg/m^3)
A	2.1	3 990	1 900
B	3.4	7 140	2 100
C	4.6	9 200	2 000
D	5.1	10 200	2 000

Discussion

1. Since all 4 solids were the same material, the densities should be the same. Were they? Why?
2. Table 24 shows the results of an experiment similar to yours but using a different type of solid.
   Which answer do you think is closest to the real density of the solid? Why?

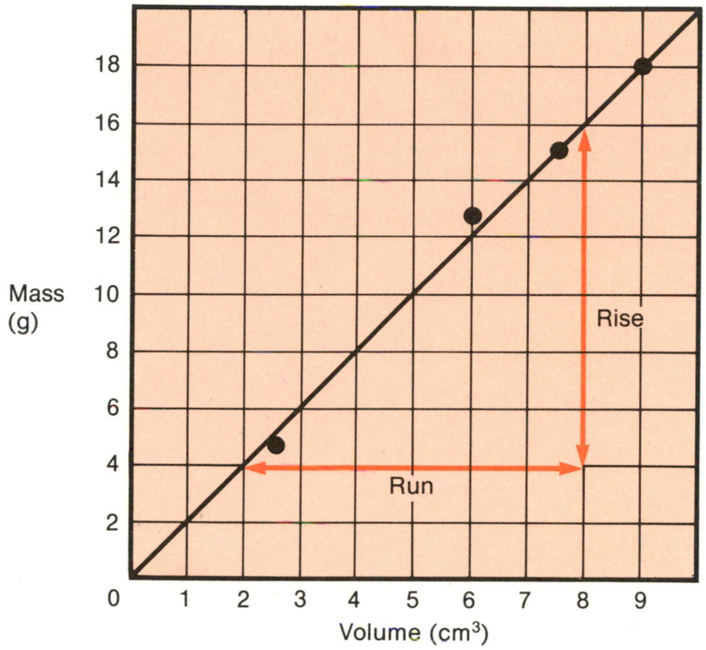

Figure 3-3 Density of a solid.

Figure 3-3 is a mass-volume graph for this substance. Let us calculate the slope ("slant") of the line. The formula for slope is:

$$\text{Slope} = \frac{\text{Rise}}{\text{Run}} = \frac{16.0\text{ g} - 4.0\text{ g}}{8.0\text{ cm}^3 - 2.0\text{ cm}^3} = \frac{12.0\text{ g}}{6.0\text{ cm}^3} = 2.0\text{ g/cm}^3$$

Note that the slope gives the density of the substance. Calculate the slope for other rises and runs. Compare your answers.

What value do you see in using the slope of the graph to calculate density? (Hint: Compare the slope to the average value.)

3. Determine the slope of your graph. What is the best answer for the density of the rectangular solid?
4. Use the graph in Figure 3-3 to find the mass of a piece of the solid that has a volume of 6.5 cm^3.
5. Use the same graph to find the volume occupied by a piece of the solid with a mass of 8.5 g.
6. Describe how you would find the density of a cylindrical solid.
7. Describe how you would find the density of a spherical solid.

3.4 INVESTIGATION: Identifying an Irregular Solid Using Density

In this investigation you are given a sample of metal pellets. The metal is one of those in Table 25. You are to find out which one it is by determining its density.

TABLE 25 Densities of Some Metals

Metal	Density (kg/m^3)
Magnesium	1 700
Aluminum	2 700
Tin	5 600
Zinc	7 100
Iron	7 900
Nickel	8 900
Lead	11 400

Materials

sample of metal pellets
graduated cylinder (100 mL)
balance

Procedure

a. Determine the mass of the metal pellets. Make sure that they are dry.
b. Determine the volume of the metal pellets using the most suitable method.
c. Calculate the density of the metal pellets. Express your final answer in kilograms per cubic metre.

Discussion

1. Compare your answer with the values in Table 25. What is the metal?
2. The density of cobalt is 8 900 kg/m^3 and the density of chromium is 7 100 kg/m^3. Compare these values with those in Table 25. Could density alone always be used to identify a substance? Explain.

3.5 INVESTIGATION: The Density of Water

"Water is a very special substance in the metric system."

Find out what this statement means by determining the density of water.

Materials

balance
graduated cylinder (100 mL)
graph paper

Procedure

a. Copy Table 26 into your notebook.

TABLE 26 Density of Water

	Column A	Column B	Column C	Column D	Column E	Column F
Trial	Volume of water (mL)	Mass of cylinder (g)	Mass of cylinder + water (g)	Mass of water (g)	Density (g/mL)	Density (kg/m^3)
1						
2						
3						
4						

b. Find the mass of the graduated cylinder. Enter your answer in Column B.
c. Add some water to the graduated cylinder. Read the volume (see Fig. 2-9, page 42). Enter your answer in Column A.
d. Find the mass of the water plus graduated cylinder. Enter your answer in Column C.
e. Calculate the mass of water by subtracting the mass of the graduated cylinder (Column B) from that of the water plus graduated cylinder (Column C). Enter your answer in Column D.
f. Calculate the density of water by dividing the mass of water (Column D) by the volume of water (Column A). Enter your result in Column E. Complete Column F.
g. Repeat steps c to f three times. Use a different volume of water each time.

h. Plot a mass-volume graph like the one you plotted in Investigation 3.3. Determine the slope of the line.

Discussion

1. Find the average of the four densities you got in Table 26.
2. What is the slope of the line?
3. What is the density of water in grams per millilitre (g/mL)? in grams per cubic centimetre (g/cm^3)? in kilograms per cubic metre (kg/m^3)?
4. What is meant by the statement "Water is a very special substance in the metric system"?

3.6 INVESTIGATION: The Density of Two Unknown Liquids

Your teacher will give you samples of two unknown liquids. Determine their densities using the same procedure that you used in Investigation 3.5.

Materials

balance
graduated cylinder (100 mL)
graph paper
samples of 2 unknown liquids

Procedure

a. Prepare two tables for the recording of the results.
b. Use a procedure like that of Investigation 3.5 to find the densities of the two liquids. Plot both graphs on the same piece of graph paper that you used for the water in Investigation 3.5. Use a different colour for plotting the graph for each liquid.

Discussion

1. What are the densities of the two liquids? Express your answers in kilograms per cubic metre.
2. What do you think makes one liquid more dense than another? (This is a hard question. Give it a try but don't worry if you cannot answer it.)

3.7 INVESTIGATION:
The Density of a Gas

Alka-Seltzer® tablets form carbon dioxide gas when they are placed in water. In this investigation you will measure the mass and volume of carbon dioxide gas produced by an Alka-Seltzer tablet. Then you will calculate the density of the gas.

Gases have such low densities that it is difficult to determine their mass. You cannot get enough of the gas on a balance to get a decent reading. Therefore, in this experiment you determine the mass by an **indirect method.** Instead of finding the mass of the gas, you find the mass of the system before and after the gas is given off. The difference is the mass of the gas.

Materials

balance
graduated cylinder (100 mL)
beaker (500 mL)
30 mL test tube
piece of wire
2 Alka-Seltzer tablets

Procedure A Mass of Gas

Follow this procedure to get the *mass* of gas given off by one Alka-Seltzer tablet:

a. Make sure the two Alka-Seltzer tablets have the same mass. If they don't, scrape material from the larger one until they do.
b. Fill the test tube one-third full of water.
c. Attach the test tube + water to the balance pan, using the wire.
d. Place one Alka-Seltzer tablet on the balance pan beside the test tube.
e. Find the mass of the test tube + water + wire + Alka-Seltzer tablet.
f. Drop the Alka-Seltzer tablet into the test tube. (You may have to break it.)
g. After fizzing has stopped, find the mass of the test tube + wire + contents.
h. Calculate the mass of gas given off. (Subtract g from e.)

Procedure B Volume of Gas

Follow this procedure to get the *volume* of gas given off by one Alka-Seltzer tablet:

a. Fill the graduated cylinder with water. Invert it in a 500 mL beaker that is half-full of water.
b. Break the Alka-Seltzer tablet into two halves.
c. Drop one half of the tablet into the beaker. Quickly put the mouth of the graduated cylinder over the tablet to collect the gas (Fig. 3-4).
d. After fizzing has stopped, read and record the volume of gas that collected in the graduated cylinder.
e. Repeat steps c and d for the other half of the tablet.
f. Add the volumes together. This gives the volume of gas formed by one tablet.

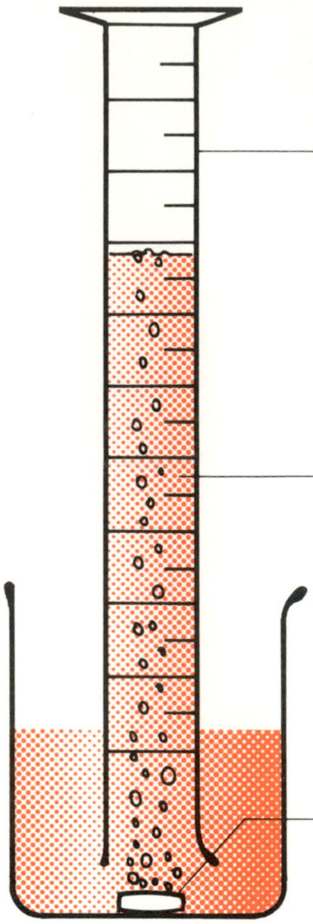

Figure 3-4 Density of carbon dioxide gas.

Discussion

1. What is the mass of gas formed?
2. What is the volume of gas formed?
3. Calculate the density of carbon dioxide gas.
4. Compare your result with the average value for the class.
5. Ask your teacher for the true value for the density of this gas. If your value differs from the true value, explain why.

3.8 Using Density to Identify Substances

Comparing Densities of Solids, Liquids, and Gases

Table 27 shows the densities of several solids, liquids, and gases. Study the table closely.

Note that, in general, solids are the most dense and gases the least dense. Note, too, that all densities were measured at a certain temperature. In addition, the gas densities were measured at a certain pressure. Why is this so? You have seen the liquid in a thermometer expand (rise) when the thermometer is warmed. Yet the mass of the liquid in the thermometer does not change; no liquid escapes. As a result, the density becomes less. Most liquids behave like this. Solids do likewise, but to a lesser extent. Gases expand even more than liquids when heated the same amount. Therefore it is important for accurate work to state the temperature at which the density was determined. For gases it is important to state the pressure also. Gases are easily compressed. A high pressure will squeeze a large mass of gas into a smaller volume. As a result, its density will increase. This happens when you pump up a bicycle tire.

TABLE 27 Some Densities

Solids (at 20°C) kg/m³		Liquids (at 20°C) kg/m³		Gases (at 0°C & standard pressure) kg/m³	
Osmium	22 500	Mercury	13 600	Carbon dioxide	1.98
Platinum	21 400	Carbon tetrachloride	1 600	Oxygen	1.43
Gold	19 300	Chloroform	1 490	Air	1.29
Uranium	18 700	Sea water	1 030	Nitrogen	1.25
Lead	11 300	Water	1 000	Helium	0.178
Nickel	8 900	Olive oil	920	Hydrogen	0.089
Copper	8 900	Turpentine	870		
Iron	7 900	Methyl alcohol	790		
Zinc	7 100	Ether	740		
Tin	5 600	Gasoline	690		
Aluminum	2 700				
Magnesium	1 700				
Ice (0°C)	920				

Density, Melting Point, and Boiling Point

In Investigation 3.4, page 58, you found out that density can often be used to identify substances—but not always. Two completely different substances can have the same density. If that happens, then we must use other characteristic physical properties such as melting point and boiling point. Look at Table 28. Chromium and zinc could not be told apart by density. However, either melting point or boiling point would serve the purpose. The same is true of tin and manganese. Look at the great difference in melting points—a candle flame will melt tin. Nickel, cobalt, and copper have the same density. You could tell copper by its colour, but the other two are shiny silvery metals. You would, again, have to use melting points and boiling points. Very careful experiments would have to be done this time. Look at how close the values are.

Methyl alcohol is a deadly poison. It attacks the nervous system causing blindness and death when drunk in even very small amounts. Ethyl alcohol is used in alcoholic beverages. Although it, too, attacks the nervous system, it is not as deadly. If people wanted to distinguish between these alcohols, they clearly could not use density to decide. What would they have to do?

Sometimes density alone can identify a substance. This is particularly true when there are only 3 or 4 possibilities to consider. However, you can almost always tell what the substance is by finding the three characteristic properties—density, melting point, and boiling point. In the next unit you will learn how to find the latter two.

TABLE 28 Some Characteristic Properties

Substance	Density (kg/m^3)	Melting point (°C)	Boiling point (°C)
Chromium	7 100	1 615	2 200
Zinc	7 100	420	907
Tin (white)	7 300	232	2 260
Manganese	7 300	1 260	1 900
Nickel	8 900	1 455	2 900
Cobalt	8 900	1 495	3 000
Copper	8 900	1 083	2 300
Methyl alcohol	790	− 98.0	64.7
Ethyl alcohol	790	− 117.3	78.5

Discussion

1. a) In general, how do the densities of solids, liquids, and gases compare?
b) How does an increase in temperature affect the density of most substances? Why is this so?
c) How does pressure affect the density of a gas? Why is this so?

2. a) Explain why density, alone, cannot often be used to identify a substance.
b) What other characteristic properties could be used, with density, to identify a substance?
c) How would you distinguish between methyl alcohol and ethyl alcohol?

3. a) Why will turpentine float on water?
b) Will a piece of iron float or sink in mercury? Why?

3.9 A Theory to Explain Density

Lead and tin look very much alike. They are both soft silvery metals. You can bend a fairly thick bar of either metal with little effort. Yet they differ greatly in density. Lead has a density of 11 300 kg/m^3; tin has a density of 5 600 kg/m^3. Lead is twice as dense as tin (Fig. 3-5). What is there about lead that makes 1 m^3 of it have twice the mass of 1 m^3 of tin? Look at and feel the samples of lead and tin your teacher has. Do they look different? Do they feel different?

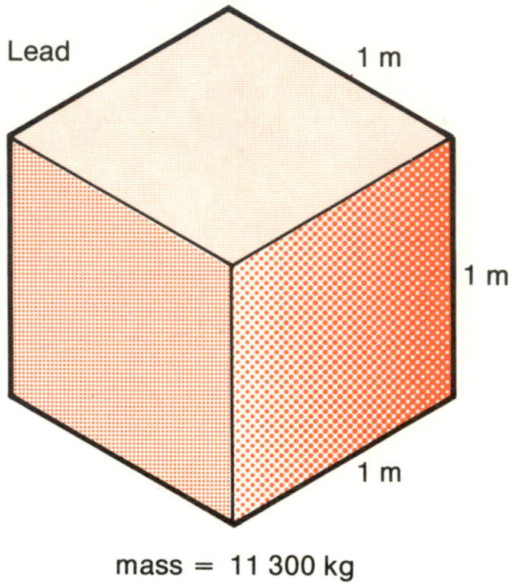

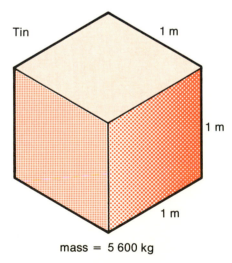

Figure 3-5 Why is lead twice as dense as tin?

More than 2 000 years ago the Greek philosophers had two theories for the structure of matter. One is called the **continuous** theory. It says that a sample of matter is all one continuous piece. Thus a pail of water is one piece of water; a glass of water is also one piece of water; so is a tiny dew-drop. The other theory is called the **particle** theory. It says that all matter is made of tiny particles held together by some kind of hooks or forces. Thus a glass of water simply has more water particles in it than a dew-drop. A pail of water has more water particles in it than a glass of water.

The continuous theory cannot explain the difference in density between lead and tin. For this (and other) reasons, it has been discarded. The particle theory can explain the difference in density, provided we assume that lead particles are different from tin particles. Maybe lead particles are more massive; maybe they are closer together and there are more of them in 1 m^3.

We really do not have enough evidence so far in this course to prove the particle theory, even though it seems to explain differences in density. Keep it in mind, however, and we will see if it explains the characteristic properties we study in the next unit.

Discussion

1. Explain the difference between the continuous theory and the particle theory.
2. How might lead particles differ from tin particles?
3. When water freezes, its density changes from 1 000 kg/m^3 to 920 kg/m^3. According to the particle theory, what has likely happened?
4. Table 28 shows us that methyl alcohol and ethyl alcohol have the same density. Suppose that the particle theory is true and that ethyl alcohol particles have a greater mass than methyl alcohol molecules. What else must be true about the particles of these liquids?

3.10 Problems Involving Density

Problems involving density are of three types. You may be given the mass and volume and be asked to find the density. You may be given the density and mass and be asked to find the volume. You may be given the density and volume and be asked to find the mass. The following are model solutions to these three types of problems.

TYPE 1. Finding Density, Given Mass and Volume

Problem

A sample of iron has a mass of 31 600 kg and a volume of 4.0 m^3. Find the density of iron.

Solution Using Statements

4.0 m^3 of iron have a mass of 31 600 kg

therefore 1 m^3 of iron has a mass of $\frac{31\,600 \text{ kg}}{4.0} = 7\,900$ kg

Therefore the density of iron is 7 900 kg/m^3.

Solution Using Formula

Density is mass per unit volume.

Therefore the formula for density is $\boldsymbol{D = \frac{m}{V}}$

Thus $D = \frac{31\,600 \text{ kg}}{4.0 \text{ m}^3} = 7\,900 \text{ kg/m}^3$

TYPE 2. Finding Volume, Given Density and Mass

Problem

Iron has a density of 7 900 kg/m^3. Find the volume occupied by 31 600 kg of iron.

Solution Using Statements

Iron has a density of 7 900 kg/m^3.

Therefore 7 900 kg of iron occupy 1 m^3.

or 1 kg of iron occupies $\left(\frac{1}{7\,900}\right)$ m^3

and 31 600 kg of iron occupy $\left(\frac{1}{7\,900} \times 31\,600\right)$ m^3 = 4.0 m^3.

Solution Using Formula

$$D = \frac{m}{V}$$

$$\text{or} \quad V = \frac{m}{D} = \frac{31\,600 \text{ kg}}{7\,900 \text{ kg/m}^3} = 4.0 \text{ m}^3.$$

TYPE 3. Finding Mass, Given Density and Volume

Problem

Iron has a density of 7 900 kg/m^3. Find the mass of a piece of iron with a volume of 4.0 m^3.

Solution Using Statements

Iron has a density of 7 900 kg/m^3.

Therefore 1 m^3 of iron has a mass of 7 900 kg.

or 4.0 m^3 of iron have a mass of (7 900 x 4.0) kg = 31 600 kg.

Solution Using Formula

$$D = \frac{m}{V}$$

or $m = D \times V = 7\,900\ \text{kg/m}^3 \times 4.0\ \text{m}^3 = 31\,600\ \text{kg}.$

Discussion

1. What is the density of a metal if 38 600 kg of the metal occupy a volume of 2.0 m^3? Look back to Table 27. What is the metal?
2. A piece of aluminum has a volume of 10.0 m^3 and a mass of 27 000 kg. What is the density of aluminum?
3. An oil storage tank contains 40 m^3 of oil with a mass of 35 000 kg. What is the density of the oil?
4. **a)** What is the mass of 50 m^3 of water?
   b) What is the mass of 50 m^3 of ethyl alcohol? (See Table 28.)
5. **a)** What volume is occupied by 40 000 kg of water?
   b) What volume is occupied by 40 000 kg of methyl alcohol?
6. **a)** Show that 1 kg/m^3 = 1 g/dm^3.
   b) Show that 1 g/dm^3 = 1 g/L.
   c) Show that 1 g/cm^3 = 1 000 kg/m^3.
7. A rectangular solid measures 8.0 cm × 20.0 cm × 10.0 cm. It has a mass of 3 040 g. What is its density in grams per cubic decimetre and in kilograms per cubic metre?
8. The density of iron is 7 900 kg/m^3 and the density of aluminum is 2 700 kg/m^3.
   a) Compare the volumes of blocks of iron and aluminum both of which have a mass of 50 000 kg.
   b) Compare the masses of blocks of iron and aluminum both of which have a volume of 50 m^3.

9. A sample of metal has a density of 10 500 kg/m^3 and a mass of 2 000 kg. What is its volume?
10. A piece of white pine wood has a density of 430 kg/m^3. What is the mass of 50 m^3 of this wood?
11. A piece of nickel has a mass of 130 g and a density of 8 900 kg/m^3. A piece of magnesium has a mass of 60 g and a density of 1 700 kg/m^3. Which piece of metal will displace more water from an overflow can?
12. **a)** Would a 100 mL graduated cylinder hold 120 g of ethyl alcohol (density = 790 kg/m^3)?
b) Would a 100 mL graduated cylinder hold 250 g of mercury (density = 13 600 kg/m^3)?
13. A few years ago North American car makers used steel (mostly iron) to construct the frame, motor, and transmission case of their cars. Today aluminum is used for much of the same parts. Explain why.
14. The density of gasoline is 690 kg/m^3 and the density of turpentine is 870 kg/m^3. Suppose you find a 5 L can that is full of one of these liquids. You do not know which, and the label has fallen off. How could you tell which one is in the can without removing the lid?

Highlights

Density is the mass per unit volume of a substance. It is a characteristic physical property and, as a result, can be used to help identify substances. In general, solids are more dense than liquids, and liquids are more dense than gases.

The preferred unit of density is the kilogram per cubic metre (kg/m^3).

The density of a substance can be determined in the laboratory by finding the mass and the volume of a sample of the substance. These measurements are then substituted in the formula $D = \frac{m}{V}$. The same formula can be used to calculate the volume if the mass and density are known and also to calculate the mass if the volume and density are known.

The particle theory partially explains why substances have different densities.

UNIT

4 The Three States of Matter

More Characteristic Physical Properties

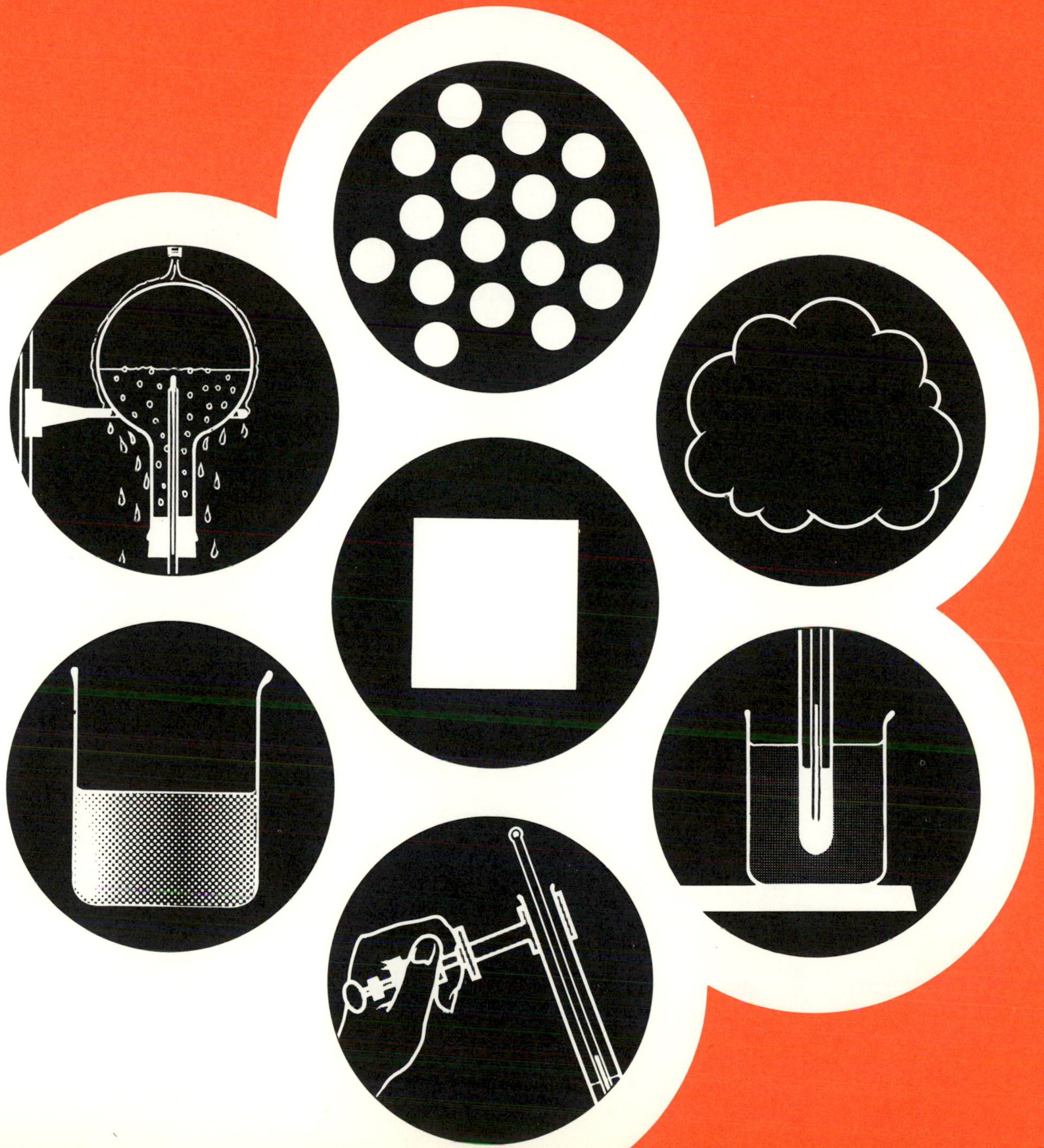

You found out in Unit 3 that all forms of matter have mass and volume. In fact, **matter** is defined as anything that has mass. You also found out that matter occurs in three states—solid, liquid, and gaseous.

In this unit you study the nature and behaviour of the three states of matter. You will examine more characteristic physical properties which, along with density, can be used to identify substances. But, most important, you will learn enough about matter so that you will be able to understand one of the most interesting and important theories of all science—the particle theory.

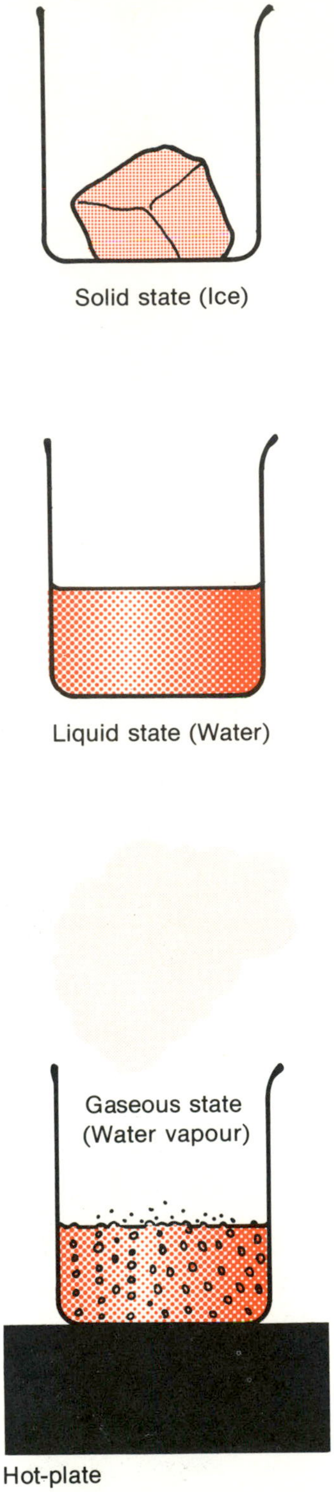

Figure 4-1 Water exists in all three states—solid, liquid, and gaseous.

4.1 The Three States of Matter

The purpose of this section is to make sure you have the knowledge needed to do the investigations that follow. Read it carefully. Pay particular attention to any terms that are new to you.

As you read these descriptions, refer to Figure 4-1. Keep in mind that water is just one of thousands of substances that can exist in three states.

The Solid State

A solid has a definite shape and a definite volume. That is, it is rigid or stiff and is not easily pressed together or compressed. (Under extremely high pressures a small change in volume can occur.) Most solids increase in volume (expand) when heated, although the change is not great.

The Liquid State

The most obvious thing about a liquid is the fact that it can be poured. It is said to be **fluid**, because it flows. The shape is not definite like that of a solid. Instead, a liquid flows to take the shape of its container. However, the volume of a liquid remains the same regardless of the shape (Fig. 4-2). Like solids, liquids are almost non-compressible. The hydraulic brakes of a car would not work if the brake fluid were compressible. Liquids increase in volume when heated. This property is used in thermometers that contain mercury or alcohol.

The Gaseous State

A gas has neither a definite shape nor a definite volume. It usually expands to fill its container. For example, if you spray

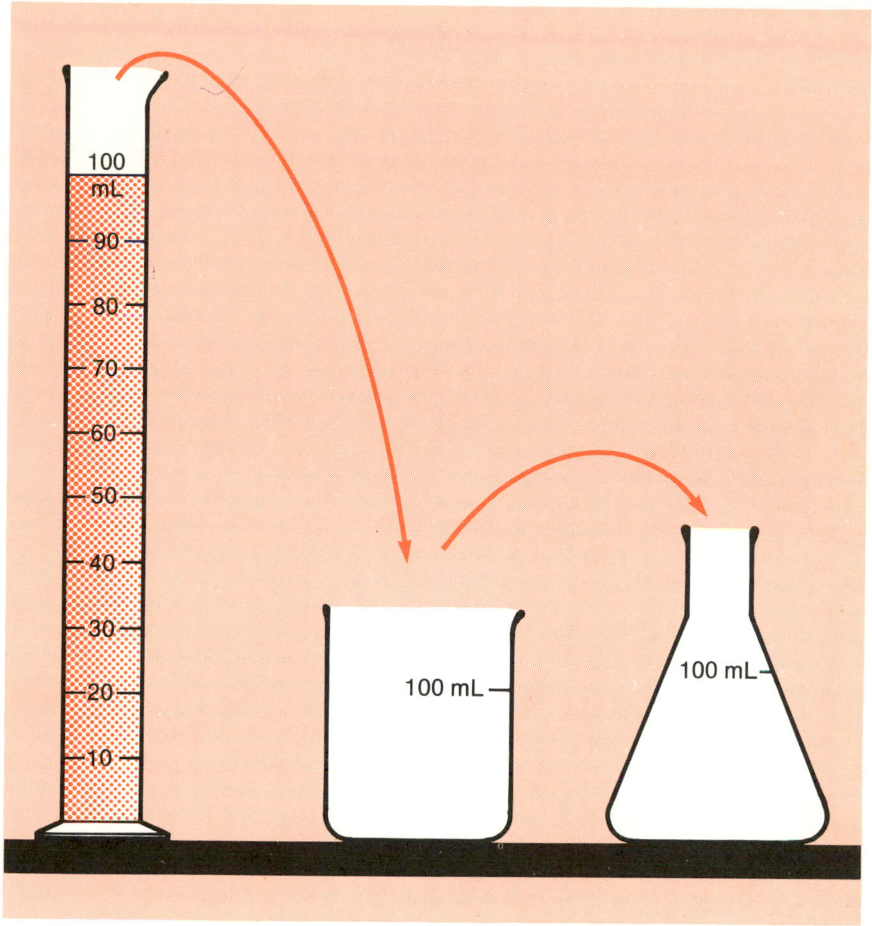

Figure 4-2 Pour 100 mL of water from a graduated cylinder to a 100 mL beaker, then to a 100 mL Erlenmeyer flask. What changes occur in the shape and the volume?

some room deodorant into one corner, the deodorant does not stay there. It gradually spreads across the room. Thus a gas has both the volume and shape of its container. Unlike solids and liquids, gases are easily compressed. You have probably seen "bottles" of compressed oxygen and other gases. A gas expands considerably when heated. The engine of a car makes use of this fact. The exploding gasoline in a cylinder heats up the air and other gases in the cylinder, causing them to expand. This, in turn, pushes the piston down.

Gases, like liquids, are fluids. They can be poured. If you were not aware of this fact, try this experiment at home. Make some carbon dioxide gas by putting vinegar on baking soda in a glass. After the mixture has fizzed for a few minutes, the glass should be full of carbon dioxide gas. Pour the gas (but not the vinegar and baking soda) into a second glass that contains a short, burning candle (Fig. 4-3).

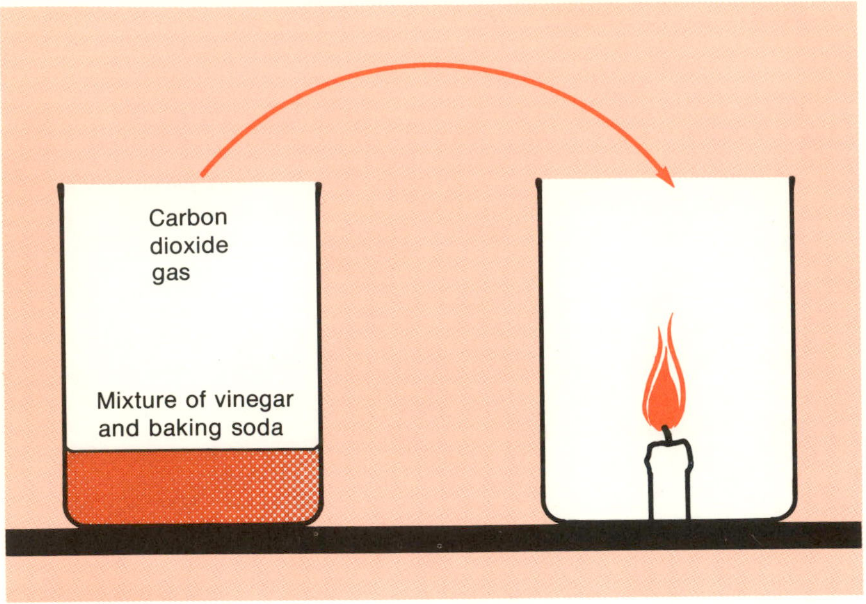

Figure 4-3 Will a gas pour?

Gas and Vapour

A vapour is the gaseous state of a substance that is in the solid or liquid state at room temperature. Thus we speak of water vapour coming out of a kettle instead of water gas. Alcohol and gasoline are liquids at room temperature. Therefore their gaseous states are called alcohol vapour and gasoline vapour. Camphor, the smelly substance in some cold remedies, is a solid at room temperature. Thus the gaseous state you inhale is called camphor vapour.

A gas is a substance that is in the gaseous state at room temperature. Oxygen, hydrogen, nitrogen, and carbon dioxide are called gases instead of vapours.

Discussion

1. Complete Table 29.

TABLE 29 Comparing Solids, Liquids, and Gases

Property	Solid	Liquid	Gas
Shape			
Volume			
Effect of compression			
Effect of heat			

2. **a)** Distinguish between a gas and a vapour.
   b) Classify the gaseous state of each of the following as either a gas or vapour: water, nitrogen, oxygen, rubbing alcohol, moth crystals, helium, turpentine, fuel oil, dry ice (frozen carbon dioxide).
3. **a)** If all matter can be classified as solid, liquid, or gaseous, how would you classify butter, margarine, lard, and shortening? (Review the properties of the solid state.) Explain your answer.
   b) Sand can be poured. Is it a fluid? Explain.
   c) Look at Figure 4-3. The carbon dioxide gas could be poured through air because it is more dense than air. Helium gas is much less dense than air. That is why it is used in weather balloons. Can it be poured through air? Explain.

4.2 Changes of State

Figure 4-4 shows the six changes of state that can occur. Study the diagram and the following definitions carefully. You need to know these terms before doing the investigations in this unit. As you think about each change of state, try to answer these questions. Your answers are the predictions for the investigations that follow. Remember that water is not the only substance that can be in three states. Even iron can be a liquid and a vapour.

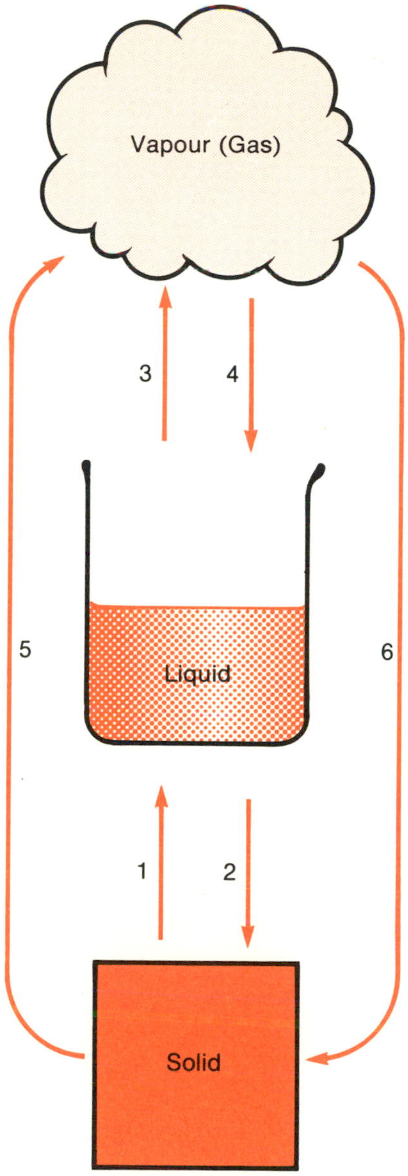

Figure 4-4 Changes of state.

Questions to Consider

1. Does this change of state require heat or must heat be removed?
2. Does the volume increase or decrease during this change of state?

Definitions

1. **Melting (liquefaction; fusion)** is the change of state in which a solid changes to a liquid.
2. **Freezing (solidification)** is the change of state in which a liquid changes to a solid.
3. **Evaporation (vaporization)** is the change of state in which a liquid changes to a vapour or gas. Sometimes evaporation is so rapid that bubbles of vapour form quickly throughout the entire mass of liquid. This is called **boiling**.
4. **Condensation (liquefaction)** is the change of state in which a vapour or gas changes to a liquid. Note that the term liquefaction applies to both 1 and 4. It means "formation of a liquid". A liquid is being formed in both changes of state.

5. and 6. Sublimation is the change of state directly from a solid to a vapour and/or from a vapour to a solid, without the appearance of a liquid.

Discussion

1. Summarize your thoughts and predictions regarding changes of state by completing Table 30. Copy the table into your notebook so that you can go back and check your predictions after the investigations.

TABLE 30 Changes of State: Predictions

	Heat		Volume	
Change of state	**Add**	**Remove**	**Increases**	**Decreases**
Melting				
Freezing				
Evaporation				
Condensation				
Sublimation (#5)				
Sublimation (#6)				

2. Give a reason for each of the predictions you made in question 1. For example, if you marked "Add" and "Decreases" for melting, explain why you did so.

4.3 INVESTIGATION: The Melting of Ice—Mass and Volume

The purpose of this investigation is to find out what happens to the mass and volume of water as it changes from a solid to a liquid. Also, it should show whether this change of state requires heat or removal of heat.

Materials

small ice cubes
graduated cylinder (100 mL)
balance
thin pointed object (dissecting needle or stiff wire)

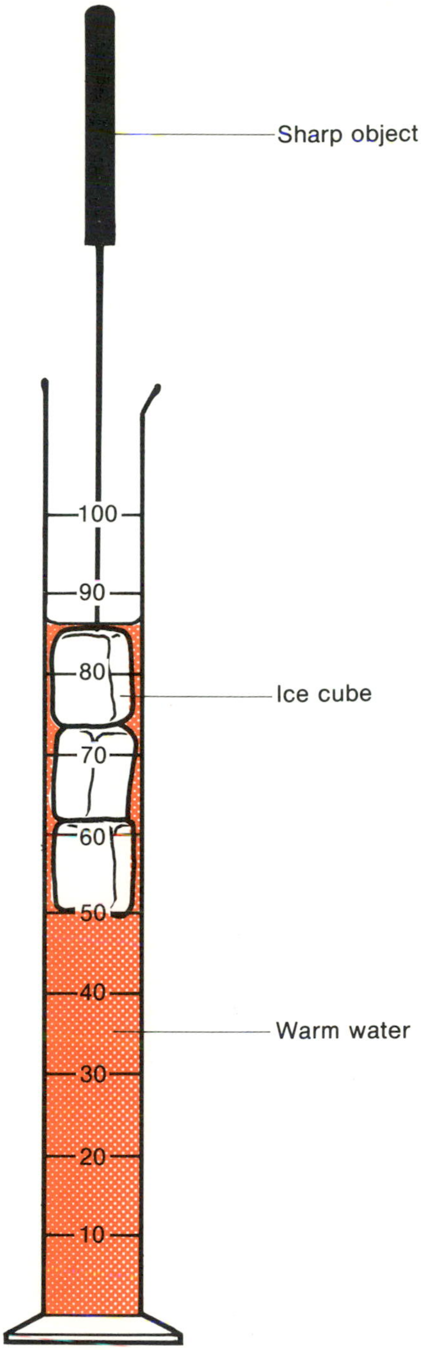

Figure 4-5 Mass and volume change during melting.

Procedure

a. Place 50 mL of warm water (60-70 °C) in the graduated cylinder.
b. Add 3 or 4 ice cubes to the water. Immediately push them under the water using the sharp object. This is most easily done if you use ice cubes that are just a little less wide than the diameter of the graduated cylinder (Fig. 4-5). Measure the volume of the water + ice cubes as quickly as you can.
c. Quickly place the graduated cylinder on the balance. Measure the mass of the cylinder + water + ice. Leave it on the balance for the rest of the experiment. Measure the mass after the ice has all melted.
d. Measure the volume of water after the ice has all melted.
e. Calculate the volume of ice and the volume of melted ice.
f. Record your results and calculations in a table like Table 31.
g. If time permits, repeat the experiment twice.

TABLE 31 Melting of Ice

Measurement or calculation	Trial 1	Trial 2	Trial 3
Volume of warm water			
Volume of water + ice			
Volume of ice			
Volume after melting			
Volume of melted ice			
Mass before melting			
Mass after melting			

Discussion

1. Give three names for the change of state that occurred during this investigation.
2. What happens to the mass of ice when it melts?
3. What happens to the volume of ice when it melts?
4. What is a possible explanation for the observed behaviour of volume in this investigation?
5. What characteristic property of water changed when the ice melted? In what way did it change? Why?
6. Do you think that all substances behave like water when they change from the solid state to the liquid state? (Hint: This is a hard question that may require a trip to the library or a discussion with a more experienced person. You might seek an answer to this question: 'What is the **anomalous behaviour** (abnormal behaviour) of water?)

4.4 INVESTIGATION: The Effect of Heat on the Temperature of Melting Ice

In this investigation you measure the temperature of ice as it melts. What do you think will happen to the temperature?

Materials

finely crushed ice (or snow)
beaker (250 mL)
thermometer
glass or plastic stirring rod
ring stand, iron ring, adjustable clamp, wire gauze
watch or clock with second hand
Bunsen burner
string

CAUTION: Wear safety goggles during this investigation.

Procedure

a. Half fill a 250 mL beaker with crushed ice.
b. Set up the apparatus as shown in Figure 4-6. Be sure that the flame is very low.
c. Stir the ice continually with the stirring rod. Do not stir with the thermometer; it may break.
d. Record the temperature every 30 s until all of the ice is gone. Keep the thermometer toward the centre of the beaker. Do not let it touch the bottom.
e. Continue to record the temperature every 30 s for 5-10 min after the ice is gone.
f. Plot a temperature-time graph, with time on the horizontal and temperature on the vertical axis.

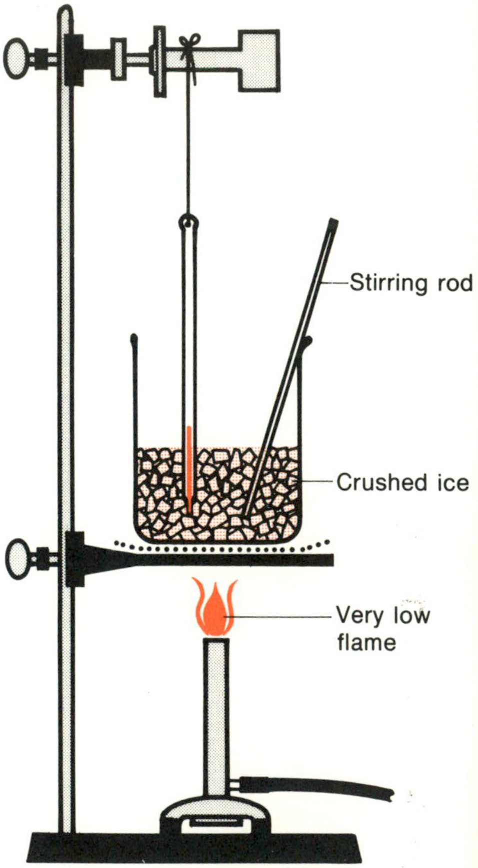

Figure 4-6 Effect of heat on the temperature of melting ice.

Discussion

1. What happens to the temperature while the ice is melting?
2. What happens to the heat that the Bunsen burner puts into the ice-water mixture?
3. Does the melting of ice require the addition of heat or does it require the removal of heat?
4. The graph you plotted is called the **heating curve** for water. What do you think the **cooling curve** would look like?
5. What is the melting point of ice? What happens to the graph at this temperature?
6. What is the freezing point of water (liquid)?
7. Imagine a lake with large chunks of ice floating at the surface. What will the temperature of the water be among the chunks of ice? Will it be the same at the bottom of the lake? Why?
8. Under what conditions would it be possible to have ice that is colder than 0°C?

4.5 INVESTIGATION: The Effects of Heating and Cooling on Paradichlorobenzene

This investigation is similar to Investigation 4.4, with two exceptions. First, the substance being used has a melting point that is above room temperature. In other words, it is a solid at room temperature. Second, you will study its cooling behaviour as well as its warming behaviour.

Materials

paradichlorobenzene (15 g)
beaker (400 mL)
test tube
ring stand, iron ring, wire gauze, adjustable clamp
Bunsen burner
marking pen
2 thermometers (-10°C to 110°C)
water bath at 70°C (on teacher's bench)
watch or clock with second hand

CAUTION: Wear safety goggles during this investigation.

Procedure A Cooling Behaviour

In this part of the procedure you melt the paradichlorobenzene and then study its behaviour as it cools.

a. Add the 15 g of paradichlorobenzene to the test tube. Put your initials on the test tube. Then place it in the water bath that your teacher has prepared. Leave it there until all of the paradichlorobenzene has melted.

b. While you are waiting for the paradichlorobenzene to melt, one partner should prepare a data table for the recording of the results. The column headings should be "Time (s)", "Temperature (°C)", and "Observations and Comments".

c. At the same time, the other partner should prepare the apparatus as shown in Figure 4-7 so that it is ready to receive the test tube. The water in the beaker should be at room temperature.

d. Decide which partner will be the timer-recorder and which one will read the temperature.

e. When the paradichlorobenzene has melted, clamp the test tube into position as shown in Figure 4-7. The part of the test tube containing paradichlorobenzene must be completely under water.

f. Record the temperature every 30 s until the temperature has dropped to 40°C. For each reading estimate the temperature to the nearest 0.2°C. Do not let the thermometer touch the

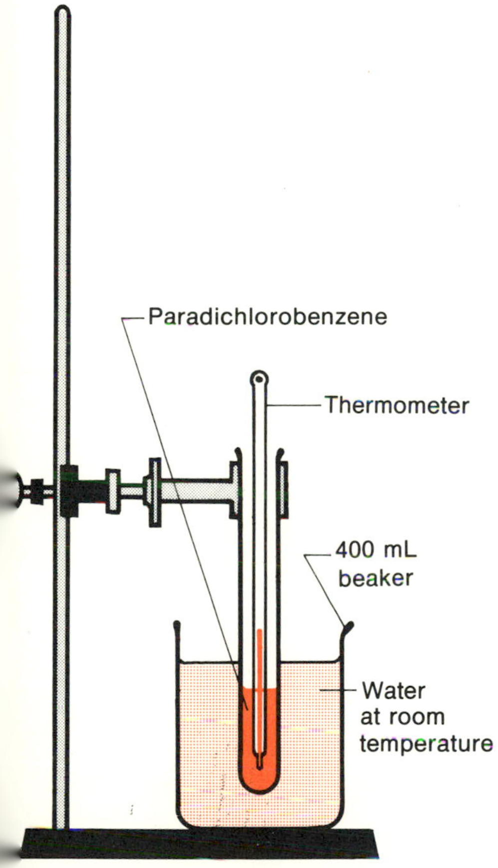

Figure 4-7 Cooling behaviour of paradichlorobenzene.

bottom. Keep the bulb near the centre of the liquid. Leave it there when the paradichlorobenzene solidifies.

g. Note in the "Observations and Comments" column of your data table the temperatures at which solidification first starts and at which it is complete.

Procedure B Warming Behaviour

In this part of the procedure you warm up a water bath. Then you use it to warm the paradichlorobenzene so you can study its warming behaviour.

a. Remove the test tube from the water bath.

b. Mount the beaker of water on a ring stand and iron gauze as shown in Figure 4-8.

c. Heat the water bath to about 70°C. (If your classroom has hot running water, start with that to save time.)

d. Turn off the burner.

e. Clamp the test tube back in the water bath in the same position it was in for Procedure A.

f. Record the temperature every 30 s until the temperature of the paradichlorobenzene is about 60°C. Move the thermometer around in the paradichlorobenzene just as soon as you can, but do not strike the bottom.

g. Note in your data table when melting starts and when it is complete.

h. Keep the second thermometer in the water bath during this entire procedure. If the water temperature drops below 60°C before the paradichlorobenzene is melted, turn on the Bunsen burner and warm the water up to 60°C or 65°C. If you don't, the experiment will take too long.

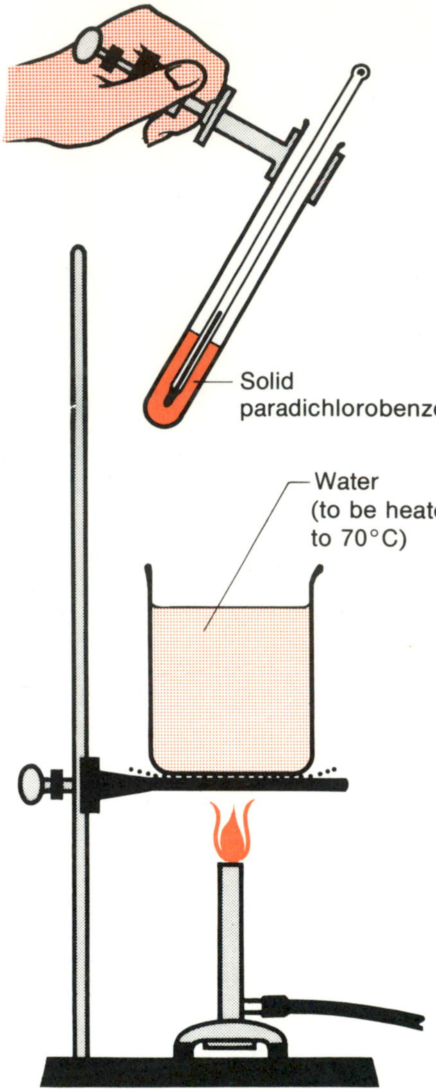

Figure 4-8 Warming behaviour of paradichlorobenzene.

Procedure C Plotting the Results

a. Use the data from Procedure A to plot a cooling curve for paradichlorobenzene. Put time on the horizontal axis and temperature on the vertical axis.

b. Use the data from Procedure B to plot a heating curve for paradichlorobenzene. Use a different colour for this curve. Plot it on the same axes as the data from Procedure A.

Discussion

1. What changes of state occurred in this investigation?
2. What is occurring during the flat part of the cooling curve?
3. What is occurring during the flat part of the warming curve?
4. What is the melting point of paradichlorobenzene?
5. What is the freezing point of paradichlorobenzene?
6. What is the relationship between the freezing point and the melting point of a substance?

7. What states of matter are present at the freezing point and melting point?
8. Compare your results for Investigations 4.4 and 4.5 to those of your classmates. What evidence do you have that melting point is a characteristic physical property?
9. The odour of paradichlorobenzene repels moths. If you looked at some paradichlorobenzene in a clothes closet, you would see only the solid state. What change of state occurs in the clothes closet? (Note: Moths do not eat paradichlorobenzene!)

4.6 INVESTIGATION: Effects of Salt and Antifreeze on the Melting Point of Ice

In this investigation you study the effects of salt and antifreeze on the melting point of ice.

Materials

beaker (250 mL)
thermometer
table salt
10 mL measure
finely crushed ice (as small as coarse sand)
antifreeze (ethylene glycol)
graduated cylinder (25 mL)
stirring rod

Procedure A Effect of Salt on the Melting Point of Ice

a. Pour crushed ice into the beaker until it is half full.
b. Take the temperature of the ice while you are stirring it. Do not stir with the thermometer. Use the stirring rod.
c. When the thermometer reads 0°C, add 1 level measure of salt to the ice. Continue stirring. Record the temperature after it stops changing.
d. Continue adding 1 level measure of salt at a time, with continuous stirring and temperature readings, until you have added 7 or 8 measures.
e. Record your results in a table.
f. Empty the beaker and wash it.

Procedure B Effect of Antifreeze on the Melting Point of Ice

The antifreeze used in this investigation is ethylene glycol. It is the permanent type that comes in new cars.

Repeat Procedure A, but use 15 mL of antifreeze each time instead of 1 level measure of salt.

Discussion

1. What evidence do you have that the ice is at its melting point, even after salt and antifreeze are added? (Look back to step 5 of the Discussion in Investigation 4.5.)
2. What effect does salt have on the melting point of ice?
3. What effect does antifreeze have on the melting point of ice?
4. Explain why salt is added to ice-covered roads.
5. Explain why antifreeze is added to the water in the radiators of cars in cold areas of the country.
6. Why is salt not used in the radiators of cars?
7. Where did the heat go in this experiment? In other words, how can something get colder when you add a warmer substance to it in a warm room?

4.7 INVESTIGATION: Effect of Heat on the Temperature of Boiling Water

In this investigation you measure the temperature of water as it boils. What do you think will happen to the temperature?

Materials

round-bottomed flask (250 mL)
thermometer (−10°C to 110°C)
distilled water
boiling chips
Bunsen burner
ring stand, iron ring, wire gauze, 2 adjustable clamps

CAUTION: Wear safety goggles during this investigation.

Procedure

a. Set up the apparatus as shown in Figure 4-9. The bulb of the thermometer must be just below the surface of the water. The boiling chips are added to ensure even boiling.
b. Note the temperature of the water.
c. Begin heating the water strongly. Record the temperature every 30 s.
d. Continue heating until the water is boiling rapidly. Keep it boiling for about 10 min. During this time continue to record the temperature every 30 s. Make sure the bulb of the thermometer is always just below the surface of the water.
e. Record all of your findings in a table. Be sure to note the temperature at which boiling began.

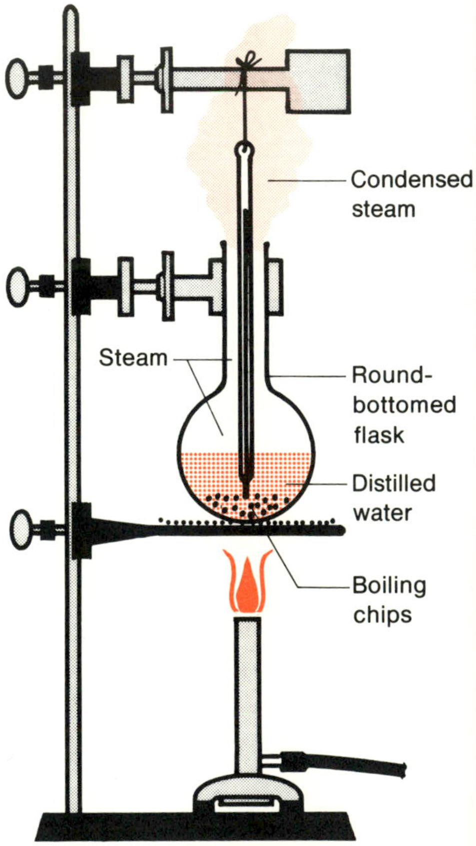

Figure 4-9 Studying boiling water.

f. Plot a temperature-time graph, with time on the horizontal and temperature on the vertical axis.

g. While the water is still boiling, take the temperature deep in the liquid and also in the steam just outside the mouth of the flask. (See Figure 4-9.)

Discussion

1. What happens to the temperature while water is boiling?
2. What happens to the heat that the Bunsen burner puts into the water?
3. What two changes of state took place in this experiment?
4. Which change requires heat and which requires loss of heat?
5. What is the boiling point of water?
6. Compare your results to those of some of your classmates. Probably each person started with a different mass of water. Does the boiling point depend on the mass of water used?
7. What similarities and what differences would you see in the results if you used a liquid other than water in this experiment?
8. Why is boiling point called a characteristic physical property?
9. What was the temperature deep in the water? Why?
10. What was the temperature in the steam? Why?
11. Figure 4-10 is a warming curve for a liquid. What is happening during part A? part B? part C? What is the boiling point of the liquid? On what factors does the length of B depend?

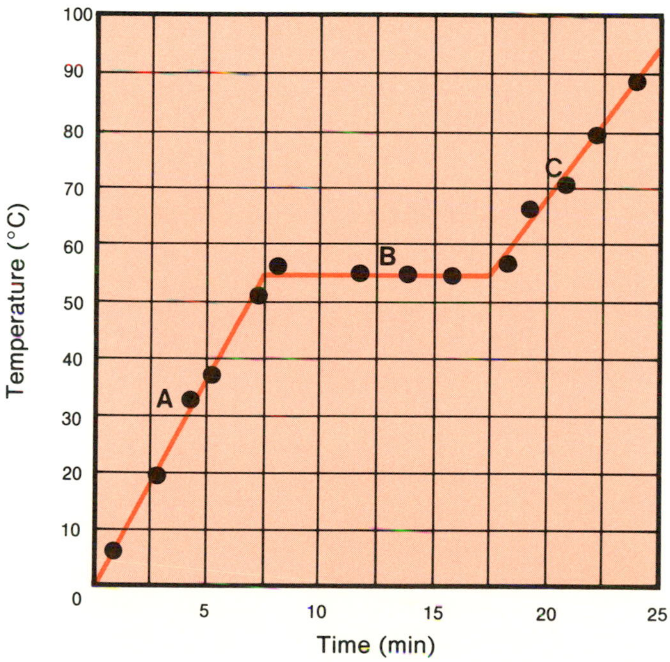

Figure 4-10 A warming curve for a liquid.

4.8 INVESTIGATION: Effect of Increased Pressure on the Boiling Point of Water

In this investigation you increase the pressure on the surface of the water as you boil it. This is done by making it hard for the vapour to escape. What do you think will happen to the boiling point?

CAUTION: This experiment must be done carefully to avoid burns. If you are unsure of any step, check with your teacher. Your teacher may choose to demonstrate this study.

Materials

distilled water
boiling chips
Bunsen burner
ring stand, iron ring, wire gauze, adjustable clamp
thick-walled distillation flask (250 mL), preferably a Franklin flask
thermometer (−10°C to 110°C)
2-holed rubber stopper
glass tubing
tall glass container

CAUTION: Wear safety goggles during this investigation.

Procedure

a. Set up the apparatus as illustrated in Figure 4-11, with this exception: Do *not* have the tall glass container in place.

b. Heat the water until it is boiling *gently*. **CAUTION:** You must not boil the water too fast or the stopper will blow out.

c. Note the temperature of the boiling water.

d. Bring the tall glass container into place. The glass tube should be no more than 10 cm under the surface of the water.

e. When the water is boiling as it was in step b, note the temperature again.

f. **CAUTION:** Before you turn off the heat, remove the tall glass container from the system. If you don't, cold water may back up into the hot distillation flask and break it.

Discussion

1. What was the boiling point of the water before you increased the pressure?
2. What was the boiling point of the water after you increased the pressure?

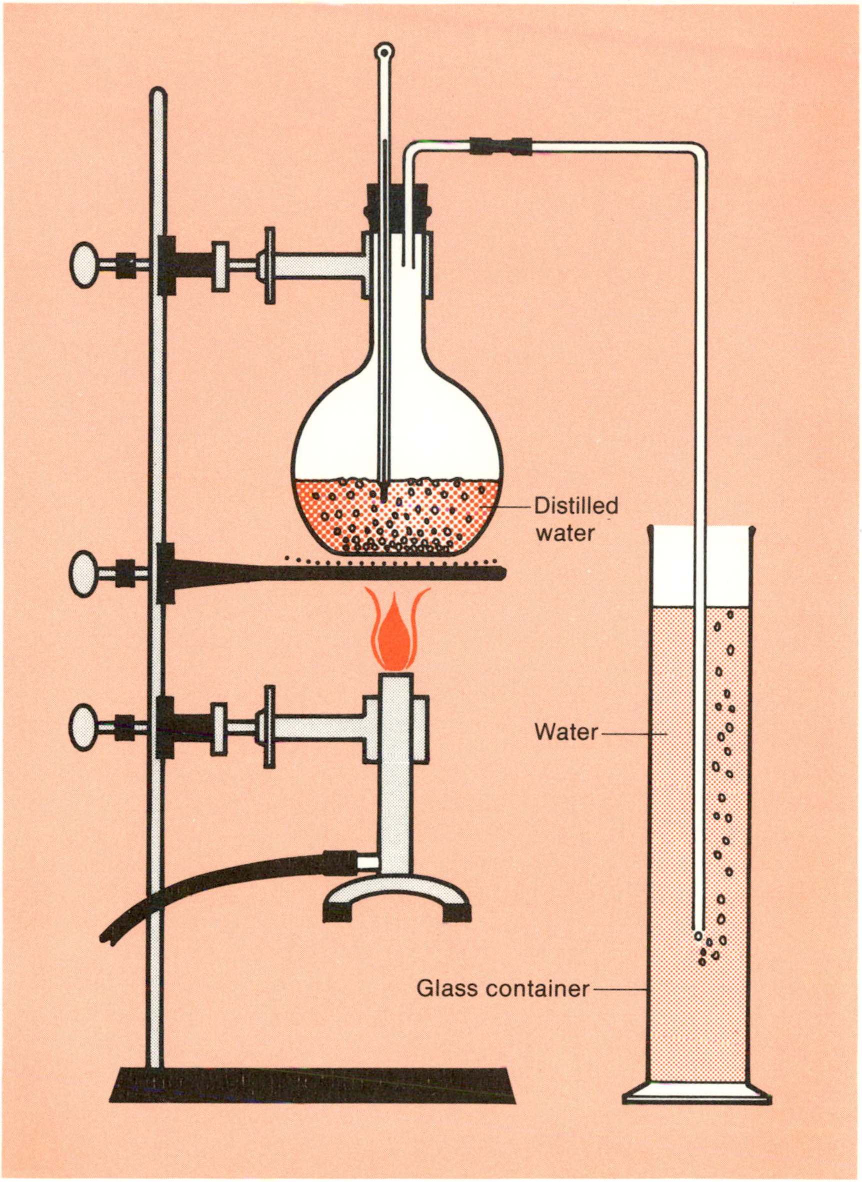

Figure 4-11 Effect of increased pressure on the boiling point of water.

3. Compare your results to those of two or three classmates. Make a general statement about the effect of increased pressure on the boiling point.
4. Why is a pressure cooker used to cook certain foods? How does it work?
5. Why are automobile radiators kept under a pressure that is above normal?
6. Will water always boil at the same temperature in your home? Explain.

4.9 INVESTIGATION: Effect of Decreased Pressure on the Boiling Point of Water

In this investigation you decrease the pressure on the surface of the water by removing vapour above it. How will this affect the boiling point?

CAUTION: This experiment must be done carefully to avoid burns and broken glass. Your teacher may choose to demonstrate this study.

Materials

thick-walled distillation flask (250 mL), preferably a Franklin flask
thermometer ($-10°C$ to $110°C$)
distilled water
boiling chips
Bunsen burner
ring stand, iron ring, wire gauze, adjustable clamp
one-holed rubber stopper
oven mitts

CAUTION: Wear safety goggles during this investigation.

Procedure A Pressure Decreased with a Vacuum Source

If you have a vacuum source such as a vacuum pump or water aspirator, the effect of decreased pressure can be studied as follows: Simply connect the vacuum source to the end of the glass tubing in Investigation 4.8. Begin with the water boiling, then turn off the heat and connect the vacuum source. Record the temperature every 30 s as long as the water is boiling.

CAUTION: A vacuum pump can be damaged by water. Therefore you should place a trap containing a dessicant (drying agent) like silica gel between the tubing and the pump.

Procedure B Pressure Decreased by Condensation

Use this procedure with the listed materials if you have no vacuum source.

a. Add water to the distillation flask until it is one-third full. Heat the water until it is boiling quickly. Use the usual ring stand assembly.

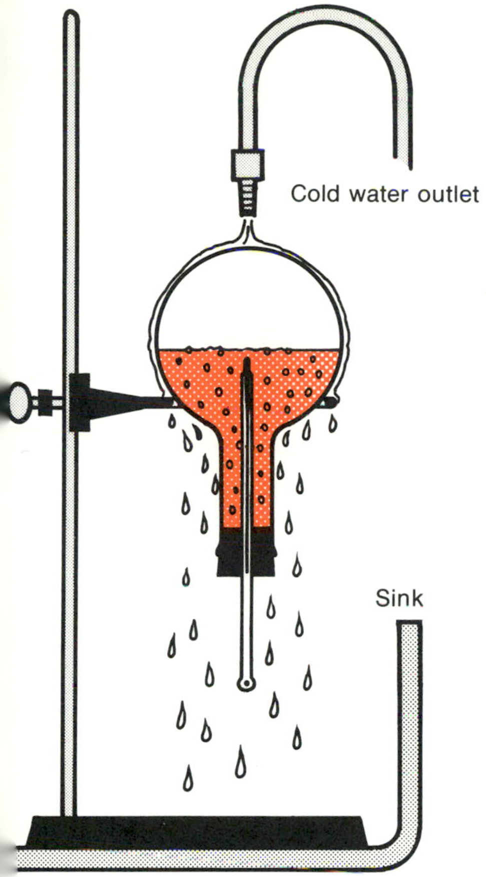

igure 4-12 Effect of decreased ressure on the boiling point of ater.

b. While you are waiting for the water to boil, fit the thermometer through the one-holed stopper. The bulb should be just inside the round part of the flask when the stopper is in place in the mouth of the flask. **CAUTION:** To avoid cuts, follow carefully the instructions given by your teacher.
c. When the water is boiling quickly, remove the source of heat and immediately insert the stopper into the flask.
d. Invert the flask. Use oven mitts to avoid burns. Note the temperature.
e. Slowly run cold water over the flask (Fig. 4-12). Note the temperature.
f. Continue with step e until the water no longer boils.

Discussion

1. What was present above the water in the flask just before the stopper was inserted?
2. What happened to this substance when cold water was run over the flask?
3. Explain why decreased pressure is produced by running cold water over the flask.
4. What was the temperature of the water before you ran cold water over the flask? Was the water boiling?
5. Describe the effect of the running water on the water in the flask. Explain the results.
6. What was the lowest boiling point that you noted?
7. The pressure on a mountain-top is less than the pressure at the foot of the mountain. Compare the boiling points of a liquid at these two locations.
8. Why does it take longer to cook potatoes in boiling water at the top of a mountain than it does at the foot of the mountain?
9. The instructions on a cake mix say: "Bake at 175°C. (At altitudes of 600 m or over, bake at 190°C.)" Why is this so?

4.10 INVESTIGATION: Effect of Antifreeze on the Boiling Point of Water

Most car makers recommend leaving the antifreeze in a car radiator during the summer. Why?

Materials

as listed for Investigation 4.7
50 mL of antifreeze (ethylene glycol)
CAUTION: Wear safety goggles during this investigation.

Procedure

a. Set up the apparatus as shown in Figure 4-9, page 80. Begin with 100 mL of water in the flask.
b. Heat the water until it is boiling. Note the temperature.
c. Add 10 mL of antifreeze to the water. Continue heating the mixture. Note the new boiling point.
d. Repeat step c until all of the antifreeze has been used.

Discussion

1. Describe the effect of antifreeze on the boiling point of water.
2. Why do car makers recommend leaving the antifreeze in a car radiator during the summer?
3. Compare the effects of antifreeze on the boiling point and freezing point of water.
4. Does antifreeze help or hinder boiling? Does antifreeze help or hinder melting?

4.11 INVESTIGATION: Effect of Gentle Heating on Iodine Crystals

Here you study the effects of heat on a solid called iodine. Pay particular attention to any changes of state that occur.

Materials

test tube (20 × 150 mm)
iodine crystals
ring stand, iron ring, wire gauze
Bunsen burner

CAUTION: Wear safety goggles during this investigation.

Procedure

a. Set up the water bath shown in Figure 4-13. Heat the water until it is boiling.
b. Place a few iodine crystals in the test tube. Describe the crystals. **CAUTION:** Iodine is very corrosive. It will damage skin and clothing. Do not touch it or smell it. If you spill any, tell your teacher.
c. Hold the bottom of the test tube in the boiling water as shown in Figure 4-13. Do not let the top half of the test tube get so warm that you cannot hold it.
d. Keep the test tube in this position until all of the crystals are gone or until you are sure you understand what is happening.
e. Make a careful record of all observations.
f. Return the test tube and contents to your teacher. Do not try to wash it.

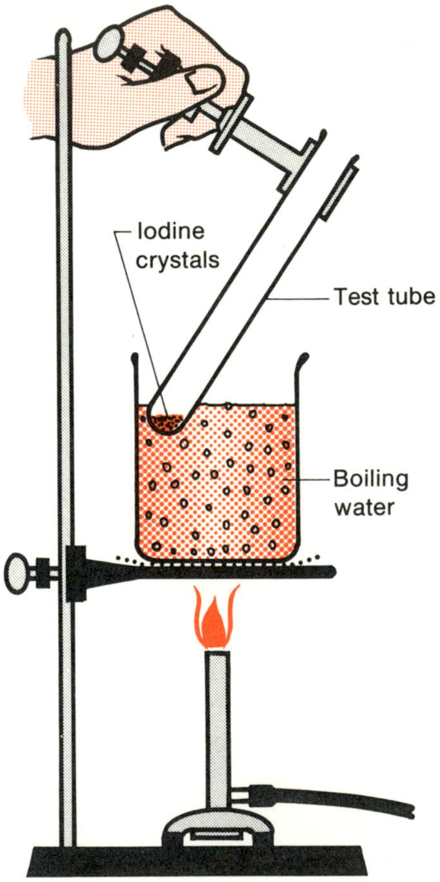

Figure 4-13 Effect of gentle heating on iodine crystals.

Discussion

1. What changes of state took place in this experiment? Give evidence to support your answer.
2. The melting point of iodine is 113.5°C. Why was the water bath used? Ask your teacher to demonstrate what happens if iodine is heated without the water bath. **CAUTION:** Use a fume-hood.
3. Compare the volume of iodine vapour to the volume of the same mass of solid.
4. List four other examples of sublimation. For each example, explain why the change of state is sublimation.

4.12 A Summary of Volume and Heat Changes during Changes of State

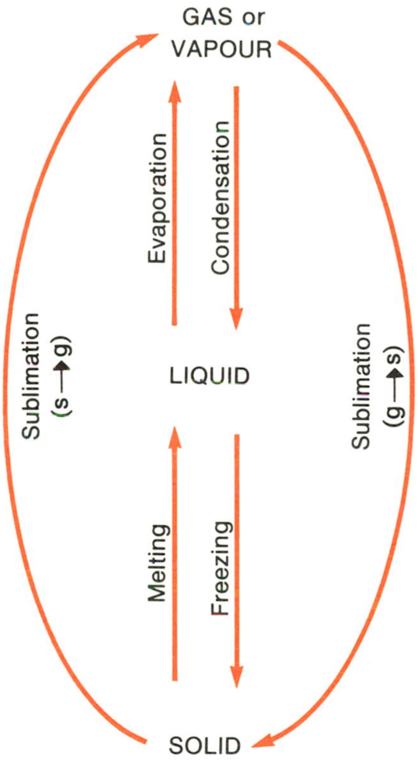

Figure 4-14 Changes of state.

You have just completed several investigations involving changes of state. Let us look back at your results to see if we can make any generalizations. Perhaps they can help us to develop and understand a theory that explains the physical properties of matter.

You noticed that heat is involved in all changes of state. It must be either added or removed for the change to occur. You also saw that the volume of a given mass of a substance changes during all changes of state. That is, the density changes. It either increases or decreases. Are there any relationships between the heat changes and the density changes that occur during changes of state? Refer to Figure 4-14 while looking for these relationships. Perhaps some useful generalizations will emerge.

You should realize that it is unwise to develop generalizations on the basis of so few experiments. But time does not permit you to do enough experiments. You will have to take our word for the fact that, except for the freezing and melting of water, most substances behave like the ones you studied.

Melting

You discovered in Investigation 4.3 that the volume decreased when ice changes to water. But you also learned that the behaviour of water is anomalous or abnormal. For most substances, including iron, copper, and paradichlorobenzene, the volume increases when a solid changes to a liquid. That is, the density decreases. Heat is required for this process.

Freezing

This process is the reverse of melting. Therefore you might expect that for most substances the density increases when a liquid changes to a solid. Your studies showed that heat must be removed during this process. In other words, freezing produces heat.

Sublimation (solid to gas)

The iodine experiment clearly showed that a decrease in density occurs during sublimation from a solid to a gas. A few tiny crystals of iodine produced large quantities of vapour. This experiment also showed that heat is required for this process.

Sublimation (gas to solid)

This change is the reverse of the preceding one. Your experiment showed that the density increased as the vapour sublimed to a solid. Large quantities of vapour were needed to form just a few crystals at the top of the test tube. The fact that crystals form only in cool regions of the test tube indicates that heat must be removed for sublimation from a gas to a solid. The process gives off heat.

Evaporation

Your studies showed that a large increase in volume occurs when a liquid is changed to a vapour. In other words, the density decreases considerably. They also revealed that heat is required for this process.

Condensation

This change is the reverse of evaporation. Thus one can deduce that the density should increase when a vapour condenses to a liquid and that heat must be removed during this process. Your observations support this deduction.

Summary

For most substances the following generalizations seem to be true:

1. *Whenever the volume increases during a change of state heat is required (must be added).*
2. *Whenever the volume decreases during a change of state heat is given off (must be removed).*

Generalizations like these don't just happen by accident. They are true because of the basic nature of matter. The next section explores that nature and introduces a theory to explain these and other generalizations.

Discussion

1. **a)** Which of the experiments that you did support the first generalization in the preceding summary?
   b) Which experiments support the second generalization?
2. Our pioneer forefathers used to store potatoes, apples, carrots, and other foods in the basements of their homes. On a cold winter night when there was a chance that these edibles could freeze and be destroyed, the pioneers filled all available containers with water and placed them in the basement. The water in the containers would often freeze but the vegetables would not. Explain why. (Hint: Think about these questions: What heat change accompanies the freezing of the water? The fruits and vegetables contain sugar in the water that is within them. What difference will this make?)
3. **a)** Tobacco plants are easily damaged by temperatures below 0°C. The tobacco farmers of southern Ontario often turn on their sprinkling systems when the temperature drops below 0°C. Layers of ice may form on the tobacco leaves. Yet the leaves do not freeze. Why?
   b) Explain this statement: "The freezing of a lake moderates the climate of the land area near it."
   c) Explain this statement: "The thawing of the ice in a lake moderates the climate of the land area near it."
4. Dry ice (solid carbon dioxide) sublimes to gaseous carbon dioxide. People often say that they "burned" their fingers when they picked up dry ice. Explain what happened.
5. If you have ever stood outside in a heavy snowfall with no wind, you probably noticed that the air got much warmer than it was before the snowfall. Why would this be so?

4.13 The Particle Theory of Matter

You saw in Section 3.9, page 64, that the particle theory explains reasonably well why some substances are more dense than others. In this section you look more closely at the particle theory to see how it explains most other physical properties of matter. We will give you each point in the theory and explain it using, where possible, examples from the experiments that you have done. Make sure you understand all of this because, in Unit 5, you will be asked to do many experiments and then to explain your results using the particle theory.

You have probably heard of atoms and molecules. They are both particles. Don't worry about the difference between them at this time. You will learn about atoms and molecules in Units 7 and 8.

The Particle Theory

Each of the points in italics is a statement that scientists generally agree upon concerning the particle nature of matter.

1. ***All matter is made of small particles***. Although we have not seen these particles, we do know that density and changes of state are easily accounted for if we assume that matter is made of particles. Try to explain them. The particle theory is a model for matter.

   The word "small" requires explanation. One scientist estimated that the particles in the head of a pin are so small and numerous that everyone in the world would have to count them for their entire lives to get them all counted! Another scientist estimated that the number of atoms (small particles) in the human body equals the number of green peas that it would take to cover 250 000 earths 1.5 m deep! When we say small, we mean small!
2. ***All particles of the same substance are identical***. We have no proof of this at the moment. But it is reasonable to assume that a particle of water (ice), a particle of water (liquid), and a particle of water (vapour) are all the same. All particles of water have the same mass and volume, regardless of state. The particles of different substances have different masses.
3. ***The particles in matter attract one another. These "attractive forces" get stronger as the particles get closer together***. This explains why solids are more rigid and more dense than liquids. It also explains why liquids are more dense than gases. The particles in a solid are closer together than in a liquid or gas (Fig. 4-15). Thus the solid will be the most dense. Since the particles are closer together, the attractive forces will be stronger. Thus a solid is rigid. Why does a liquid offer more resistance to your passing through it than a gas does?

   If you want to change a solid to a liquid, you have to give the particles energy to overcome the attractive forces. This is why a change of state from a solid to a liquid requires heat. A similar explanation accounts for the fact that heat is needed to vaporize a liquid.
4. ***The spaces between the particles are very large compared to the sizes of the particles themselves***. Our diagram in Figure 4-15 is misleading in this respect. The spaces are much larger than the diagram suggests. One researcher figured out that if the empty spaces between the particles are removed, all of the buildings in New York City would fit in a match box! How heavy would the match box be?

   Our studies show that the spaces are smallest in a solid and greatest in a gas. Because the spaces are large in a gas, the gas can be compressed (reduced in volume) by applying pressure to it. The pressure pushes the particles closer together. Solids and liquids cannot be easily compressed since

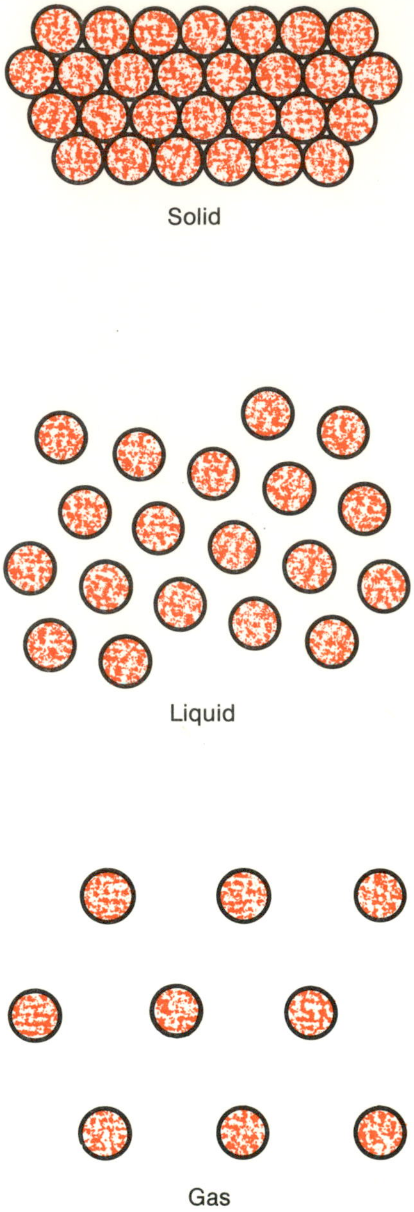

Figure 4-15 The particle nature of solids, liquids, and gases.

the spaces between their particles are much smaller than those of a gas.

5. ***The particles in matter are in constant motion***. The extent of this motion depends on the temperature. As the temperature goes up, the particles have more energy. Therefore they can move more rapidly. How does this explain why water evaporates faster at its boiling point than it does at room temperature?

   The motion of the particles causes them to bump into the walls of the container. These collisions cause pressure. When you put air in a bicycle tire, the collisions of the particles with the inside of the tire cause the pressure that inflates the tire.

NOTE: As you learn more and more, you will find exceptions to this theory. You will also find that some of the statements are only true within certain limits. A better theory may be needed to go on from that point. Recall that theories may change as new evidence is discovered. (See Section 1.11, page 16.) In spite of this, you should find that this theory will explain most of what you do in the next unit. Be sure you understand it.

Discussion

Use the particle theory to answer these questions:

1. Why is it easier to compress a gas than a liquid?
2. Why is iron (solid) more dense than iron (liquid)?
3. Why must heat be removed to change liquid water to ice?
4. What happens to the heat energy that must be used to melt a solid? (Recall that the temperature does not change during melting.)
5. How does air inside a balloon hold the rubber out? That is, what causes pressure?
6. Why are melting point, boiling point, and density characteristic physical properties? That is, why are they generally different for every substance? Explain each one separately.
7. Why does decreasing the pressure lower the boiling point of a liquid?
8. Explain how substances like salt and ethylene glycol lower the freezing point of water.
9. Explain how antifreeze raises the boiling point of water.
10. The air pressure inside an automobile tire increases when the tire becomes warm. Why is this so?

Highlights

Matter is anything that has mass. It occurs in three states: solid, liquid, and gaseous. Most substances can exist in all three states and can be changed from one state to another by the removal or addition of heat. When the volume increases during a change of state, heat is usually required. When the volume decreases during a change of state, heat is usually given off.

The temperature stays constant while a substance is undergoing a change of state. Thus the heating curve for a solid and the cooling curve for its liquid state have a common plateau at the melting (or freezing) point.

Melting point and boiling point are characteristic physical properties that are used to help identify substances.

Impurities like salt and antifreeze lower the freezing point of water and raise its boiling point.

The boiling point of a liquid depends on the pressure above the liquid.

The particle theory of matter explains many of the properties of matter. It says that all matter is made up of small particles that are attracted to one another. It says, further, that large spaces exist between these particles and that the particles are in constant motion.

UNIT

5 Using the Particle Theory

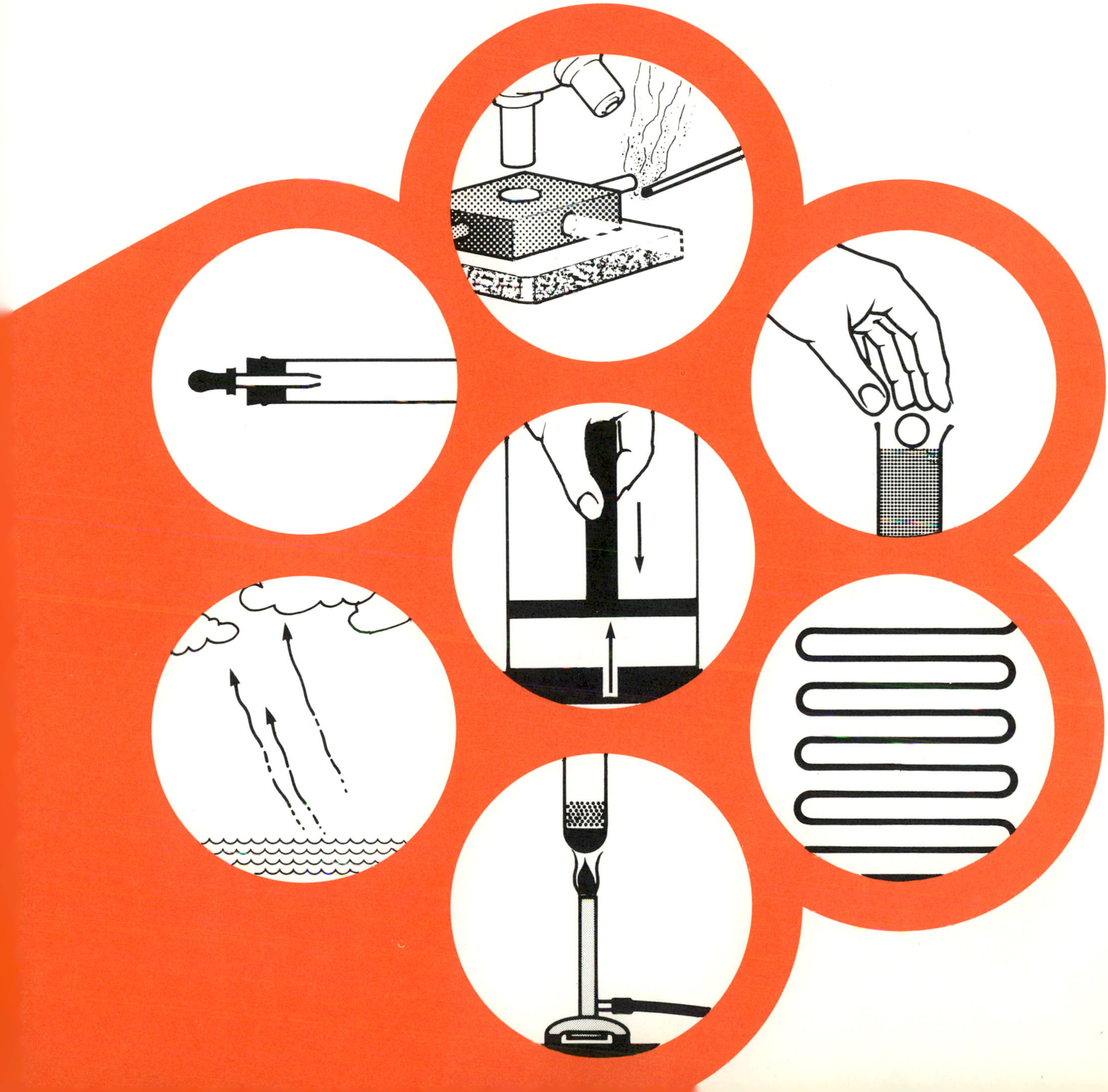

At the end of Unit 4 we introduced the particle theory. This theory explains most of the physical properties of matter that we have met so far. It accounts for density as a characteristic physical property (Unit 3). It also accounts for most other physical properties of the three states of matter (Unit 4).

The real test of a theory is its ability to predict with success the results of experiments before you do them. And that is what this unit is all about. For most of the investigations in this unit you will use the particle theory to predict the results before you begin the investigation. In some cases you will be asked to design your own procedure, based on your predictions. Then, after you have done the experiment, you will, if possible, relate your findings to your prediction and the particle theory.

The investigations in this unit have been chosen with care. They were picked because they teach you basic principles that make it possible for you to explain everyday happenings. Therefore the unit consists mainly of investigations and practical applications of what you learn in the investigations.

5.1 INVESTIGATION: The Particle Demonstration Tube

Here is the first test of the particle model of matter.

The particle demonstration tube shown in Figure 5-1 contains liquid mercury with light chips of coloured glass floating on it. The tube has been evacuated (all air removed). Use the particle theory to predict what will happen if the mercury is heated strongly. Then do the experiment to test your prediction.

Since this is the first time you have had to use the theory to make a prediction we will give you some help. Think about these questions and look back to the particle theory as you make up your prediction: What is liquid mercury made of? What usually happens when a liquid is heated strongly for a long time? What happens to the heat that goes into a liquid? What will happen to the glass chips?

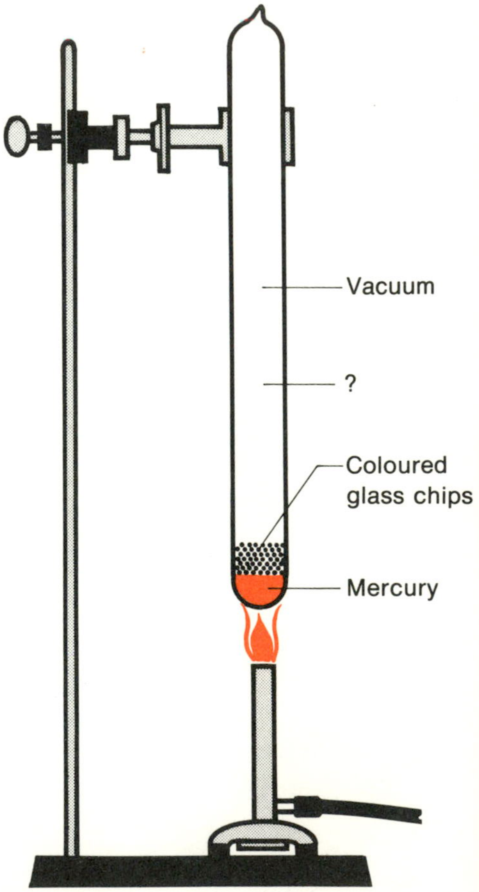

Figure 5-1 The particle demonstration tube.

Materials

particle demonstration tube
ring stand, adjustable clamp
Bunsen burner
CAUTION: Wear safety goggles during this investigation.

Procedure

a. Write in your notebook a prediction of what you think will happen if the mercury is heated strongly.

b. Set up the tube as shown in Figure 5-1. The tube is expensive. Make sure you clamp it securely.
c. Heat the bottom end only. Use a very hot flame.
d. Make a careful list of every change that occurs.

Discussion

1. Explain why you made the prediction that you did.
2. What changes of state took place in the tube?
3. While the tube was being heated, what was present at the position indicated by a question mark in Figure 5-1? Could you see it? What proof do you have that it was there?
4. Why is the tube called a particle demonstration tube?
5. What points in the particle theory are illustrated by this experiment?

5.2 INVESTIGATION: Diffusion of Gases (Teacher Demonstration)

If you release some air freshener or perfume into one corner of the room, the odour does not stay there. It gradually spreads, or **diffuses**, throughout the room. This happens even if the room is tightly sealed to prevent air currents. The odour seems to move all by itself. Such movement is called **spontaneous.**

Diffusion *is the process by which a substance spreads spontaneously in all directions from a region where there is a high concentration of the substance to a region where there is a lower concentration (perhaps none) of the substance.*

This investigation studies the diffusion of gases. You predict the results of some studies, using the particle theory. You and your teacher then test your predictions with experiments.

Background Information

a. When concentrated sulfuric acid is added to a mixture of sodium bromide and manganese dioxide, bromine gas is formed.
b. When concentrated nitric acid is added to copper, nitrogen dioxide gas is formed.

CAUTION: Both gases are brown and much denser than air. They are also poisonous. Do not inhale them. A fume-hood should be used.

Materials

2 tall glass containers
2 glass plates

sodium bromide, manganese dioxide,
sulfuric acid (concentrated)
copper, nitric acid (concentrated)
2 droppers

CAUTION: Wear safety goggles during this investigation.

Procedure

a. Use the definition of diffusion and the particle theory to predict what will happen when bromine gas is produced at the bottom of one container and nitrogen dioxide gas at the bottom of the other.

b. Mix a pinch of sodium bromide with a pinch of manganese dioxide. Place the mixture in the bottom of container A (Fig. 5-2). Add one or two drops of concentrated sulfuric acid to the mixture. Immediately cover the container with a glass plate.

c. Place a piece of copper in the bottom of container B. Add one or two drops of concentrated nitric acid to the copper. Immediately cover the container with a glass plate.

d. Make a careful record of all observations.

Discussion

1. Where are the gases most concentrated? least concentrated? Account for this observation using the particle theory.
2. Explain the process of diffusion using the particle theory.
3. Is there any limit to the distance through which diffusion will occur? Use the particle theory and the definition of diffusion to explain your answer.
4. A possible source of error in this demonstration is the presence of convection currents. Find out what they are (try a dictionary). How might they affect the results of this demonstration?

5.3 INVESTIGATION: Diffusion of Gases

This investigation also deals with the diffusion of gases. There are two basic differences between it and Investigation 5.2. First, the gases are invisible this time; therefore you cannot see what is going on. You will have to rely more on your imagination and the particle theory. Second, you, not your teacher, do the work.

Materials

glass tube (l = 50 cm; d = 2 cm)
2 one-hole rubber stoppers

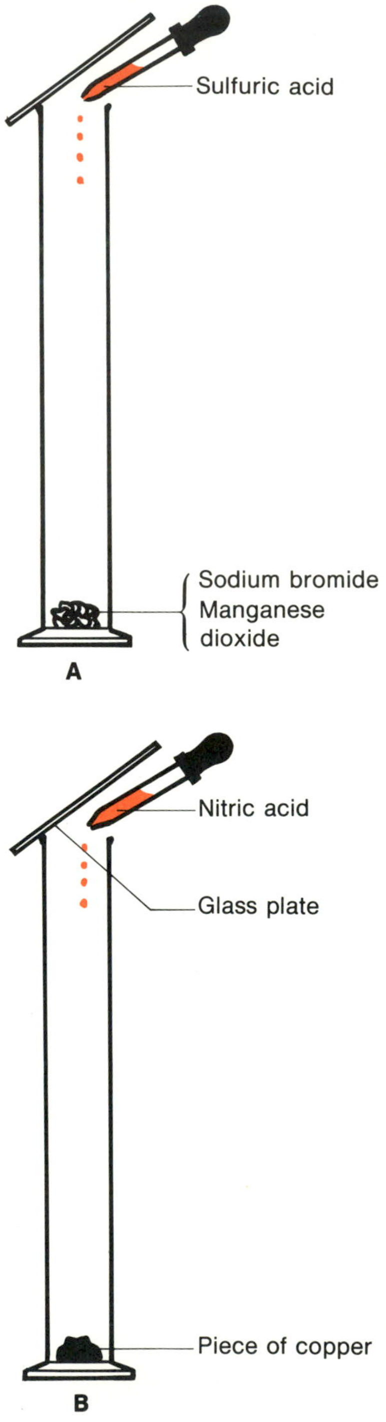

Figure 5-2 Studying the diffusion of gases.

2 droppers
concentrated ammonium hydroxide
concentrated hydrochloric acid
ruler
watch or clock
ring stand, adjustable clamp

Background Information

a. Hydrochloric acid is made of a gas called hydrogen chloride dissolved in water. Ammonium hydroxide is made of a gas called ammonia dissolved in water. Your teacher will show you proof of these two points. **CAUTION:** Handle these materials with care. Do not smell them except as directed by your teacher.

b. Scientists have shown that a particle of hydrogen chloride gas has a mass that is over twice that of a particle of ammonia gas.

c. Ammonia gas and hydrogen chloride gas react with each other to form a white solid. Your teacher will demonstrate this point.

Figure 5-3 Studying the diffusion of gases.

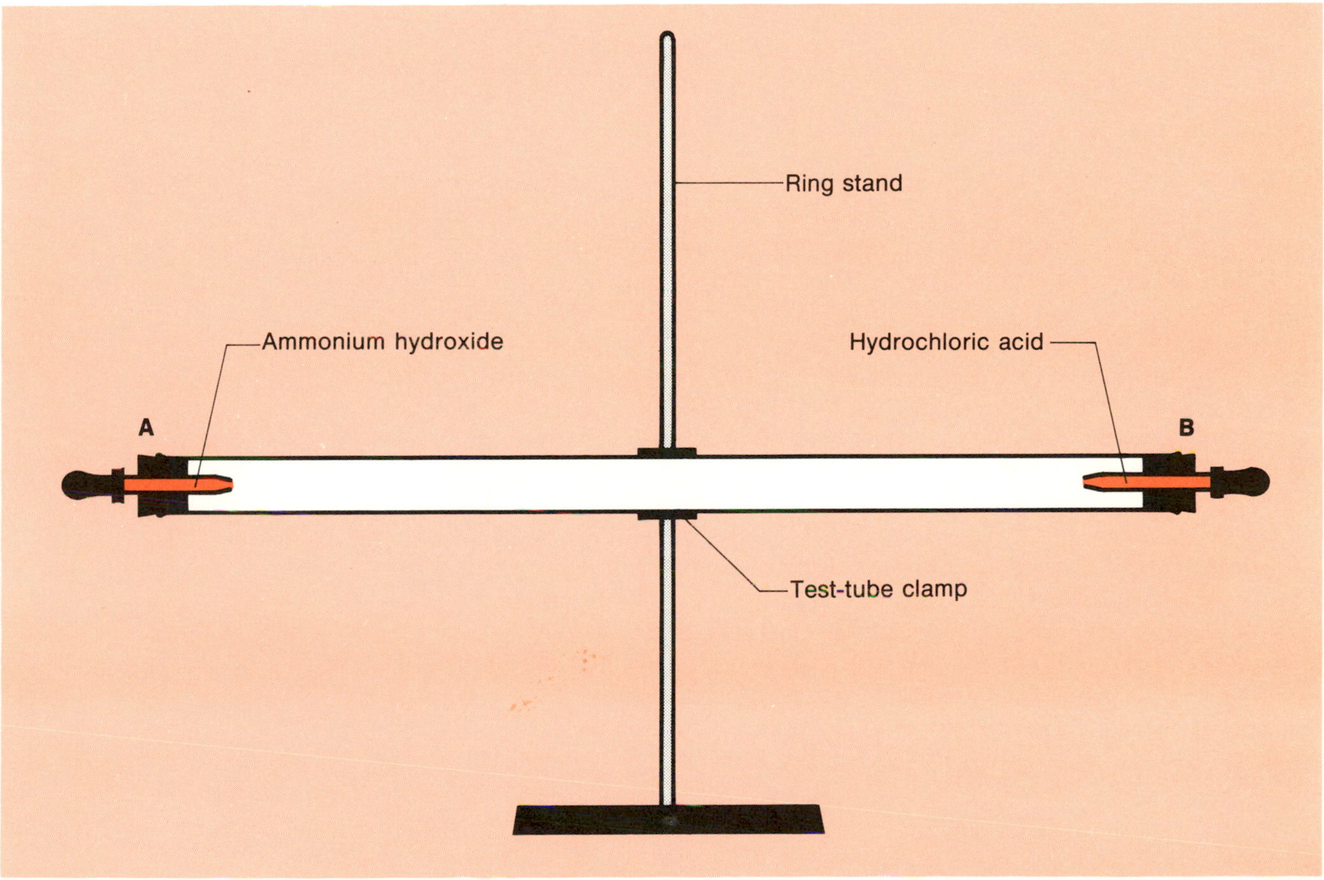

Procedure

a. Set up the apparatus as shown in Figure 5-3, but with the droppers empty. Make sure that the tube is perfectly horizontal.

b. Use the particle model to predict what will happen if ammonia gas is released at end A at the same time that hydrogen chloride gas is released at end B.

c. Fill one dropper with ammonium hydroxide and the other with hydrochloric acid.

d. At the same time, insert the ammonium hydroxide dropper into position A and the hydrochloric acid dropper into position B.

e. Also at the same time, squeeze 4 drops of solution from each dropper. Record the time that you did this.

f. Observe the tube closely for several minutes. Record any change that occurs and the time of that change.

g. Remove the tube and wash it thoroughly. Then dry it completely on the inside.

h. Repeat the entire experiment with one change: Use the particle theory to predict what would happen if the experiment was done at a higher temperature. Then play a low flame along the glass tube to warm it up until it is uncomfortable to touch, **before** you insert the stoppers. Observe the same factors as before.

Discussion

1. Which gas diffuses faster, ammonia or hydrogen chloride? Use the particle theory to explain this observation.
2. How does temperature affect the rate of diffusion of a gas? Use the particle theory to explain this result.
3. The tube had air in it. What effect might it have on the results of your experiment? Why?
4. Hydrogen sulfide gas (often called "rotten egg" gas) is used in the making of heavy water for nuclear reactors. This gas is a deadly poison. It is also more dense than air. The electric company that runs one plant using hydrogen sulfide has sent all residents of the immediate area a notice which reads, in part: "If a warning is given that hydrogen sulfide has escaped from our plant, head for the top floor of your home." Use the results of Investigations 5.2 and 5.3, and your knowledge of the particle theory to explain the purpose of this warning.

5.4 INVESTIGATION: Diffusion in Liquids

Would you expect a substance to diffuse more rapidly through a liquid or through a gas? Refer to the particle theory as you make this prediction. Then try this Investigation and compare your results to Investigations 5.2 and 5.3.

Materials

2 petri dishes
water
ethyl alcohol
iodine crystals
potassium permanganate crystals

Procedure

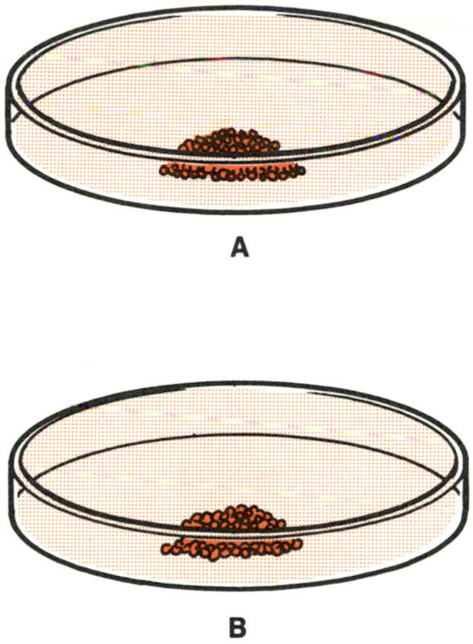

Figure 5-4 Diffusion in liquids.

a. Fill one petri dish (A) with water and the other (B) with ethyl alcohol.
b. Place a few crystals of potassium permanganate in the centre of dish A. Place approximately the same volume of iodine crystals in the centre of dish B. Do not disturb the dishes from this point on (Fig. 5-4).
c. Observe each dish carefully for 10 min. Make notes and sketches of the changes that occur. (If your classroom has an overhead projector, you could place the dishes on the projector and observe the change on the projection screen. The enlargement will make it easier for you to see what is happening. Step b must, of course, be performed with the dishes on the projector.)
d. Clean out both petri dishes. Fill one with tap water and the other with hot water. Place a few crystals of potassium permanganate in the centre of each dish. Note any differences between the observations of the two dishes.

Discussion

1. Describe the appearance as a solid diffuses through a liquid.
2. Use the particle theory to explain this appearance, right from the moment the solid is added to the liquid.
3. Does diffusion appear to be faster through a liquid or through a gas? (Compare the results of this investigation to those of the previous two investigations.)
4. What variable did you ignore when you answered part 3? In other words, how good are the controls when you compare the results of this investigation with those in Investigations 5.2 and 5.3?
5. Will a teaspoon of sugar dissolve faster in a glass of iced tea or in the same volume of hot tea? Why? (Assume no stirring.)
6. Explain why stirring usually increases the rate of dissolving of a solid in a liquid. In your explanation, use your knowledge of diffusion and the particle theory.
7. Sewage plants have been putting phosphates into Lake Erie for decades. A great deal of this phosphate is now in the solid state in the bottom ooze. Phosphates are not good for lakes.

They cause algal blooms which can eventually kill most of the fish. Use the particle theory and the results of this investigation to explain why Lake Erie may still have algal blooms long after sewage plants stop putting phosphates into the lake.

5.5 INVESTIGATION: The Motion of Particles in a Gas

You cannot see the particles of a gas, even with a microscope. They are far too small. However, you can see smoke particles since they are millions of times larger than gas particles. If you enlarge and brighten the smoke particles by bouncing light off them, they become very easy to see. This makes it possible for you to study their motion.

The particle theory says that the particles of a gas such as air are in motion. We can guess that, if smoke particles are mixed with the air particles, the air particles will bump into the smoke particles and start them moving. Therefore, by watching the motion of smoke particles, we can get an idea of how the particles of a gas move.

In this investigation you mix smoke and air. Then you shine a light on the smoke particles and watch their motion with a microscope. The motion you see is called **Brownian Motion**, since it was first observed by Robert Brown in 1827.

Materials

smoke chamber
wax taper or other source of fine smoke
microscope
bright light source (e.g. projector)

Procedure

a. Set up the apparatus as shown in Figure 5-5. The smoke chamber shown is a commercial one. It uses a squeeze-bulb to suck smoke into the chamber. It has a viewing window for the microscope and an entry window for the light. You can make your own chamber, if you give it some thought.
b. Put smoke into the chamber.
c. Turn on the light.
d. Put the microscope on low power and observe the smoke particles. Try powers up to 200 ×.
e. Observe closely the way one particle moves. Does it move in straight lines or circles? how far? how fast?

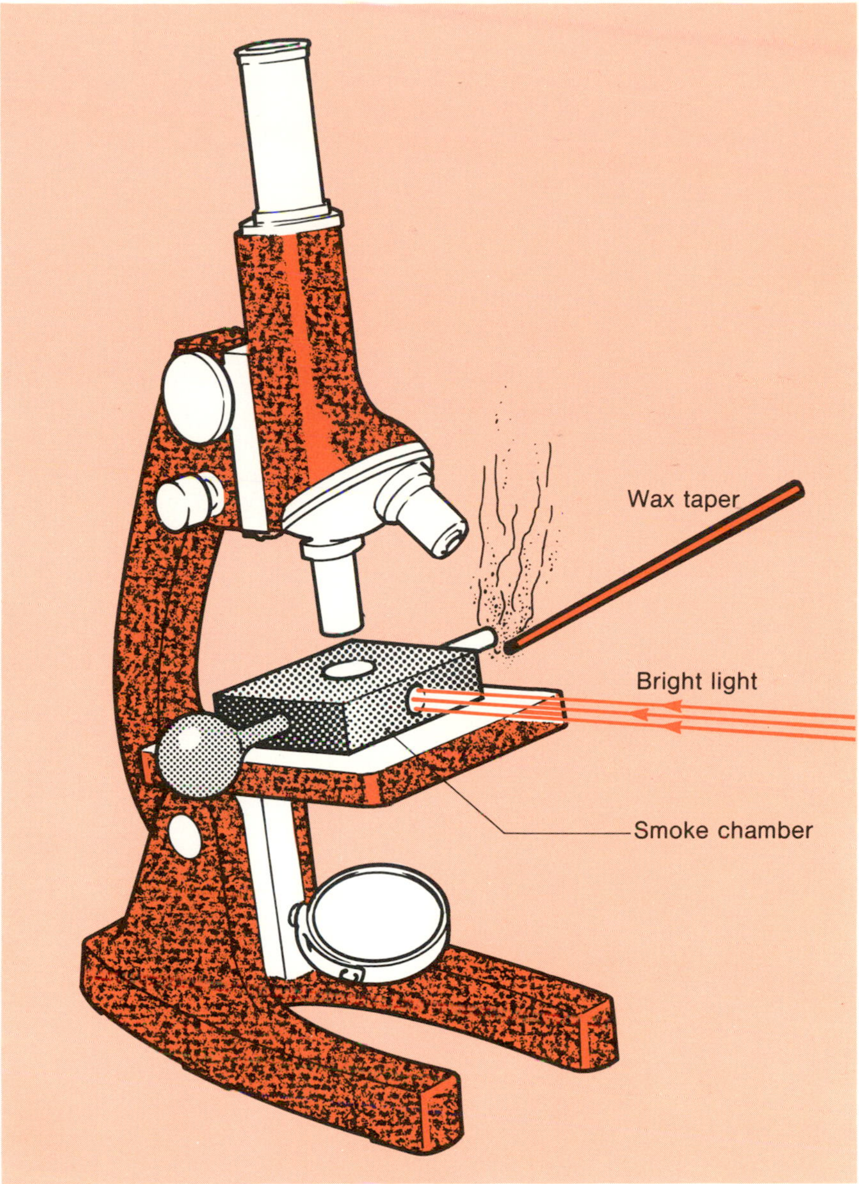

Figure 5-5 Observing the motion of smoke particles.

Discussion

1. Describe the overall movement of the smoke particles.
2. Describe the motion of one smoke particle.
3. What causes the smoke particle to move?
4. Why is this motion considered indirect evidence for the particle theory of matter?
5. Would the particles always move at the speed you observed? Explain.

6. Theatre owners usually assume that non-smokers will not be bothered by smoke if smokers are seated at the back of the theatre. Use your knowledge of the particle theory, diffusion, and the results of this investigation to prepare a scientific argument for or against the theatre owners' assumption.

5.6 INVESTIGATION: The Motion of Particles in a Liquid

The purpose of this investigation is to compare the motion of the particles of a liquid to that of the particles of a gas. You cannot see the particles of a liquid, either. So to do this investigation you will observe the motion of tiny particles of carbon in water and the motion of small butterfat droplets in homogenized milk. Does the particle theory suggest any differences between their motion and the motion of smoke particles in air?

Background Information

a. India ink is made of tiny particles of carbon suspended in water.
b. Homogenized milk has in it, among other things, tiny droplets of butterfat suspended in water.

Materials

microscope
microscope slides (2) and cover slips
dropper
pin
India ink
homogenized milk

Procedure

a. Mix 5 mL of homogenized milk with 5 mL of water.
b. Place one drop of this mixture on a microscope slide. Cover it with a cover slip.
c. Observe the mixture with a magnification of at least $400\times$.
d. Place a drop of water on the second microscope slide.
e. Dip the tip of the pin in the India ink. Touch the drop of water with this tip.
f. Cover the mixture with a cover slip. Observe under the microscope, again at $400\times$ or greater.

Discussion

1. Describe the motion of the butterfat droplets.
2. Describe the motion of the carbon particles.
3. What do the motions described in 1 and 2 tell you about the motion of water particles?
4. Compare the motion of air particles to that of water particles. What relationship is there between water pressure and the movement of water particles?
5. A deep-sea diver knows that the water pressure increases with depth. Explain why this is so using the particle theory.

5.7 INVESTIGATION: Evaporation of Liquids

Our studies in Unit 4 showed that the evaporation of water requires heat. They also showed that the volume increases (density decreases) during evaporation of water. Now that you understand the particle theory, we will take a closer look at the evaporation of water. We will also study the evaporation of other liquids.

Many important household devices make use of the heat and volume changes that occur during evaporation. Therefore, in this investigation you will learn all you can about this change of state. Then, in the next three sections you will apply what you have learned to some practical situations. Keep the particle theory in mind as you think about the results of this investigation.

Materials

glass plate
ether
cotton wick
ring stand, adjustable clamp
beaker
watch glass
2 thermometers ($-10°C$ to $110°C$)
2 petri dishes (or other flat containers)
rubbing alcohol (50 mL)
dropper
graduated cylinder (50 mL)
electric fan (optional)

Procedure A Evaporation of Rubbing Alcohol

a. Use the dropper to place a few drops of the alcohol on the back of your hand. Record your observations. Add more alcohol, if necessary. Then fan the wet spot vigorously with your hand. Record your observations.

b. Place 25 mL of alcohol in each of the 2 petri dishes. Take the air temperature and the temperature of the alcohol in each dish.
c. Set one dish aside as a control. Fan the other vigorously for at least 10 min. (If possible, set it in front of an electric fan.) Take the temperature of the alcohol in each dish every minute. Record your results in a table.
d. Pour the alcohol from each dish back into the graduated cylinder and note the volume from each dish.

Procedure B Evaporation of Ether (*Teacher Demonstration*)

CAUTION: Ether is flammable. Turn off all flames. Also, perform this demonstration in a well-ventilated area.
a. Place 10-20 drops of water in the centre of the glass plate.
b. Set the watch glass in this pool of water (Fig. 5-6).
c. Place about 10 mL of ether in the watch glass.
d. Fan the ether vigorously with your hand or an electric fan.
e. After a few minutes, pick up the watch glass.
f. Record your observations.

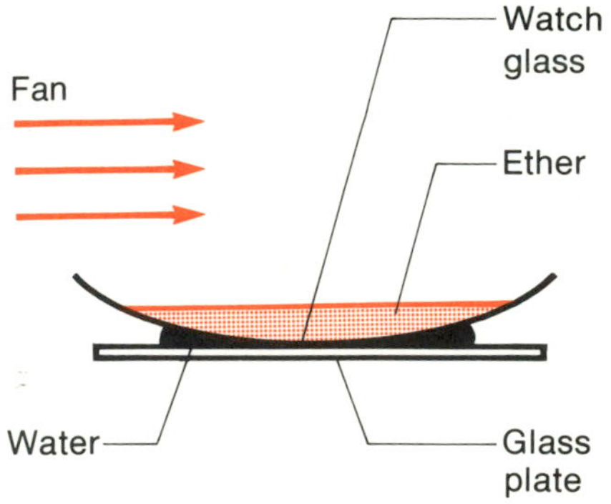

Figure 5-6 Studying the evaporation of ether.

Procedure C Evaporation of Water

You have already studied the evaporation of water at its boiling point. Here you study its evaporation at room temperature. Water does not evaporate as fast at room temperature as does ether or alcohol. Thus a different procedure is needed to study its evaporation. The cotton wick is used to get the fastest possible rate of evaporation.
a. Place one end of the cotton wick over the bulb of the thermometer. Suspend the thermometer so that the other end of the wick is in a beaker of water as shown in Figure 5-7. Hang the second thermometer beside the first one. Make sure its bulb is dry.
b. Fan the thermometers with a book for at least 5 min.
c. Record the temperatures of the wet and dry bulb thermometers.

Discussion

1. Describe the results when you fanned the alcohol on your hand. Why did this happen?
2. Why is rubbing alcohol used on the skin of bed-ridden persons?
3. Describe the differences observed between the 2 containers of alcohol in Procedure A under these headings: temperature

change, rate of evaporation, amount of evaporation. Use the particle theory to explain these differences.

4. Describe and explain, using the particle theory, the results of Procedure B.
5. Compare the rates of evaporation of ether and rubbing alcohol. Use the particle theory to explain the difference.
6. Describe and explain the results of Procedure C.
7. If a fan is used to blow air on a dry thermometer on a hot summer day, the temperature reading will not change. Yet, if the fan is used to blow air on you, you will feel cooler. Why?
8. When you come out of a swimming pool on a windy day, you often feel cold, even though the air and water are quite warm. Why?
9. In some warm, under-developed countries, drinking water is stored in porous jugs. Why?
10. A desert nomad often carries his drinking water in a leather bag that leaks slowly over its entire surface. Why does he do this?
11. Explain how sweating helps a human to maintain a constant body temperature on a hot day.
12. Dogs do not sweat. What adaptation do they have to help maintain a constant body temperature on a hot day? Explain how it works.

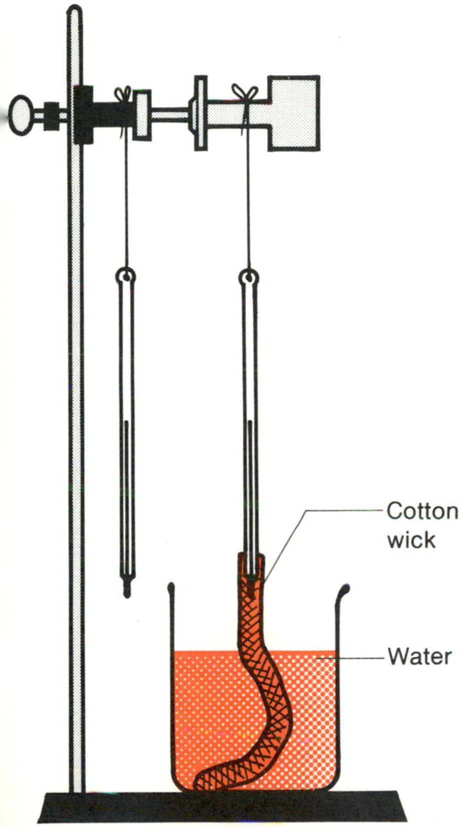

'igure 5-7 Studying the 'vaporation of water.

5.8 Humidity

You know that air contains water vapour and that it contains more on some days than on others. Investigation 5.7 and several investigations in Unit 4 showed how this water gets into the air.

In order to forecast the weather, a weather station needs to know the moisture content, or humidity, of the air. In this section you meet some terms used by weather stations. In the next two sections you use those terms.

Absolute Humidity

The **absolute humidity** *is the mass in grams of water vapour in one cubic metre of air at a certain temperature.* Thus the absolute humidity could be 6.5 g/m^3 at 20°C. This means that 1 m^3 of air at 20°C contains 6.5 g of water vapour. Absolute humidity is usually measured by passing one cubic metre of air through a drying agent like silica gel. The silica gel is massed before and after to get the mass of water vapour in the cubic metre of air.

Relative Humidity.

Absolute humidity values are not of much use to the average person. If the morning news reported that the absolute humidity was 6.5 g/m^3, most of us really would not know what kind of a day it was going to be. For that reason, the weather office usually reports the water vapour content of the air as the relative humidity. You have probably heard the weather report say "The relative humidity is 90%." What does this mean?

A given volume of air can only hold a certain amount of water vapour at a given temperature. Beyond that amount, the water vapour changes to rain or some other form of precipitation. When the air is holding its maximum amount of water vapour, it is said to be **saturated**. Its relative humidity is 100%. If the air is holding only half the amount of water vapour required for saturation, it is 50% saturated and its relative humidity is 50%. *The* **relative humidity** *is the ratio of the mass of water vapour in a given volume of air to the mass of water vapour present when the same volume of air is saturated at the same temperature*. It is usually expressed as a percentage. The following formula is used to calculate relative humidity:

$$\textbf{Relative humidity} = \frac{\textbf{Mass of water vapour in a given volume of air}}{\textbf{Mass of water vapour in the same volume, at saturation, and at the same temperature}} \times \textbf{100\%}$$

Here is an example of the use of this formula: The weather office knows that, at 20°C, the saturation point is 17.1 g/m^3. Suppose the absolute humidity at a certain time is 4.3 g/m^3 and the temperature is 20°C. Then, $\text{R.H.} = \frac{4.3\ g/m^3}{17.1\ g/m^3} \times 100\% = 25\%$.

The air is holding 25% as much water vapour as it can at 20°C.

Relative humidities between 40% and 60% are the most comfortable for humans. They are also the best for human health. When the relative humidity is very low, the skin, eyes, nose, and throat dry out. Not only does this make you uncomfortable, but it also increases your chances of catching many diseases. The protective mucus lining of your nose and throat dries up, making invasion by bacteria and viruses more likely. When the relative humidity is very high, you also feel uncomfortable. You found out in Investigation 5.7 that evaporation of water from the skin causes a cooling effect. However, if the air next to the skin is close to saturation with water vapour, little evaporation of sweat can occur. There is little room for it in the air. Therefore the cooling effect is slight. Sweat remains on the skin; you feel clammy and hot.

Relative Humidity and Temperature

Simply stated, warm air can hold more water vapour than the same volume of cold air can. For example, saturated air at 0°C has an absolute humidity of 4.8 g/m^3. This means that 1 m^3 of air at 0°C can hold no more than 4.8 g of water vapour. But at 20°C (room temperature), the absolute humidity of saturated air is 17.1 g/m^3. The same volume of air holds 3.6 times as much water vapour! This fact causes several problems in areas with cold winters. To see why, we must compare the *relative* humidity of the outside and inside air. The 0°C air outside will likely be saturated with water vapour. Therefore its relative humidity is $\frac{4.8\ \text{g}}{4.8\ \text{g}} \times 100\% = 100\%$. But when this air enters the home at 20°C, its relative humidity becomes $\frac{4.8\ \text{g}}{17.1\ \text{g}} \times 100\% = 28\%$. This dry air removes water from furniture, causes static electricity on rugs, and affects the comfort and health of humans. As a result, many homeowners install humidifiers to raise the relative humidity in the winter.

Discussion

1. Explain the difference between absolute humidity and relative humidity.
2. **a)** 4.5 m^3 of air contain 27.0 g of water vapour at 20°C. What is the absolute humidity at 20°C?
   b) Air at 20°C is saturated when it contains 17.1 g/m^3 of water vapour. What is the relative humidity of the air in part a?
3. **a)** Explain why a person feels uncomfortable at a low relative humidity.
   b) Explain why a person feels uncomfortable at a high relative humidity.
4. **a)** Explain why you feel hotter on a humid day than you do on a non-humid day at the same temperature.
   b) If a home has a high relative humidity, water will condense on the windows on a cold day. Why?
5. **a)** What happens to the absolute humidity when air at 0°C enters a home at 20°C? Explain.
   b) What happens to the relative humidity when air at 0°C enters a home at 20°C? Explain.
6. Why should the temperature be included when a relative humidity value is being given?
7. Use the particle theory to explain why warm air can hold more water vapour than the same volume of cold air.
8. **a)** Homeowners usually find that, if they humidify their homes in the winter, they feel just as warm with the thermostat set at 19°C as they used to feel with it set at 21°C. Why is this so?
   b) Makers of humidifiers often use the evidence given in part a to convince homeowners that they can save fuel by using humidifiers. What do you think of this?

9. Homeowners are warned to adjust the humidity levels in their homes during winter as the outside air temperature changes. Typical recommendations are shown in Table 32.

TABLE 32 Humidity Recommendations

Outside air temperature (°C)	Recommended inside humidity at 20°C
−30 or below	15%
−30 to −24	20%
−24 to −18	25%
−18 to −12	35%
−12 and above	40%

a) Describe the change in the recommended inside humidity as the outside air temperature decreases.
b) What happens to the relative humidity of a sample of air as the temperature decreases?
c) What might happen if the recommended relative humidities are exceeded? Where would the problem first be noticed? What damage might result if the recommendations are exceeded?
d) Why is an upper limit of 40% recommended?

10. Many homeowners are adding insulation to the walls and ceilings of their homes to lessen heat transfer. The Department of Energy, Mines, and Resources of Canada recommends that vapour barriers also be added.
a) Account for this recommendation in terms of what you know about the effect of temperature on relative humidity.
b) Should the vapour barrier be installed next to the inside or next to the outside wall? Why?

5.9 INVESTIGATION: Measuring Relative Humidity

You saw in Section 5.8 that the relative humidity can be obtained from the absolute humidity and the saturation value. However, this involves a great deal of work. Fortunately, scientists developed a much simpler method. It uses the apparatus shown in Figure 5-7 (page 105) which you used in your studies of evaporation in Investigation 5.7. This apparatus is called a **wet and dry bulb hygrometer**.

When water on the wet bulb evaporates, it cools the bulb and lowers the temperature reading below that of the dry bulb. If the air has a high relative humidity, little evaporation occurs from the wet bulb. Thus it will have a reading only slightly less than that of the dry bulb. If the air has a low relative humidity, much evaporation occurs from the wet bulb. It will now have a reading several degrees less than that of the dry bulb. Scientists have used this cooling effect, along with absolute humidity measurements, to develop Table 33. Let us see how it is used to find the relative humidity.

Materials

Wet and dry bulb hygrometer as shown in Figure 5-7. (You may choose to build a more portable one.)
Table 33

Procedure

a. Practise using Table 33 to get the relative humidity. Table 34 gives you some examples to try. Check your answers with a classmate. Then consult your teacher if you are having problems.

TABLE 34 Determining Relative Humidity

Dry bulb temp.	Wet bulb temp.	Relative humidity
20°C	10°C	24%
25°C	22°C	77%
18°C	8°C	
28°C	25°C	

b. Use your hygrometer to find the relative humidity of the classroom. Fan it vigorously for 2 min before taking the temperature readings.
c. Find the relative humidity of at least 5 other locations. Here are some suggestions: the sunny side of the school, the shady side of the school, the playing field, the basement, a confined storeroom, the classroom at the beginning and at the end of the period. Record all of your results in a table.

Discussion

1. Account for any variations in results you observe among the locations you chose.
2. Why must the hygrometer be fanned?
3. Compare your outdoor readings to those of the official weather report. Account for any differences. (Check the newspaper for the official figures.)
4. The wet and dry bulb hygrometer used by a weather station is called a sling psychrometer. Find out why it has that name.

TABLE 33 Relative Humidity

Wet bulb temperature (°C)	Dry bulb temperature (°C)																				
	10	11	12	13	14	15	16	17	18	19	20	21	22	23	24	25	26	27	28	29	30
1	10																				
2	15	9																			
3	24	18	12	7																	
4	34	27	21	15	10	6															
5	44	36	29	23	18	13	8														
6	55	46	39	32	26	20	15	11	7												
7	66	56	48	41	34	27	23	18	14	10	6										
8	77	67	58	50	42	36	30	25	20	16	12	9	6								
9	88	78	68	59	51	44	38	32	27	22	18	14	11	8	5						
10	100	91	78	69	60	53	46	40	34	29	24	20	17	13	10	8					
11		100	89	79	70	61	54	47	41	36	31	26	22	19	15	12	10				
12			100	89	79	71	63	55	49	43	37	32	28	24	20	17	14	12	5		
13				100	90	80	71	64	57	50	44	39	34	30	26	22	19	16	13	7	
14					100	90	81	72	65	58	51	46	40	36	31	28	24	21	18	15	9
15						100	90	81	73	65	59	53	47	42	37	33	29	26	22	19	17
16							100	90	82	74	66	60	54	48	43	39	34	31	27	24	21
17								100	91	82	74	67	61	55	49	44	40	36	32	28	25
18									100	91	83	75	68	62	56	50	46	41	37	33	30
19										100	91	83	76	69	62	57	51	47	42	38	35
20											100	91	83	76	69	63	58	52	48	43	39
21												100	92	84	77	70	64	58	53	49	44
22													100	92	84	77	71	65	59	54	50
23														100	92	84	78	71	65	60	55
24															100	92	85	78	72	66	61
25																100	92	85	78	72	67
26																	100	92	85	79	73
27																		100	93	86	79
28																			100	93	86
29																				100	93
30																					100

5. A dial-type hygrometer does not use the wet and dry bulb principle. Examine one closely. "Huff" on it or hold it in a source of water vapour. Then make up a theory explaining how it works.

5.10 Home Appliances Using Evaporation

Three common home appliances make use of the cooling effect of evaporation. These are the refrigerator, air conditioner, and dehumidifier. We will explain how the first one works. You can figure out how the other two work as you do the "Discussion" questions.

The Refrigerator

The refrigeration unit of a refrigerator contains a substance called a freon. When a freon is in the gaseous state, it can be easily compressed to a liquid. Also, when it is in the liquid state, it can be easily evaporated. A condensation-evaporation cycle of the freon is used to produce the cooling of materials in the refrigerator.

Basically all a refrigerator does is move heat from within itself to the air in the room. Inside the refrigerator are cooling coils (Fig. 5-8). Liquid freon is forced through a small opening called a needle valve into these coils. As it passes through the valve, the freon expands from a liquid to a vapour; that is, it evaporates. Evaporation requires heat. The heat comes from the cooling coils themselves and the air in the refrigerator. These, in turn, get heat from the materials stored in the refrigerator, thereby cooling them.

The freon vapour must now be condensed to a liquid so it can be forced through the needle valve again. A compressor is used to force the freon particles closer together, forming a liquid. Condensation produces heat and we do not want that in the refrigerator. Therefore the compression is done outside the refrigerator box. You have probably felt the heat rising from the condenser coils on the back of the refrigerator when the compressor is running. The quantity of heat given off by the condenser coils is the same as that absorbed from materials inside the refrigerator during evaporation.

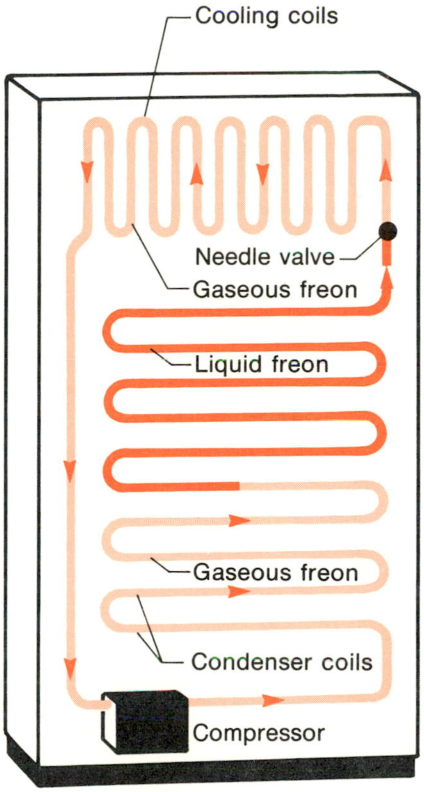

Figure 5-8 The electric refrigerator uses the cooling effect of evaporation.

Discussion

1. Locate the following parts on the refrigerator in your home: cooling coils, compressor, condenser coils.
2. Would it be possible to cool the kitchen of your home by leaving the refrigerator door open? Why?
3. An air conditioner is basically a refrigeration unit that is partly outdoors. What part is outdoors? Why?
4. Some people with damp basement rooms use a dehumidifier to lower the humidity in that part of the house. A dehumidifier

is basically a refrigeration unit with a tank on it to collect water. How does it work? What will happen to the air temperature in a room containing a dehumidifier that is operating? Why?

5. Oranges do not have to be kept in a refrigerator to prevent them from spoiling if you are only going to keep them for 2 or 3 weeks. Because of this, Ian says that they should be stored in a cupboard to save energy. Sandra says that they might as well be kept in the refrigerator. It has to be kept cold for the other things, so no energy will be wasted on the oranges. Whose argument do you support? Why?

6. **a)** Make a list of the benefits of an air conditioner to the owner.
b) Make a list of the "costs" of an air conditioner to society.
c) "Air conditioners are essential in homes in Florida, in most large department stores during July, and in cars that must be driven in cities." What do you think of this statement? Why?

5.11 Changes of State in Nature

The Water Cycle

In nature, water occurs in all three states. It is constantly changing from one state to another through a continuous process called the water cycle (Fig. 5-9).

Figure 5-9 The water cycle involves all changes of state.

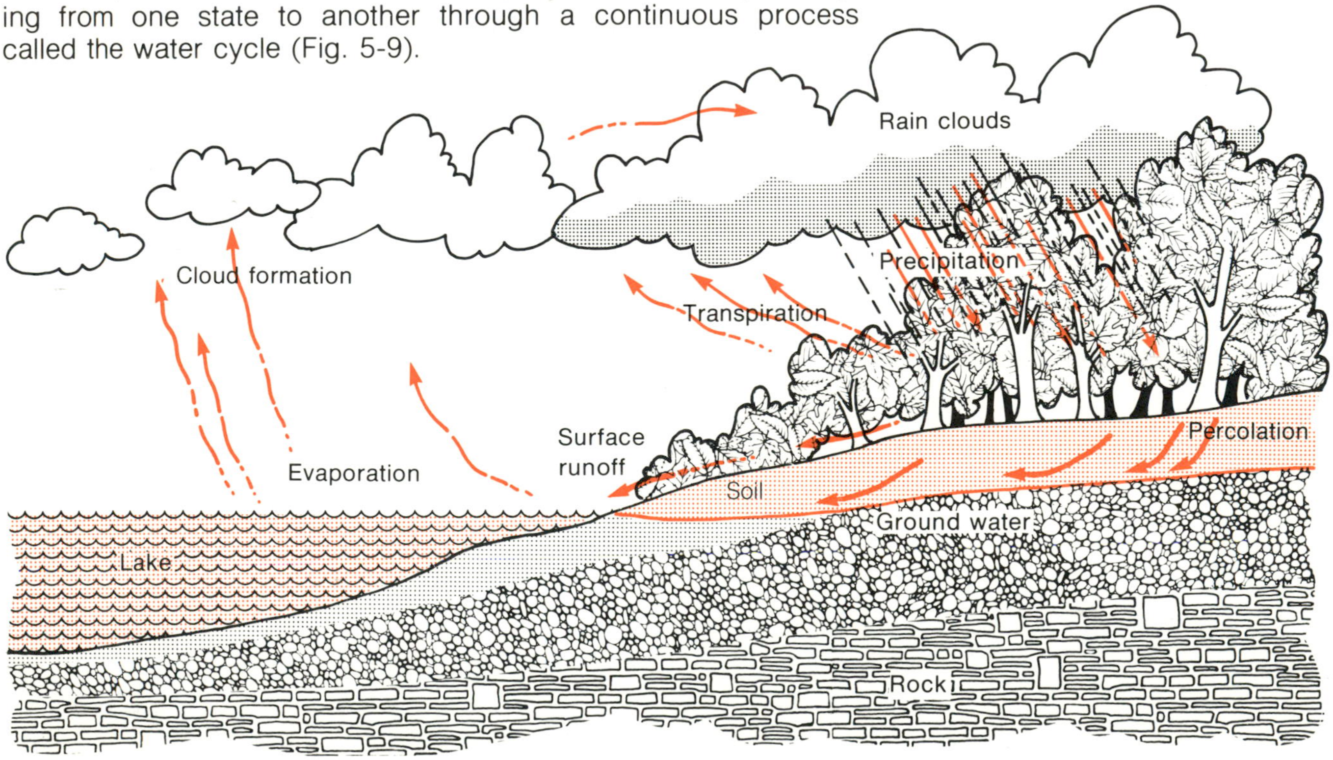

Evaporation occurs from the oceans, lakes, rivers, and soil. Plants also give off water vapour (transpire) through tiny pores in their leaves. Warm air currents carry the water into the cooler upper atmosphere. The vapour condenses into tiny water droplets or sublimes into small ice crystals. These are the condensation nuclei, or starting spots, for the formation of rain drops and snowflakes. These condensed forms of water return to the earth and are then put through the water cycle again.

We will now look more closely at some changes of state that occur in nature. Most of these are familiar to you. For each one we will give you some basic information. You then have to use your knowledge of the particle theory and the heat changes that take place during changes of state to answer some questions.

Dew

Dew consists of tiny droplets of water. It forms most often on cloudless nights when there is little wind. The "fall" of dew is particularly heavy when a cool night follows a warm humid day.

Questions

1. What change of state takes place during the formation of dew?
2. Why does dew form on cars, grass, and rocks instead of forming in the air (like a mist)?
3. Why does a cloudless night help form dew?
4. Why does a windless night help form dew?
5. Explain why the fall of dew is particularly heavy when a cool night follows a warm humid day. (Recall your work on relative humidity.)

Frost

Frost consists of tiny crystals of ice. But it is not just frozen dew. However, like dew, it forms most often on cloudless nights when there is little wind. A very heavy coating of frost appears on cars and grass when a very cold night follows a warm humid day.

Questions

1. How cool must the night become if frost is to form?
2. What change of state occurs during the formation of frost?
3. What changes of state occur as the temperature rises in the morning?
4. Frost damages most garden plants. When is it most likely to do the damage, when it is forming or when it is disappearing? Why?

Clouds, Fog, Mist, Rain, and Snow

Sometimes one small part of the earth's surface gets warmer than the area around it. For example, the black soil of a bog usually gets much warmer on a sunny day than the land and water around it. Warm air is less dense than cold air. Therefore it begins rising through the colder air. As this less dense air rises into the colder upper atmosphere, it is cooled. It also expands as it rises, because the pressure is lower in the upper atmosphere. This causes still more cooling of the air. Usually these rising air currents contain a fair amount of water vapour. If the temperature of the air becomes so low that the saturation point is reached, the water vapour in the air condenses into tiny water droplets. If the air is very cold, the water vapour may sublime into ice crystals. Usually the condensation or sublimation begins on dust particles in the air.

Rain or snow forms when the water droplets or ice crystals get so large that the rising air currents can no longer hold them up.

A fog or mist is simply a cloud that has formed at or near the surface of the earth.

Questions

1. Why is warm air less dense than cold air?
2. Why does the less dense air rise through the denser air?
3. Explain why the rising air cools.
4. How cold must the rising air become before sublimation of water vapour to ice crystals will occur?
5. In what places do fogs (mists) usually form? At what time of day do they usually form? Why is this so?
6. The particle theory says that particles have attractive forces between them. If you read an advanced book on the particle theory, it would tell you that these forces act more strongly in certain directions than they do in others. Does the appearance of a snowflake support this statement? Explain.
7. Describe how a large island in a large lake might cause the formation of clouds on a calm, hot summer day.
8. To answer this question, you will have to go to the library.
   a) What are the meanings of the terms warm front, cold front, high pressure area, and low pressure area?
   b) What kinds of clouds are usually formed by each type of front?
   c) How do low and high pressure areas affect the weather? Why?
   d) Make a table with the column headings "Type of Cloud", "Location", and "Weather". Place in the first column the names of the clouds in b. Then describe in the second column

where each cloud type is usually located. Finally, explain in column three what each cloud type tells us about the weather.

Hail

Hailstones are usually formed during violent storms. Raindrops on their way to the ground are caught in fast rising air currents. They are carried high into the cold upper atmosphere where they freeze into small hailstones. Layers of frost may sublime onto their surfaces at this time. As the hailstones fall or are blown through the lower humid air and clouds, water condenses on them. Again, updrafts may throw the hailstones back up into the cold region. Freezing and sublimation take place, adding more layers to the hailstones. This cycle may be repeated many times. The more violent the storm, the more often the hailstones are cycled, and the larger they get. Finally they become too heavy to be blown upwards and they drop or are blown to the earth.

If you ever get the chance, cut open a hailstone to see how many layers it has. This will give you an idea of how often it made the round-trip from upper to lower atmosphere.

Questions

1. **a)** Describe the damage hailstorms often cause.
   b) Describe the weather that normally precedes a hailstorm.
   c) How large do hailstones get?
2. How could you tell the difference between a layer of a hailstone that was formed by sublimation and one that was formed by freezing of condensed water?

Sleet

Sleet is formed when raindrops fall through air that has a temperature below 0°C. If the air does not have enough dust particles, the raindrops can cool below 0°C without freezing. In this state they are called **supercooled**. When they strike an object or the ground, they immediately leave the supercooled condition and freeze. This produces a coating of ice called sleet.

Questions

1. How does sleet on the window of a car differ in appearance from frost on the window of a car? Why?
2. What will happen to the temperature of a car window when sleet forms on it? Explain.
3. Describe the damage that is normally caused by a sleet storm.

5.12 INVESTIGATION: The Viscosity of Liquids

The term **viscosity** describes how easily a liquid "runs" when it is poured. Some liquids such as corn syrup are thick and run slowly when poured. They are called viscous, and are said to have a high viscosity. Other liquids such as gasoline and water are thin and splashy. They have a low viscosity. As a result, they run easily through a rubber hose. Imagine trying to run corn syrup through a garden hose!

What makes corn syrup more viscous than gasoline? According to the particle theory, both liquids are made of particles. Also, the particles have attractive forces that tend to pull them together. Knowing this, make up a hypothesis to explain why corn syrup is more viscous than gasoline. You know that viscous liquids such as corn syrup thin out when heated. How does the particle theory explain this?

Discuss your hypothesis and your explanation of the thinning with your lab partner. Then proceed with this investigation.

Here you study how fast an object falls through samples of motor oil. Each sample has a different viscosity. You are told the grades of motor oil used; the grades give you an idea of the relative viscosities. The results of this investigation should help you decide why one liquid has a different viscosity from another. You also study the effect of temperature on viscosity.

Materials

tall glass containers (at least 20 cm)
single grade motor oil of several viscosities (SAE 10; SAE 20; SAE 30; SAE 40; SAE 50)
multigrade motor oil (SAE 10W-40; SAE 20W-50)
hot-plate (on teacher's bench)
plastic sphere
thermometer (−10°C to 110°C)
large tin can
watch or clock with second hand

Procedure A Comparing Different Grades of Motor Oil

Your teacher has placed several tall glass containers at various stations in the lab. Each container contains motor oil of a certain grade. The grade is marked on the container. At each station you are to do the following:

a. Take the temperature of the oil.

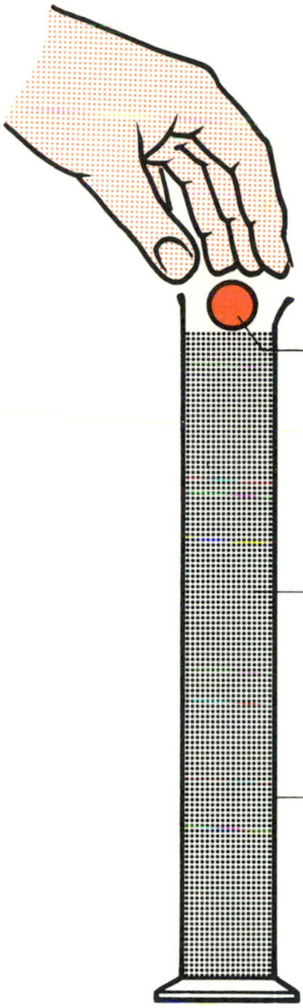

Figure 5-10 Investigating the viscosity of motor oil.

b. Hold a plastic sphere so it just touches the surface of the oil (Fig. 5-10). Release it and time how long it takes for the sphere to reach the bottom.
c. Record your results in a table.
d. When you have finished, plot a graph that summarizes your results.

Procedure B Viscosity and Temperature

a. Your teacher has samples of SAE 50 motor oil at several temperatures on the demonstration desk. Form a group of 6 to 8 students and go to the demonstration desk. Repeat steps a and b of Procedure A using each of the samples. Your teacher will remove the spheres.
b. Record your results in a table.
c. Plot a graph that summarizes your results.

Procedure C What is Multigrade Motor Oil?

a. A multigrade motor oil is described by two numbers, for example, SAE 10W-40. Write in your notebook what you think this means. If you have no idea at all, discuss the matter with a classmate who does. What you write in your notebook is a hypothesis concerning the nature of multigrade motor oil.
b. Design an experiment to test your hypothesis. Write the procedure in your notebook.
c. Discuss your procedure with your teacher. Then get the materials to do the experiment.
d. Record your results in a table.
e. Plot a graph that summarizes your results.

Discussion

1. Make a general conclusion that summarizes the results of Procedure A. Keep in mind that the oil has to flow around the sphere as it drops.
2. Compare the strengths of the attractive forces between the particles in each of the oil samples used in Procedure A.
3. Compare the strength of the attractive forces between the particles of corn syrup with that between the particles of water.
4. How does temperature affect the viscosity of a single grade motor oil? Explain your answer using the particle theory.
5. Compare the graphs from Procedures B and C. What are the advantages of a multigrade motor oil?
6. If you are using a single grade motor oil in your car, why must you change the oil when the season changes?

7. Check the owner's manual of the family car. What grades of oil does it recommend be used in different seasons?
8. Find out from a service station what happens to the oil that is drained from a car during an oil change. What do you think should be done with it? Why?

5.13 INVESTIGATION: Effect of Temperature on the Volume of a Gas

What does the particle theory say about the effect of temperature changes on the particles? As a result, what do you think will happen to the volume of a given mass of gas if its temperature is increased? Does your prediction agree with common sense (your everyday experiences)?

This investigation is qualitative. No measurements of volume are made. Its purpose is simply to help you decide whether or not the particle theory applies to the study of the behaviour of gases.

Materials

250 mL flask (Erlenmeyer or distillation)
2-hole rubber stopper
thermometer (−10°C to 110°C)
glass tubing (50 cm)
Bunsen burner
marking pen
CAUTION: Wear safety goggles during this investigation.

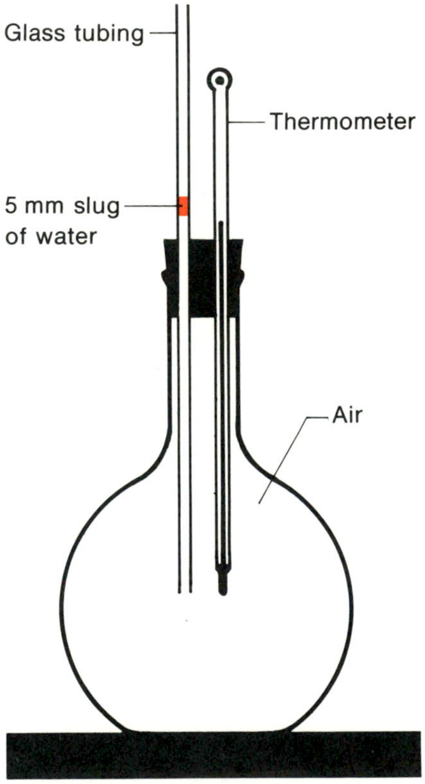

Figure 5-11 Temperature and volume of a gas.

Procedure

a. Set up the apparatus as shown in Figure 5-11. The glass tubing and thermometer should reach halfway into the flask. The slug of water should be about 5 mm long. At the start of the investigation it should be just above the stopper. Your teacher will tell you how to get it there if you have trouble.
b. Warm the flask gently by wrapping your hands around it for a few minutes. Mark the new position of the slug of water. Note the temperature.
c. Warm the flask still more by lightly playing a low flame over it. Keep touching the flask from time to time so you won't overheat it. (**CAUTION:** The thermometer will break if the temperature goes above its range.) Continue marking new positions of the slug and the corresponding temperatures until the temperature is in the 70°C-80°C range.
d. Record all of your data in a table.
e. Observe closely the movement of the slug of water as the flask cools.

Discussion

1. Make a generalization concerning the effect of temperature on the volume of a gas.
2. Explain your generalization using the particle theory.
3. Does the particle theory suggest that your generalization may be true for gases other than air? Explain.
4. Why can this investigation not be used to study *quantitatively* the effect of temperature on volume? (Would a graph of temperature vs. volume be valid? What are the major sources of error?)

5.14 INVESTIGATION: Effect of Temperature on the Volume of a Liquid

Three things should help you to predict the results of this investigation:

a. your everyday experiences
b. the results of Investigation 5.13
c. the particle theory.

Write in your notebook a prediction of the effect of temperature on the volume of a liquid. Explain why you made this prediction. Will all liquids behave in exactly the same manner? Why?

Now, try the experiment to test your prediction.

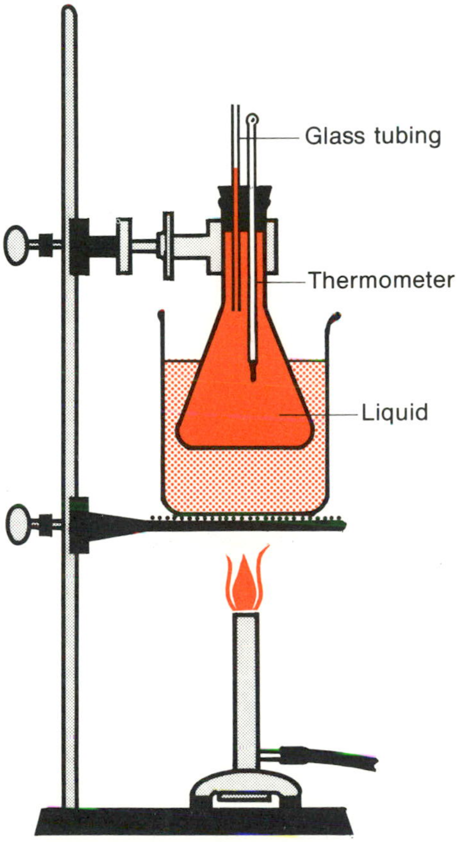

Figure 5-12 Apparatus for studying the effect of temperature on the volume of a liquid.

Materials

1 000 mL beaker
2-hole rubber stopper
thermometer (−10°C to 110°C)
250 mL flask (Erlenmeyer or distillation)
ring stand, iron ring, wire gauze, adjustable clamp
water, glycerine, ethylene glycol
marking pen
glass tubing (50 cm)
Bunsen burner

CAUTION: Wear safety goggles during this investigation.

Procedure

a. Set up the apparatus as shown in Figure 5-12. Make sure the liquid in the glass tube is just above the rubber stopper when you start. Use the liquid suggested by your teacher.
b. Before you begin heating, mark the position of the liquid and note the temperature. Record your observations in a data table with the headings "Change in volume" and "Temperature". (In this case, the change in volume is 0.)

c. Begin heating the liquid slowly. Record the change in volume at the following temperatures: 30°C, 35°C, 40°C, 45°C, 50°C, 55°C, 60°C, 65°C. Do not heat the liquid beyond 65°C.
d. Plot a graph with change in volume on the vertical axis and temperature on the horizontal axis.
e. If time permits, repeat the experiment using another liquid that your teacher will supply. Plot your results on the same graph. If time does not permit you to try a second liquid, get the results for another liquid from a classmate and plot them on your graph.

Discussion

1. Describe the relationship between the temperature and volume of a liquid. How does this agree with your prediction?
2. Use the particle theory to explain the results of this experiment. Your explanation should include the general behaviour of liquids when heated and also any differences you observed between different liquids.

5.15 INVESTIGATION: Effect of Temperature on the Volume of a Solid

You have discovered that both gases and liquids expand when heated. According to the particle theory, it is reasonable to assume that solids will do the same. However, solids are likely to show less change in volume per unit temperature change than gases or liquids.

In this investigation you heat a metal rod and observe any change in length. Since the change may be too small to see directly, a volume change indicator is used to magnify any change in length. (See Figure 5-14.)

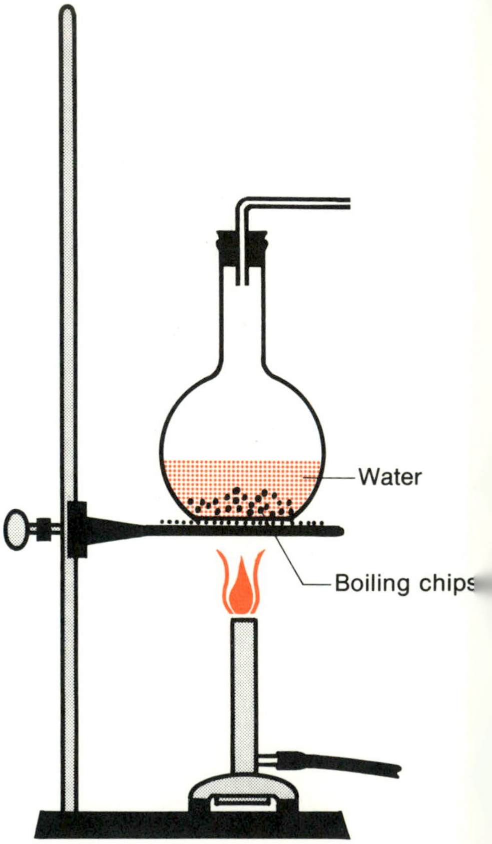

Figure 5-13 A steam generator.

Materials

steam generator as shown in Figure 5-13
hollow metal rods (3 different metals of the same length)
volume change indicator
ring stand and adjustable clamp
rubber hose
short plastic ruler
thermometer (−10°C to 110°C)
CAUTION: Wear safety goggles during this investigation.

Procedure

a. Set up the steam generator as shown in Figure 5-13.
b. Set up the rest of the apparatus as shown in Figure 5-14. Use transparent tape to fasten the ruler in the required position.

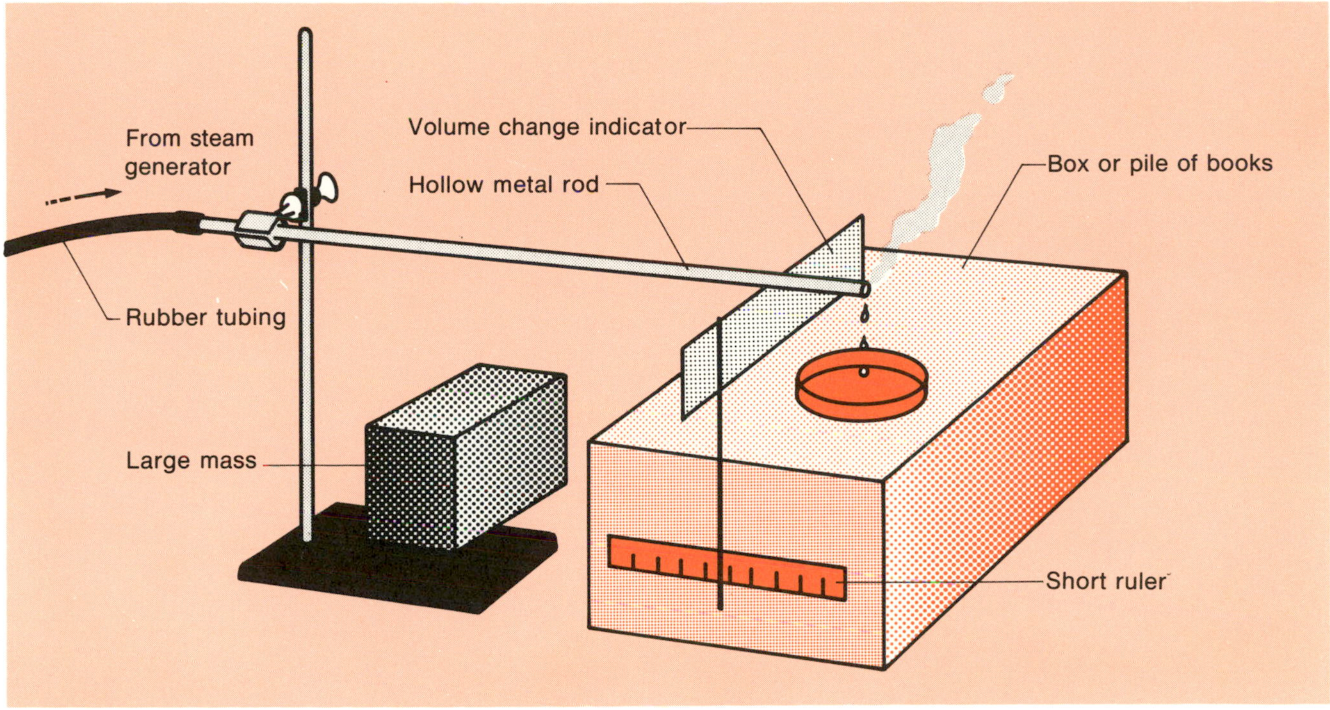

Figure 5-14 Apparatus for studying the effect of temperature on the volume of a solid.

c. Connect the steam generator to the rest of the apparatus and begin heating the water. You may heat strongly at first. When the water is close to boiling, turn down the flame so that the water will boil gently. **CAUTION:** If the water boils vigorously, it may blow the stopper out of the flask.
d. Continue passing steam through the tube until the volume change indicator no longer moves. Record the position of the indicator. Measure the temperature of the steam going through the tube by loosely inserting the bulb of the thermometer into the end of the metal tube. Keep it there until it shows no further change.
e. Stop heating the steam generator. When the metal tube has cooled, replace it with one of the other metal tubes. Repeat steps c and d.
f. Repeat step e for the third tube, if time permits.

Discussion

1. Compare the effect of temperature on the volume of solids, liquids, and gases.

2. Explain any differences using the particle theory.
3. How does the same temperature change affect the same volume of different metals? Why is this so?
4. The metal tube could get longer without an increase in volume, provided its diameter decreased. Is this likely to happen? Use the particle theory to support your answer.
5. Will power transmission lines hang lower in the summer or in the winter? What precaution must be taken by a line construction crew that builds a power transmission line in the summer?
6. Describe and explain any device you have seen on bridges that allows for expansion and contraction of the metal in the bridge.
7. What allowance is made in railway tracks for the expansion and contraction of the tracks?
8. Review the results of Investigation 1.8 on page 13. What effect would an increase in temperature have on the accuracy of a pendulum clock? Explain your answer.
9. A metal screw-type top comes off a glass jar more easily if you run hot water over the top for a few seconds. Why is this so?

5.16 INVESTIGATION: Effect of Temperature on the Pressure of a Gas

As the particles of a gas move around, they strike the walls of the container. As a result, they tend to push the walls outward. This "push" is what we call gas pressure. If the pressure is too great, the container may break or explode.

Temperature is one factor that can affect the pressure that a gas exerts on its container. Use the particle theory to predict the effect of temperature on the pressure of a gas. Then try this experiment to check your prediction.

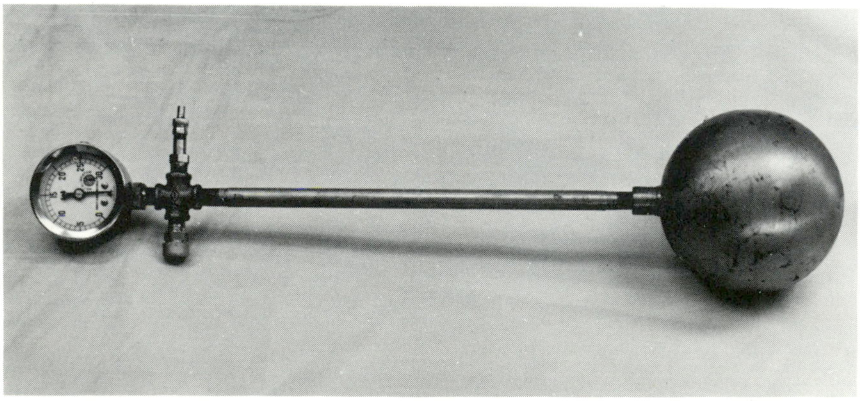

Figure 5-15 Apparatus for studying the effect of temperature on the pressure of a gas.

Materials

pressure-temperature apparatus (Fig. 5-15)
Bunsen burner
CAUTION: Wear safety goggles during this investigation.

Procedure

a. Note the pressure reading on the dial of the apparatus.
b. Gently heat the bulb and note the reading.
c. Heat the bulb strongly and note the reading.

Discussion

1. Explain the results of this investigation using the particle theory.
2. Aerosol cans have a warning "DO NOT INCINERATE" on them. Why?
3. A car manual tells the owner to check the tire pressure when the tires are cool. Why?

5.17 INVESTIGATION: Relationship between the Pressure and Volume of a Gas

In this investigation you use Figure 5-16 and the particle theory to predict the relationship between the pressure and volume of a gas. You then do an experiment to check the prediction.

Look at Part A of Figure 5-16. This diagram shows a cylinder with a piston in it. The piston can be moved up and down. A volume of gas, V, is trapped by the piston. The piston is being held down by a pressure, p_d. Since the piston is not moving, the gas inside must be pushing up with a pressure that equals p_d. It is called p_u. The red dots represent the particles of the gas. These particles cause the pressure, p_u. As they move around, they strike the piston (and the walls of the cylinder), causing the pressure.

Your Prediction

Answer the following questions to make your prediction of the relationship between the pressure and volume of a gas.

1. Part B of Figure 5-16 suggests that, in order to decrease the original volume by half, the pressure on the piston would have to be doubled. Does this agree with the particle theory? Explain.

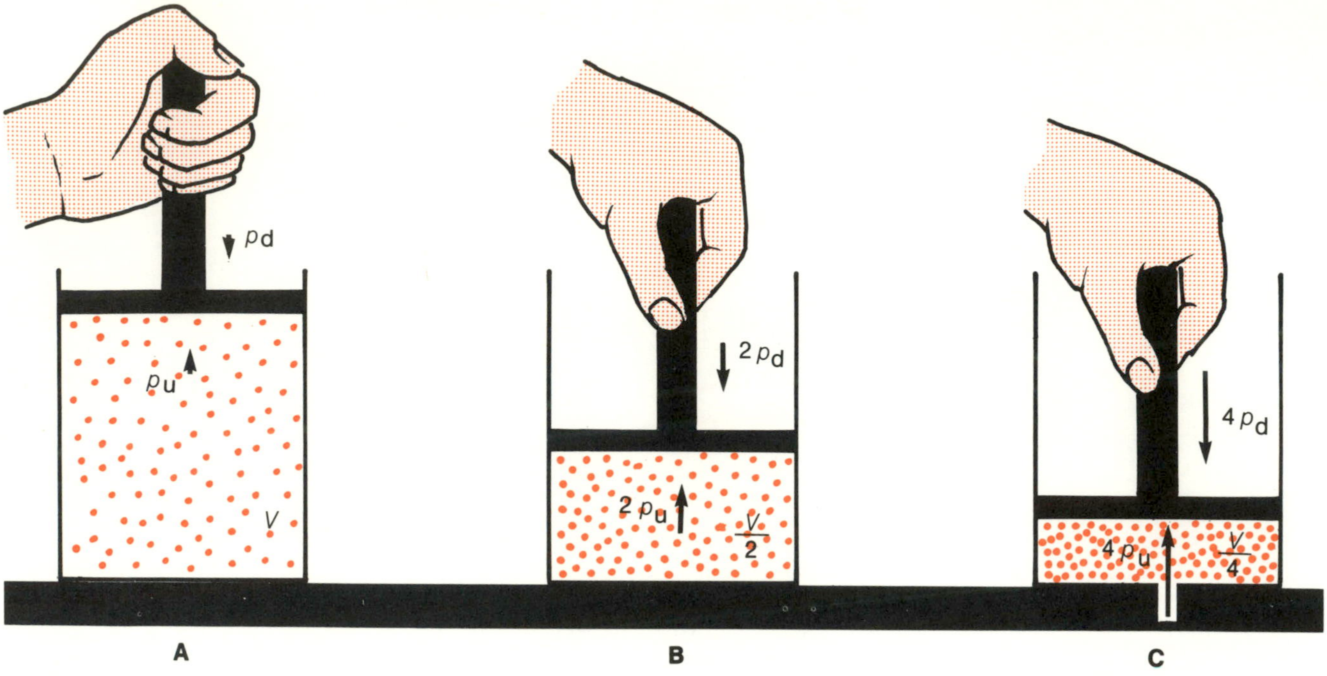

Figure 5-16 The particle model suggests that the volume of a gas decreases as the pressure increases.

2. What does Part C of Figure 5-16 suggest? Does this agree with the particle theory? Explain.
3. What is your prediction of the relationship between the pressure and volume of a gas? Now try the following experiment to check your prediction.

Materials

syringe
ring stand
adjustable clamp
six identical textbooks
rubber stopper with hole part way through
(The tip of the syringe must fit *tightly* into this hole.)

Procedure A A Qualitative Experiment

a. Set the piston of the syringe at the 30 mL mark.
b. Place the rubber stopper over the tip of the syringe.
c. Hold the syringe and stopper in one hand and push the piston with the other.
d. Remove the rubber stopper and push the piston down as far as it will go. Put your finger over the tip of the syringe and pull up on the piston.

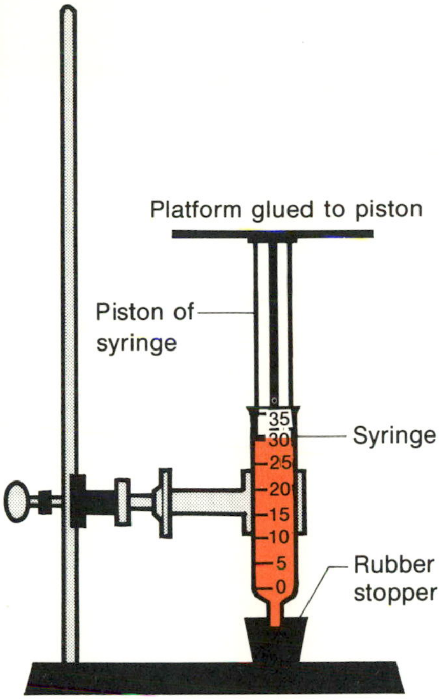

Figure 5-17 Apparatus for studying the pressure-volume relationship for a gas.

Procedure B A Quantitative Experiment

a. Repeat steps a and b of Procedure A.
b. Set up the apparatus as shown in Figure 5-17.
c. Carefully place one textbook on the platform of the piston. Record the new volume in Column A of a table like Table 35.
d. With the book still on the platform, press the piston until the volume is a few millilitres less than the one you just recorded. Release the piston and record the volume at which it stops in Column B.
e. Average the volumes in Columns A and B. Record your answer in Column C.
f. Repeat steps c to e with 2 books on the platform. Repeat again using 3, 4, 5, and finally 6 books.
g. If time permits, repeat the entire procedure, starting again with one book, to check your values.
h. Plot a graph with the average volume on the horizontal axis and the pressure (in books) on the vertical axis.

TABLE 35 Pressure-Volume Data

Number of books	A	B	C
0	30.0 mL	28.0 mL	29.0 mL
1			
2			
3			
4			
5			
6			

Discussion A

1. Describe and explain the results of step c in Procedure A.
2. Describe and explain the results of step d in Procedure A.

Discussion B

1. Explain how we obtained the volumes inserted in Table 35 opposite 0 books.
2. What is the purpose of averaging the volumes obtained in steps c and d of Procedure B?
3. Does your graph agree with your prediction? Explain.
4. Robert Boyle did an experiment in 1662 from which he obtained this law: *At constant temperature, the volume of a*

given mass of gas is inversely proportional to the pressure. Look up the meaning of "inversely". Does your graph agree with **Boyle's Law?** Explain. Why must the temperature be constant in this experiment?

5. Account for the graph using the particle theory.

5.18 INVESTIGATION: The Crystallization of Salol

Crystals are solids with regular shapes and smooth flat surfaces. Many solids exist in the form of crystals. Each solid has a certain shape of crystal. For example, common salt is made of crystals that are small cubes. The surfaces of each crystal are always flat and the corners are always 90°.

What causes crystals to have regular shapes? You know that the particles in a sample of matter exert attractive forces on one another. Also, as a liquid is cooled, the particles are pulled together to form a solid. If the solid is always a crystal with the same shape, it seems logical to hypothesize that the attractive forces are strongest in certain directions.

The purpose of this investigation is to let you watch crystals of a solid form under a variety of conditions. Perhaps you can verify the hypothesis just made.

Materials

salol (phenyl salicylate)
glass petri dish
Bunsen burner
ice
ring stand, iron ring, wire gauze
hand lens (10 ×)
test tube
beaker (250 mL)

CAUTION: Wear safety goggles during this investigation.

Procedure

a. Pour a thin layer of solid salol into the petri dish. Place the top on the dish. Gently heat the dish until all of the salol has melted.

b. Turn off the Bunsen burner as soon as the salol has melted.

c. Let the petri dish cool until it is warm to the touch but not hot.

d. Drop a small piece of salol crystal into the liquid salol. Using the hand lens, observe what happens. (A stereo microscope is better if it is available.) Write a description of what you observe and draw several sketches. What shape are the crystals? Are they all alike? Note: You can remelt the salol and repeat steps b to d to check your results, if you wish.

e. Melt the salol again and let it cool as described in step c. This time sprinkle some powdered salol over the entire surface of the liquid salol. (Prepare the powdered salol by crushing a few small crystals of salol.) Describe what happens to the salol. Compare the time required for crystals to form with the time required for crystals to form in step d. Compare the size and shape of the crystals with those in step d.

f. Melt 2-3 cm^3 of salol in a test tube. Immediately plunge the lower half of the test tube into a beaker of ice water. Describe what happens. Compare the time required for crystals to form with the times required in steps d and e. Compare the size and shape of the crystals with those in steps d and e.

Discussion

1. Do the results of step d of the procedure support the hypothesis made in the introduction? Explain.
2. Account for what you observed in step e of the procedure.
3. Account for what you observed in step f of the procedure.
4. Ryolite is a fine-grained igneous rock that forms when a certain type of liquid rock erupts from a volcano and cools quickly in the air (Fig. 5-18). If the same liquid cools slowly below the ground, a coarse-grained granite (also an igneous rock) is formed. Using the results of this investigation, explain the formation of these two types of rocks from the same liquid.

A

B

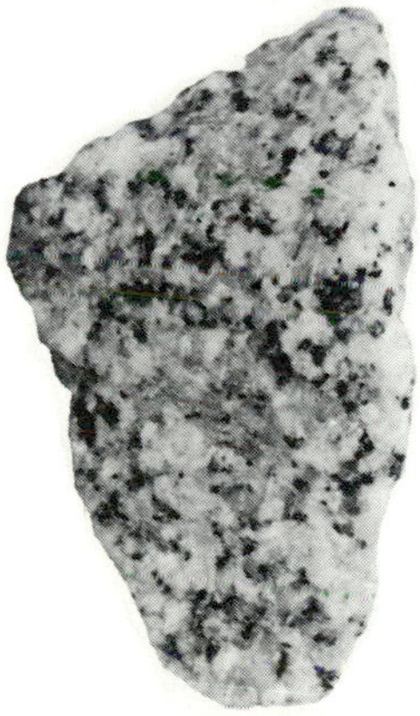

Figure 5-18 Two types of igneous rock: ryolite (A) and granite (B). Compare the size of the crystals.

Highlights

The particle theory explains such phenomena as diffusion, Brownian motion, and the cooling effect of evaporation.

Evaporation of a liquid requires heat. Humidity gauges, the refrigerator, and the air conditioner all use this fact in their operation.

Changes of state occur in nature due to the heating and cooling of air masses that contain water vapour. A decrease in temperature tends to form a more condensed state of water, whereas an increase in temperature tends to form a less condensed state.

Viscosity is a measure of how easily a liquid runs when it is poured. The particle theory explains why many substances differ in viscosity. It also explains why most liquids become less viscous when the temperature is increased.

The volume of a gas, liquid, or solid generally increases when the temperature is increased. The change in volume is most marked in the case of a gas. Water is anomalous; its liquid state is more dense than its solid state. Liquid water is most dense at 4°C. Most other liquids are most dense at their freezing points.

The volume of a gas decreases when the pressure is increased. The particle theory explains this behaviour.

UNIT

6 Water, Solutions, and Mixtures

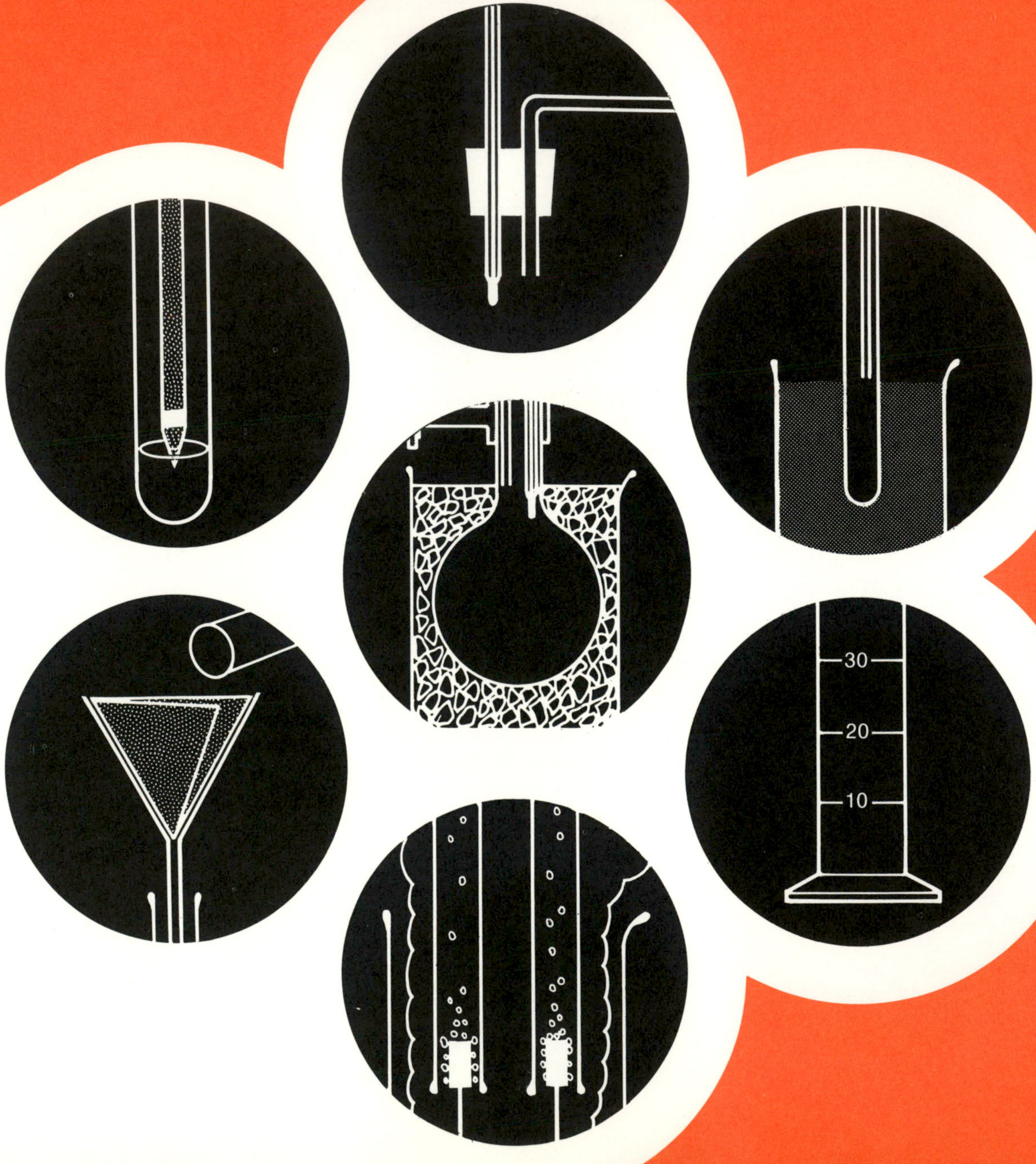

In Units 3, 4, and 5 you studied the properties of matter and used the particle theory to explain those properties. In this unit you combine what you have learned with new studies to investigate the properties of water, solutions, and mixtures. Most of this material will be useful to you in your everyday life. Try your best to understand all of it.

A. Water

6.1 Water: Occurrence, Importance, and Properties

Water is so common that you probably have not thought much about its occurrence, importance, and properties. Yet, because it is so common, it affects you every day in many ways. Therefore you should have some basic knowledge about this substance.

Occurrence

In Section 5.11, page 112, you learned that water exists in nature in all three states—solid, liquid, and gaseous. You also learned that it is always changing from one state to another through a series of processes that together are called the water cycle.

Over 70% of the earth's surface is covered by water. Most of this, of course, is made up of the oceans. In fact, 97% of the world's water is salt water. The remaining 3% is fresh water of which 98% is frozen in the ice caps of Antarctica and Greenland. If these ice caps were to melt, the sea level would go up by about 60 m. That would be enough to flood many coastal cities such as New York and London, England. The 2% of fresh water that is not frozen is in lakes and rivers. Much of this is in big lakes like the Great Lakes. But a surprisingly large amount is in rivers. The Amazon River of South America, with about 1 100 tributaries and a length of over 6 000 km, contains at least 65% of the world's river water. From its mouth each day comes 20% of all the river water that enters the oceans. The importance of such a vast reservoir of fresh water cannot be ignored.

Plants contain large amounts of water in their tissues. Lettuce, cabbage, and tomatoes are over 90% water. Animal tissue also contains much water. Eggs and lean beef are about 75% water. Your body is almost 80% water. Even some minerals in rocks contain water. Water is so tightly held within the rocks that the rocks appear dry. Yet some of these rocks may be as much as half water. All types of soil contain water. You can see and feel it in the peat of a northern bog. But even the dryest desert sand contains some water. Last, but not least, the atmosphere contains vast amounts of water. If this were not so, the water cycle could not move water from the oceans to inland areas.

Importance

All living things, plant and animal, need water. Some of these are aquatic and must live entirely in water. Fish and algae are examples. Others such as dogs, most birds, and the grass of a lawn are terrestrial. Yet they, too, require water. They would die if water were not present in the air and soil.

Of vital concern to humans today is the need for water to produce food. Growing just 1 kg of wheat requires 600 L of water; 1 kg of rice 2 000 L; 1 kg of meat 25 000 – 60 000 L; 1 L of milk 9 000 L. A single corn plant absorbs over 200 L of water from the soil in one growing season.

Industry also uses large amounts of water. At least 500 000 L of water are necessary to produce one car!

Each one of us requires directly or indirectly about 9 000 L of water a day. This includes water for drinking and bathing as well as a share of that used by agriculture and industry. In 1900 each person in North America required only 2 400 L. Industrialization and irrigation of farmland caused the increase. Since only 3% of the earth's water is fresh, care must be taken not to use large amounts needlessly. Global studies have shown that humans are now removing fresh water from the land faster than the water cycle can replace it. In North America we use about twice as much water as the water cycle returns. Every time we need more water, we pump it from a lake or river or we drill a new well. Sooner or later a limit will be reached. Some scientists say that, in the next decade, we will no longer be able to find all the fresh water we need. Perhaps it would be more accurate to say we will not be able to find all we want.

This situation may cause some worry in North America. But in a country such as India that is short of food, has a large population, and a low average rainfall, the situation is desperate. In an attempt to find more water to grow more food, India drilled in just one year about 80 000 wells. In the same year it installed about 250 000 pumps to bring water from lakes and rivers! How long can this continue?

Physical Properties

Water in the liquid state is clear and colourless when in thin layers. In thick layers it has a bluish-green colour. Pure water is also odourless and tasteless. The presence of dissolved substances often gives water an odour or taste. The freezing point of water is 0°C, the boiling point 100°C. Its density is about 1 000 kg/m^3 in the liquid state and about 900 kg/m^3 in the solid state. It is unusual for the solid state of a substance to be less dense than the liquid state. Thus water is said to be **abnormal** or **anomalous**.

Discussion

1. List five ways in which water affects your daily life.
2. **a)** Some desert animals like gerbils seldom drink water. Where do they get the water they need?
   b) The Amazon River valley contains over 65% of the world's river water. The Brazilian government is cutting trees from the Amazon forest to produce farmland for settlers. How will this affect the amount of water the valley can hold? Explain your answer.
3. Explain why more water is required to produce 1 kg of meat than is needed to produce 1 kg of any grain crop.
4. **a)** If you live in a house, read the water meter at the beginning and end of a week. Then calculate the water used in your home per person per day.
   b) Suggest methods that could be used to save water in your home.
   c) Compare the amount used directly by you in your home with the amount used indirectly by you for industrial and agricultural purposes. Are your economy methods listed in b worth the effort?
5. What is the difference between needing a certain amount of water and wanting the same amount of water?

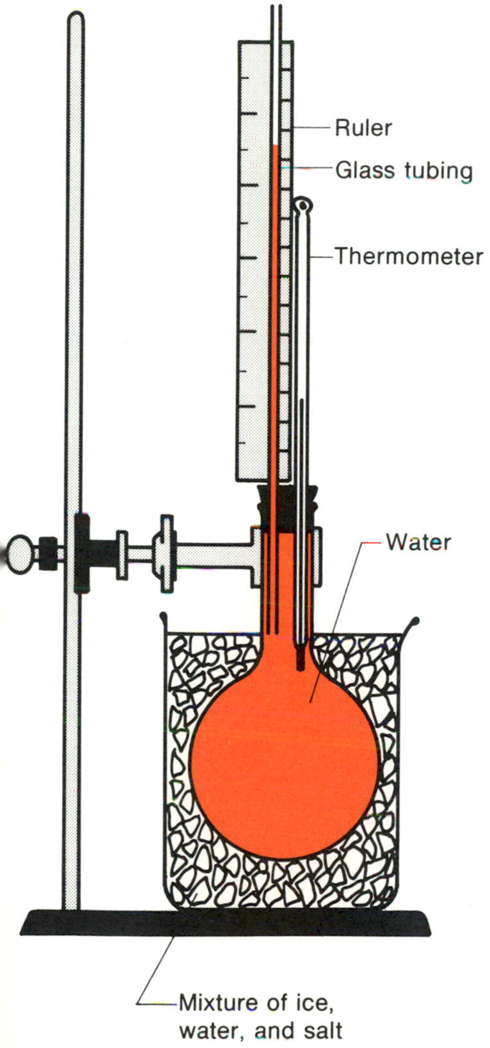

Figure 6-1 Studying the effect of cooling on the volume of liquid water.

6.2 INVESTIGATION: The Abnormal Behaviour of Water

In the previous section we referred to the abnormal or anomalous behaviour of liquid water as it is cooled. Here we study this interesting and important property of water.

Materials

large beaker (1 000 mL)
crushed ice
salt
water
250 mL flask (Erlenmeyer or round-bottomed)
thermometer (−10°C to 110°C)
2-hole rubber stopper
glass tubing
ruler
ring stand, adjustable clamp

Procedure

a. Set up the apparatus as shown in Figure 6-1. The water should be at least 15 cm up the glass tube.
b. Note in a table the level of the water in the tube every minute until the temperature of the water is 0°C. Stir the ice-water-salt mixture from time to time to keep it as cold as possible.
c. Remove the flask from the mixture. Allow it to warm up. Note the water level in the tube every minute until the temperature of the water is near room temperature.
d. Plot graphs with temperature on the horizontal axis and water level on the vertical axis.

Discussion

1. Carefully describe what happens to the volume of water as the temperature drops.
2. What happens to the density of water as the temperature drops?
3. At what temperature does the water have its minimum volume or maximum density?
4. Why is the behaviour of the water called abnormal or anomalous?
5. Figure 6-2 shows a pond in mid-winter. Account for the various temperature readings.
6. If water did not have this anomalous behaviour, fish could not live in ponds and lakes in areas where freezing occurs. Explain this statement.

Figure 6-2 The temperature profile of a pond in mid-winter.

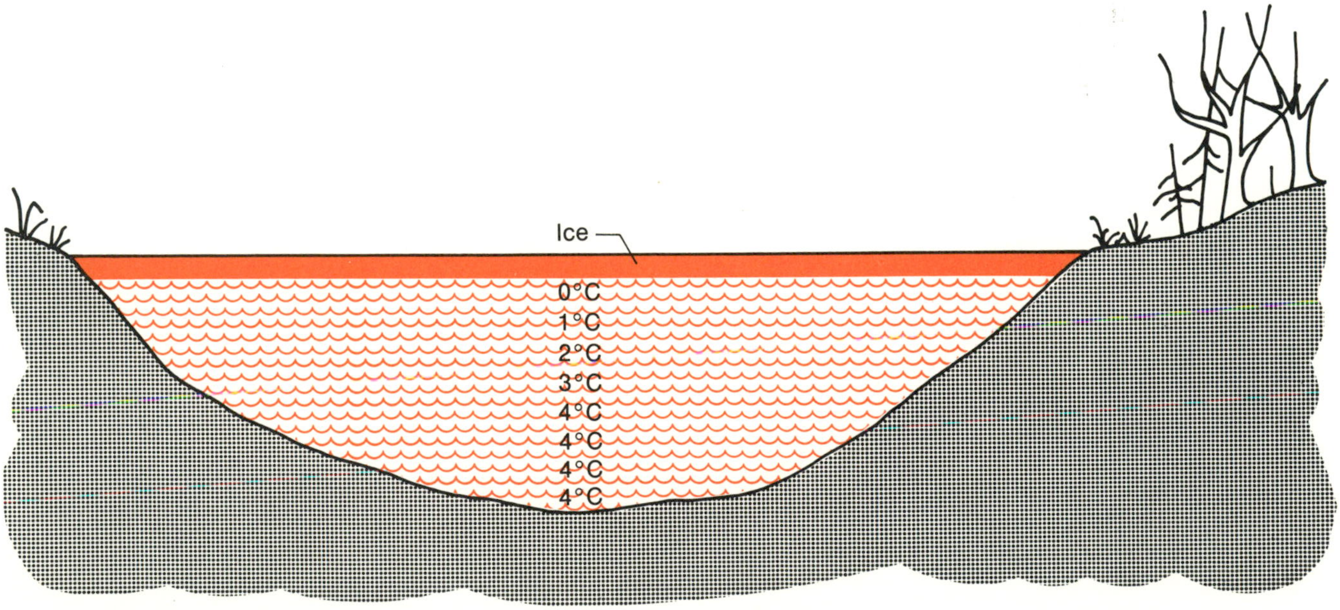

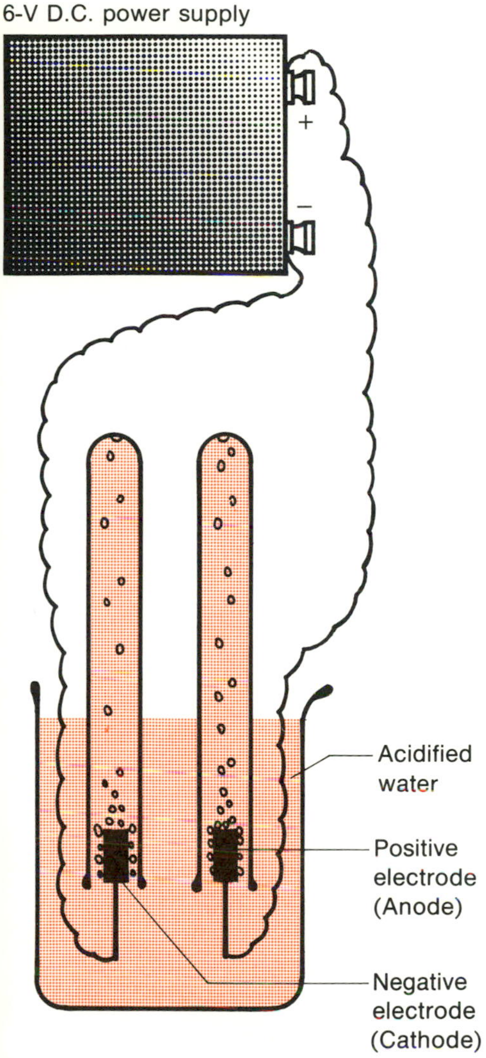

Figure 6-3 Apparatus for determining the composition of water.

6.3 INVESTIGATION: Electrolysis of Water

The purpose of this investigation is to find out what water is made of. Water can be decomposed by passing an electric current through it. This is called **electrolysis.** You are not expected to be able to explain this process. Later in this course you will learn more about the nature of the particles of matter. Then you may be able to explain how electrolysis works.

Materials

electrolysis apparatus (Fig. 6-3)
6-V power supply and connectors
acidified water
wooden splints
Bunsen burner
graduated cylinder (25 mL)
marking pen

Procedure

a. Fill the electrolysis apparatus with acidified water as shown in Figure 6-3. Acidified water contains a few drops of sulfuric acid in each litre of water. The purpose of the acid is to speed up the decomposition of the water by the electricity.

b. Connect one electrode of the electrolysis apparatus to the positive terminal of the power supply and the other electrode to the negative terminal. By doing this, you have made the first electrode positive (an *anode*) and the second electrode negative (a *cathode*).

c. Turn on the power until several millilitres of gas have collected in each tube. Turn off the power.

d. Use a marking pen to mark the level of the gas in each tube.

e. Find out what gas is in each tube by doing the following: Lift the tube out of the apparatus and let the remaining water escape. Keep the tube upside down. Insert a burning splint into the tube. Use Table 36 to decide what gas is in each tube. (Both gases are in this table.)

TABLE 36 Effect of a Flame on Some Gases

Gas	Effect of flame
Oxygen	Flame burns more brightly
Carbon dioxide	Flame goes out
Hydrogen	A "pop" occurs
Sulfur dioxide	Flame goes out

f. Find out the volume of the gas that was in each tube by filling the tube with water to the mark you made. Pour this water into the graduated cylinder.

Discussion

1. What gas collected at each electrode?
2. Compare the volumes of each gas that formed.
3. Electrolysis of water produces only the substances you discovered. What, then, is water made of?

6.4 INVESTIGATION: Amount of Water in Foods

All foods contain water. Earlier you were told the approximate amount of water in some foods. Here you find out experimentally how much water is in several other foods.

Materials

evaporating dish
ring stand, iron ring, wire gauze
Bunsen burner
balance
food material (e.g. oatmeal, potato, apple, or carrot)
cobalt chloride test paper (for water)
glass plate
CAUTION: Wear safety goggles during this investigation.

Procedure

a. Find the mass of the evaporating dish to the nearest 0.01 g.
b. Add about 10 cm^3 of the food to the dish. If you are testing a food like potatoes or carrots, cut it into small pieces.
c. Find the mass of the evaporating dish plus the food.
d. Place the glass plate over the dish. Heat the dish gently as shown in Figure 6-4.
e. Test the liquid that collects on the glass plate with cobalt chloride paper. (Recall from Section 1.4, page 7 , that water turns cobalt chloride from blue to red.)
f. Leave the glass plate off the dish. Continue to heat the food gently for a least 15-20 min. *Do not burn it.* Just try to evaporate all of the water from it.
g. Let the dish cool. Find the mass of the dish plus its contents.

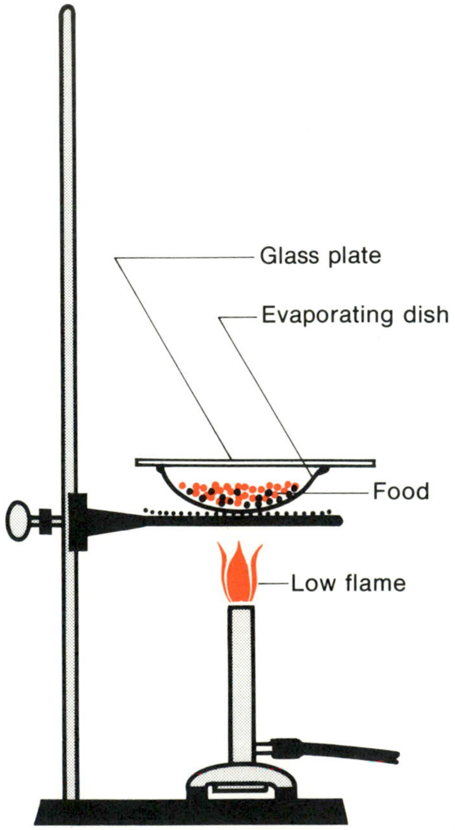

Figure 6-4 Apparatus for determining the percentage of water in a food.

Discussion

1. Calculate the mass of the food.
2. Calculate the mass of the food after the water was removed.
3. Calculate the mass of water that was in the food.
4. Calculate the percentage of water that was in the food.
5. Compare your answer to those of other students who studied the same food.
6. Where does the water in the food come from?

6.5 INVESTIGATION: Testing a Mineral for Water

Many of the minerals found in rocks contain water, even though they appear dry. The water is so strongly held within the minerals that the heat of the sun cannot drive it off. Even your Bunsen burner could not do so for most minerals. In this investigation you heat a mineral called bluestone, or copper sulfate crystals, to study the water contained in it.

Materials

several bluestone crystals
2 test tubes
Bunsen burner
ring stand, 2 adjustable clamps
cobalt chloride test paper
CAUTION: Wear safety goggles during this investigation.

Procedure

a. Place bluestone crystals to a depth of 1-2 cm in a test tube. Use a paper towel to remove any bluestone particles that stick to the sides of the test tube above the crystals. Clamp the test tube at an angle as shown in Figure 6-5.
b. Gently heat the bluestone and observe closely what happens.
c. Heat the bluestone strongly until no further changes occur.
d. Test the liquid formed with cobalt chloride paper.
e. Gradually move the Bunsen flame toward the mouth of the test tube to drive out all of the liquid.
f. Pour the solid onto a piece of paper and examine it closely. Add a few drops of water to the solid.

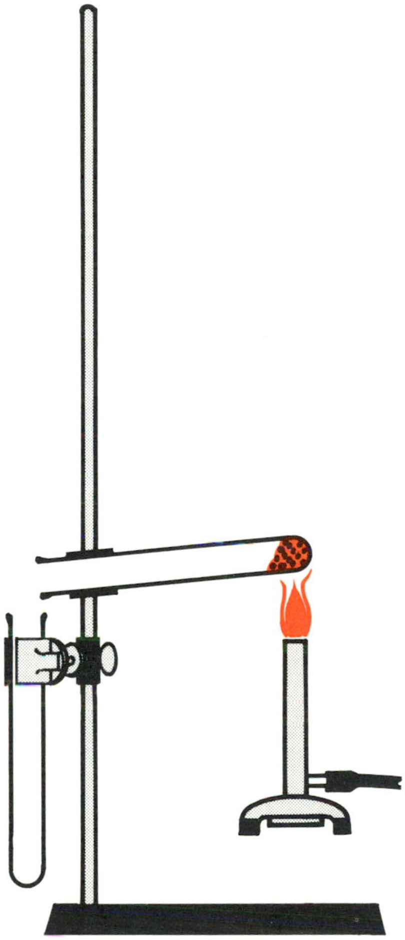

Figure 6-5 Studying the effect of heat on a mineral.

Discussion

1. Describe what happens when bluestone is heated.

2. The water in a mineral like bluestone is often called "water of crystallization". Do you think this is a good name? Why?
3. Describe and explain what happens when water is added to the dry solid.
4. Describe how you would determine the percentage of water in bluestone. Note: Your teacher may choose to give you an unknown mineral to study quantitatively at this time.

B. Solutions

6.6 INVESTIGATION: Characteristic Properties of Solutions

You are familiar with many solutions. Soft drinks, sea water, tea, and vinegar are solutions. Solutions are mixtures with special properties. The purpose of this investigation is to find out what those properties are. Ketchup, salad dressing, and tomato soup are also mixtures. But they are not solutions. How do they differ from solutions?

When two substances are mixed together to form a solution, one is usually called the solute and the other the solvent. The **solute** is the substance which dissolves; the **solvent** is the substance that does the dissolving. The solute is said to be **soluble** in the solvent. For example, if a sugar cube is placed in a glass of water and stirred, the sugar dissolves. Sugar is soluble in water. The sugar is the solute; the water is the solvent.

Materials

5 test tubes
samples of copper sulfate, cobalt chloride, common salt, ammonium dichromate, nickel sulfate

Procedure

a. Place about 10 mL of water in each of the 5 test tubes.
b. Add a few crystals of copper sulfate to one test tube. Shake it vigorously for 2-3 min. Then let it stand for a few minutes. Describe the results carefully.
c. Repeat step b for each of the other solids.

Discussion

1. Is colour a characteristic property of solutions? Explain.

2. Is clarity a characteristic property of solutions? Explain. ("Clear" is the opposite of "turbid" or "cloudy".)
3. A substance has one **phase** in it if you can see only one type of matter in it. How many phases does a solution have?
4. **Homogeneous** matter has only one phase. **Heterogeneous** matter has two or more phases. Is a solution homogeneous or heterogeneous?
5. Make up a definition of a solution.
6. Give an example of a solution in which the solute is a gas and the solvent a liquid.
7. Give an example of a solution in which the solute is a liquid and the solvent a liquid.

6.7 INVESTIGATION: Temperature Changes during Dissolving

In this investigation you are to find out whether heat is required or given off when a solid dissolves in a liquid. Let us begin by making a prediction.

The particle theory tells us that the particles in a solid are held together by strong attractive forces. In order for the solid to dissolve, these particles must break away from one another and move into the liquid. Will this require heat or release heat? Why? Will the liquid get warmer or cooler?

Materials

test tube
thermometer ($-10°C$ to $110°C$)
ammonium chloride, potassium nitrate, sodium thiosulfate, sodium sulfate, and other soluble solids

Procedure

a. Add 5 mL of water to the test tube. Take the temperature of the water.
b. Add 1 g of ammonium chloride to the water. Shake the mixture vigorously for about 30 s. Take the temperature of the resulting solution.
c. Repeat steps a and b with the other solids provided by your teacher. Try to begin each test with water at the same temperature.
d. Record all of your data in a table.

Discussion

1. Make a conclusion regarding heat changes that occur during the dissolving of a solid in a liquid.
2. Explain your conclusion using the particle theory.
3. Are the heat changes the same for all solids? Explain your answer.
4. Does the melting of ice require or give off heat? Use the particle theory to explain your answer. Compare this explanation to the one you wrote in 2.

6.8 INVESTIGATION: Volume Changes during Dissolving

The particle theory tells us that there are spaces between the particles in any type of matter. Solid ammonium chloride has spaces between its particles; so does liquid water. Imagine that 10 mL of ammonium chloride is in a graduated cylinder with 90 mL of water as shown in Figure 6-6. The total volume is 100 mL. Use the particle theory to predict what would happen to the total volume if you shook the cylinder and dissolved the ammonium chloride in the water. If you have trouble making this prediction, imagine what would happen to the total volume if 50 mL of sand were poured into 50 mL of water. Would the final volume be 100 mL? Why?

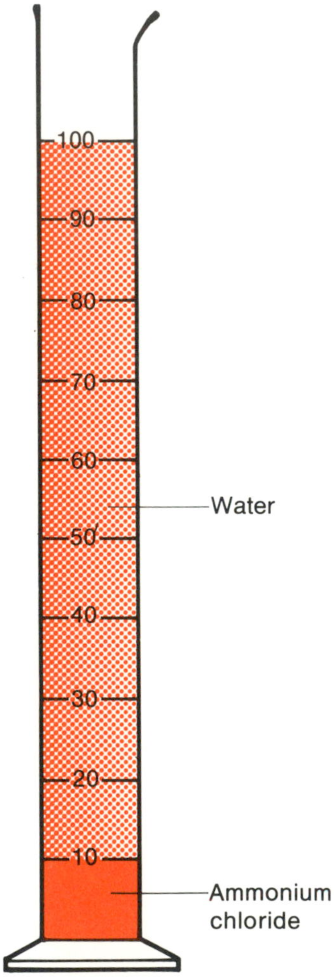

Figure 6-6 What will happen to the total volume when these substances are mixed?

Materials

2 graduated cylinders (100 mL)
sand
ammonium chloride
alcohol
100 mL gas measuring tube (if available)

Procedure A Sand and Water

a. Place 50 mL of water in one graduated cylinder and 50 mL of sand in the other.
b. Pour the sand into the water and mix them thoroughly. Measure the final volume.

Procedure B Ammonium Chloride and Water

a. Place 50 mL of water in a graduated cylinder.
b. Add 5 g of ammonium chloride to the water without stirring.
c. Note the total volume of water plus ammonium chloride.

d. Shake the mixture until the ammonium chloride dissolves. Be careful not to lose any of the solution.
e. Note the volume of the resulting mixture.

Procedure C Alcohol and Water

a. Place 50 mL of water in a graduated cylinder. (Use a gas-measuring tube if one is available. It will give a more accurate result.)
b. Carefully and slowly pour about 50 mL of alcohol down the side of the graduated cylinder so that it forms a layer on top of the water.
c. Accurately measure the total volume of alcohol and water.
d. Mix the two liquids, being careful not to lose any of the solution.
e. Note the volume of the resulting solution.

Discussion

1. What was in the 50 mL of sand, other than particles of sand? What was in the 50 mL of water, other than particles of water? Describe and explain the change in volume when sand and water are mixed. What left the graduated cylinder during this experiment? How much of it left?
2. Use the particle theory to explain the results of your experiment in Procedure B.
3. Use the particle theory to explain the results of your experiment in Procedure C. Are alcohol particles and water particles the same size? (If they were, what would happen to the volume when they were mixed?)

6.9 Concentration, Saturation, and Solubility

The purpose of this section is to make sure you understand a few terms before we continue the investigations.

Dilute and Concentrated

If you dissolve just 2 or 3 small crystals of table salt in 100 mL of water, the solution tastes slightly salty. You have made a **dilute** solution. But, if you dissolve a spoonful of table salt in the same amount of water, the solution tastes very salty. Compared to the first solution, this one is **concentrated.** Dilute and concentrated are relative terms only. If all of you were asked to make up dilute solutions of salt, you would not be likely to use the same amount of salt. But all of you would probably put only a few crystals of salt in 100 mL of water. Also, if all of you were asked to make up concentrated solutions of salt, you would not all add the same amount. But each of you would use more salt than you did to make a dilute solution.

Concentration

Since dilute and concentrated are relative terms only, they are of little help in scientific studies. Thus a quantitative way of describing the amount of solute in the solvent was developed. It is called the concentration of the solution. *The* **concentration** *of a solution is the mass of solute dissolved in one litre of solvent.* For example, a solution of ammonium chloride may have a concentration of 120 g/L. This means that someone has dissolved 120 g of ammonium chloride in 1 L of water.

Unsaturated, Saturated, and Supersaturated

An **unsaturated** solution is one in which more solute can be dissolved in the same amount of solvent at the same temperature. For example, suppose you made up a solution containing just 1.0 g of ammonium chloride in 1 L of water at 20°C. This solution is unsaturated because studies have shown that 360 g of ammonium chloride can dissolve in 1 L of water at 20°C.

A **saturated** solution is one in which no more solute can dissolve in the same amount of solvent at the same temperature. Thus a solution of 360 g of ammonium chloride in 1 L of water at 20°C is saturated. No more ammonium chloride can dissolve.

Under special conditions a supersaturated solution may form. A **supersaturated** solution contains more solute than can normally be dissolved in the solvent at that temperature. In other words, it contains more solute than a saturated solution at the same temperature. How can this be? You will find out later how this is possible.

Solubility

360 g of ammonium chloride in 1 L of water makes a saturated solution at 20°C. Here is another way of saying the same thing: The solubility of ammonium chloride in water at 20°C is 360 g/L.

At 20°C only 110 g of potassium sulfate will dissolve in 1 L of water. In other words, its solubility is 110 g/L at 20°C. Ammonium chloride is more **soluble** than potassium sulfate at 20°C. It takes more of it to form a saturated solution.

Solubility *is the mass of solute in grams that dissolves in one litre of solvent to form a saturated solution at a given temperature.*

Discussion

1. **a)** Distinguish between a dilute and a concentrated solution.
   b) A person with whom you are having lunch dissolves a teaspoon of sugar in a cup of tea. She asks you if the solution is dilute or concentrated. What would you say?
2. **a)** Define the term concentration.
   b) A solution has a concentration of 45 g/L. What does this mean?
   c) 350 g of table sugar were dissolved in 3 L of water. What is the concentration of this solution?
3. **a)** Distinguish between an unsaturated and a saturated solution.
   b) A saturated solution of potassium nitrate at 60°C has a concentration of 1.1 kg/L. What does this mean?
4. **a)** Define the term solubility.
   b) The solubility of sodium nitrate at 25°C is 900 g/L. What does this mean?
   c) Suppose you were making up a solution of sodium nitrate at 25°C and that you added 1.1 kg of sodium nitrate to 1 L of water. What would this mixture be like after you had stirred it for several minutes?

6.10 INVESTIGATION: Factors Affecting Rate of Dissolving

In order for a solid to dissolve in a liquid, its particles must break away from one another and move among the particles of the liquid. In this investigation you study the effect some factors have on the rate of this process. Before you begin each part of the procedure, use the particle theory to predict the effect on the rate of dissolving of the factor named.

Materials

2 test tubes
copper sulfate crystals
Bunsen burner
CAUTION: Wear safety goggles during this investigation.

Procedure A Effect of Agitation

a. Write in your notebook a prediction of the effect of shaking or stirring on the rate of dissolving of a solid in a liquid.
b. Half fill the two test tubes with water.
c. Add 1 g of copper sulfate crystals to each test tube.
d. Let one tube stand undisturbed. Shake the second one vigorously for 2-3 min.

Procedure B Effect of Subdivision

a. Write in your notebook a prediction of the effect of grinding the solid into small pieces on the rate of dissolving of a solid in a liquid.
b. Half fill the two test tubes with water.
c. Add 2 g of copper sulfate crystals to one test tube. Then add 2 g of powdered copper sulfate to the other.
d. Shake each tube vigorously for 2 min.

Procedure C Effect of Temperature

a. Write in your notebook a prediction of the effect of an increase in temperature on the rate of dissolving. (You discovered in Investigation 6.7 that dissolving of a solid in a liquid usually requires heat. This fact should help you make your prediction.)
b. Half fill the two test tubes with water.
c. Add 2 g of copper sulfate crystals to each test tube.
d. Let one test tube stand at room temperature. Heat the other one carefully with a Bunsen flame for 2-3 min.

Discussion

1. State and explain the effect of agitation on the rate of dissolving. Will the copper sulfate in the undisturbed test tube eventually dissolve? Explain.
2. State and explain the effect of state of subdivision on the rate of dissolving.
3. State and explain the effect of temperature on the rate of dissolving.

6.11 INVESTIGATION: Factors Affecting Solubility

In the preceding investigation you found out what factors affect *how fast* a solute dissolves in a given volume of solvent. In this investigation you will find out what factors affect *how much* solute can dissolve in a given volume of solvent. Be sure to predict the result in each case before you try the experiment.

Materials

5 test tubes
Bunsen burner
adjustable clamp
sugar, common salt, calcium carbonate, kerosene, alcohol, iodine crystals, ammonium chloride

CAUTION: Wear safety goggles during this investigation.

Procedure A Nature of the Solute

a. Write in your notebook a prediction of the effect of the nature of the solute on the solubility of that solute. (Hint: One possibility is that one solute may have stronger attractive forces between its particles than another solute. Therefore it would be harder to push them apart and squeeze solvent particles between them. How would this affect the solubility?)
b. Half fill the 5 test tubes with water. To each test tube add one of the following solutes: 1 g common salt, 1 g sugar, 1 g calcium carbonate, 2 mL kerosene, 2 mL alcohol.
c. Shake each tube vigorously for 2-3 min. Let it stand for 2-3 min and note the results.

Procedure B Nature of the Solvent

a. Write in your notebook a prediction of the effect of the nature of the solvent on the solubility of the solute. Here is an example to help you with your prediction: Imagine a room full of basketballs (a solvent) and a room the same size full of baseballs (another solvent). Try to "dissolve" a few hundred ping-pong balls in each of these solvents.
b. Half fill 1 test tube with water and the other with alcohol.
c. Add an iodine crystal of the same size to each test tube.
d. Shake each tube vigorously for 2-3 min. Note the results.

Procedure C Effect of Temperature

a. Write in your notebook a prediction of the effect of an increase in temperature on the solubility of a solute.
b. Place 10 mL of water in a test tube.
c. Obtain 10 g of ammonium chloride from your teacher. Divide it into 5 portions that are about 2 g each.

d. Add one portion of the ammonium chloride to the water. Shake the test tube vigorously until the ammonium chloride dissolves.
e. Repeat step d with successive portions of ammonium chloride until some of the ammonium chloride will not dissolve.
f. Heat the mixture to 50°-60°C. Note the effect on the undissolved ammonium chloride.
g. Add another portion of ammonium chloride and shake the test tube vigorously.
h. Heat the mixture to the boiling point. Note the effect.
i. Add any remaining ammonium chloride, shaking the tube vigorously after each addition.
j. Cool the solution by running cold water over the outside of the test tube.

Discussion

1. Which of the solutes in Procedure A were soluble? Which were insoluble? Are you satisfied with your prediction?
2. Describe and explain the results of Procedure B.
3. Describe and explain the effect of increased temperature on the solubility of ammonium chloride. What happened when the hot concentrated solution was cooled? Why?

6.12 INVESTIGATION: Solubility of a Solid

In this investigation you find out the solubility of common salt (sodium chloride) in water. Before you begin, review the definition of solubility given in Section 6.9, page 139.

Materials

common salt
2 beakers (150 mL)
balance
ring stand, iron ring, wire gauze
Bunsen burner
thermometer (−10°C to 110°C)

CAUTION: Wear safety goggles during this investigation.

Procedure

a. Place about 70-80 mL of water in a beaker. Add salt to this water in small amounts, with continuous stirring, until a saturated solution is formed. (An excess of salt will remain on the bottom no matter how long you stir.)
b. Find the mass of the other beaker.

c. Pour about 50 mL of the solution from the first beaker into the second one. Be sure to let the undissolved salt settle out. Do not pour any of it into the second beaker.

d. Find the mass of the beaker plus solution.

e. Take the temperature of the solution.

f. Evaporate the water from the salt solution. Heat the solution slowly and carefully to prevent spattering, particularly near the end of this step. Be sure all of the water has been driven off.

g. Find the mass of the beaker and its contents after it has cooled.

h. Record your results and calculations in this manner:

1. Mass of beaker .... _______ g
2. Mass of beaker + saturated solution .... _______ g
3. Mass of beaker + salt .... _______ g
4. Temperature .... _______ °C
5. Mass of saturated solution (2-1) .... _______ g
6. Mass of salt in saturated solution (3-1) .... _______ g
7. Mass of water in saturated solution (5-6) .... _______ g

i. Calculate the solubility of the salt.

Discussion

1. In this investigation you determined the solubility at one temperature only. Table 37 shows the results obtained when the solubility of a solid is determined at several temperatures. The solid used in this case was ammonium chloride. Plot a graph of these results, with temperature on the horizontal axis and solubility on the vertical axis. This is the **solubility curve** for ammonium chloride.

TABLE 37 Solubility of Ammonium Chloride

Temperature (°C)	Solubility (g/L of water)
0	300
20	372
40	458
60	550
80	658
100	771

2. Figure 6-7 shows the solubility curves for sodium chloride and sodium nitrate. Compare your value for the solubility of sodium chloride with the value the graph indicates for the same temperature.

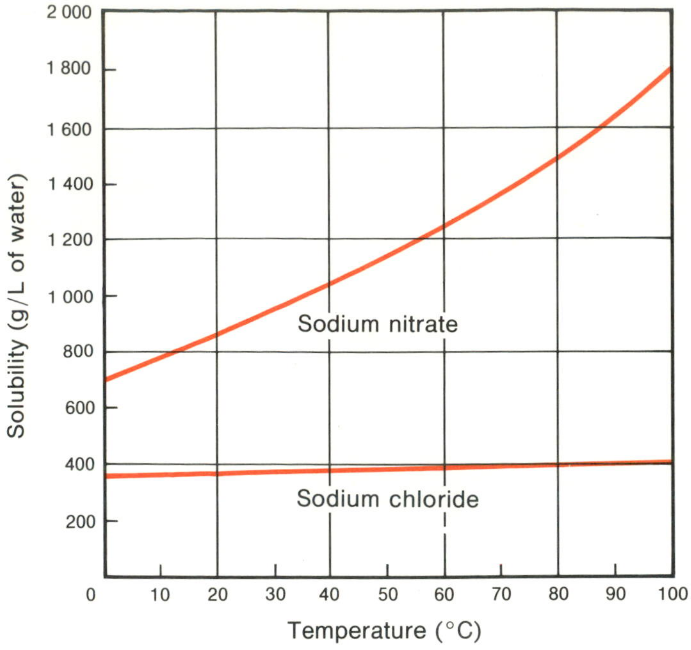

Figure 6-7 Solubility curves for sodium chloride and sodium nitrate in water.

3. List sources of error that could account for the difference between your value and the "true value".
4. What is the lowest temperature that is needed to completely dissolve 1.4 kg of sodium nitrate in 1 L of water? (See Figure 6-7.)
5. Will 1.2 kg of sodium nitrate in 1 L of water at 80°C form an unsaturated or a saturated solution? Explain.
6. If a saturated solution of sodium nitrate is cooled from 80°C to 20°C, how many grams of sodium nitrate will settle out from the solution?

TABLE 38 Solubility of Carbon Dioxide

Temperature (°C)	Solubility (g/L)
0	3.4
20	1.8
40	1.2
60	0.86

7. Table 38 shows the solubility of carbon dioxide gas in water at atmospheric pressure for several temperatures. Plot a graph of these data, with temperature on the horizontal axis and solubility on the vertical axis. Carbon dioxide is a typical gas. Sodium chloride and sodium nitrate (Fig. 6-7) are typical solids. How do the solubilities of gases and solids at various temperatures differ? Soft drinks contain carbon dioxide. That is

what gives them the stinging taste. Explain why a soft drink goes "flat" in taste if you leave it in a glass on the table for a long time.

6.13 INVESTIGATION: Properties of a Supersaturated Solution

In Investigation 6.11, page 142 you discovered that the solute settles out of a solution as the solution is cooled. This happens because the solute is not as soluble at the lower temperatures as it is at the higher temperatures.

All solutes do not always behave in this manner when cooled. Some solutes, under special conditions, stay in the solvent and form supersaturated solutions. In this investigation you discover the nature of a supersaturated solution.

Materials

test tube (20 × 150 mm)
adjustable clamp
hypo (sodium thiosulfate)
Bunsen burner
CAUTION: Wear safety goggles during this investigation.

Procedure

a. Make sure the test tube is clean and dry.
b. Fill the test tube half full of hypo crystals.
c. Add 10 drops of water to the hypo. Shake the mixture. Note any change in temperature with your hand.
d. Gently heat the mixture until all of the hypo has dissolved.
e. Hold the test tube in a stream of running water from a tap until the solution has been cooled to about room temperature. If no crystals form, you have made a supersaturated solution. If crystals do form, repeat steps d and e, being more gentle with your handling of the test tube.
f. Add one crystal of hypo to the supersaturated solution. Describe what happens. Note any changes in temperature with your hand.

Discussion

1. Is the solution after step c unsaturated or saturated?
2. Which word best describes the solution after step d—concentrated or saturated?
3. Is the solution saturated or unsaturated after step f? Explain.

4. Is heat required or given off when hypo dissolves? Explain your answer using the particle theory.
5. Is heat required or given off when hypo crystallizes out of solution? Explain your answer using the particle theory.
6. What evidence do you have from your observations that supports the idea that hypo is made of small particles that exert attractive forces on one another?
7. Does the melting of ice require or give off heat? Does your explanation in part 4 apply to the melting of ice?
8. Does the freezing of water require or give off heat? Does your explanation in part 5 apply to the freezing of water?

6.14 INVESTIGATION: Some Other Mixtures

Not all mixtures of solids in liquids are solutions. Nor are all mixtures of liquids in liquids solutions. To be a solution, the mixture must be clear and homogeneous. Only one phase can be visible. In this investigation you compare a solution you know well (ammonium chloride) with some other mixtures. See if you can find out some of the properties of these mixtures.

Materials

4 test tubes (20 × 150 mm)
ammonium chloride, liquid laundry starch, fine soil, kerosene
light source (e.g. flashlight)

Procedure

a. Fill all 4 test tubes about two-thirds full of water.
b. Add one of the following to each test tube: 1 g of ammonium chloride, 1 g of fine soil, 1 drop of liquid laundry starch, 1 mL of kerosene.
c. Make a mixture in each case by shaking each test tube vigorously for 30 s or more. Describe each mixture.
d. Shine a beam of light through each mixture. Note the appearance of the beam.
e. Wait 10 min then write a new description of each mixture and repeat step d for each one.

Discussion

This background material will help you to answer the questions that follow.

A mixture is said to show the **Tyndall effect** when it makes a beam of light visible. You have probably seen the Tyndall effect when sunlight shines through a window into a dusty room or when a movie projector shines through smoky air. The Tyndall effect is used to help distinguish among different types of mixtures.

A **solution** is a mixture that is clear and homogeneous. The particles of the solute are small and will not settle out on standing (unless the solution is cooled or evaporated). They are too small to cause the Tyndall effect.

A **suspension** is a mixture in which some of the particles are large enough to be seen with the naked eye. It is cloudy, nontransparent, and heterogeneous. Some of the particles will settle out on standing. A suspension will show the Tyndall effect. A suspension of tiny droplets of one liquid in another is called an **emulsion.**

A **colloidal dispersion** is a mixture that is part way between a solution and a suspension. It is slightly cloudy, semi-transparent, and heterogeneous, although it may appear homogeneous. Its particles are large enough to show the Tyndall effect. But they are not large enough to settle out on standing. The particles cannot be seen with the naked eye or even with your microscope. Yet they are much larger that the particles of the solute in a solution. Note: It is possible for a mixture to be two or more of these types at the same time.

1. Which of the four mixtures is a solution? Why?
2. Which mixtures are suspensions? Why?
3. Which mixture is an emulsion? Why?
4. Which mixtures are colloidal dispersions? Why?
5. Explain how your studies of these mixtures support the particle theory.
6. What type of mixture or mixtures are each of the following: vinegar, vanilla extract, ketchup, house paint, shaving lotion, hand cream, salad dressing, freshly squeezed orange juice, homogenized milk?

C. Separating Substances from Mixtures

6.15 INVESTIGATION: Filtration

In the preceding investigation you learned that solids form three types of mixtures in water—solutions, colloidal dispersions, and suspensions.

In this investigation you use filter paper to try to remove the solid from each of these types of mixture. Filter paper is paper that has been made with a very small pore size. If the pore size is smaller than the particles of solid in the mixture, the solid will remain on the paper. Anything that passes through the filter paper is called the **filtrate.** Any solid that remains on the filter paper is call the **residue.**

Materials

funnel
filter paper
8 test tubes
4 microscope slides
copper sulfate solution
salt solution
starch dispersion
clay suspension
dropper
hot plate

Procedure

a. Prepare the filter paper and funnel for filtration as directed by your teacher.
b. Obtain a test tube full of each of the 4 mixtures.
c. Filter each mixture as shown in Figure 6-8.
d. In each case examine the filter paper to see if any residue is on it.
e. In each case compare the filtrate to the original mixture.
f. Place a drop or two of each filtrate on a microscope slide. Carefully evaporate the water from the slides by placing them on a hot plate.

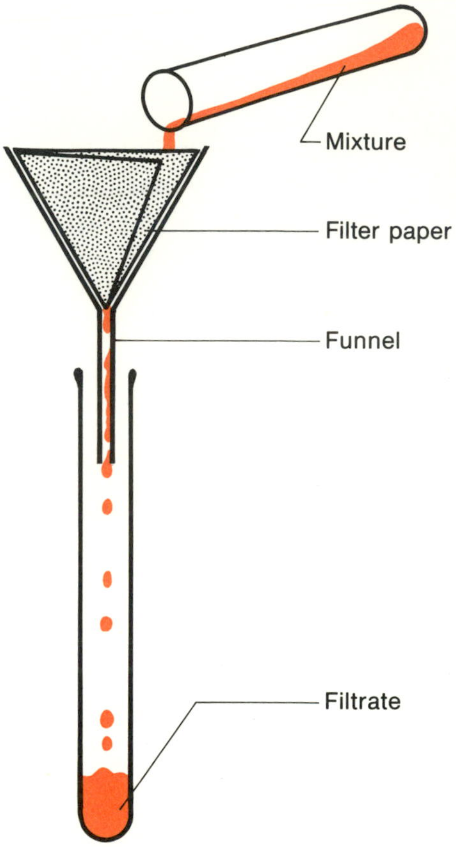

Figure 6-8 What types of mixtures can filtration separate?

Discussion

1. Explain the terms filtrate and residue.
2. Describe the effect of filtration on the solutions.
3. Describe the effect of filtration on the colloidal dispersion.
4. Describe the effect of filtration on the suspension.
5. In general, what kind(s) of mixtures can be separated using filter paper? Why can they be separated while the others cannot?
6. Explain how this investigation supports the particle theory.
7. Find out how filtration is used in the water treatment plant to help purify your drinking water.
8. Name two places in an automobile where filters are used. In each case name the mixture being filtered and the residue that is likely to be found on the filter.
9. Name two other common examples of the use of filtration.

6.16 INVESTIGATION: Distillation

Distillation can be used to separate many mixtures. It is necessary only that the substances in the mixture have different boiling points. The substance with the lowest boiling point vaporizes first when the mixture is heated. It can then be condensed in another container as shown in Figure 6-9. This condensed substance is called the **distillate**. The material that remains in the distilling flask is called the **residue**.

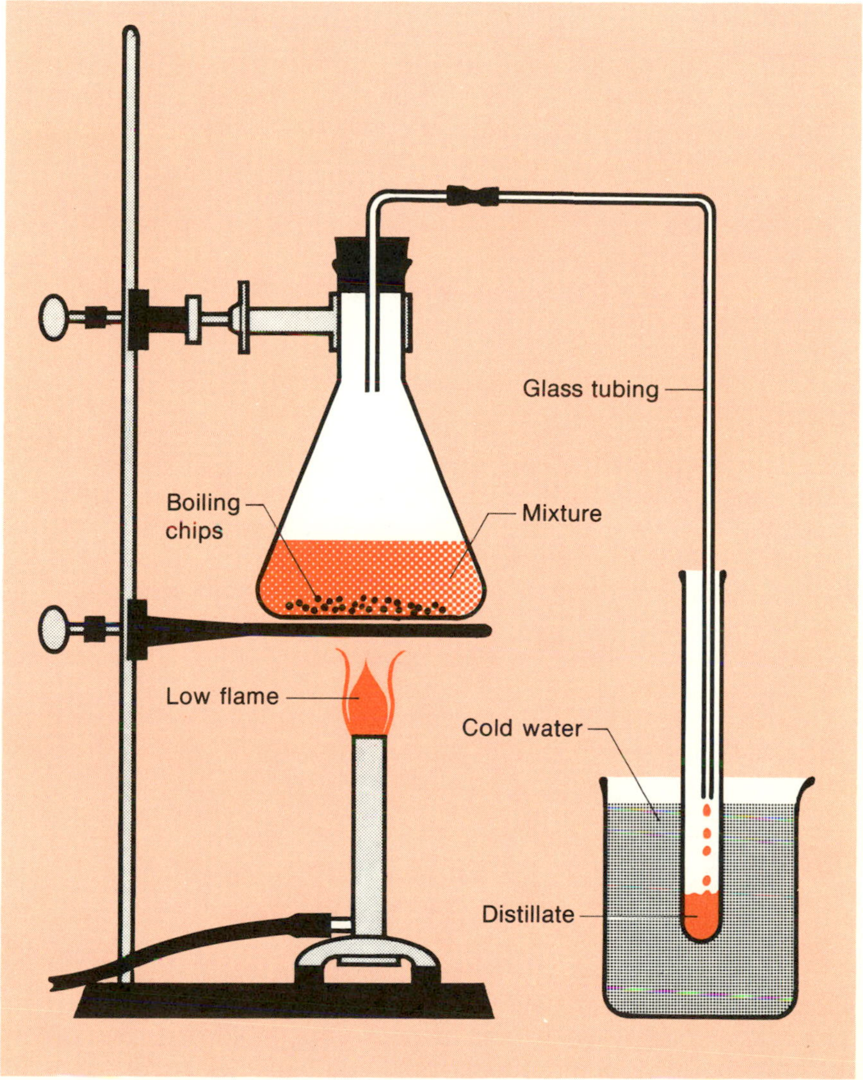

Figure 6-9 A simple distillation apparatus.

Materials

ring stand, iron ring, wire gauze
Bunsen burner
boiling chips
adjustable clamp
one-hole rubber stopper
U-shaped glass tubing
test tube
large beaker
solution of copper sulfate in water
suspension of clay in water
Erlenmeyer or round-bottom flask
CAUTION: Wear safety goggles during this investigation.

Procedure

a. Set up the apparatus as shown in Figure 6-9. Your teacher will give you 30-40 mL of a solution of copper sulfate in water or a suspension of clay in water. Place the mixture that you are given in the Erlenmeyer flask.
b. Heat the mixture slowly and carefully until it is boiling gently. **CAUTION:** Do not heat too strongly or the vapour may blow the stopper from the flask.
c. Turn off the Bunsen burner when a few millilitres of distillate have been collected.

Discussion

1. Explain the terms distillate and residue.
2. Describe the appearance of the original mixture, the distillate, and the residue. Exchange this information with a student who distilled the other mixture.
3. Does distillation separate the substances in both solutions and suspensions?
4. Explain how distillation works. Be sure to identify any changes of state that occur.
5. Explain how drinking water can be obtained from sea water. Where is this being done in the world?

6.17 INVESTIGATION: Fractional Distillation

In the preceding investigation distillation was used to separate a liquid and a solid. It works well because the liquid and solid differ greatly in boiling points. The boiling point of the liquid is much lower than that of the solid. Therefore the liquid vaporizes easily and the solid remains behind.

Distillation can also be used to separate a mixture of two or more liquids, provided they have different boiling points. The same principle applies. When the mixture is heated, the liquid with the lowest boiling point distils off first. When it is nearly all gone, the liquid with the next lowest boiling point distils off, and so on. Each separate liquid that distils off is called a **fraction**. The process is called **fractional distillation**.

The mixture that you use in this investigation has only two substances in it—water (b.p. = 100°C) and alcohol (b.p. = 78.5°C). See if you can separate these substances by fractional distillation.

Materials

Same as Investigation 6.16, with these exceptions:

a. the one-hole stopper is replaced by a two-hole stopper, equipped as shown in Figure 6-10.

b. the mixture to be used is alcohol and water

CAUTION: Wear safety goggles during this investigation.

Procedure

a. Set up the apparatus as shown in Figure 6-9, with the modification described in part a of the Materials. Make sure the thermometer bulb is well above the liquid and opposite the beginning of the glass tube. Place 50 mL of the alcohol-water mixture in the Erlenmeyer flask.

b. Heat the mixture slowly and carefully until it is boiling gently. Record the temperature every minute. When the temperature changes quickly, begin collecting the distillate in another test tube. Continue heating and recording temperatures for about 5 min.

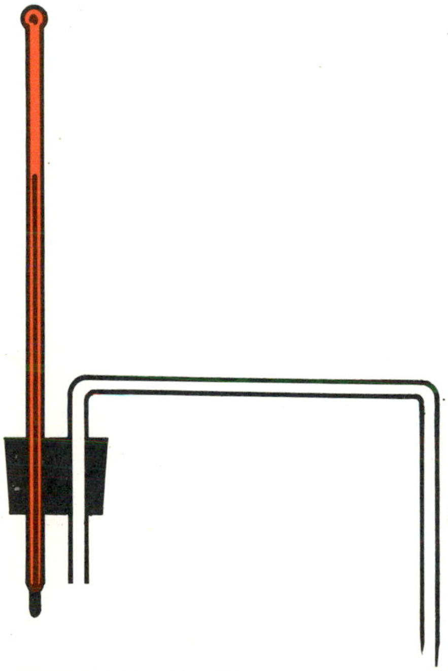

Figure 6-10 Modification of distillation apparatus for fractional distillation.

Discussion

1. Explain the terms fraction and fractional distillation.
2. In what characteristic physical property must the liquids in a mixture differ before they can be separated by fractional distillation?
3. What liquid is most of the first fraction made of? Why is this so?
4. What liquid is most of the second fraction made of? What proof do you have of this?
5. Why will the alcohol fraction always contain some water? How could you make the alcohol fraction more pure?

6.18 Fractional Distillation of Petroleum

The most important use of fractional distillation is the separation of petroleum (crude oil) into its components. From petroleum, fractional distillation produces gasoline, kerosene, furnace oils, diesel fuel, lubricating oils, greases, and hundreds of other useful substances. In this section we investigate the fractional distillation of petroleum and several related topics.

Origin of Petroleum

The word petroleum means rock oil. It was given this name because its source is sedimentary rock layers. Millions of years ago, as ancient seas evaporated, sediment settled to the bottom. This sediment was made of minerals such as sand and clay, as well as dead organisms, both plant and animal. By repeated flooding and evaporation for millions of years, the sediment became hundreds of metres thick in places. It eventually became so heavy that the lower layers were compressed into sedimentary rock. The sand became sandstone; the clay became shale. The dead organisms became petroleum and natural gas. The petroleum and natural gas tend to collect in the sandstone since sandstone is more porous than shale. (The spaces between the particles of sandstone are much larger than those between the particles of shale.) But they remain in the sandstone only if they are covered by a dome of non-porous rock such as shale. Figure 6-11 shows one way in which a dome of non-porous rock traps oil and gas in a layer of porous rock. The porous rock is called the **reservoir rock**, since it holds the oil and gas. The non-porous rock is called the **cap rock**.

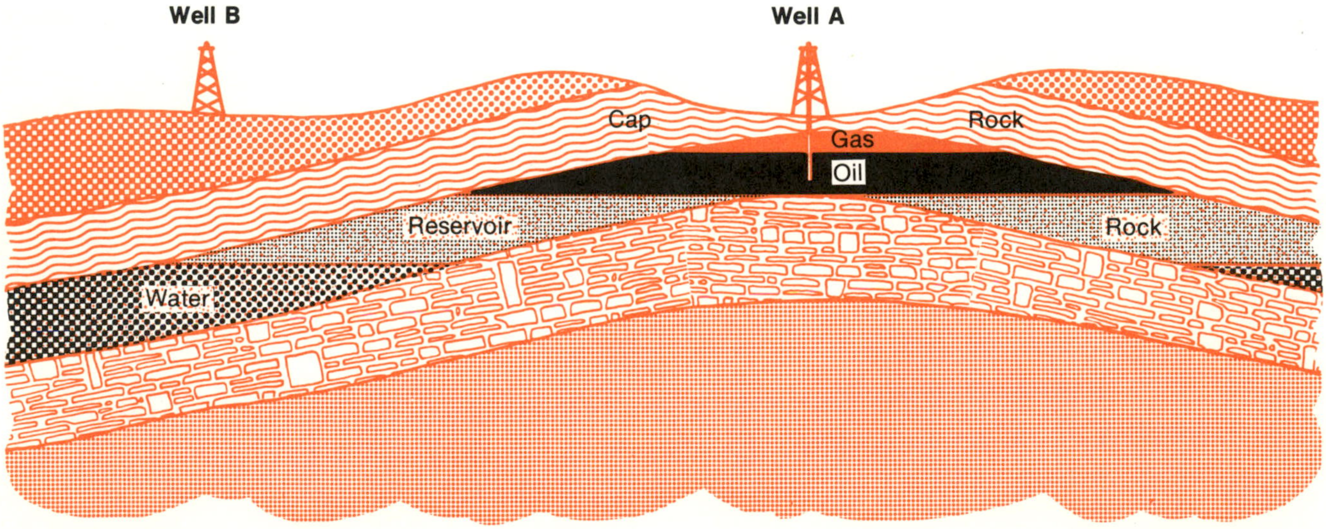

Figure 6-11 A common method by which petroleum is trapped in the earth.

Figure 6-12 A simplified diagram of a bubble tower for the fractional distillation of petroleum. An actual tower has 20 – 30 trays and several bubbles per tray.

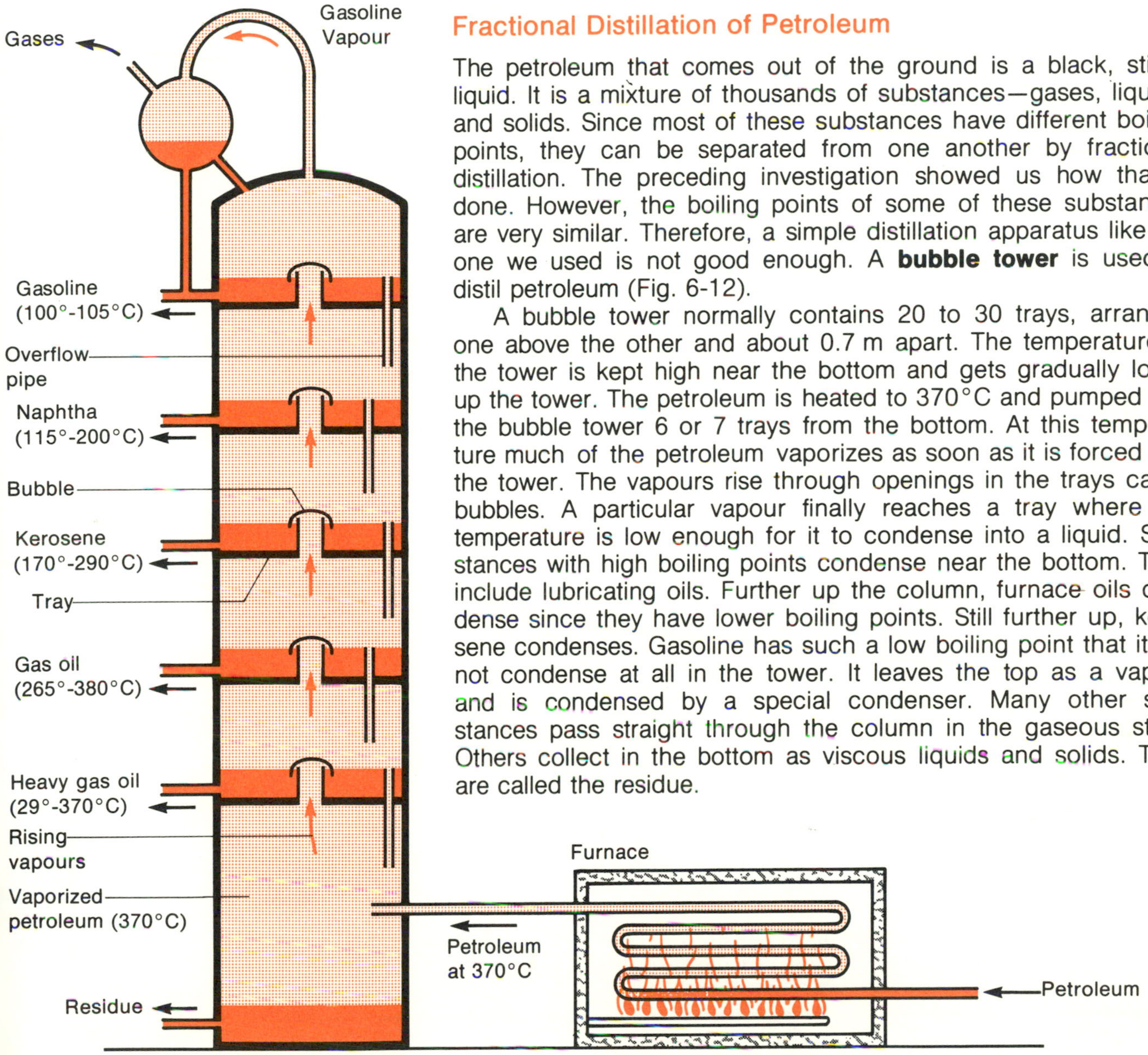

Well A in the diagram has struck what geologists call an **oil pool**. The name is somewhat misleading, as is our diagram. Little of the oil is in an actual pool. Most of it is in the reservoir rock. The pressure of the cap rock forces it to the drilling site. Several oil pools, either beside one another or on top of one another, make up an **oil field**.

If Well B is completed, the drilling company will be disappointed. This well will bring mostly sea water to the surface. Large pockets of sea water are often trapped by the sedimentary rock.

Fractional Distillation of Petroleum

The petroleum that comes out of the ground is a black, sticky liquid. It is a mixture of thousands of substances—gases, liquids, and solids. Since most of these substances have different boiling points, they can be separated from one another by fractional distillation. The preceding investigation showed us how that is done. However, the boiling points of some of these substances are very similar. Therefore, a simple distillation apparatus like the one we used is not good enough. A **bubble tower** is used to distil petroleum (Fig. 6-12).

A bubble tower normally contains 20 to 30 trays, arranged one above the other and about 0.7 m apart. The temperature of the tower is kept high near the bottom and gets gradually lower up the tower. The petroleum is heated to 370°C and pumped into the bubble tower 6 or 7 trays from the bottom. At this temperature much of the petroleum vaporizes as soon as it is forced into the tower. The vapours rise through openings in the trays called bubbles. A particular vapour finally reaches a tray where the temperature is low enough for it to condense into a liquid. Substances with high boiling points condense near the bottom. They include lubricating oils. Further up the column, furnace oils condense since they have lower boiling points. Still further up, kerosene condenses. Gasoline has such a low boiling point that it will not condense at all in the tower. It leaves the top as a vapour and is condensed by a special condenser. Many other substances pass straight through the column in the gaseous state. Others collect in the bottom as viscous liquids and solids. They are called the residue.

Note that the system of bubbles and overflow pipes allows part of each condensed liquid to flow down to a lower tray. This system gives substances a "second chance" to stop at the right tray. Often the upward rush of vapour carries some substances too far up the tower.

Uses of Petroleum

Table 39 shows just a few of the substances found in petroleum. Note that gasoline contains a mixture of several of these substances. Likewise, kerosene contains several of these substances. After the substances in petroleum have been completely separated from one another, they can be used to make a wide variety of materials. Among these are rubbing alcohol, antifreeze, synthetic fibers like Dacron, Orlon, Acrilan, and Polyesters, resins, rayon, polyethylene, detergents, paints, explosives, rubber, lacquers, insecticides, herbicides, and even fertilizer. Nearly all medicines and drugs are also petroleum by-products. Of course, most of our petroleum is used for transportation, heating, and the operation of our industries. Clearly, petroleum is important to all of us.

TABLE 39 Some Substances in Petroleum

Substance	Usual state at room temperature	Boiling point (°C)	Major use(s)
Methane	Gas	−161	(As gas for industrial and domestic fuel; to make other substances)
Ethane	Gas	−88	
Propane	Gas	−43	
Butane	Gas	−0.5	In gas and gasoline
Pentane	Gas & liquid	36	Gasoline
Hexane	Liquid	69	Gasoline
Heptane	Liquid	98	Gasoline
Octane	Liquid	125	Gasoline
Nonane	Liquid	151	Gasoline
Decane	Liquid	174	Gasoline & kerosene
Undecane	Liquid	196	Kerosene
Dodecane	Liquid	216	Kerosene
Tridecane	Liquid	236	Kerosene
Tetradecane	Liquid	254	Kerosene
Pentadecane	Liquid	271	Kerosene
Hexadecane	Liquid & solid	287	Dissolved in heavy oils
Pentatriacontane	Solid	very high	Waxes

Petroleum Reserves

Petroleum is a non-renewable resource. Once it has been used it is gone forever. Many warnings have been issued regarding an upcoming shortage of petroleum. Yet North Americans continue to use it as though the supply were unending.

Scientists tell us that the known reserves of petroleum in the world are about one hundred billion cubic metres ($100 \times 10^9\ m^3$). The world used about $5 \times 10^9\ m^3$ in 1976. Therefore the known reserves will last only 20 years. This assumes, of course, that the rate of consumption will not increase —an unlikely possibility. Most of the known reserves are in the Middle East. Saudi Arabia has 17% of the reserves; Kuwait has 15%. The greatest user in the world is the United States. It accounts for 33% of the world's total consumption even though it has only about 6% of the world's population.

Optimists tell us that new reserves will be found. But they are not being found as fast as scientists predicted. In the meantime, several disastrous oil spills have occurred as giant tankers haul petroleum from the mid-East to North American refineries. And oil companies are searching far out to sea and in the northern barrens for more petroleum. But discoveries of new reserves are not keeping up with the increase in demand. Can you think of anything that we should do to curb our demand before we exhaust this important reserve?

Other Sources of Petroleum

An important source of petroleum in North America is the Athabasca **tar sands** in northern Alberta. In the tar sands, petroleum is found in the solid state. The tar sands consist of asphalt ("tar") in the pores of sandstone. Since the mixture is solid, it must be mined and then heated to drive off the asphalt. Refining, including fractional distillation, produces gasoline and other products from the asphalt. The tar sands cover $75\,000\ km^2$ and are, in places, 60 m deep. The petroleum reserves of the tar sands have been estimated at $120 \times 10^9\ m^3$—more than the known reserves of liquid petroleum in the world! But the recovery problems are tremendous. The first plant to begin operating in the tar sands mined 100 000 t of tar sands per day. From this it produced $10\,000\ m^3$ of petroleum, a negligible amount when compared to the amount used in Canada per day.

Another important source is the Green River **oil shales** in Colorado, Wyoming, and Utah. The reserves of petroleum present in the oil shales are estimated at $200 \times 10^9\ m^3$. Pilot plants have successfully obtained petroleum from this shale. But commercial production has not yet started. It is still cheaper to buy the oil abroad. Oil company officials argue that it is economically impossible to extract petroleum from the shale. Yet Russia and China have been doing it economically for years. As long ago as 1972, 35% of China's petroleum came from oil shale.

A final source of petroleum is **coal**. Coal can be heated to produce some petroleum products. It can also be burned directly, in place of petroleum. The world's recoverable coal reserves are equal to about $7\,600 \times 10^9$ m^3 of oil.

When you consider all of the petroleum in tar sands, oil shales, and coal, you may think that there is no need to worry about a shortage of petroleum. But the engineering problems involved in extracting petroleum from these sources are enormous; also, the pollution produced is at present unacceptable.

Discussion

1. a) How is petroleum formed?
b) Describe how it is obtained from the earth.

2. Describe the fractional distillation of petroleum, referring to Figure 6-12 as you do so.

3. Explain why petroleum is important to all of us.

4. a) What is a non-renewable resource?
b) Why is petroleum a non-renewable resource?
c) Name a renewable resource that can be used instead of petroleum in certain instances.

5. Several topics are listed here that should concern you. Your teacher will give you further information on some of these as well as a resolution to be debated. Select a topic that interests you and decide whether you are for or against the resolution. Your teacher will place you in a team with 3 or 4 other students who feel the same as you do. As a team you are to prepare a defence of your position. Then you will conduct a debate with a team that holds the opposite view.
a) Gasoline consumption of automobiles
b) Super-tankers and oil spills
c) Gas for California
d) Rationing of petroleum products
e) Fuel consumption by supersonic passenger jet aircraft
f) Oil exploration in the Arctic
g) Offshore drilling for oil
h) Trans-Arctic pipelines for gas and oil

6.19 INVESTIGATION: Paper Chromatography

In this investigation you use a procedure called paper chromatography to separate substances that are not easily separated by other methods. The way it works is easy to understand. You have seen water soaked up by a paper towel. You have probably seen alcohol move up the wick of an alcohol burner or kerosene move up the wick of a kerosene lamp. Liquids appear to move well through some materials like paper and cloth.

Suppose a substance is dissolved in the kerosene of a kerosene lamp. This substance will be carried up the wick by the kerosene. But how far it is carried up depends on several factors. One factor is the mass of the particles of that substance. Another factor is the solubility of the substance in the kerosene.

Paper chromatography works in much the same way. Paper replaces the wick of the kerosene lamp, and a solvent that will dissolve the substances to be separated replaces the kerosene. Very complex substances can be separated using this method.

Both biologists and chemists have used this procedure for years to separate substances. A biological and a chemical study are outlined here. If you prefer chemistry, select Procedure A; if you prefer biology, select Procedure B.

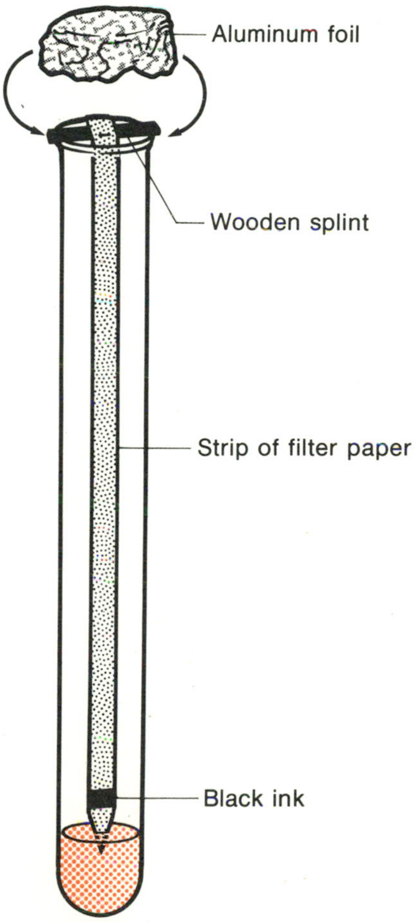

Figure 6-13 Making a chromatograph of black ink.

Materials

large test tube (20 × 150 mm)
filter paper (or chromatography paper, if available)
fine-tipped dropper
piece of wooden splint
aluminum foil
black ink (Procedure A)
green leaves, mortar & pestle, alcohol (Procedure B)

Procedure A Separating the Substances in Black Ink

a. Place 3-4 mL of water in the test tube. Then dry the inside wall of the test tube above the water line.

b. Prepare a strip of filter paper (or chromatography paper) that is about 1 cm wide and 20 cm long. Taper one end as shown in Figure 6-13.

c. Use the dropper to "paint" a strip of black ink across the filter paper about 2 cm above the tip of the paper. Make the strip about 0.5 cm wide.

d. Hang the filter paper over the wooden splint as shown in Figure 6-13. Only the tip of the filter paper should be in the water. The ink strip should be about 1 cm above the surface of the water. Make sure the filter paper hangs in the centre; it must not touch the wall of the test tube.

e. Cover the top of the test tube with aluminum foil to prevent escape of water vapour.

f. Let the system stand until the colour has risen almost to the top of the filter paper. Remove the paper and hang it in a safe place to dry.

Procedure B Separating the Substances in Leaf Pigment Extract

a. Collect a few green leaves. Tear them into pieces. Put the pieces in a mortar. Add about 5 mL of alcohol. Grind the mixture with the pestle until a deep green (almost black) solution is obtained. This is the pigment extract.
b. Prepare a strip of filter paper as described in part b of Procedure A.
c. Using a pencil (not pen), draw a line across the filter paper about 2 cm above the tip of the paper. Use the dropper to spread one drop of the pigment extract along the line. Allow the paper to dry completely. (Gentle fanning will help.) Now spread a second drop of the pigment extract along the line. Allow it to dry. Keep repeating this until a dark green strip is present across the filter paper.
d. Place 3-4 mL of alcohol in the test tube.
e. Hang the filter paper in the test tube as described in part d of Procedure A.
f. Cover the top of the test tube with aluminum foil to prevent escape of alcohol vapour.
g. Let the system stand until the colour has risen almost to the top of the filter paper. Remove the paper and hang it in a safe place to dry.

Discussion A

1. What kind of force pulled the water up the filter paper?
2. How many different pigments (colored substances) were in the black ink?
3. Explain why some pigments move further up the paper than others.
4. What colour would you get if you combined the pigments that you just separated? You can try this by cutting out each coloured section. Put each one in a separate test tube. Add 1 mL of water and shake the mixture. Pour all of the mixtures together.

Discussion B

1. What kind of force pulled the alcohol up the filter paper?
2. How many different pigments were in the extract? List the colours in order from the top of the filter paper down. Name as many of these as you can.

6.20 INVESTIGATION: Sedimentation and Floc Formation

You saw in Investigation 6.14 that some solids which are suspended in water tend to settle out when the water is allowed to stand undisturbed. This settling is called **sedimentation**. You also saw that some solids, such as fine earth, or clay, appear to stay suspended indefinitely in water. Such solids can be removed by distillation. If the solid particles are large enough, they can be removed by filtration, provided filter paper with very small pores is used. But a cheaper and better method for commercial use is a process called **floc formation**. In this method, alum (aluminum sulfate) is added to the suspension. If the water is acidic, lime (calcium hydroxide) must also be added. A **floc** forms. It consists of globules (tiny drops) of a sticky substance called aluminum hydroxide. These globules settle from the water. The suspended particles stick to the globules as they settle. The globules can be easily filtered from the water if settling takes too long.

In this investigation you study sedimentation. You also observe floc formation and its effect on the rate of sedimentation. Then you are asked to find practical applications of this process by visiting the school resource centre, community library, and other places.

Materials

100 mL graduated cylinder
loam soil
jar of motor oil (from Investigation 5.12)
Plastic beads (diameters of 12 mm, 7 mm, and 4 mm)
alum (aluminum sulfate) solution
lime (calcium hydroxide) solution

Procedure A Sedimentation of Soil from Water

a. Add 2-3 cm^3 of loam soil to 100 mL of water in a graduated cylinder or other tall glass container.
b. Shake the mixture for 15-20 s.
c. Let the mixture stand undisturbed for 24 h. Observe it several times during the remaining part of the period. Observe it again in your next class. Each time record the appearance of the suspension and the appearance of the sediment.

Procedure B Rate of Sedimentation and Particle Size

a. Use the materials and procedures of Investigation 5.12 to discover how long it takes plastic beads of several different diameters to settle through motor oil of one viscosity.
b. Plot a graph with rate of sedimentation on the vertical axis and bead diameter on the horizontal axis.
c. Extrapolate the graph to the smallest bead diameter your graph permits.

Procedure C Floc Formation

a. After you have examined the mixture in Procedure A for the last time, carefully pour off about half of the suspension into a clean graduated cylinder (or other tall glass container). Do not disturb the sediment.
b. Add 5-10 mL of alum solution and 5-10 mL of lime solution to the suspension.
c. Let the mixture stand undisturbed for the rest of the lab period.

Discussion A

1. Describe the mixture of soil and water after it was first formed.
2. Describe the changes that you observed in the suspension and in the sediment.
3. Which particles settle out first, large ones or small ones?

Discussion B

1. What is the relationship between bead diameter and rate of sedimentation?
2. Use this relationship to account for the appearance of the sediment in Procedure A.
3. Use the results of the extrapolation of your graph to discuss why clay particles stay suspended for a long time in water.

Discussion C

1. Describe the aluminum hydroxide floc that formed in this experiment.
2. Use the results of Procedure B to explain why floc formation increases the rate of sedimentation.

Practical Applications

Form a group of 3 or 4 students. Select one of the following topics. Collect information on the topic from the resource centre of your school, the community library, and other sources. Prepare a group presentation for the class as directed by your teacher.

a. Use of sedimentation in water treatment plants.
b. Use of sedimentation in sewage treatment plants.
c. Use of the centrifuge to speed up sedimentation; application of this procedure in the separation of blood fractions.
d. Use of floc formation in water treatment plants.
e. Use of floc formation in swimming pools.
f. Use of sedimentation or floc formation by any industry.
g. Use of the particle theory to explain the relationship between rate of sedimentation and particle size.

6.21 INVESTIGATION: Fractional Crystallization

You discovered in Investigation 6.15 that an insoluble solid can be separated from water by filtration. The insoluble solid stays on the filter paper and the water goes through the paper. You also discovered that, if the water has a soluble solid dissolved in it, the solid goes through the filter paper with the water. Clearly, before one can separate a dissolved solid from water by filtration, the dissolved solid must first of all be made insoluble. That is, it must be made to crystallize from the solution.

Examine the solubility curve of potassium nitrate in Figure 6-14. A saturated solution at 50°C contains 830 g of potassium nitrate in 1 L of water. But at 0°C a saturated solution contains only 120 g of potassium nitrate in 1 L of water. Therefore, if a saturated solution was cooled from 60°C to 0°C, all but 120 g of the original 830 g would crystallize out and could be recovered by filtration.

Suppose that the solution of potassium nitrate was also saturated with sodium chloride. From Figure 6-14 you can see that, at 50°C, a saturated solution contains 370 g/L of sodium chloride. At 0°C a saturated solution contains 350 g/L of sodium chloride. Therefore, when 1 L of the mixture is cooled from 50°C to 0°C, only 20 g of sodium chloride would crystallize out. It is possible, then, to separate most of two solids which are both soluble, provided the solubility of one solid changes with temperature more than does the solubility of the other solid. You simply cool the mixture. Much of the solid with the greater change in solubility crystallizes out. The mixture is then filtered to recover the

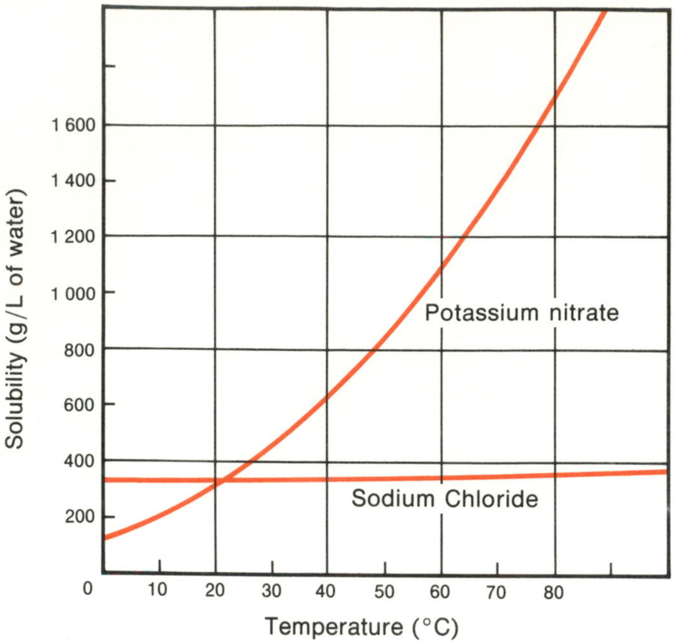

Figure 6-14 Solubility curves for potassium nitrate and sodium chloride.

crystals. The solution that passes through the filter paper contains most of the other solid. This solid can be recovered by evaporating the water.

The process by which a solid is crystallized from a solution that contains a mixture of solids is called **fractional crystallization**. In this investigation you use fractional crystallization to separate potassium nitrate and sodium chloride from a solution.

Materials

test tube (20 × 150 mm)
solution saturated at 50°C with potassium nitrate and sodium chloride
large beaker (400 mL)
ring stand, iron ring, wire gauze
thermometer (−10°C to 110°C)
ice
funnel
filter paper
small beaker (150 mL)
Bunsen burner

CAUTION: Wear safety goggles during this investigation.

Procedure

a. Your teacher has prepared a solution that is at 50°C and is saturated with both potassium nitrate and sodium chloride. Obtain half a test tube of this solution from your teacher.

b. Make notes on any changes that take place as the solution cools in air from 50°C to about 30°C.

c. Place the test tube in an ice-water mixture in a large beaker. Make notes on any changes that take place as the solution cools to about 0°C.

d. Filter the mixture to recover the crystallized solid.
e. Evaporate the filtrate to dryness to recover the second solid.

Discussion

1. What is the solid that crystallized from the solution on cooling?
2. What is the solid that was recovered by evaporating the filtrate?
3. Use the solubility curves to explain why fractional crystallization could be used to separate these two solids.
4. Sodium chloride is common salt. It occurs in the ground in some parts of Canada. Sometimes other solids are mixed in with it. Suggest a method that could be used to purify this salt for use by humans.

6.22 INVESTIGATION: Can You Separate These Mixtures?

In this investigation you have to design your own procedure. You are given very brief information about several mixtures. You are to select one of these and write a procedure for separating it. Have your teacher check your procedure. If it is acceptable, he/she will give you what you need to try your procedure.

Prepare a report on the separation of the mixture that includes procedure, results, and a discussion that points out how the results prove that your procedure was successful.

You should be able to do 2 or 3 mixtures.

MIXTURE A Iron and Sulfur

This mixture contains iron filings mixed with powdered sulfur. How can you separate the iron and sulfur?

MIXTURE B Salt and Sand

This mixture contains table salt mixed with fine sand. How can you separate the salt and sand? The salt should be ready for use on the table after the separation.

MIXTURE C Peanuts

A peanut is a mixture of several things, one of which is peanut oil. You are to separate the peanut oil from the rest of the mixture. Hint: Peanut oil is soluble in methyl alcohol.
CAUTION: Methyl alcohol is flammable.

MIXTURE D Iodine and Salt

This mixture contains iodine crystals mixed with salt crystals. Review Investigation 4.11, page 86 and Investigation 6.11, page 142, before designing your procedure.

MIXTURE E Tomato Juice

Tomato juice is both a suspension and a solution. You are to separate the suspended substances from the dissolved substances. What gives tomato juice most of its taste, the suspended or the dissolved substances?

MIXTURE F Copper Turnings and Wood Charcoal

This mixture contains small pieces of wood charcoal and small pieces of copper metal. Can you separate them? The density of copper is 8 900 kg/m^3 and the density of charcoal (including the air in the pores of the charcoal) is 300-400 kg/m^3.

Highlights

Water is one of the most important substances on earth. It is composed of hydrogen and oxygen. It occurs almost everywhere and is required by all living things. Even apparently dry foods and minerals contain water.

A solution is a homogeneous mixture that consists of a solute and a solvent. Temperature and volume changes occur when a solute and a solvent are mixed. A solution may be described as dilute or concentrated; it may also be described as unsaturated, saturated, or supersaturated.

The rate of dissolving can usually be increased by agitation, subdivision of the solute, and an increase in temperature.

The solubility of a substance is the mass of the substance in grams that dissolves in 1 L of solvent to form a saturated solution at a given temperature. The solubility depends on the nature of the solute, the nature of the solvent, and the temperature.

Suspensions, emulsions, and colloidal dispersions are heterogeneous mixtures that can sometimes be mistaken for solutions.

Most mixtures can be separated by one or more of these methods: filtration, distillation, fractional distillation, paper chromatography, fractional crystallization, sedimentation and floc formation. In each case the separation depends on the fact that the components of the mixture differ in some physical properties such as solubility and boiling point.

UNIT

7 The Chemical Properties of Matter

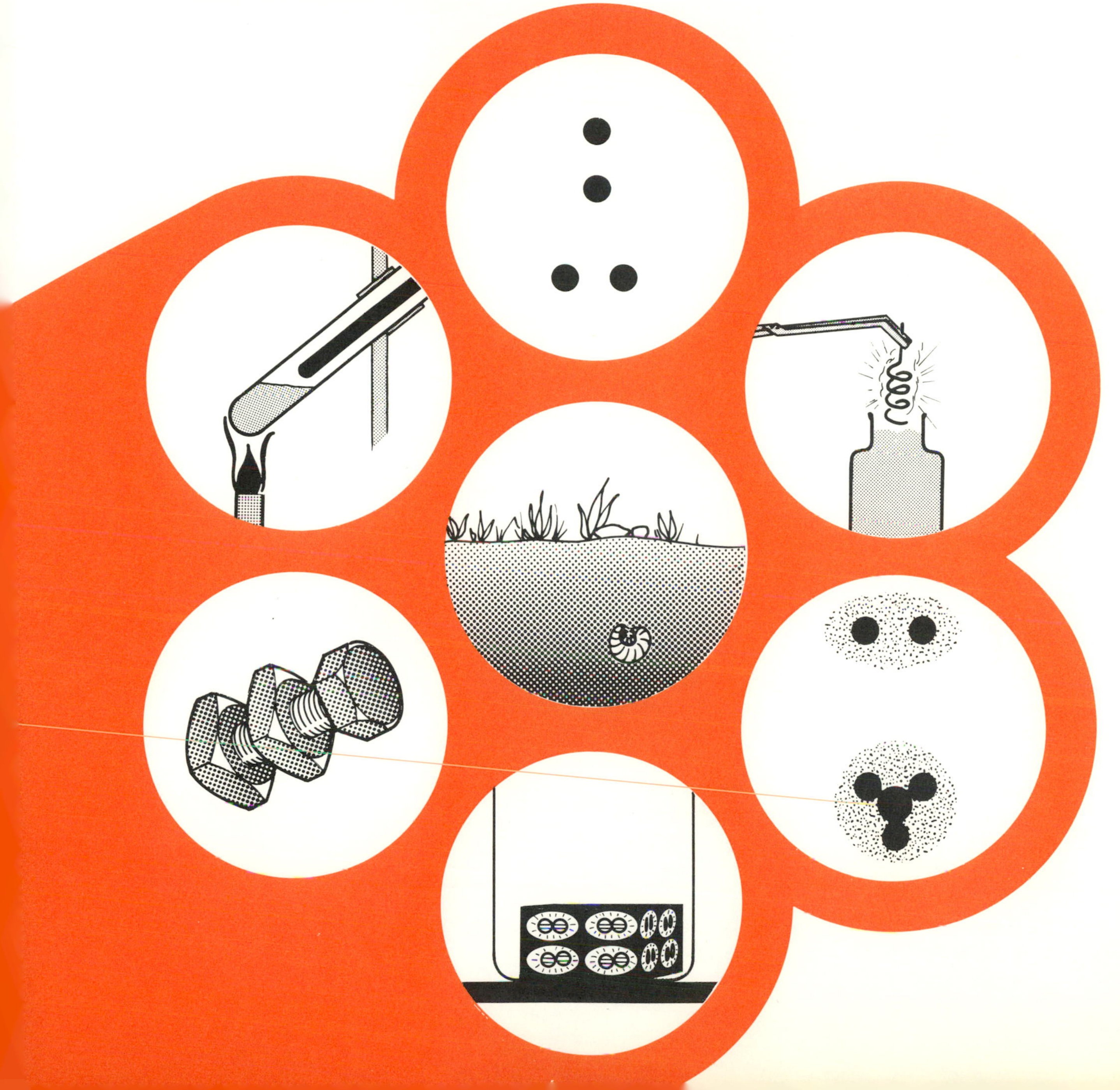

As you will recall from Unit 3, chemical properties involve the formation of a new substance. In order to determine what the chemical properties of a substance are, we must observe what that substance does when it comes in contact with other substances or energy. The amount of time it takes for one substance to react with another varies. For example, gold is not very reactive. It tarnishes slowly when exposed to air. On the other hand, nitroglycerine is very reactive. A rise in temperature or a shock can cause nitroglycerine to explode. Thus the chemical properties of gold and nitroglycerine are very different.

Because there are so many substances, it is not possible in this course to study the chemical properties of all of them. In this unit you investigate the chemical properties of the two classes of substances, elements and compounds. Then an atomic model for matter is developed to explain the properties observed.

7.1 A Classification System for Matter

Why Classify Matter?

What do the substances in Table 40 have in common? How are they different?

TABLE 40 Some Examples of Matter

Substance	Brief description
Fertilizer	solid granules of different colours
Air	clear, colourless gas
Lye (sodium hydroxide)	white flakes
Bluestone (copper sulfate)	blue crystals
Milk	white, opaque liquid
Ethyl alcohol	clear, colourless liquid
Diamond (carbon)	clear, sparkling solid
Tea	clear, yellow-brown liquid
Sugar (sucrose)	white crystals
Ketchup	red, opaque liquid
Granite	white-and-black-speckled solid
Baking powder	white powder
Aluminum	shiny, grey solid
Oxygen	clear, colourless gas
Sulfur	yellow powder

All of these substances have mass and take up space, but they differ in appearance.

You know that matter is anything that has mass and takes up space. If you place each of the substances of Table 40 on a balance pan, the pan will go down. Each substance has mass. With the aid of a ruler or graduated cylinder you can measure the volume of each substance. Each substance takes up space. Thus each substance is a form of matter. In fact, everything you see, and some things you do not see, like air, are forms of matter. They all take up space and have mass.

If you wish to remember the properties of matter, it is necessary to group substances with common properties together. The process of grouping substances according to common properties is called **classification**. Imagine the task of memorizing Table 40! Now rearrange the table so that all of the white solids are together. The task becomes much easier because one description serves for several substances. When substances with common properties are grouped together, you reduce the amount of information that must be memorized, without losing any of the information. Thus a **classification system** makes the study of a subject easier.

In Unit 4 you classified matter into three groups—solids, liquids, and gases. This method of classifying matter was very useful for keeping track of physical properties such as density. However, it is not sufficient for chemical properties.

Classifying by Composition

A classification system that will work for chemical properties is shown in Figure 7-1. It is based on the composition of matter and can be used to classify the substances in Table 41. But before you can classify matter according to this scheme you must understand the following terms:

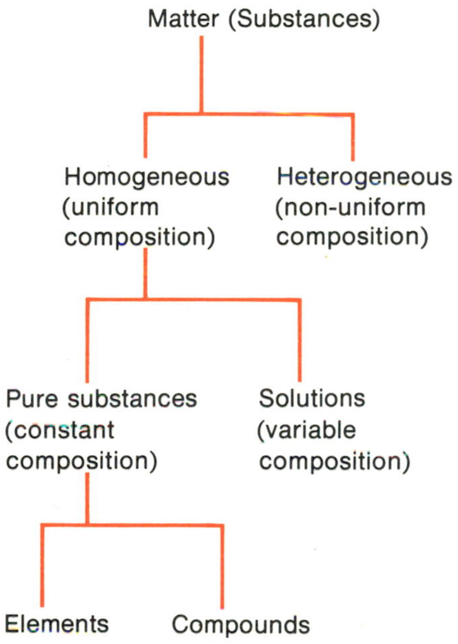

Figure 7-1 A classification system for matter.

Composition

The **composition** *of a substance is a description of the quantitative or qualitative make-up of the substance.* A quantitative description of a fertilizer is given by the numbers 14-4-8. This means that 14% of the fertilizer is nitrogen, 4% is phosphoric acid, and 8% is potash. A qualitative description of the composition of a fertilizer is the list of the ingredients used to make the fertilizer. Such a list includes ammonium sulfate, potassium chloride, sodium nitrate, calcium dihydrogen phosphate, and calcium sulfate.

Phase

A **phase** *is a visibly distinct portion of matter.* There are definite boundaries that separate one phase from another. The granite rock shown in Figure 7-2 is an example of a substance made up of two phases.

Granite is a rock formed deep in the earth's crust. The light-coloured parts are one phase, the dark-coloured parts another phase. The light-coloured parts are composed of the minerals feldspar and quartz. The dark-coloured parts are composed of hornblende and biotite.

Figure 7-2 Granite, an igneous rock. The light-coloured phase is feldspar and quartz; the dark-coloured phase is horneblende and biotite.

Homogeneous

Homogeneous *means that the composition is uniform throughout the phase*. The dark-coloured phase of granite is homogeneous. The percentage of hornblende would not change within the boundaries of the phase.

Homogeneous Matter

Homogeneous matter *is any substance that has only one phase and a uniform composition*. Diamond, sugar, and after-shave lotion are examples of substances with only one phase. Their composition is the same throughout the phase.

Heterogeneous Matter

Heterogeneous matter *is any substance that has more than one phase and a non-uniform composition*. Fertilizer and granite are examples. In both cases colour can be used to tell the phases apart. You can see more than one colour of material. Table 41 shows the composition of some common fertilizers.

TABLE 41 The Composition of Some Common Fertilizers

Fertilizer	Total nitrogen	Available phosphoric acid	Soluble potash
Nutrite, Turf Special (14-4-8)	14%	4%	8%
Nutrite, Garden Special (4-12-8)	4%	12%	8%
So-Green (10-6-4)	10%	6%	4%
Golden Vigoro (12-6-3)	12%	6%	3%

The table shows the variation in the composition of common fertilizers. The variation is made possible by changing the amount of each phase used. Heterogeneous matter has a variable composition.

Pure Substances

Pure substances are homogeneous forms of matter whose properties you study later in this unit. There are two types of pure substances: elements and compounds. For example, sugar is a pure substance that is a compound, and diamond is a pure substance that is an element. Sugar is 100% sucrose; diamond is 100% carbon.

It is not always possible to tell a pure substance by just looking at it. Ketchup, for example, appears to be homogeneous. The red colour is uniform throughout. However, the label lists the ingredients as: tomatoes, sugar, vinegar, salt, and spices. Thus ketchup is a mixture. Container labels can often help you classify the contents. If the label lists only one ingredient, the contents may be a pure substance. If two or more ingredients are listed, the contents must be a mixture.

Solutions

Solutions *are homogeneous mixtures of two or more pure substances*. After-shave lotion is a mixture of alcohol, a coloured substance, and some perfume. Only one phase is visible. The colour is uniform throughout. Thus after-shave lotion is an example of a solution. You have already studied the properties of solutions in Unit 6.

Elements

Elements are pure substances. They are the building blocks of matter. All forms of matter are built from elements. Among the 106 known elements are: oxygen, nitrogen, iron, aluminum, and tin.

Compounds

Compounds are also pure substances. There are about two million pure substances known, and new ones are discovered every day. Since there are only 106 elements, then most pure substances must be compounds. Compounds, then, are pure substances formed from the union of elements.

Because of sheer numbers it is not possible to list all the known compounds in a book. However, a systematic way of naming compounds has been developed that makes it easier for you to tell when you are working with a compound. The chemical names of compounds often use the names of the elements that make up the compound. Thus the name provides a qualitative description of the composition of the compound. Table 42 shows the composition for five compounds. Note that in some cases the name tells you what elements are in the compound.

TABLE 42 Names of Compounds and Their Composition

Common name	Chemical name	Elements used to build the compound
Table salt	Sodium chloride	sodium, chlorine
Lye	Sodium hydroxide	sodium, hydrogen, oxygen
Bluestone	Copper sulfate	copper, sulfur, oxygen
Ethyl alcohol	Ethanol	carbon, hydrogen, oxygen
Table sugar	Sucrose	carbon, hydrogen, oxygen

Discussion

1. **a)** Define the following terms: matter, composition, phase.
   b) Define and give an example of the following forms of matter: homogeneous, heterogeneous, pure substance, solution, element, compound.
   c) What property is used in Figure 7-1 to classify matter?
2. Explain what is meant by the term "pure orange juice". Why would a chemist not consider orange juice to be pure?

7.2 INVESTIGATION: Elements—the Building Blocks of Matter

There are 92 naturally-occurring elements. Another dozen or so have been made by scientists. Some elements, like oxygen and nitrogen, are found in the atmosphere. Others are found in the earth's crust. Figure 7-3 shows the percentage composition by mass of elements in the earth's crust. A few elements—gold, silver, copper, carbon, and sulfur—can exist in the earth's crust uncombined with other elements. However, most elements exist in combination with one or more other elements. Elements are the building blocks of matter. They combine with one another to produce compounds. Almost everything around you is in this form. The salt you use to season foods is a combination of the elements sodium and chlorine. The water you drink is a combination of the elements hydrogen and oxygen. Gasoline is a combination of the elements carbon and hydrogen. Beach sand contains quartz granules. Quartz is a combination of the elements silicon and oxygen. Thus the elements carbon and hydrogen propel the family car as it makes its way toward a lakeshore of silicon, oxygen, and hydrogen. In this investigation you study the combination of several pairs of elements. Pay particular attention to observations that suggest a new pure substance is formed. Copy Table 43 into your notebook. Record all of your observations in the table.

TABLE 43 Combining of Elements

Elements	Appearance before reaction of element 1	Appearance before reaction of element 2	Observations that suggest elements combine	Observations that suggest a new pure substance formed
Procedure A 1. Magnesium 2. Oxygen				
Procedure B 1. Copper 2. Sulfur				
Procedure C 1. Iron 2. Chlorine				

CAUTION: Because of possible dangers in this investigation, your teacher may decide to demonstrate some of the reactions.

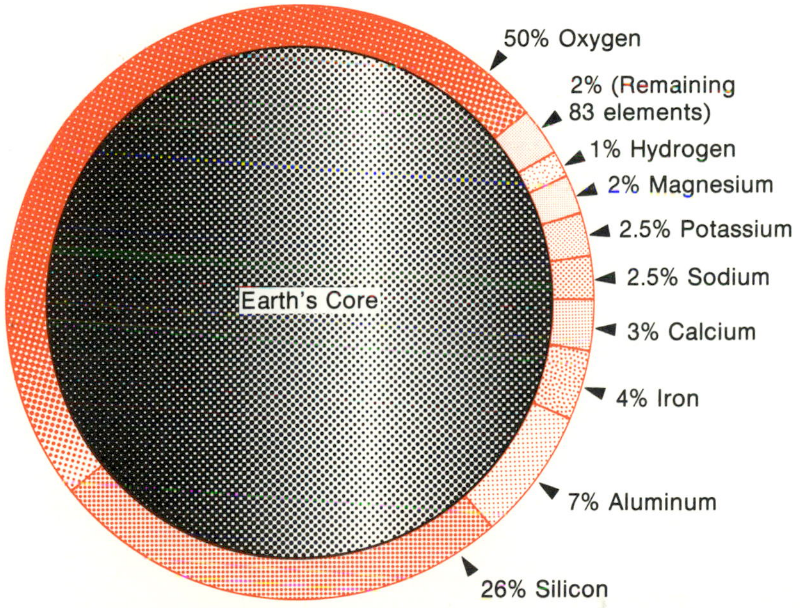

Figure 7-3 Composition of the earth's crust.

Materials

spiral of magnesium ribbon (the spiral is formed by winding 8 cm of magnesium ribbon around a pencil.)
oxygen
copper (10 cm x 1 cm strip of foil or coil of copper wire)
sulfur (crushed or powdered)
iron (steel wool; degrease with alcohol and twist into a compact cylinder 10 cm long and 2 cm in diameter)
chlorine
2 gas bottles and 2 glass plates
tongs
adjustable clamp
Bunsen burner and asbestos pad
CAUTION: Wear safety goggles during this investigation.

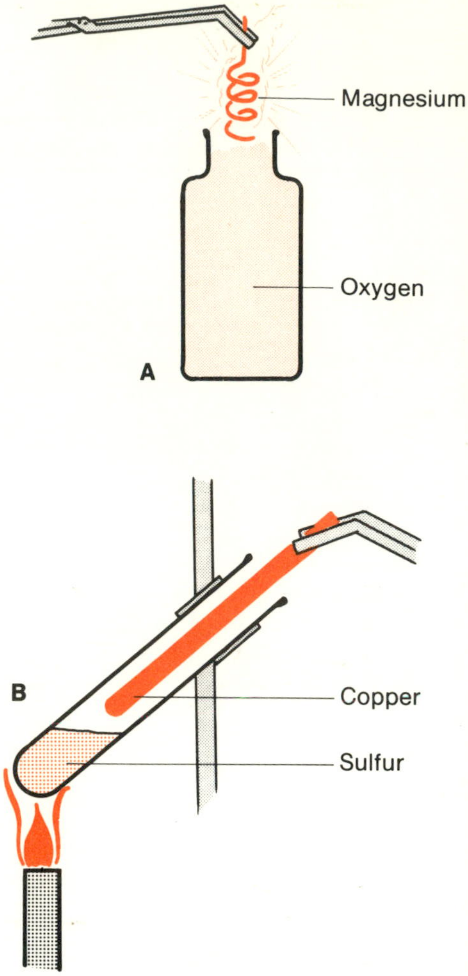

Figure 7-4 Combining elements: A—magnesium and oxygen; B—copper and sulfur; C—iron and chlorine.

Procedure A Magnesium and Oxygen

a. Obtain a gas bottle filled with oxygen from your teacher. Keep the mouth of the gas bottle covered with the glass plate until you are ready to use it.
b. Place the gas bottle close to the Bunsen burner on the asbestos pad.
c. Hold the magnesium spiral firmly with the tongs.
d. Ignite the tip of the magnesium spiral in the hottest part of the burner flame. **CAUTION:** Work over an asbestos pad. Burning magnesium could burn the desk top if it falls from the tongs.
e. Remove the plate from the gas bottle and lower the burning magnesium into the oxygen. Be careful not to let the burning magnesium touch the sides of the bottle (Fig. 7-4,A). **CAUTION:** Do not stare at the brilliant magnesium flame.

Procedure B Copper and Sulfur

a. Fill a test tube ¼ full of sulfur.
b. Use an adjustable clamp to hold the test tube. Clamp the test tube near its mouth.
c. Heat the sulfur until it boils. **CAUTION:** Irritating and poisonous gases are produced. Adequate ventilation is required.
d. Hold the copper firmly with the tongs.
e. Lower the copper into the sulfur vapour until the foil is almost touching the boiling sulfur (Fig. 7-4, B).

Procedure C Iron and Chlorine

a. Obtain a gas bottle filled with chlorine from your teacher. Keep the mouth of the gas bottle covered with the glass plate until

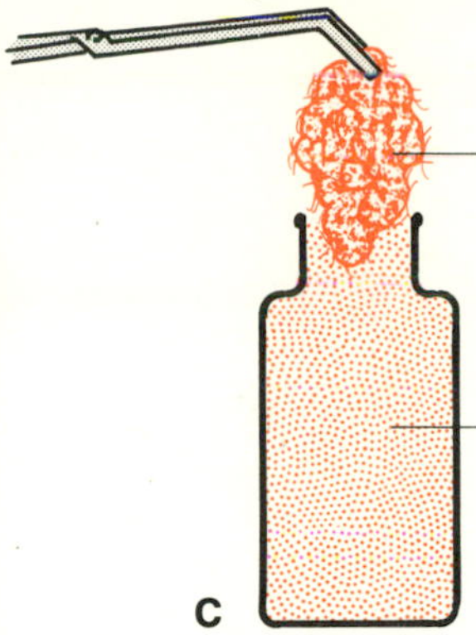

Figure 7-4 Combining elements: A —magnesium and oxygen; B— copper and sulfur; C—iron and chlorine.

you are ready to use it. **CAUTION:** Chlorine is an irritating and poisonous gas. Adequate ventilation is required.

b. Hold the iron (steel wool) firmly with the tongs.
c. Heat the tip of the steel wool in the hottest part of the Bunsen burner flame until it begins to glow.
d. Remove the plate from the gas bottle and lower the glowing iron into the chlorine (Fig. 7-4,C).

Discussion

1. Define an element.
2. What are the two most abundant elements found in the earth's crust?
3. Which elements do you think were discovered first? Explain why you think they would have been discovered first.
4. How were the three reactions in this investigation the same? How were they different?
5. Each reaction is an example of a combination reaction.
   a) What observations suggested that the elements are used up in these reactions?
   b) What combined in these reactions?
6. What observations suggested that a new pure substance was formed?
7. What type of pure substance was formed? Explain how you decided whether it was an element or a compound.
8. Use the information provided by your teacher to determine the name of the product of each combination reaction.
9. What observations suggest that energy (heat and light) is a product of a combination reaction?
10. Write a word equation to describe each combination reaction. A word equation for a combination reaction has the following form:
    Name of element 1 ( ) + Name of element 2 ( ) → Name of compound ( ) + energy
    The "( )" is used to indicate the state the substance is in when the reaction occurs. The solid state is represented by (s), the liquid state by (l), the gaseous state by (g), and an aqueous (water) solution by (aq). Here is an example:

    Sodium (l) + oxygen (g) → sodium oxide (s) + energy.

    This word equation is read: "liquid sodium plus gaseous oxygen combine to produce solid sodium oxide plus energy." The reactants are sodium and oxygen; the products sodium oxide and energy.
11. Use your answers to the previous questions to explain what is meant by the term "combination reaction".
12. What chemical property of elements is demonstrated in this investigation?

7.3 INVESTIGATION: Can an Element Be Recovered from Its Compound?

The name of the compound magnesium oxide tells us that the element magnesium is part of the compound. But how is this possible? In Investigation 7.2 the silvery magnesium ribbon became a white powder. The product did not look like the starting materials. Has the magnesium been destroyed? Can magnesium be recovered from the compound, magnesium oxide? This investigation provides observations that you can use to answer these questions. In Procedure A a compound of carbon is produced. In Procedure B magnesium is used to try to recover the carbon from the compound.

CAUTION: Because of possible dangers in this investigation, your teacher may decide to demonstrate some of the reactions.

Materials

oxygen
carbon (a small lump of charcoal)
spiral of magnesium ribbon
gas bottle and glass plate
tongs
Bunsen burner and asbestos pad

CAUTION: Wear safety goggles during this investigation.

Procedure A Producing a Compound of Carbon and Oxygen

a. In this investigation you will need to record observations that will enable you to decide two things: first, that a compound of carbon is produced and, second, that carbon is recovered from the compound. Construct a table that will make it easier for you to record such observations.

b. Obtain a gas bottle filled with oxygen from your teacher. Keep the mouth of the gas bottle covered with the glass plate until you are ready to use it.

c. Place the gas bottle near the Bunsen burner on the asbestos pad.

d. Hold the piece of carbon (charcoal) firmly with the tongs.

e. Heat the charcoal in the hottest part of the burner flame until it glows red.

f. Remove the plate from the gas bottle and lower the glowing charcoal into the oxygen.

g. When the reaction ceases, remove the remaining charcoal and quickly replace the glass plate.

Procedure B Recovering the Element Carbon

a. Hold the magnesium spiral firmly with the tongs.
b. Ignite the tip of the magnesium spiral in the hottest part of the burner flame. **CAUTION:** Work over an asbestos pad to protect the desk.
c. Remove the plate from the gas bottle and lower the burning magnesium into the bottle. **CAUTION:** Do not stare at the magnesium flame.

Discussion A

1. What observations suggest that carbon and oxygen combine?
2. Use your observations and the information provided by your teacher to name the product.
3. Write a word equation for this combination reaction.
4. What chemical property of elements is demonstrated in Procedure A?

Discussion B

1. What observations suggest that magnesium reacts with the gas produced in Procedure A?
2. Use your observations to support or reject the conclusion that "carbon can be recovered from its compound".
3. Use your observations and the information provided by your teacher to name both products of this reaction.
4. Write a word equation for the reaction.
5. What chemical property of compounds is demonstrated in Procedure B?

7.4 INVESTIGATION: The Effect of a Combination Reaction on the Properties of an Element

You have seen a grey solid become red-brown, a black solid become a clear, colourless gas, and a copper-coloured solid turn black. In all of the combination reactions studied so far, the appearance of the elements changed a great deal. Are other physical and chemical properties, such as melting point, density, and ability to react with an acid changed as much? In this investigation you study the changes in the magnetic and chemical properties of iron when it combines with sulfur. The densities, melting points, and boiling points of the reactants and product are provided.

Materials

iron filings (degrease with alcohol)
sulfur powder
test tube
2 wooden splints
dilute hydrochloric acid and medicine dropper
magnifying glass
adjustable clamp
permanent magnet
porcelain spot plate

CAUTION: Wear safety goggles during this investigation.

Procedure

a. Copy Table 44 into your notebook. Record your observations in this table.

TABLE 44 Properties of an Element and Its Compound

Substance	Density (kg/m^3)	Melting point (°C)	Boiling point (°C)	Appearance	Effect of magnet	Effect of dilute hydrochloric acid
Iron	7 860	1 539	2 887			
Sulfur	1 960	119	445			
Mixture	9 820	Glows when heated				
Compound	4 740	Decomposes at 1 195				

b. Place 3 measures of iron filings in a test tube. (A measure is the amount of solid that will cover 1 cm of a wooden splint.)
c. Touch the end of the magnet to the bottom of the test tube. Move the magnet up and down the side of the test tube.
d. Place ½ measure of iron filings in one section of the spot plate.
e. Add two drops of dilute hydrochloric acid to the iron filings on the spot plate.
f. Gently wave some of the gas produced toward your nose.
g. Add 6 measures (use a different wooden splint) of powdered sulfur to the test tube containing the iron filings. Shake the test tube until the colour of the mixture in the test tube appears uniform.
h. Repeat step c. Remove the magnet and shake the test tube until the colour of the mixture in the test tube appears uniform.
i. Transfer ½ measure of the mixture to a section of the spot plate. Examine the mixture with a magnifying glass.
j. Repeat steps e and f for the mixture.
k. Gently heat the test tube containing the rest of the mixture. Heat a corner of the mixture with the hottest part of the Bunsen burner flame, as shown in Figure 7-5. When the mixture begins to glow, remove the test tube from the flame.

CAUTION: Irritating and poisonous gases are produced. Adequate ventilation is required.

l. When the product cools, repeat step c.
m. Transfer some of the product to a section of the spot plate. Examine the product with a magnifying glass.
n. Repeat steps e and f for the product.

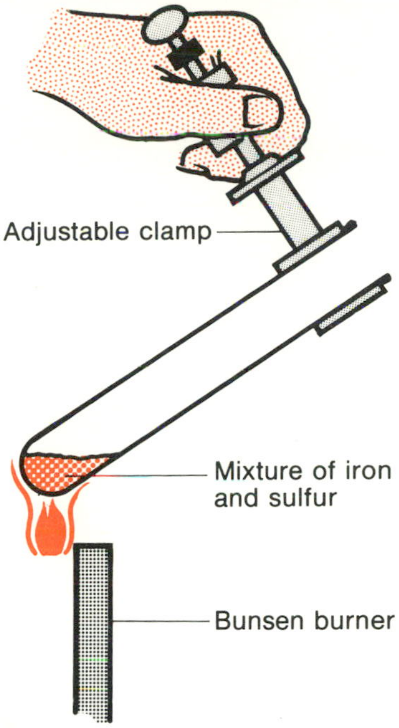

Figure 7-5 Effect of heat on a mixture of iron and sulfur.

Discussion

1. What observations suggest that iron combines with sulfur?
2. Use your observations and the information provided by your teacher to identify the compound.
3. Write a word equation for the combination reaction.
4. Compare the properties of the elements with those of the mixture. How are they the same? How are they different?
5. Compare the properties of the elements with those of the compound. How are they the same? How are they different?
6. Compare the properties of the mixture and the compound. How are they the same? How are they different?
7. What physical properties are compared in this investigation?
8. What chemical properties are compared in this investigation?
9. What chemical properties of elements are demonstrated in this investigation?

7.5 Atoms and Molecules

Developing a Model

In Section 1.11 you learned that a model must explain observations and predict successfully the results of further experiments. The following is a summary of the results of the preceding studies in this unit:

1. There are 106 elements or building blocks of matter.
2. Compounds are produced by combining two or more different elements.
3. The physical and chemical properties of elements change a great deal when elements combine to form compounds.
4. Elements can be recovered from their compounds, but only by chemical reactions.

Will the particle theory of Unit 4 explain these results? The answer to this question is not obvious. There are at least three ways of combining particles that should be considered. Figure 7-6 shows the three models. Each circle represents a particle of a substance.

Model A explains the combination of elements as a mixing of particles. In this model the particles are not changed. The model

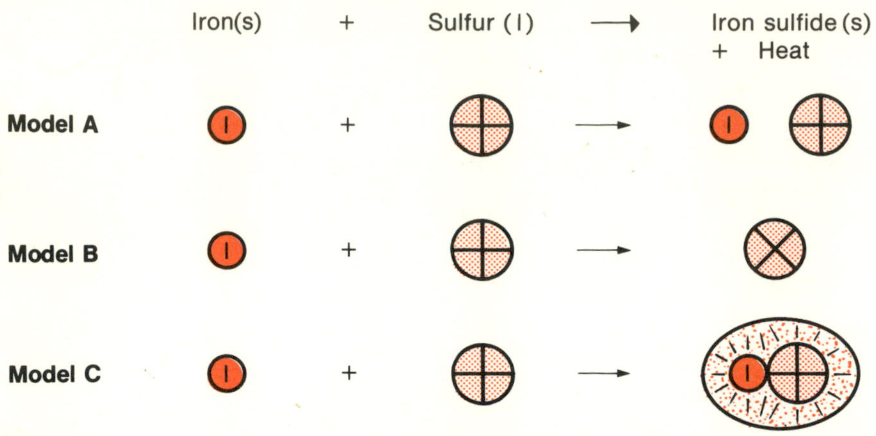

Figure 7-6 Three models for the formation of iron sulfide.

is similar to mixing red and green jellybeans. In the case of the jellybeans and model A, the particles retain their properties. No new substance is formed.

With model B a new particle is produced. The new particle has different properties than those of the two elements. Model B explains the changes in properties observed when two elements combine. Model B is similar to a tomato combining with an apple to form a grapefruit. With this model it is difficult to imagine how an element could be recovered from its compound.

Models A and B do not explain the observations. Then why consider them? Because that is the way of science. Scientists use their imaginations to create as many models as possible. Models that do not explain the observations are discarded. Models that are successful are used to predict the results of new experiments.

Model C, like model B, shows the formation of a new particle. Thus model C explains the change in properties. With this model the new particle is formed by joining a particle of iron to a particle of sulfur. Model C is similar to joining the two jellybeans with a toothpick or glue. Since the new particle is made up of the particles of the two elements, it is easier to imagine how an element could be recovered from its compound. Thus model C best explains the observations.

Model C shows a single particle of a compound as being much more complex than a single particle of an element. To distinguish between these two types of particles, the words atom and molecule are used. **Atom** *is the name given to a single particle of an element*. An atom of iron is a single particle of the element iron. *A single particle of a compound is called a* **molecule**. A molecule is really a compound atom. A molecule of iron sulfide is made up of one atom of iron and one atom of sulfur. *A molecule is composed of two or more atoms*. The two or more atoms can be the same or different. If the atoms are different, the substance is a compound. If the atoms are the same, the substance is an element. Figure 7-7 shows the molecules of some pure substances.

Figure 7-7 The molecules of some pure substances.

Molecule	Composition	Element	Compound
Nitrogen	2 atoms of nitrogen	✓	
Magnesium Oxide	1 atom of magnesium 1 atom of oxygen		✓
Hydrogen	2 atoms of hydrogen	✓	
Carbon dioxide	1 atom of carbon 2 atoms of oxygen		✓
Ammonia	1 atom of nitrogen 3 atoms of hydrogen		✓

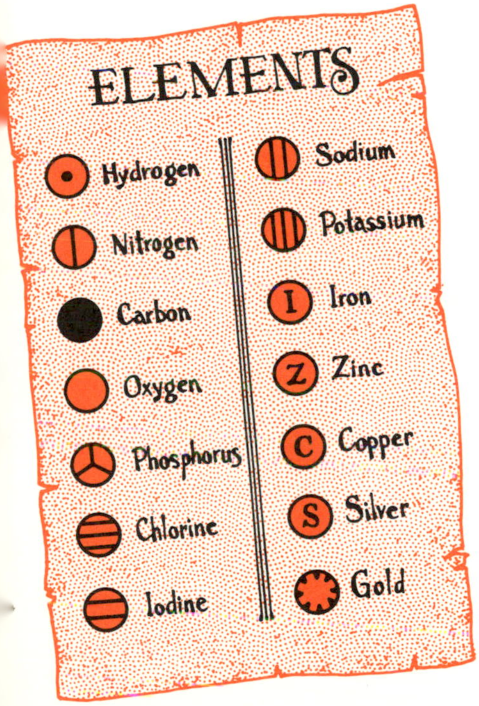

Figure 7-8 Symbols for the atoms of some elements.

Dalton's Atomic Theory

John Dalton (1766-1844) first used circles to represent atoms. Figure 7-8 shows his way of representing atoms of some of the elements.

Dalton was very interested in the study of weather. He measured barometric pressure, humidity, and rainfall. His aim was to predict the weather and thus help fishermen and farmers. His observations led him to investigate the properties of the gases of the atmosphere. As a result of his work with gases, he stated the atomic theory. Dalton wrote his ideas about atoms in his notebook in 1803. This is Dalton's Atomic Theory:

1. Matter is made up of small particles or atoms.
2. Atoms are indestructible.
3. Atoms of an element are identical. They have the same mass.
4. Atoms of different elements have different masses.

5. Atoms can attract or hold onto other atoms.
6. Compounds are formed by combining atoms of different elements.

Dalton was not the first to speak of atoms. The Greeks, as early as 430 B.C., had proposed that a sample of matter was made up of a very large number of tiny indestructible particles. These particles were called atoms. Dalton's main contributions to the development of a model for matter were:

1. Atoms of different elements differ in mass.
2. Compounds are formed by combining atoms of different elements.

Chemical and Physical Changes

Changes that involve the formation of new substances are called **chemical changes**. A chemical change is accompanied by a change in properties. The properties of the new substance formed are different from the properties of the elements that formed it. Recall the changes that occurred when iron and sulfur formed iron sulfide. Dalton's Atomic Theory provides a picture on a small scale of what we think is happening. According to the model, atoms of different elements combine to form a new substance. Thus *chemical reactions involve a rearrangement of atoms of different elements*.

Silver nitrate (aq) colourless solution + Sodium chloride (aq) colourless solution ⟶ Sodium nitrate (aq) colourless solution + Silver chloride (s) white solid

Figure 7-9 A chemical reaction involves the rearrangement of different atoms.

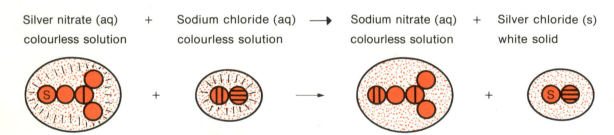

Figure 7-9 illustrates a reaction between two compounds. The appearance of the white solid indicates the change in properties caused by the rearrangement of atoms. In this example the atoms have formed new combinations. The result is two new pure substances.

Changes that involve a rearrangement of identical particles (atoms or molecules) are called **physical changes** *or* **changes of state**. You studied changes of state in Unit 4. Figure 7-10 illustrates the rearrangement that occurs during changes of state.

Discussion

1. List three characteristics of combination reactions.
2. Explain the difference between an atom and a molecule.
3. Compare the following terms using examples to illustrate your answers: Atom and element; molecule and compound; physi-

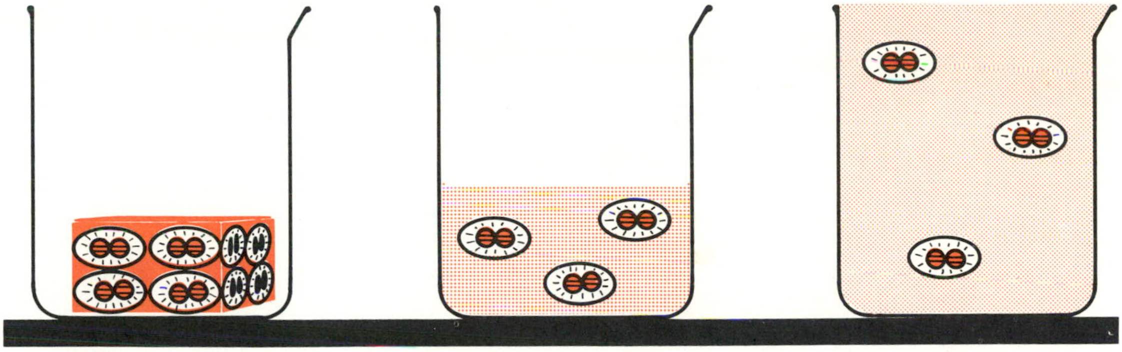

Figure 7-10 A physical reaction involves the rearrangement of identical particles (atoms or molecules).

cal properties and chemical properties; physical change and chemical change.

4. Summarize Dalton's Atomic Theory in your own words.

5. What were Dalton's main contributions to the development of a model for matter?

6. Use Dalton's Atomic Theory to explain the four results summarized on page 179.

7. We no longer accept two of Dalton's ideas about atoms. What are those ideas? Why are they no longer accepted?

8. The following are brief descriptions of some actions that result in changes. Classify each change as either chemical or physical. Indicate the observations you used to make your choice. If the change is physical, name it.

a) baking a cake
b) making ice cream
c) toasting bread
d) firing a cap pistol
e) burning fondue fuel
f) drying clothes
g) adding sugar to cereal cream
h) allowing pop to go flat
i) adding alka-seltzer to water
j) using battery-powered toys or a flashlight

9. The following are brief descriptions of investigations you did in the laboratory. Classify each as either a chemical or physical change. Indicate the observations you used to make your choice. If the change is physical, name it.

a) a candle was burned;
b) ammonium dichromate was heated;
c) lumps of bluestone were ground to a fine powder;
d) hot steel wool was placed in chlorine gas;
e) solid iodine was warmed;
f) sulfuric acid was added to a mixture of sodium bromide and manganese dioxide.

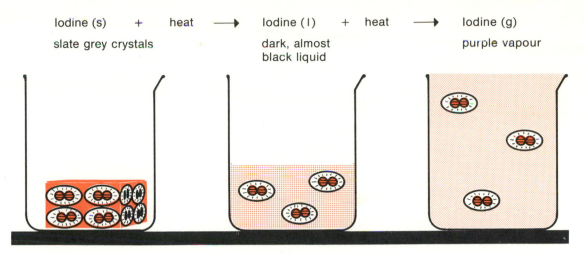

10. Dalton's reason for studying gases was to help fishermen and farmers predict the weather. Do you think scientific research should only be done if it benefits society? Discuss this question with your classmates.

7.6 INVESTIGATION: Using the Dalton Model to Predict

Scientists test a model by making predictions with it. In this investigation you use Dalton's atomic model to predict a characteristic of chemical reactions and a property of compounds. Nuts and bolts are used to represent Dalton atoms. Table 45 shows that nuts and bolts are a good analogy for Dalton's atomic model. (An **analogy** is a tool often used by scientists. It explains by comparing one thing to another whose properties are known. In this case we are comparing Dalton's model with nuts and bolts. Since the statements of Dalton's model agree well with the properties of nuts and bolts, then nuts and bolts are a good analogy for the Dalton model.)

To simplify instructions, B will be used for one bolt and N for one nut. Thus BN represents a molecule of one nut threaded on one bolt (Fig. 7-11, A); BN_2 represents a molecule with two nuts threaded on one bolt (Fig. 7-11, B); B_2N represents a molecule with two bolts held together by one nut (Fig. 7-11, C).

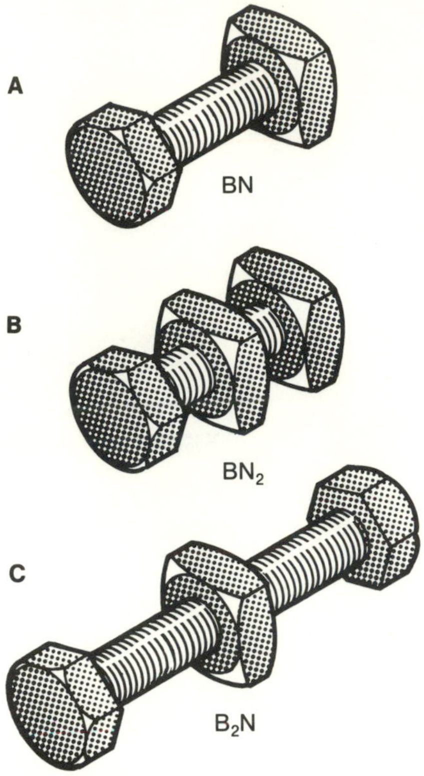

Figure 7-11 Nut and bolt molecules.

TABLE 45 A Nut and Bolt Analogy for Dalton's Atomic Model

Dalton's Atomic Model	Nuts and Bolts
Matter is made up of small particles or atoms.	1 nut represents 1 atom of an element. 1 bolt represents 1 atom of another element.
Atoms are indestructible.	Nuts and bolts are indestructible (under normal conditions).
Atoms of an element are identical. They have the same mass.	The bolts are identical; they have the same mass.
Atoms of different elements have different masses.	The mass of 1 bolt is greater than the mass of 1 nut.
Atoms can attract or hold onto other atoms.	Nuts can be threaded on a bolt.
Compounds are formed by combining atoms of different elements.	Compounds are formed by threading nuts on bolts.

Materials

container
nuts and bolts
balance

Procedure A Mass of Combined and Uncombined Nuts and Bolts

a. Copy Table 46 into your notebook.

TABLE 46

	Trial 1	Trial 2
Mass of Container plus B and N atoms, m_1(g)		
Mass of Container plus BN molecules, m_2(g)		

b. Trial 1: Find the mass of the container plus 5 bolts and 5 nuts (m_1). The bolts and nuts must be separated.
c. Thread one nut on each bolt. Find the mass of the container and the BN molecules (m_2).
d. Trial 2: Repeat steps b and c using 10 bolts and 10 nuts.

Procedure B Percent Composition of Bolt-Nut Molecules

a. Copy Table 47 into your notebook.

TABLE 47

	Trial 1	Trial 2	Trial 3
Mass of container + BN molecules, m_1(g)			
Mass of container + B atoms, m_2(g)			
Mass of container, m_3(g)			
Mass of B atoms, $m_4 = m_2 - m_3$(g)			
Mass of BN molecules, $m_5 = m_1 - m_3$(g)			
Percent B = $\frac{m_4}{m_5} \times 100\%$			
Percent N = 100% − %B			

b. Find the mass of the empty container (m_3).
c. Trial 1: Add 5 bolts to the container. Find the mass of the 5 bolts and the container (m_2).
d. Add 15 nuts to the container. Thread 1 nut on each bolt. Remove any excess bolts or nuts.

e. Find the mass of the container and the BN molecules (m_1).
f. Calculate the mass of the B atoms (m_4) and the mass of the BN molecules (m_5).
g. Calculate the percent B and percent N in the compound BN.
h. Trial 2: Repeat steps c to g using 10 bolts.
i. Trial 3: Repeat steps c to g using 15 bolts.

Discussion A

1. Compare m_1 and m_2 for each trial. Compare your results with those obtained by your classmates.
2. Does the mass change as a result of threading a nut onto a bolt? Use the properties of nuts and bolts to explain why this is reasonable.
3. What mass in this investigation represents the mass of the container and the reactants? What mass represents the mass of the container and the products?
4. In Investigation 7.7 you will find the mass of the reactants of a chemical reaction and then allow them to react to form products. Assume that matter behaves like nuts and bolts. Then predict the relationship between the mass of the reactants and the mass of the products in the chemical reaction.

Discussion B

1. Compare the percent compositions from the three trials. Calculate the average percent B and percent N. Compare your average with that obtained by your classmates.
2. Does the percent B or percent N depend on the number of bolts used? Use the properties of nuts and bolts to explain why this is reasonable.
3. In Investigation 7.8 you will carry out the following chemical reaction: zinc (s) + hydrochloric acid (aq) → zinc chloride (aq) + hydrogen (g). Different masses of zinc will be used. The percent composition of zinc chloride will be determined for each trial. Assume that matter behaves like nuts and bolts. Then predict the relationship between the percent composition that you will get for each trial.

7.7 INVESTIGATION: Conservation of Mass

In Investigation 7.6 you used nuts and bolts to make predictions about how atoms behave. Predictions made in this way are little more than educated guesses. It is absolutely necessary to do experiments to test your predictions. And that is what you are going to do here.

In this investigation you compare the total mass of the products formed in a chemical reaction with the total mass of the reactants.

Materials

balance
Erlenmeyer flask and stopper
small test tube
solutions of each of the following: lead nitrate, potassium iodide, copper sulfate, sodium hydroxide, barium chloride, sodium sulfate

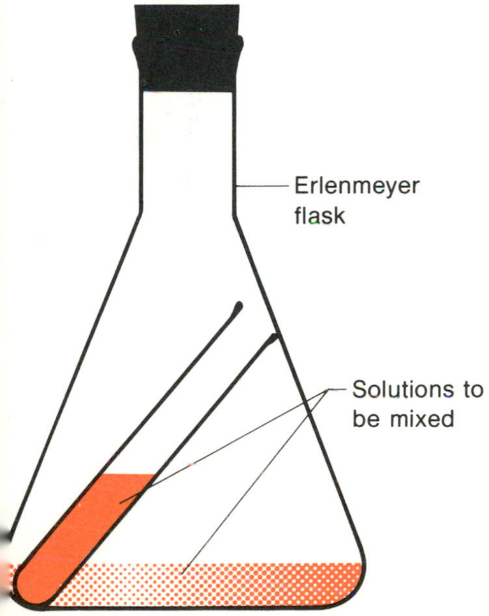

Figure 7-12 Conservation-of-mass assembly.

Procedure

a. Prepare a copy of Table 48 for recording your results. The procedures for all three reactions are the same. Your teacher will tell you which reaction to do. Obtain half a test tube of each of the two solutions you need.

b. Half-fill the small test tube with one of the solutions.

c. Pour about 10 mL of the other solution into the Erlenmeyer flask.

d. Carefully slide the small test tube and contents into the flask (Fig. 7-12).

e. Put the stopper in the flask. Record a description of the reactants.

CAUTION: Avoid contact of solutions with eyes, skin, and clothing. Rinse thoroughly with water if an accident does happen.

f. Find the mass of the entire assembly. Record the mass in the "Mass of reactants" column.

g. Remove the assembly from the balance pan. Do not adjust the balance. Invert the assembly and swirl the contents to mix the solutions.

h. Return the assembly to the balance and find its mass.

i. Record a description of the reaction and the products. Note the mass and record it in the "Mass of products" column.

TABLE 48 Conservation of Mass

Reaction number	Names of reactants	Descriptions of reactants	Mass of reactants (g)	Description of reaction and products	Mass of products (g)
1	Lead nitrate (aq) Potassium iodide (aq)				
2	Copper sulfate (aq) Sodium hydroxide (aq)				
3	Barium chloride (aq) Sodium sulfate (aq)				

Discussion

1. For each reaction explain how you know that a chemical change was involved. Identify the products with the aid of the information provided by your teacher. Write a word equation for each reaction.
2. For each reaction compare the total mass of the reactants with the total mass of the products. Compare your results with those obtained by your classmates. Make a general statement comparing the total mass of the products formed in a chemical reaction with the total mass of the reactants.
3. The **Law of Conservation of Mass** states that *the total mass of the products formed in a chemical reaction is equal to the total mass of the reactants.* Do the data gathered by the class support the Law of Conservation of Mass? Do the data prove the law? Why?
4. In Investigation 7.6 you found that mass did not change as a result of rearranging nuts and bolts. Does mass change as a result of rearranging atoms in a chemical reaction? Use Dalton's model to explain how mass is conserved in a chemical reaction.
5. What chemical property was measured in this investigation?
6. According to Dalton's model, if mass is conserved then atoms are neither destroyed nor created in a chemical reaction. Thus man cannot use chemical reactions to dispose of substances. Chemical reactions can only be used to rearrange atoms. Nature has always operated within this law and has used chemical reactions to cycle elements so that they can be used by all the organisms that need them. Figure 7-13 shows how carbon passes through the soil and into and

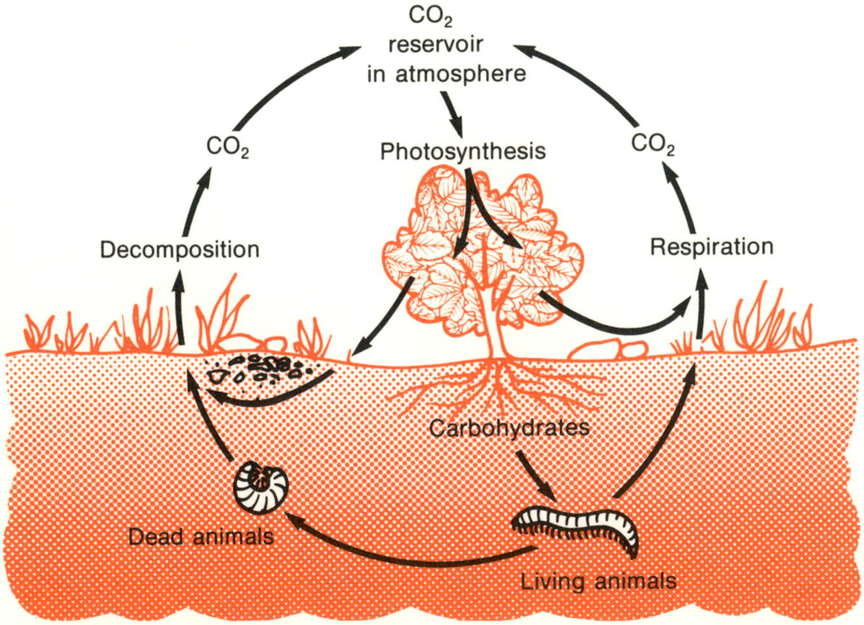

Figure 7-13 The carbon cycle in and above the soil.

out of the atmosphere. In our handling of garbage we have operated for many years as if we were unaware of the Law of Conservation of Mass. We have buried or burned garbage in the belief that this would dispose of it. List some of the harmful effects of such action. Suggest better ways of operating, particularly ways that use the Law of Conservation of Mass.

7. How could you prove that mass is conserved when a flash bulb flashes? You may wish to try this experiment.

8. It seems obvious to us that the total mass of the products formed in a chemical reaction is equal to the total mass of the reactants. Yet, when magnesium is heated, a white ash forms and the mass increases. Also, when a candle burns, it grows shorter and its mass decreases. Account for these observations which apparently violate the Law of Conservation of Mass.

9. Predict the mass of sulfur trioxide produced when 100 g of sulfur react with 150 g of oxygen. Write the word equation for the reaction.

10. When 80 g of sulfur trioxide react with water, 98 g of sulfuric acid are formed. What mass of water is required? Write the word equation for the reaction.

7.8 INVESTIGATION: Constant Composition

Recall the prediction made about the percent composition of zinc chloride in Investigation 7.6. To test this prediction, members of your class will use different amounts of zinc to produce zinc chloride. The amount of hydrochloric acid used will be the same in each case. After the zinc and hydrochloric acid have reacted, you will remove any remaining zinc and determine the mass of zinc used up. The solution that remains contains the zinc chloride. By evaporating the solvent you can determine the amount of zinc chloride produced. Then you can calculate the ratio of the mass of zinc that reacted to the mass of zinc chloride that was produced.

Materials

balance
10 mL graduated cylinder
250 mL beaker
test tube (20 × 150 mm)
zinc (1 cm^3 pieces)
hydrochloric acid
heat lamp
evaporating dish
alcohol
ring stand, iron ring, wire gauze, Bunsen burner

CAUTION: Wear safety goggles during this investigation.

Procedure A Making Zinc Chloride

a. Design a data table for recording your results. If you have trouble, look back to Investigation 7.6, Procedure B.
b. Find the mass of the test tube.
c. Your teacher will tell your group how many zinc pieces to use. Place them in the test tube. Find the mass of the test tube and the zinc pieces.
d. Add 10 mL of hydrochloric acid to the test tube.
CAUTION: Avoid contact of acid with eyes, skin, and clothing. If an accident happens, rinse thoroughly with water.
e. Place the test tube in a beaker half full of water.
f. On a blank piece of paper put your name, your partner's name, your class, and the date.
g. Place the beaker and test tube on the paper with your name on it in a place where it will not be disturbed. Allow the mixture of zinc and hydrochloric acid to stand overnight.

Procedure B Separating Zinc Chloride

a. Find the mass of the evaporating dish.
b. Pour the solution from the test tube into the evaporating dish. Keep any zinc that has not reacted in the test tube.
c. Add 5 mL of water to the test tube. Shake the water in the test tube to wash the test tube and the zinc. Pour this wash water into the evaporating dish.
d. Set up the apparatus as shown in Figure 7-14.
e. Heat the evaporating dish with a hot flame. If the solution begins to spatter, move the flame back and forth under the dish.
f. Heat the material until the residue appears dry. Then continue to heat the solid in the dish until it melts and a small pool of liquid forms in the bottom of the evaporating dish. Turn off the Bunsen burner.
g. When the evaporating dish has cooled so you can handle it with your fingers, find the mass of the dish and its contents.
h. If there is any unreacted zinc in the test tube, add 5 mL of alcohol to the test tube. **CAUTION:** Alcohol is flammable. Stay away from flames.
i. Shake the alcohol in the test tube to wash the test tube and the zinc. Pour off the alcohol. Keep the zinc in the test tube. Let the test tube stand under a heat lamp for 10-15 min. Then find the mass of the test tube and its contents.

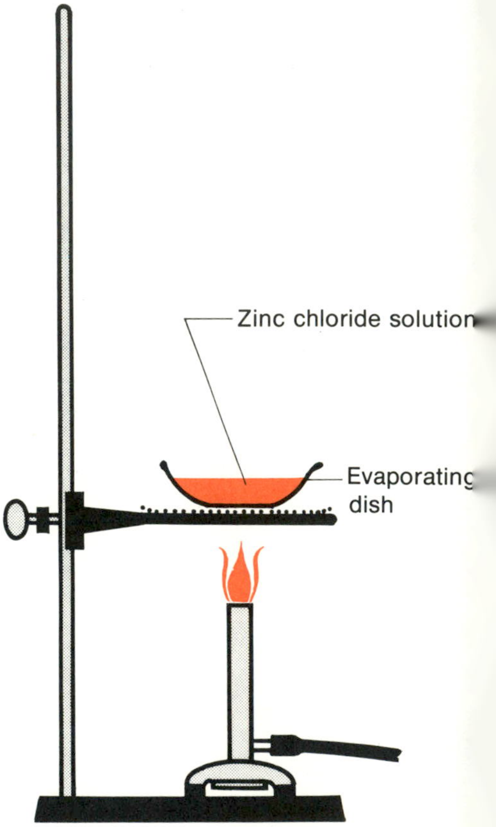

Figure 7-14 Separating zinc chloride.

Discussion

1. Write a word equation for the making of zinc chloride. Explain how you know that a chemical reaction took place.
2. Why was the test tube placed in a beaker half full of water?

3. Why was the mixture of zinc and hydrochloric acid allowed to stand overnight?
4. Why was the water added to the test tube after the solution has been poured into the evaporating dish?
5. What part did alcohol play in this investigation?
6. Calculate the percent composition of zinc chloride. Compare your results with those of your classmates.
7. If the zinc was not thoroughly dry when you found its mass, how would this affect the percent zinc? If the zinc chloride was not thoroughly dry when you found its mass, how would this affect the percent zinc?
8. The **Law of Constant Composition** states that *the percent composition of a compound is always the same*. Do the data gathered by your class support the Law of Constant Composition? Do the data prove the law? Why?
9. In Investigation 7.6 you found that the percent composition of a nut-bolt molecule did not change with an excess of nuts nor did it change with an excess of bolts. Did an excess of either zinc or hydrochloric acid affect the percent composition of zinc chloride? Use Dalton's model to explain the Law of Constant Composition.
10. What evidence suggests that the product, zinc chloride, is a pure substance and not a mixture?
11. What chemical properties were studied in this investigation?
12. When dry hydrogen was passed over 7.5 g of copper (II) oxide, 6.0 g of copper were produced. In a second experiment 4.5 g of the copper were burned in oxygen to form 5.6 g of copper (II) oxide. Do these data support the Law of Constant Composition? Show calculations to support your answer.

Highlights

Elements are the building blocks of matter. They combine to produce compounds and energy. An element is made up of identical atoms. A compound is made up of atoms of different elements. A molecule is the name given to a single particle of a compound.

Chemical reactions involve the rearrangement of atoms of different elements. A chemical reaction produces a new pure substance. Physical changes or changes of state involve a rearrangement of identical particles (atoms or molecules). No new pure substance is produced.

In a chemical reaction, the total mass of the reactants is equal to the total mass of the products. This is called the Law of Conservation of Mass. The Law of Constant Composition states that the present composition of a compound is always the same. Dalton's Atomic Theory explains both of these laws.

UNIT

8 Using the Atomic Model

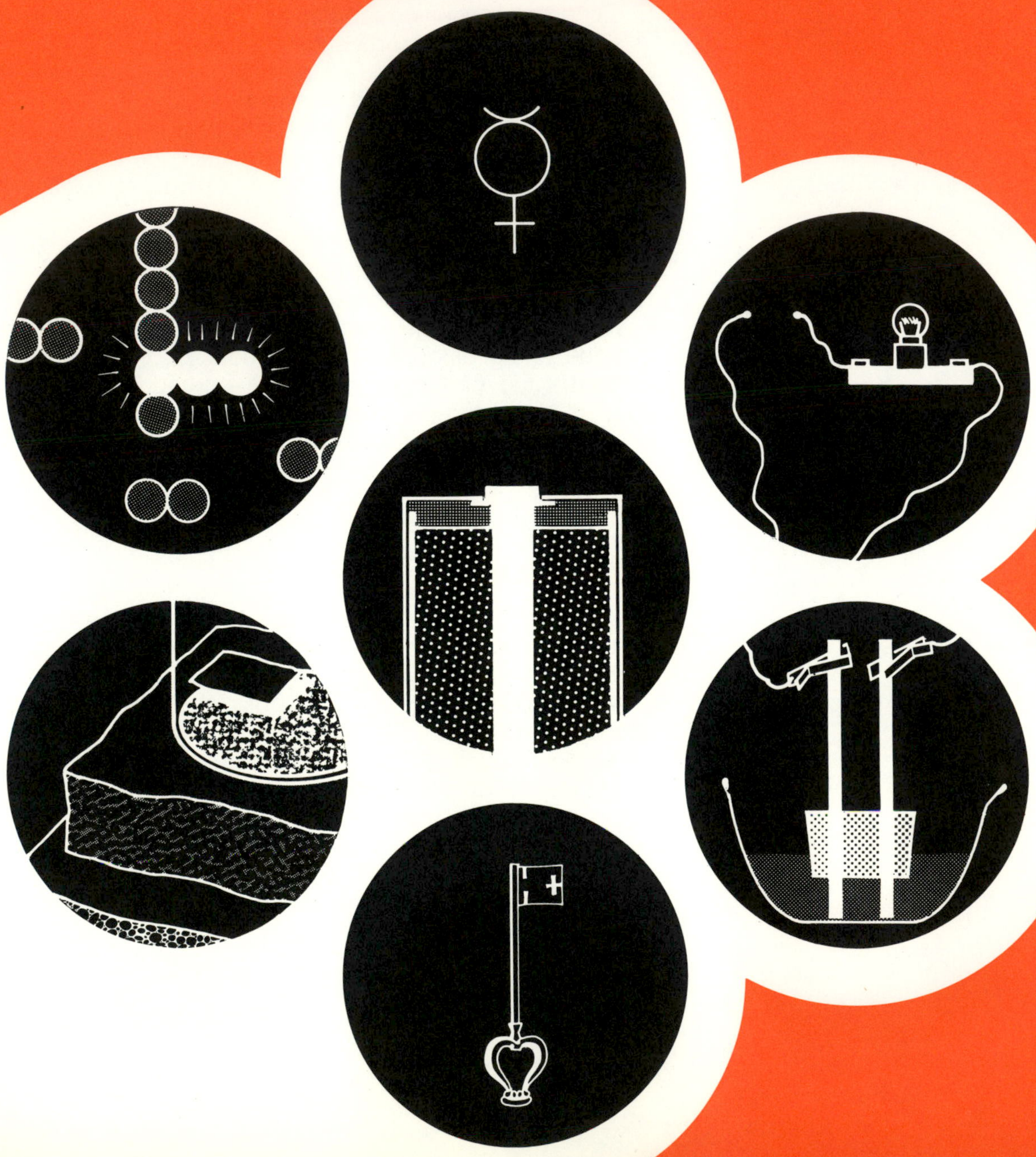

Unit 7 introduced Dalton's atomic model of matter. The model explains most of the chemical properties you have met so far. It accounts for the Law of Conservation of Mass and the Law of Constant Composition. It also provides a description at a microscopic level of chemical and physical changes.

You may recall that Section 1.9 defined a model as a mental picture. Dalton's atomic model is what we *think* matter looks like on a very small scale. A model is created to explain observations. Thus it is reasonable to expect that, as more observations are made, it may become necessary to modify the model.

In this unit you use Dalton's atomic model to explain observations. You also use it to predict results before you begin an investigation. However, in some investigations, you will make observations that may force you to modify Dalton's atomic model.

8.1 INVESTIGATION: The Law of Multiple Proportions

Dalton's atomic model explained the Law of Constant Composition. But can it explain the fact that certain pairs of elements combine to form two or more compounds? If it is a good model for matter it should be able to explain this new information. In this investigation you test Dalton's atomic model by using it to explain the composition of compounds formed from the same two elements. In order to test the model you will need to do some calculations.

Study the following example using the data for two compounds of copper and sulfur.

Copper (I) Sulfide

This compound was found to be 80.0% copper and 20.0% sulfur by mass. What mass of sulfur combines with 100 g of copper?

80.0 g of copper combines with 20.0 g of sulfur

Therefore 1 g of copper combines with $\frac{20.0}{80.0}$ g of sulfur

and 100 g of copper combines with $\frac{20.0}{80.0} \times 100\ \text{g}$

$= 25.0$ g of sulfur (m_1)

Copper (II) Sulfide

This compound was found to be 66.7% copper and 33.3% sulfur. What mass of sulfur combines with 100 g of copper?

66.7 g of copper combines with 33.3 g of sulfur

Therefore 1 g of copper combines with $\frac{33.3}{66.7}$ g of sulfur

and 100 g of copper combines with

$$\frac{33.3}{66.7} \times 100\text{ g} = 49.9\text{ g of sulfur } (m_2)$$

The Ratio of the Masses of Sulfur Combining with 100 g of Copper

$$\frac{m_1}{m_2} = \frac{49.9\text{ g}}{25.0\text{ g}} = \frac{2}{1}$$

Thus, when copper and sulfur combine to form two compounds, the ratio of the masses of sulfur that combine with 100 g of copper is given by two small whole numbers. Is this a coincidence or is it an important generalization? If this result is a generalization, then ratios determined by this method for other pairs of elements that form two or more compounds should also be given by two small whole numbers. Table 49 provides experimental data which you will use to test whether the ratios are always small whole numbers. You will then use the nut and bolt analogy for Dalton's atomic model to account for your findings.

Materials

balance
nuts and bolts
container for nuts and bolts

Procedure A Mass Ratios for Some Compounds

a. Copy Table 49 into your notebook. Do the calculations needed to complete the table.

Procedure B Mass Ratios for Bolt-Nut Compounds

a. Copy Table 50 into your notebook.
b. Your teacher will tell you which of the following pairs of nut and bolt compounds to use in this investigation: BN, BN_2/BN_2, BN_3/BN_2, BN_4/B_2N, BN/B_2N, B_2N_3.
c. Construct 5 molecules for each compound you have been assigned.
d. Design an experiment to determine the percent composition of each nut and bolt compound.

e. Write the method in your notebook. Have it checked by your teacher before you begin to find the mass of the nuts and bolts.

f. Experimentally determine the percent composition of each nut and bolt compound you have been assigned. Record the percent composition beside the appropriate pair of compounds in the table.

g. Do the appropriate calculations to complete the table for the pair of nut and bolt compounds you were assigned.

h. Complete the remainder of the table by exchanging information with your classmates. Your teacher will explain how to do this.

TABLE 49 Composition Data for Pairs of Elements That Form More Than One Compound

Composition / Compound	Percent (%)		Mass of Y (g) that combines with 100 g of X		Ratio of the Masses of Y
	Element X	**Element Y**	**Element X**	**Element Y**	
	Carbon	Oxygen	Carbon	Oxygen	
Carbon monoxide	42.9	57.1	100	$m_1 = 133$	$\frac{m_2}{m_1} = \frac{2}{1}$
Carbon dioxide	27.3	72.7	100	$m_2 = 266$	
	Iron	Chlorine	Iron	Chlorine	
Iron (II) chloride	44.1	55.9	100	$m_1 = 127$	$\frac{m_2}{m_1} =$
Iron (III) chloride	34.5	65.5	100	$m_2 = 190$	
	Tin	Chlorine	Tin	Chlorine	
Tin (II) chloride	62.5	38.5	100	$m_1 =$	$\frac{m_2}{m_1} =$
Tin (IV) chloride	45.5	54.5	100	$m_2 =$	
	Nitrogen	Oxygen	Nitrogen	Oxygen	
Nitrogen (I) oxide	63.6		100	$m_1 =$	$\frac{m_2}{m_1} =$
Nitrogen (II) oxide	46.7		100	$m_2 =$	
Nitrogen (III) oxide	36.8		100	$m_3 =$	$\frac{m_3}{m_1} =$

Discussion A

1. Compare the ratios of the masses of element Y in Table 49. How are they the same? How are they different?
2. The **Law of Multiple Proportions** states: *If two elements X and Y form two or more compounds, then the ratio of the masses of Y combining with* 100 g *of X will be given by two small whole numbers.* Use the data of Table 49 to support or reject the Law of Multiple Proportions.

TABLE 50 Composition Data for Some Nut and Bolt Compounds

Compound \ Composition	Percent (%) B	Percent (%) N	Mass of N (g) that combines with 100 g of B	Ratios of the masses of N
BN BN_2			$m_1 =$ $m_2 =$	$\frac{m_2}{m_1} =$
BN_2 BN_3			$m_1 =$ $m_2 =$	$\frac{m_2}{m_1} =$
BN_2 BN_4			$m_1 =$ $m_2 =$	$\frac{m_2}{m_1} =$
B_2N BN			$m_1 =$ $m_2 =$	$\frac{m_2}{m_1} =$
B_2N B_2N_3			$m_1 =$ $m_2 =$	$\frac{m_2}{m_1} =$

Discussion B

1. Compare the ratios of the masses of N in Table 50. How are they the same? How are they different?
2. Does the Law of Multiple Proportions apply to nut and bolt compounds? Use the data of Table 50 to support your answer.
3. Determine the number of nuts per bolt in each of the two compounds you were assigned. Call these ratio 1 and ratio 2.

$$\text{ratio 1} = \frac{\text{number of nuts in first compound}}{\text{number of bolts}}$$

$$\text{ratio 2} = \frac{\text{number of nuts in second compound}}{\text{number of bolts}}$$

4. Determine ratio 3, where ratio 3 $= \frac{\text{ratio 2}}{\text{ratio 1}}$
5. Compare ratio 3 with the ratio of the masses of N. How are they the same? How are they different?
6. When bolts and nuts combine to form two or more compounds, the ratio of the masses of N that combine with 100 g of B is given by two small whole numbers. Why is this so?
7. When two elements X and Y form two or more compounds, the ratio of the masses of Y that combine with 100 g of X is given by two small whole numbers. Use Dalton's Atomic Theory to explain why this is so.
8. Explain why the Law of Multiple Proportions increases our confidence in Dalton's atomic model.

9. Samples of two nut and bolt compounds were decomposed. The experiment provided the data in Table 51.

TABLE 51 Decomposition of Some Nut and Bolt Compounds

Nut and Bolt compound	Composition		Mass of N (g) that combines with 100 g of B	Ratio of the masses of N
	%B	%N		
BN	77.6	19.4	$m_1 =$	$\frac{m_2}{m_1} =$
BN_2	66.6	33.3	$m_2 =$	

Determine m_1, m_2, and m_2/m_1. Account for the ratio of the masses of N in terms of the number of B and N atoms in the molecules.

10. Assume that a molecule of each of the oxides of carbon in Table 49 contains 1 atom of carbon. Use the ratio of the mass of oxygen that combines with 100 g of carbon to determine how many more oxygen atoms there are in one molecule of carbon dioxide than in one molecule of carbon monoxide. If one molecule of carbon monoxide contains one atom of carbon and one atom of oxygen, what is the composition of a molecule of carbon dioxide?

11. Both of the tin chloride molecules in Table 49 contain one atom of tin. The tin (II) chloride molecule contains 2 atoms of chlorine. Describe a molecule of tin (IV) chloride.

12. A molecule of nitrogen (I) oxide contains 2 atoms of nitrogen and 1 atom of oxygen. Describe the molecules of the other nitrogen oxides.

13. Analysis of four compounds of lead and oxygen supplied the data in Table 52.

TABLE 52 Composition of Some Lead Oxides

Compound	Mass of oxygen (g) that combined with 100 g of lead	Ratio of the masses of oxygen
Lead (II) oxide	$m_1 = 7.72$	$\frac{m_2}{m_1} =$
Lead (I) oxide	$m_2 = 3.86$	
Lead (III) oxide	$m_3 = 11.58$	$\frac{m_3}{m_1} =$
Lead (IV) oxide	$m_4 = 15.44$	$\frac{m_4}{m_1} =$

Show that the data support the Law of Multiple Proportions. Then account for the ratio of the masses in terms of the number of lead and oxygen atoms in the molecules of the four compounds.

8.2 A Collision Model for Chemical Reactions

The rates of chemical reactions vary a great deal. It takes hundreds of years for stalactites to form in a cave. A light bulb often "burns out" in less than a year. To fry an egg on the stove requires only a couple of minutes. The explosion in the cylinder of a gasoline engine takes only a fraction of a second. What causes some reactions to be faster than others? What can be done to change the rate of a chemical reaction? Can Dalton's atomic model account for such a variety of rates? This section and the investigations that follow will help you find answers to these questions.

TABLE 53 A Marble Analogy for the Atomic and Collision Models

Dalton's atomic model	Analogous property of marbles
Matter is made up of small particles or atoms.	Each marble represents one atom.
Atoms are indestructible.	Marbles are indestructible under normal conditions.
Atoms of an element are identical. They have the same mass.	Marbles of the same size and design have the same mass.
Atoms of different elements have different masses.	Marbles of different sizes have different masses.
Atoms can attract or hold onto other atoms.	Plasticene can be used to stick marbles together.
Compounds are formed by combining atoms of different elements.	Plasticene can be used to stick marbles of different sizes together.
Collision Model	
Atoms or molecules must collide before a chemical reaction can occur.	Marbles can roll and collide.

The Collision Model

To combine chlorine and iron in Investigation 7.2, the iron was placed in the chlorine gas. On a large or macroscopic scale the elements were in contact. Thus on a small or microscopic scale, we would expect that the atoms of iron and chlorine would have to come in contact before a reaction could occur. (See Figure 8-1.) The basic assumption of the **collision model** is that *atoms or*

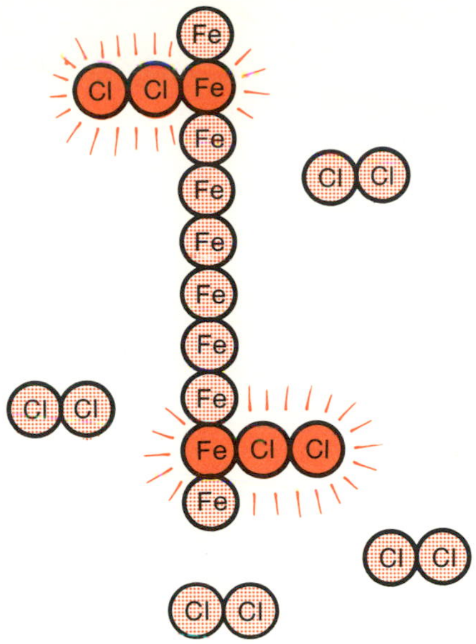

Figure 8-1 Atoms or molecules must collide before a chemical reaction can occur.

molecules must collide before a chemical reaction can occur. If this is so, then the rate of a chemical reaction will depend on the frequency of the atomic or molecular collisions. This frequency is the number of collisions of atoms or molecules that occurs per second. An understanding of how the frequency of atomic or molecular collisions can be varied is needed before you can predict the variables that change the rates of chemical reactions. Since we cannot see atomic or molecular collisions, we must use an analogy.

A Marble Analogy for Atoms

In Section 7.6 nuts and bolts were used as an analogy for Dalton's atomic model. The nuts and bolts enabled you to make predictions about combining masses. But nuts and bolts are not a suitable analogy for studying collisions because they do not roll easily. Marbles are a better choice. But are the properties of marbles analogous to Dalton's atomic model and to the basic assumption of the collision model? Table 53 illustrates which properties of marbles are analogous to the atomic and collision models.

Discussion

1. **a)** What is the basic assumption of the collision model?
   b) Explain why the assumption is reasonable.
2. What is meant by the expression "frequency of atomic or molecular collisions"?
3. In a sample of air at 0°C the average speeds of the oxygen and nitrogen molecules are 1 500 km/h and 1 600 km/h respectively. (The maximum speed of a jet airliner is about 1 000 km/h.) What will happen to the average speeds of the oxygen and nitrogen molecules if the air is warmed to 20°C?
4. The frequency of the collisions between methane and oxygen molecules at 20°C is 10^{36} collisions per second. What will happen to the frequency of the collisions if the gaseous mixture is cooled to 0°C? At which temperature would an explosion be least likely? Why?
5. Name two variables that you think will determine the frequency of collisions. Predict the effect on the rate of a chemical reaction if the value of each factor is increased.
6. Marbles are placed in a box and the box is shaken. Explain how the box and marbles could be used to demonstrate the effect of an increase in temperature.
7. As the concentration of a solution increases, what happens to the number of solute molecules in a litre of solution?
8. Explain how an increase in concentration could be demonstrated with a box and marbles. You should be able to think of two ways.

9. Explain how testing predictions based on the marble analogy could increase our confidence in Dalton's atomic model of matter.

10. Your teacher performed a demonstration that used the marble analogy to make predictions about the factors that affect the rates of chemical reactions.

a) What are the four factors that may affect the rates of chemical reactions?

b) State your prediction for the effect of each of these factors on the rates of chemical reactions. Explain in each case how you arrived at that prediction.

8.3 INVESTIGATION: Variables that Change the Rate of a Chemical Reaction

In this investigation you study the effect of each of these variables on the rate of a chemical reaction: concentration, nature of the reactants, temperature, and the amount of surface exposed. Reactions A and B are used in this study.

Reaction A.

pink solution + X (aq) → colourless solution

Reaction B.

calcium carbonate (s) + Y (aq) → carbon dioxide (g)

The disappearance of the pink colour is used to measure the rate of reaction A. The shorter the time it takes the pink colour to disappear, the faster the reaction. The appearance of the gas is used to measure the rate of reaction B. The faster the bubbles appear, the faster the reaction.

Before you begin this investigation, use your answer to question 10, page 200, to predict the effect on the rate of each reaction when each of the variables is changed.

Materials

100 mL and 10 mL graduated cylinders
2 Erlenmeyer flasks
Bunsen burner, striker, asbestos pad
4 test tubes and rack
adjustable clamp
masking tape and marking pen
clock with second hand

calcium carbonate—chips and powder
pink solution (dilute potassium permanganate solution)
X_1-acidified oxalic acid
X_2-acidified iron (II) chloride
Y_1-dilute hydrochloric acid
Y_2-dilute acetic acid

Procedure A

a. Copy Table 54 into your notebook.

TABLE 54 Rates of Reaction A

	Description of Trial							
Trial	**Variable changed**	**X_1 (mL)**	**X_2 (mL)**	**Water (mL)**	**Temperature**	**Pink solution (mL)**	**Predicted time (s)**	**Actual time (s)**
1	control	40		5	room	5	*t*	
2		10		35	room	5		
3			40	5	room	5		
4		40		5	above room	5		
5	Add 5mL of pink solution to the control					5		

b. Compare trials 2, 3, and 4 with 1. How are they the same? How are they different? Record in Table 54 the variable changed in trials 2, 3, and 4.

c. Assume that the control required "*t*" seconds for the pink colour to disappear. Predict the times required for trials 2, 3, and 4 relative to "*t*". Record your predictions in Table 54. Use words such as "same", "shorter", "longer".

d. Label one of the Erlenmeyer flasks "control".

e. Use the graduated cylinders to measure the volumes of X and water for trial 1. The volumes are recorded in Table 54. Pour the volumes of X and water into the Erlenmeyer flask. Swirl the contents of the flask.

f. Use the 10 mL graduated cylinder to transfer 5 mL of the pink solution to the Erlenmeyer flask. Start timing as soon as you begin to pour the pink solution into the flask. Swirl the contents of the flask. Stop timing when the pink colour disappears. Record the time taken for the pink colour to disappear in Table 54.

g. Do *not* pour out the contents of the first trial. Set the "control" flask aside; you use it again in trial 5.

h. Repeat steps e and f for trials 2, 3, and 4. Rinse the flask well between trials. In trial 4 warm the mixture of water and X with a Bunsen flame. Use the adjustable clamp to hold the Erlenmeyer flask in the flame (Fig. 8-2). *Warm*, do *not* boil the liquid. Add the 5 mL of pink solution to the warm mixture.

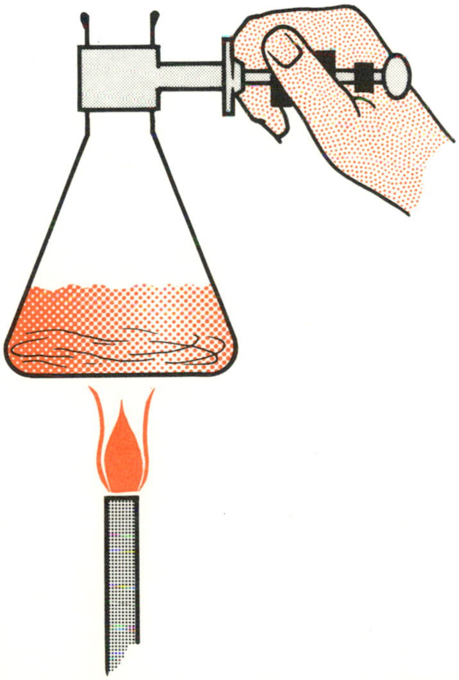

Figure 8-2 Rates of Reaction A.

i. Before doing trial 5, answer these questions. The contents of the "control" flask have already decolourized 5 mL of the pink solution. As a result, what has happened to the concentration of X? Explain your reasoning. Predict the time required to decolourize a second 5 mL of pink solution. Record your prediction in Table 54.

j. Add 5 mL of the pink solution to the flask labelled "control". Record the time taken for the pink colour to disappear.

Procedure B

a. Copy Table 55 into your notebook.

TABLE 55 Rates of Reaction B

	Description of Trial							
Trial	**Variable changed**	**Y_1 (mL)**	**Y_2 (mL)**	**Water (mL)**	**Temperature**	**Calcium carbonate**	**Predicted speed**	**Actual speed**
1	control	10			room	chips	slow	
2		5		5	room	chips		
3			10		room	chips		
4		10			room	powder		
5		10			above room	chips		

b. Compare trials 2, 3, 4, and 5 with 1. How are they the same? How are they different? Record in Table 55 the variable changed in trials 2, 3, 4, and 5.

c. Assume that the rate of the control is slow. Predict the rate of reaction in the other trials relative to the control. Record your predictions in Table 55. Use the words "same", "slower", "faster".

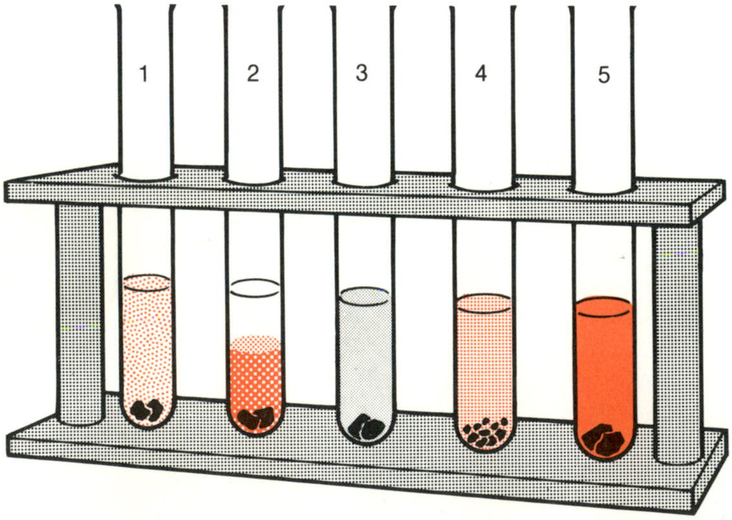

Figure 8-3 Rates of Reaction B.

d. Number 5 test tubes.
e. Use the 10 mL graduated cylinder to measure the volumes of Y and water. The volumes are recorded in Table 55. Add the liquids to the correct test tubes.
f. In trial 5 warm the liquid with a Bunsen flame. Use the adjustable clamp to hold the test tube in the flame. *Warm*, do *not* boil the liquid.
g. Set the test tubes in the rack as shown in Figure 8-3.
h. Add the calcium carbonate to all test tubes at the same time. Two chips of calcium carbonate are added to test tubes 1, 2, 3, and 5. An equal amount of powdered calcium carbonate is added to test tube 4.
i. Record the rate of the reactions in test tubes 2, 3, 4, and 5 relative to 1. Use the words "slow", "fast", "slower", "faster".

Discussion A

1. Compare the times for trials 2, 3, and 4 with the time for trial 1. What changes increased the rate of Reaction A? What changes decreased the rate of Reaction A?
2. List the variables that changed the rate of Reaction A.
3. Describe what happens to the rate of Reaction A when the value of the variable is increased.
4. Use the collision model to describe how the rates of Reaction A were changed.
5. Why was it necessary to keep the total volume of liquid the same in each trial?
6. Why was the volume of the pink solution the same in each trial?
7. For trial 5 compare your predicted and actual times. Compare the time for trial 5 with the time for trial 1. Why is this result surprising?
8. Suggest an explanation for the surprising results. Design an experiment to test your explanation. Check with your teacher before carrying out the experiment.

Discussion B

1. Compare the rates of trials 2, 3, 4, and 5 with trial 1. What changes increased the rate of Reaction B? What changes decreased the rate of Reaction B?
2. List the variables that changed the rate of Reaction B.
3. Describe what happens to the rate of Reaction B when the value of the variable is increased.
4. Use the collision model to describe how the rates of Reaction B were changed.
5. Explain why it was necessary to keep the amount of calcium carbonate the same in each trial.

General Discussion

1. Describe how these variables were changed in this investigation: concentration; temperature; nature of reactants; amount of surface exposed.
2. A **homogeneous chemical reaction** takes place in only one phase. Reaction A is a homogeneous chemical reaction. Reaction B is a heterogeneous chemical reaction. Define **heterogeneous chemical reaction**.
3. List 3 variables that change the rates of homogeneous and heterogeneous chemical reactions. What is the fourth variable that changes the rate of a heterogeneous chemical reaction?
4. Describe on a molecular scale how each variable changes the rate of a chemical reaction.
5. Explain how the results of this investigation increase our confidence in Dalton's atomic model and the basic assumption of the collision model.
6. Explain the following observations:
   a) A bicycle chain coated with grease rusts more slowly than one that is not coated.
   b) A lump of coal burns slowly in air. Coal dust burns explosively.
   c) Milk and other foods remain fresh longer in a refrigerator.
   d) Fanning a fire causes glowing coals to burst into flames.
   e) Foods cook more rapidly in a pressure cooker.
   f) It requires longer to hard boil an egg than to fry one.
7. Instructions for using a film developer are:
   a) Dilute 1 part of the developer with 3 parts water.
   b) Use at 20°C.
   c) Development time is 14 min for 125 ASA film.
   Predict the effect on the negatives if you forgot to dilute the developer; if the temperature of the developer was 35°C.

8.4 INVESTIGATION: Flame Tests

An electric current passing through a light bulb heats the tungsten filament. The heat causes the filament to glow and it appears white hot. When an electric current is passed through a glass tube containing neon, a red glow is produced. A wide variety of coloured neon lights is produced by adding small amounts of other elements to the neon and by tinting the glass tube. Table 56 shows that different gases produce different colours. Is the colour produced a characteristic property of the gas? Do other elements produce a characteristic colour? Does the colour of the light change when the element combines with other elements?

In this investigation you heat elements and compounds in a Bunsen flame until they glow or give off light. The light produced colours the Bunsen flame. This procedure is called a **flame test**. Before you perform the flame tests use Dalton's atomic model to make the following predictions:

a) Is the colour of light produced when an element is heated a characteristic property of the element?

b) Will the colour of light produced change when the element is part of a compound?

Record your predictions in your notebook.

TABLE 56 Colours of Neon Lights

Gas in tube	Colour of glass tube	Colour of light emitted
Neon	colourless	red
Neon	soft red	deep red
Helium	colourless	white
Helium	amber	golden yellow
Argon, neon and mercury	colourless	blue
Argon, neon and mercury	purple	dark blue
Argon, neon and mercury	green	light green
Argon, neon and mercury	amber	dark green

Materials

elements: copper wire, calcium turnings

compounds: copper (II) nitrate, copper (II) chloride, calcium acetate, calcium oxide, potassium carbonate, potassium nitrate, strontium chloride, strontium nitrate, sodium chloride, sodium hydrogen carbonate, barium nitrate, lithium chloride, unknown potassium or barium compound

dilute hydrochloric acid

nichrome wire and handle

Bunsen burner, striker, asbestos pad, tongs

CAUTION: Wear safety goggles during this investigation.

Procedure

a. Copy Table 57 into your notebook.

b. Light the Bunsen burner. Turn the gas on full. Open the air inlet until the flame appears to consist of two cones. The tip of the inner blue cone is the hottest point in the flame.

c. *Cleaning the wire.* This must be done after each test. Heat the wire in the hottest part of the Bunsen flame. The wire is clean when it glows white hot but no longer colours the flame.

TABLE 57 Flame Colours

Trial	Pure substance	Flame colour	Trial	Pure substance	Flame colour
1	Copper		8	Potassium nitrate	
2	Copper (II) nitrate		9	Strontium chloride	
3	Copper (II) chloride		10	Strontium nitrate	
4	Calcium		11	Sodium chloride	
5	Calcium acetate		12	Sodium hydrogen carbonate	
6	Calcium oxide		13	Barium nitrate	
7	Potassium carbonate		14	Lithium chloride	

d. *Flame tests for compounds*. Dip the hot, clean wire into dilute hydrochloric acid. Then dip the wire into the solid you wish to test. The moistened loop in the wire will pick up a small amount of the solid. Hold the loop of the wire in the hottest part of the Bunsen flame (Fig. 8-4). Record the colour of the flame in Table 57. Repeat for all compounds provided.

e. *Flame tests for elements*. Use the tongs to hold the element in the hottest part of the Bunsen flame. Record the colour of the flame in Table 57. Repeat for all elements provided.

f. Your teacher will give you an unknown substance. The unknown is a potassium or barium compound. Identify the metallic element of the compound.

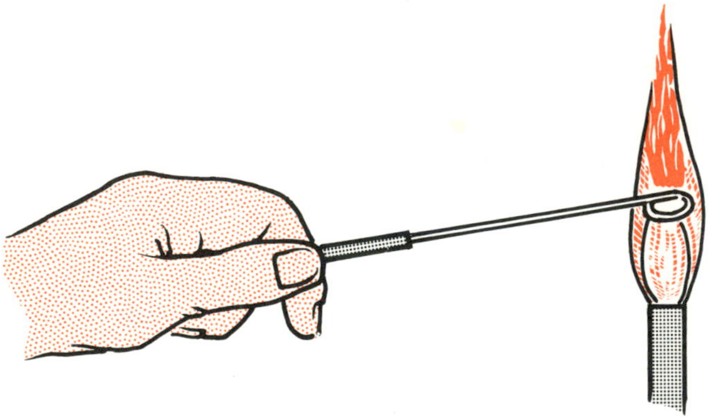

Figure 8-4 Flame tests.

Discussion

1. What do the pure substances of trials 1 to 3 have in common? Compare the flame colours of trials 1 to 3. How are they the same? How are they different?

2. What do the pure substances of trials 4 to 6 have in common? Compare the flame colours of trials 4 to 6. How are they the same? How are they different?

3. Explain the pattern of flame colours for trials 1 to 6.

4. What evidence suggests that the flame colours are produced by the first or metallic element of a compound?

5. What is a positive flame test for each of copper, calcium, potassium, strontium, sodium, barium, and lithium?
6. What is the definition of "characteristic property"? What evidence suggests that flame colour is a characteristic property?
7. What metallic element is present in the unknown substance?
8. Both the element copper and the compound copper (II) nitrate colour the flame emerald green. Which of Dalton's ideas about atoms explains this observation?
9. Rolled newspaper and pine cones were soaked in a solution and then allowed to dry. When the newspaper logs and pine cones were burned in a fireplace, the flame was tinged with green and crimson. What elements had been present in the solution?

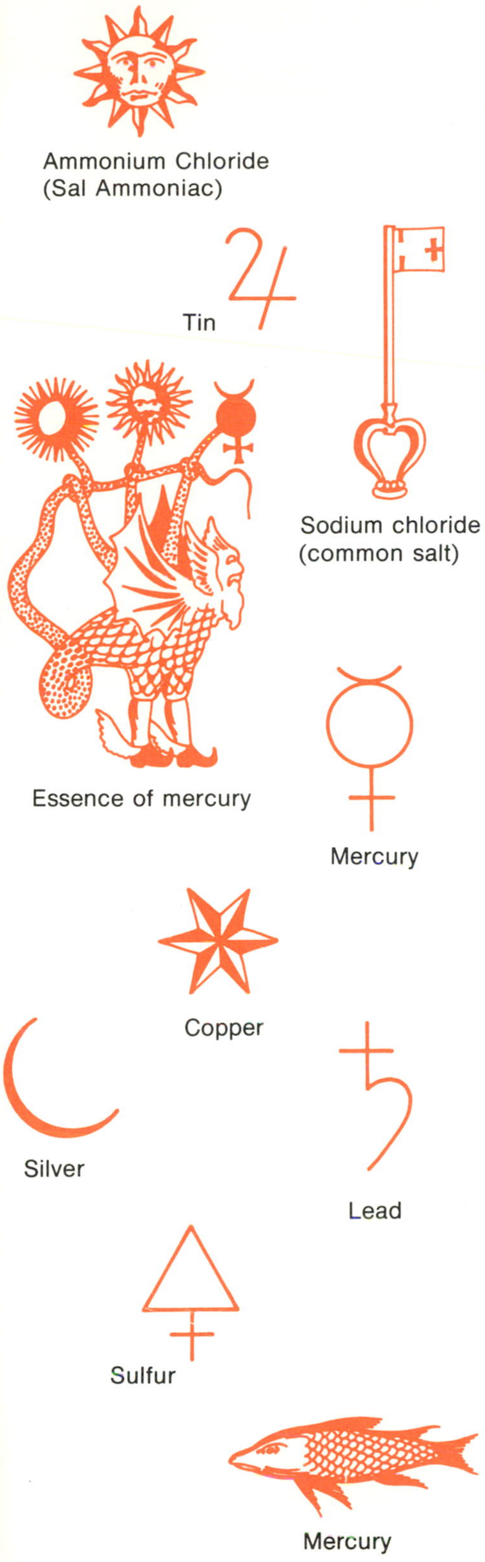

Figure 8-5 Alchemy's symbols.

8.5 Symbols and Formulas

History of Symbols

In the sixteenth and early seventeenth centuries the men who investigated matter were known as alchemists. Their chief aim was to turn cheap metals into gold. They used symbols and signs to prevent their competitors from stealing their secrets. Figure 8-5 shows some symbols used by the alchemists. In some cases the symbols illustrated a property of the substance. In other cases the symbol represented some demon or god. More than 3 000 symbols are recorded in the history of alchemy.

Modern Symbols

Dalton was the first chemist to use symbols to represent the atoms of the elements. You used his symbols in Unit 7. Because they were difficult to draw and to print in textbooks they were never widely used. Berzelius, in 1813, simplified Dalton's symbols. He used letters to represent the atoms of each element. Berzelius suggested that the first letter of the name of the element would make a suitable symbol. Because there were more than 26 elements known, a problem arose immediately, Carbon, cobalt, calcium, chlorine and copper could not all have the same symbol. To overcome this difficulty C is used for carbon, Co for cobalt, Ca for calcium, Cl for chlorine, and Cu for copper (Latin, cuprum). *Thus a* **chemical symbol** *is a capital letter, or a capital letter followed by a small letter.* In most cases the letters are the first or the first and second letters of the name of the element. When this is not so, the symbol has come from the name of the element in another language such as Latin. Remember that a chemical symbol represents *ONE* atom of the element. Thus Cl represents one atom of the element chlorine, Ag one atom of silver, and C one atom of carbon. Table 58 shows the symbols of some common elements.

TABLE 58 Symbols of Common Elements

Element	Symbol	Element	Symbol
Aluminum	Al	Lead (plumbum)	Pb
Antimony (stibium)	Sb	Magnesium	Mg
Arsenic	As	Manganese	Mn
Barium	Ba	Mercury (hydrargyrum)	Hg
Bromine	Br	Nitrogen	N
Calcium	Ca	Oxygen	O
Carbon	C	Phosphorus	P
Chlorine	Cl	Potassium (kalium)	K
Chromium	Cr	Silicon	Si
Copper	Cu	Silver (argentum)	Ag
Fluorine	F	Sodium (natrium)	Na
Hydrogen	H	Sulfur	S
Iodine	I	Tin (stannum)	Sn
Iron (ferrum)	Fe	Zinc	Zn

Formulas of Molecules

The symbols of the elements are used to show the composition of a molecule. The method used is exactly the same as the one we have used to represent nut and bolt molecules. (See Table 59.)

TABLE 59 Formulas of Molecules

Name of compound	Formula	Corresponding bolt-nut molecule
Sodium chloride (common salt)	NaCl 1 atom sodium 1 atom chlorine	BN 1 bolt 1 nut
Sulfur (IV) oxide (sulfur dioxide)	SO_2 1 atom sulfur 2 atoms oxygen	BN_2 1 bolt 2 nuts
Nitrogen (III) hydride (ammonia)	NH_3 1 atom nitrogen 3 atoms hydrogen	BN_3 1 bolt 3 nuts
Hydrogen peroxide	H_2O_2 2 atoms hydrogen 2 atoms oxygen	B_2N_2 2 bolts 2 nuts

A **chemical formula** *is made up of symbols placed side by side.* Each symbol represents one atom of the element. A formula represents a molecule of a compound. A formula shows the elements and the number of atoms of each element in the molecule of a compound. A numeral is placed at the lower right hand corner of the symbol to show the number of atoms of the element. If no numeral is placed there, then it is understood that there is only one atom of that element in the molecule. Thus H_2O shows that 2 atoms of hydrogen are joined with 1 atom of oxygen to form 1 molecule of water.

Both compounds and elements have formulas. The formulas of compounds have two or more different symbols. The formulas of elements have only one symbol. Table 60 shows the formulas of some common compounds and elements.

TABLE 60 Formulas of Some Elements and Compounds

Name	Formula of molecule	Composition
Sodium	Na_2	2 atoms of sodium
Platinum	Pt	1 atom of platinum
Sulfur	S_8	8 atoms of sulfur
Phosphorus	P_4	4 atoms of phosphorus
Iron (III) oxide (rust)	Fe_2O_3	2 atoms of iron 3 atoms of oxygen
Sodium hydroxide (lye)	NaOH	1 atom of sodium 1 atom of oxygen 1 atom of hydrogen

Discussion

1. How are Dalton's and the alchemists' symbols the same? How are they different?
2. Why do you think Berzelius's symbols were eventually accepted by international chemical societies and are now used throughout the world?
3. **a)** Define "chemical symbol".
   b) What does a chemical symbol represent?
4. What do the following symbols represent?
   Al, As, Cu, H, Pb, N, O, Sn, Zn
5. Write the chemical symbols for the following elements: antimony, barium, carbon, fluorine, iron, magnesium, mercury, potassium, silicon, sodium, sulfur.
6. **a)** Define "chemical formula".
   b) What does a chemical formula represent?
7. What is the composition of each of the following molecules? H_2 (hydrogen gas), C_2H_2 (acetylene gas), O_3 (ozone), Na_2CO_3 (washing soda), $(NH_4)_2Cr_2O_7$(ammonium dichromate).

8. The compounds in Table 61 are found in common household products. Copy the table into your notebook. List the elements in the composition column. With your parents' help, identify household products that contain the compound. Some have been done for you.

8.6 INVESTIGATION: Can Dalton's Atomic Model Explain These Observations?

This investigation shows 4 phenomena that cannot be explained by Dalton's atomic model. Should the model be discarded? No, not entirely. But it will need to be modified. You have used Dalton's idea that matter is made up of small particles called atoms to explain successfully both chemical and physical properties. It would be foolish to reject this idea. However, Dalton's description of atoms will have to be changed.

The investigation has 4 stations. Your teacher will assign you to a group of 5 or 6 students. Your group will visit each station in the order assigned by your teacher.

STATION A Static Electricity

You have probably seen sparks jump between your fingers and a metal object, especially during the winter months. A flash of light marks the path of electricity travelling between the finger and the metal object. The "zap" you hear is produced by the rapid expansion of the air. The air expands because the electricity passing through it heats it. Where does the electricity come from? Does all matter contain electrical charges? Can Dalton's atomic model explain electricity?

At this station you build up an electrical charge on various rods and strips by rubbing them with fur and cloth. To test for the electrical charge, you bring the rod or strip close to very light objects. If the objects are attracted to the rod or strip, you can conclude that there is an electrical charge. A more detailed study of electric force is found in Unit 10.

Materials

ebonite rod and dry cat's fur
acetate strip and dry cotton cloth
vinylite strip and dry wool cloth
glass rod and dry silk cloth
light objects such as flowers of sulfur, talc, sawdust, small scraps of paper, and small pieces of acetate

TABLE 61 Composition of Household Compounds

Name of compound	Formula	Composition	Household Product
Ammonia	NH_3		
Sodium hypochlorite	$NaClO$		
Phenol	C_6H_6O		acne preparations
Aluminum chlorhydroxide	$Al(OH)Cl_2$		
Talc	$MgSiO_3$		baby powder
Sodium bisulfate	$NaHSO_4$		
Toluene	C_7H_8		plastic wood
Borax	$Na_2B_4O_7 \cdot 10H_20$		
Isobutane	C_4H_{10}		
Trichlorofluoromethane	CCl_3F		propellant in spray cans
Monocalcium phosphate	$CaHP0_4$		
Sodium bicarbonate	$NaHCO_3$		
Isopropyl alcohol	C_3H_8O		rubbing alcohol
Acetic acid	$C_2H_4O_2$		vinegar
Sodium tripolyphosphate	$Na_5P_3O_{10}$		
Tricalcium phosphate	$Ca_3(PO_4)_2$		jello
Sodium metasilicate	Na_2SiO_3		
Glycerin	$C_3H_8O_3$		
Phenolphthalein	$C_{20}H_{14}O_4$		laxative
Acetylsalicylic acid	$C_9H_8O_4$		
Caffeine	$C_6H_9O_2N_4$		
Nicotine	$C_{10}H_{13}N_2$		
Ascorbic acid (vitamin C)	$C_6H_8O_6$		
Orlon	$(-C_3H_3N-)_n$		
Teflon	$(-C_2F_4-)_n$		coating for pans
Monosodium glutamate	$C_5H_8NO_4Na$		
Sodium ascorbate	$C_6H_7O_6Na$		
Potassium sorbate	$C_6H_7O_2K$		
Stannous fluoride	SnF_4		

Procedure

a. Make a table with two headings: Procedure and Results.
b. Touch the gas or water tap with the ebonite rod.
c. Bring the rod near the light objects. Record the procedure and results in the table.
d. Repeat steps b and c for the glass rod and the acetate and vinylite strips.
e. Rub the ebonite rod with the cat's fur. Repeat step c.
f. Rub the acetate strip with the cotton cloth. Repeat step c.
g. Rub the vinylite strip with the wool cloth. Repeat step c.

h. Rub the glass rod with the silk cloth. Repeat step c.
i. If you have a comb, bring it near the objects. Run the comb through your hair and then bring it near the light objects. Record the procedure and results in the table.

Discussion

1. What had to be done to the ebonite, acetate, vinylite, and glass before they would attract the objects?
2. What observations suggest that something is exchanged between the cat's fur and the ebonite rod when they are rubbed together? Do the same observations apply to the other materials that were rubbed together?
3. *Hypothesis: Atoms are exchanged when the ebonite rod is rubbed with the cat's fur.* If atoms were exchanged what would eventually happen to the rod and the cat's fur? Do you accept this hypothesis? Why?
4. What observations suggest that whatever is exchanged is common to all matter?
5. *Hypothesis: Something smaller than an atom (part of an atom) is exchanged when the ebonite rod is rubbed with the cat's fur.* Do you accept this hypothesis? Why?
6. Why is a spark sometimes produced when you walk across a rug and touch a door knob or other metallic object?
7. Which of Dalton's ideas about atoms should be modified? Why?

STATION B Conductivity

All matter conducts electricity to some extent. But some substances are better conductors than others. At this station you classify substances as good or poor conductors of electricity. Can Dalton's atomic model explain how matter conducts electricity?

Materials

2 conductivity apparati (2-hole rubber stopper, 2 carbon rods, connecting wire, flashlight bulb, d.c. power supply)
ring stand, iron ring, clay triangle, adjustable clamp
evaporating dish
Bunsen burner, striker, asbestos pad
elements: copper, iron, carbon, sulfur
compounds: sodium chloride, zinc chloride, acetic acid, distilled water, tap water, sugar
solutions: sodium chloride, zinc chloride, acetic acid, sugar
CAUTION: Wear safety goggles during this investigation.

Procedure

a. Copy Table 62 into your notebook.

TABLE 62 Conductivity of Matter

Substance	Glow of light bulb			Other observations
	Bright	Dim	None	
Copper (s)	x			
Iron (s)				
Carbon (s)				
Sulfur (s)				
Sodium chloride (s)				
Zinc chloride (s)				
Sugar (s)				
Distilled Water (l)				
Tap water (l)				
Acetic acid (l)				
Sodium chloride (aq)				
Zinc chloride (aq)				
Sugar (aq)				
Acetic acid (aq)				
Zinc chloride (l)				

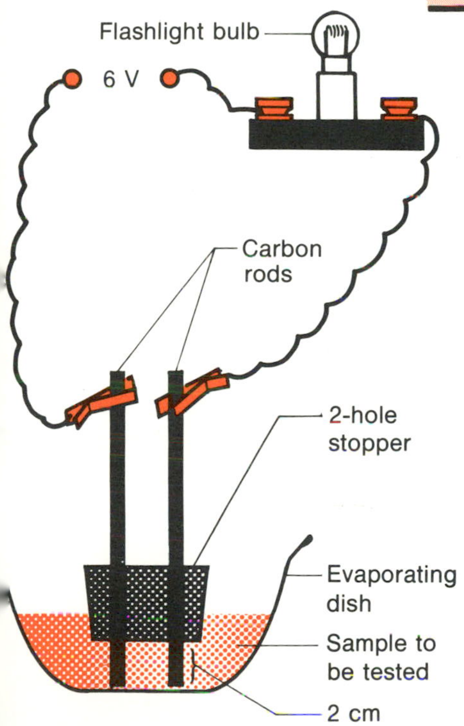

Figure 8-6 Conductivity apparatus.

b. Set up the conductivity apparatus as shown in Figure 8-6. Test the apparatus by placing a piece of copper connecting wire across the two carbon rods. If the bulb does not glow brightly ask your teacher for help.

c. Arrange the substances in the same order as in Table 62. Test the solids first.

d. Test each substance by placing the short ends of the carbon rods into the substance. Record the glow of the light bulb and any changes that occur near the carbon rods in Table 62.

Note:

1. The carbon rods must be dry to test solids.
2. To test a powdered solid, there should be no gaps in the solid between the two rods.
3. After testing a liquid or solution, rinse the carbon rods with tap water and dry them with a paper towel. **CAUTION:** Disconnect the leads to the power supply before drying the rods.

e. Half-fill an evaporating dish with zinc chloride. Set up the apparatus as shown in Figure 8-7. **CAUTION:** Adequate ventilation is required.

f. Heat the evaporating dish strongly. Record a description of the contents of the evaporating dish when the light glows. Turn off the heat as soon as the light glows.

Discussion

1. How is light produced in a light bulb?
2. In order for the light bulb to glow what must pass between the two carbon rods?
3. Which substances are good conductors of electricity? Which are poor conductors of electricity?
4. Account for the difference in the conductivity of tap water and distilled water.
5. Compare the conductivity of an aqueous solution of acetic acid with that of pure acetic acid. What evidence suggests that acetic acid is made up of electrical charges? What does the water enable the electrical charges to do?
6. Electrical current is the movement of electrical charge. The d.c. source adds electrical charges to one end of the wire and takes electrical charges from the other end of the wire. Thus, to be a conductor, a substance must be made up of electrical charges and the charges must be able to move. Zinc chloride is a good conductor when it is molten or when it is dissolved in water. But solid zinc chloride does not conduct. Why is this so?
7. What evidence suggests that sodium chloride is made up of electrical charges? Predict the conductivity of molten sodium chloride.
8. What substances do not conduct? Suggest two possible explanations. One explanation should support the idea that matter is made up of electrical charges; the other will not. Which explanation is the better? Why?
9. Which of Dalton's ideas about atoms should be modified? Why?

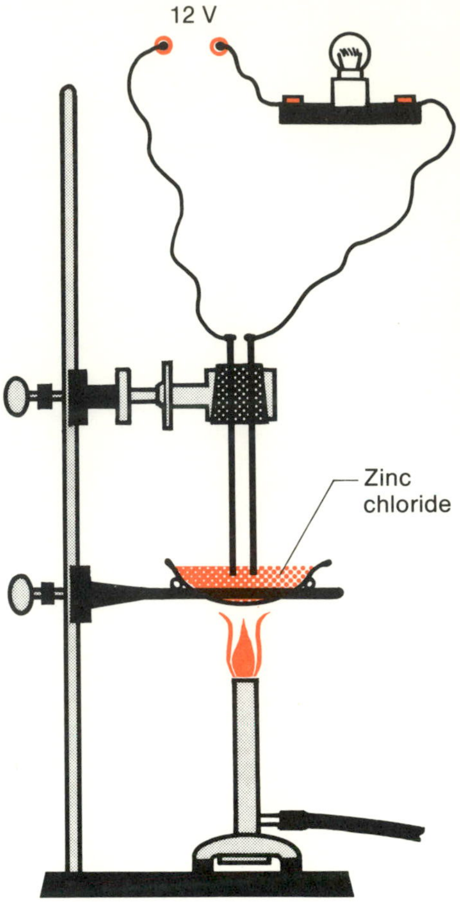

Figure 8-7 Conductivity of molten zinc chloride.

STATION C Chemical Reactions and Electrical Charges

When cells are placed in a flashlight and the switch turned on, the bulb glows. For this to happen, electrical charges must flow through the filament of the bulb. Figure 8-8 shows the main parts of a flashlight cell. When the switch is turned on, a chemical reaction occurs between the zinc case and the black mixture. The zinc case becomes thinner and electrical charges move out from the base of the cell through the circuit. The charges move through the bulb causing it to glow. They return through the cap to the carbon rod. Then another chemical reaction occurs with the black mixture. Thus chemical reactions can produce electrical charges. Can Dalton's atomic model explain the production of electrical charges by a chemical reaction?

This part of the investigation uses the cell shown in Figure 8-9 to light a flashbulb. The advantage of this cell is that you can observe the appearance and disappearance of the products and reactants.

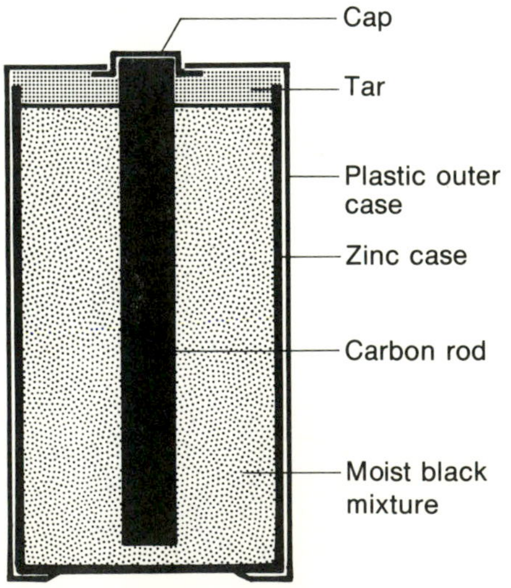

Figure 8-8 A flashlight cell.

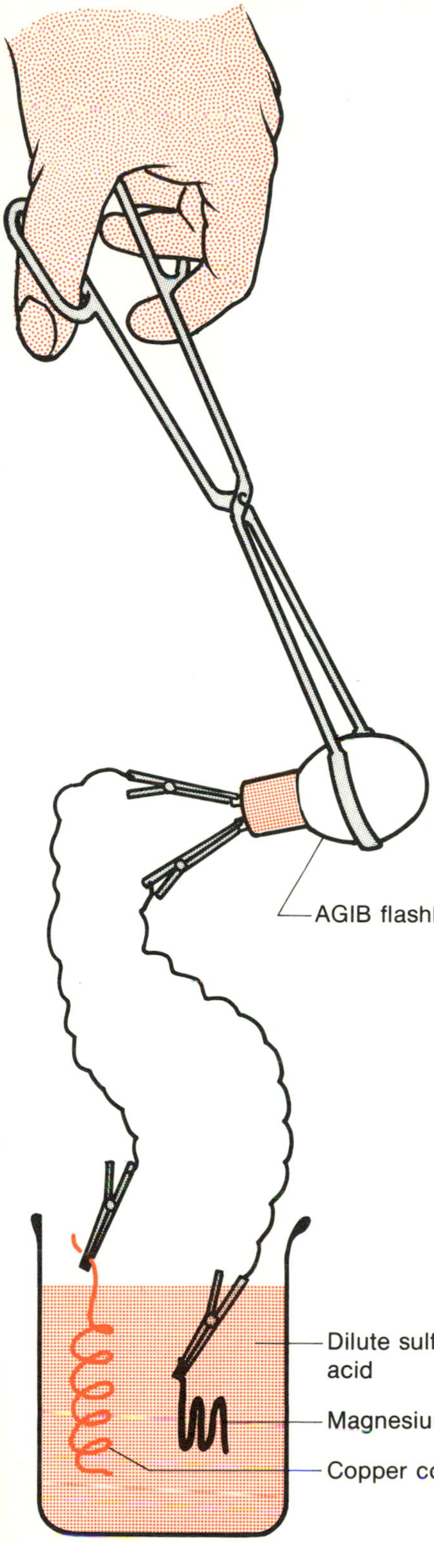

Figure 8-9 Generating an electric current.

Materials

flashbulb (AGIB type)
3 test tubes and rack
250 mL beaker
dilute sulfuric acid solution
connecting wires with alligator clips
scissors
coil of copper wire
magnesium ribbon

CAUTION: Wear safety goggles during this investigation.

Procedure

a. Cut 3 pieces of magnesium ribbon 10 cm long. Fold each piece into a bunch 2 cm long.
b. Wrap one of the bunches of magnesium with copper wire. Leave 1 cm of wire sticking out at right angles to the bunch of magnesium.
c. Half-fill 3 test tubes with water. Add 5 mL of acid to each of the test tubes. Label the test tubes 1, 2, and 3.
d. To test tube 1 add one bunch of magnesium. Record any changes that occur.
e. To test tube 2 add a short piece of copper wire. Record any changes that occur.
f. To test tube 3 add the bunch of magnesium wrapped with copper wire. Record any changes that occur. Watch closely the end of the copper wire sticking out from the bunch of magnesium.
g. Use the connecting wires to join a copper coil and a bunch of magnesium to the flashbulb (Fig. 8-9).
h. Add 200 mL of dilute sulfuric acid to the beaker.
CAUTION: Avoid contact of acid with eyes, skin, and clothing. If an accident happens, rinse thoroughly with water.
i. Place the copper coil and magnesium bunch in the sulfuric acid. The magnesium and copper must not touch. Record what happens to the flashbulb. **CAUTION:** Do not look directly at the flashbulb when the copper coil and magnesium are placed in the acid. Record any changes that occur near the magnesium and copper.

Discussion

1. What observations suggest that a chemical reaction occurs when magnesium is placed in dilute sulfuric acid?
2. What evidence suggests that copper does not react with dilute sulfuric acid?
3. Compare the reactions in test tubes 1 and 3. How are they the same? How are they different?
4. Magnesium reacts with sulfuric acid to produce magnesium sulfate and hydrogen. Write a word equation for this reaction.
5. To produce hydrogen gas, electrical charges must be exchanged between the acid and the magnesium. What

observations suggest that electrical charges are carried by the copper wire wrapped around the magnesium to the sulfuric acid?

6. A flashbulb has a magnesium filament. A small electric current passing through the filament is enough to ignite the magnesium and produce the brilliant flash. Explain how the flashbulb was flashed when the copper coil and magnesium were placed in the dilute acid.

7. Which of Dalton's ideas about atoms should be modified? Why?

STATION D Radioactivity

Uranium compounds, when left sitting on a photographic film enclosed in black paper, expose the film. When the film is developed, the image of the crystals of the compound can be seen on the negative. Elements that affect a film in this way are said to be radioactive. The discovery of radioactivity was made by Henri Becquerel in 1896, 93 years after Dalton proposed his atomic model.

Scientists who were studying the radioactive element radium noticed that it changed into two other elements. One had properties different from any gas known at that time. The other element was helium. The unknown gas was called radon because it came from radium. Can Dalton's atomic model explain how atoms of one element can change into atoms of two other elements?

In this investigation you use a cloud chamber to make visible the radiation produced by radioactive elements. Small particles accompany any change from an atom of one element to an atom of another element. These small particles escape from the solid element with considerable energy. Thus they can penetrate materials like black paper and glass. As the small particles pass through the alcohol vapour in the cloud chamber they cause the alcohol to condense, forming a trail of alcohol droplets. When the droplets are illuminated from the side you see a white track in the chamber.

Materials

cloud chamber	methyl alcohol
radioactive source	dry ice and styrofoam sheet
light source (slide projector or flashlight)	

Procedure

a. Wet the blotting paper with methyl alcohol. Put the top back on the chamber (Fig. 8-10).

b. Insert the radioactive source in the side of the cloud chamber.

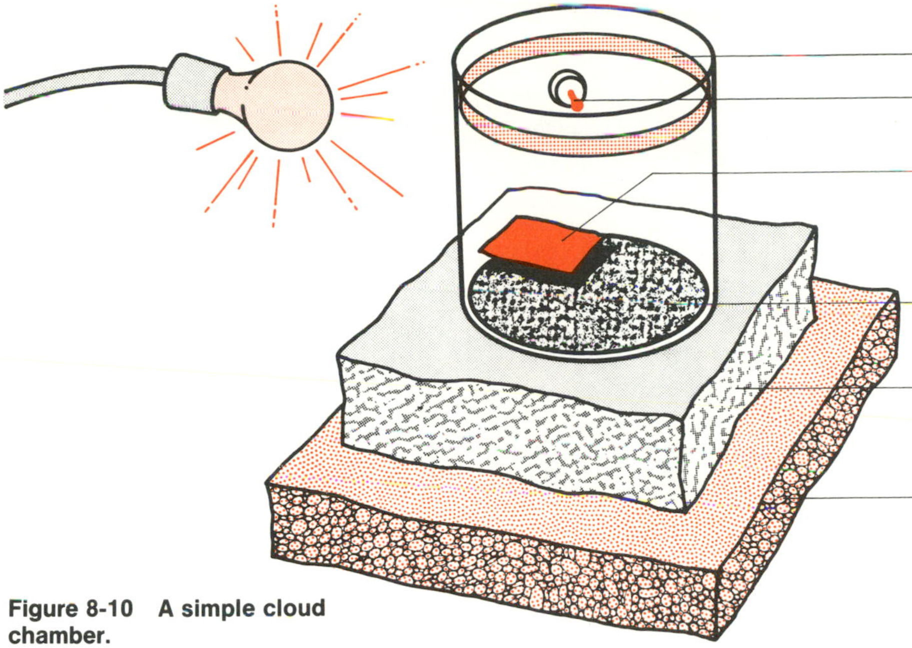

Figure 8-10 A simple cloud chamber.

c. Place the cloud chamber on the block of dry ice.

d. Use the light source to illuminate the cloud chamber from the side. Do not be disappointed if the tracks are not visible immediately. It may take 5 to 10 min for the cloud chamber to become cold enough to supersaturate the alcohol vapour and allow the cloud tracks to form. Draw a sketch in your notebook to illustrate the tracks formed.

e. If the cloud chamber has movable barriers, erect a barrier. In your notebook describe any changes in the cloud tracks.

Discussion

1. Define a radioactive substance.
2. What observations suggest that light and the radiation from the radioactive source are different.
3. What observations suggest that the radiation from a radioactive element is in the form of small particles?
4. Explain how the tracks are formed. Why is the chamber called a cloud chamber?
5. Which of Dalton's ideas about atoms should be modified? Why?

Highlights

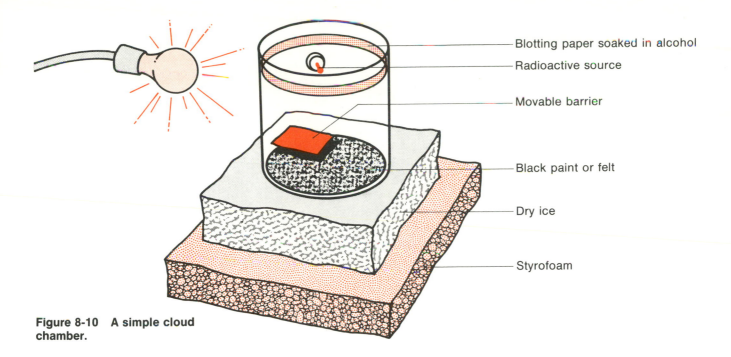

Dalton described atoms as indestructible. He also said that all atoms of an element had the same mass. These two ideas enabled him to predict the Law of Multiple Proportions. Observations supported his prediction.

When atoms form molecules, the atoms are not destroyed. Thus some of the properties of the atoms are not changed. Flame colour is such a property.

The rate of reaction is determined by the frequency of atoms colliding. The factors that alter the rate of a reaction are concentration, temperature, surface area, and the nature of the reactants. When concentration is increased or a solid is crushed, there is a greater chance of atoms colliding and the rate of the reaction increases. When the temperature is lowered, the number and strength of the collisions decrease, and the rate slows down.

Dalton's atomic model enables us to explain the Law of Conservation of Mass, the Law of Constant Composition, the Law of Multiple Proportions, the rates of chemical reactions, and, in part, flame colours. However, it will not explain static electricity, conductivity of metals, production of electricity by a chemical reaction, and radioactivity. Dalton's atomic model must be modified to explain these observations.

UNIT

9 Air and Its Components

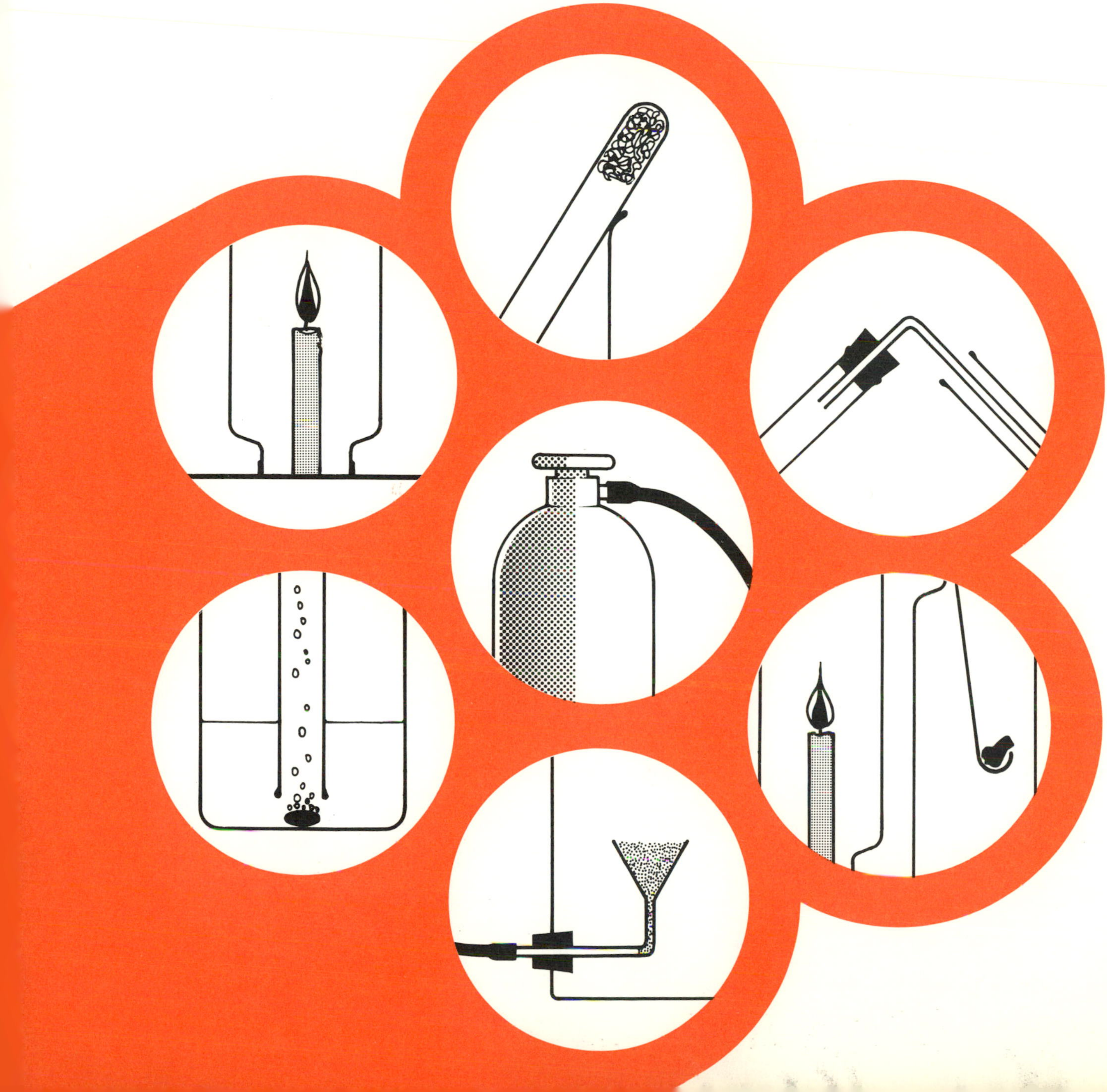

In this unit you study the atmosphere that surrounds the earth. You also consider the properties of some of the gases that make up the atmosphere. Special attention is given to oxygen, since it is very important to us in many ways. Finally you examine the nature and effects of some of the pollutants that are found in the air.

9.1 Structure of the Atmosphere

You may find it difficult to think of air as a form of matter because air is invisible and has no definite volume or shape. Yet about six thousand billion tonnes of air surround the earth. This air gets thinner and thinner (less dense) as the altitude increases. There is no definite outer edge to the atmosphere. It goes on and on, eventually becoming part of outer space. However, after a point it becomes so thin that it makes little sense to call it the atmosphere. As a result, scientists usually say that the upper edge of the atmosphere is about 1 500 km above the earth.

About 95% of all life is supported by the first 3 km of the atmosphere. At a height of 10 km, a person cannot get enough oxygen to live. At 20 km there is not enough oxygen to keep a candle burning.

Layers of the Atmosphere

The atmosphere (or air) is a mixture of several gases. Close to the earth this mixture is almost homogeneous. You would not be very wrong if you called that part of the atmosphere a solution. As the altitude increases, the composition of the air changes. For example, near the ground the most abundant gas is nitrogen; 1 000 km up, it is hydrogen. Since the composition changes as the altitude increases, scientists find it convenient to divide the atmosphere into layers (Fig. 9-1).

The **troposphere** is the lowest layer. It contains close to 90% of the mass of the atmosphere. It has an average thickness of about 15 km; it is 7 – 12 km thick at the poles and 16 – 17 km thick at the equator. Since the troposphere touches the earth, most of our weather is caused by disturbances in this layer. For every 150 m rise in altitude, the temperature drops 1 °C. At the top of this layer the temperature averages – 55 °C.

The next layer is the **stratosphere.** It extends from the troposphere to an altitude of about 40 km. It contains about 9% of the mass of the atmosphere. It is slightly warmer than the upper part of the troposphere. Jet aircraft normally fly in this layer since the very thin air offers less resistance to motion. Winds often blow at speeds of hundreds of kilometres per hour in the stratosphere.

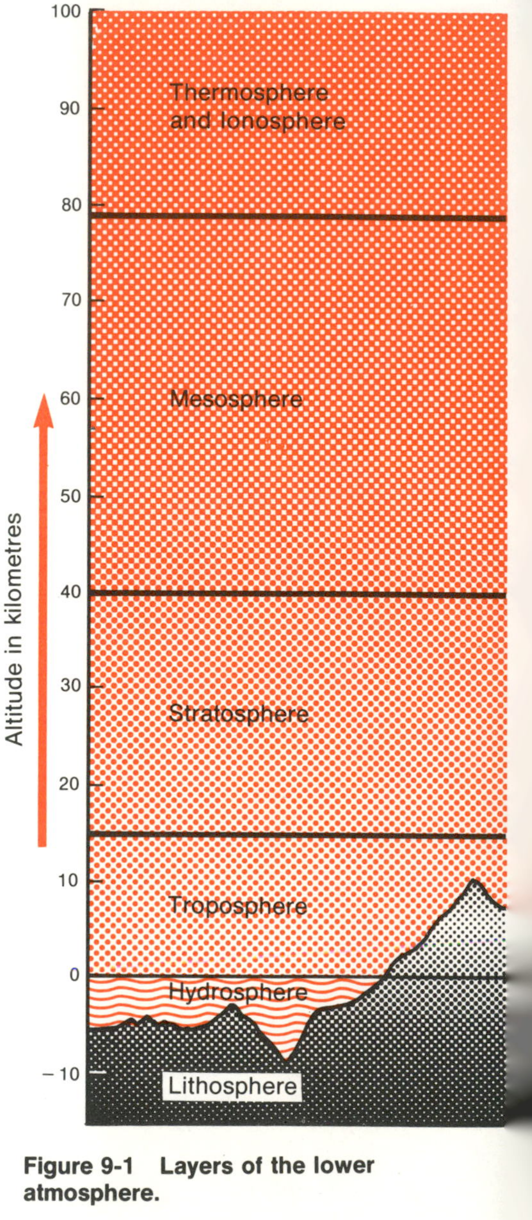

Figure 9-1 Layers of the lower atmosphere.

They may even move in directions different from those in the troposphere below. Jet aircraft often save fuel and time by flying with these winds. Within the stratosphere is the **ozone layer** which is almost 6 – 7 km thick. The molecules of ozone gas absorb a type of ultraviolet radiation from the sun that can cause skin cancer.

Above the stratosphere is the **mesosphere.** This layer extends from the stratosphere to an altitude of about 80 km. It has a relatively warm temperature of −10°C near the bottom and a relatively cool temperature of −90°C near the top. It is less dense than the stratosphere.

The next layer is the **thermosphere.** It extends from 80 km to about 500 km. In this layer the temperature rises a great deal. It is about 20°C near the bottom, 100°C at an altitude of 150 km, and 1 000°C near the top. The high temperature in this layer is due to the absorption of some of the sun's energy by the molecules of air. Some of this energy causes the molecules to break up, forming electrically-charged atoms called **ions.** This process takes place mainly in the lower part of the thermosphere between 80 km and 400 km, forming a layer called the **ionosphere.** The ionosphere is a good reflector of radio waves. Broadcasting stations use this fact to send programs thousands of miles away. They beam the radio waves into the ionosphere at an angle. The waves reflect back to the earth. They continue to bounce back and forth between the ionosphere and the earth as they travel around the earth.

The final layer of the atmosphere is the **exosphere.** This layer includes the thermosphere and extends from the top of the mesosphere to an altitude of about 1 500 km. Its density is extremely low. It consists mainly of helium gas and hydrogen gas. The helium layer reaches to an altitude of about 1 000 km. The hydrogen layer continues beyond the exosphere to about 10 000 km. Occasional molecules of hydrogen gas are found as far out in space as 50 000 km.

Discussion

1. **a)** Why is air called a mixture?
b) Explain why air near the ground can be called "almost a solution".

2. **a)** What percentage of the mass of the atmosphere is in the troposphere and stratosphere together?
b) Which layer of the atmosphere has the lowest temperatures?
c) Which layer of the atmosphere has the highest temperatures? Where does the energy come from to heat this layer?

3. **a)** How is the ionosphere formed?
b) Explain how the ionosphere is used in radio communications.

4. In 1975 scientists warned us that the propellants used in aerosol cans for deodorants and hairsprays were slowly destroying the ozone layer of the atmosphere.
   a) Why should we be concerned about this?
   b) Are aerosol sprays still available for deodorants and hairsprays?
   c) Are such sprays necessary? Explain your answer.

9.2 INVESTIGATION: Some Gases in Air

You do not have the equipment or the knowledge to find out exactly what air is made of. However, this investigation will give you an idea of the nature of some of the main gases in air. It will also give you a rough idea of the relative amounts of these gases.

Materials

candle
wide-mouth jar (about 1 000 mL size)
graduated cylinder (100 mL)
limewater
glass plate
pan (5-6 cm deep)

Procedure

a. Use the candle to make a small pool of molten wax in the centre of the pan. Stick the candle in it.
b. Add water to a depth of 4-5 cm in the pan.
c. Light the candle. Quickly lower the jar over the burning candle as shown in Figure 9-2. When it appears as though nothing further will happen, record your observations.
d. Find out how much water entered the jar as follows: Tip the bottle slightly and slide the glass plate over the mouth. Keep the mouth under the water at all times. Poke the candle with the glass plate so it will break away and enter the jar. Keep the glass plate over the mouth of the jar and lift the jar from the water. Sit it upright on the bench. Use the graduated cylinder to measure the volume of water in the jar.
e. Repeat steps a to d four times. Average your results.
f. Repeat steps a to c once more. This time lift the jar and let the water run out. Quickly sit the jar upright on the bench and cover the mouth with the glass plate.
g. Test the gas in the jar for carbon dioxide as follows: Add 5-10 mL of limewater. Cover the jar with the glass plate and shake the contents vigorously. If carbon dioxide is present, the limewater will turn milky.

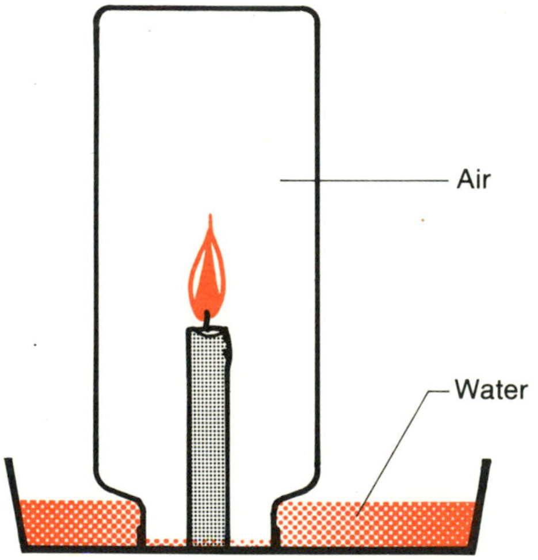

Figure 9-2 What happens when a candle burns in a jar that is inverted in water?

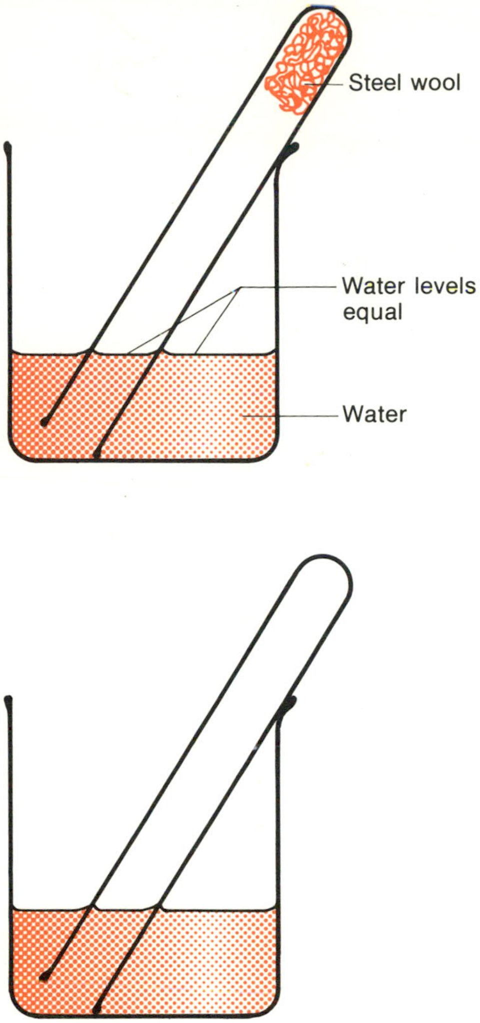

Figure 9-3 Using the rusting of iron to study the composition of air.

h. Fill the jar completely with water to get rid of all of the gases in it. Pour the water out. The jar should now be full of air only. Try the limewater test for carbon dioxide on this air.

Discussion

1. When a jar containing only air is inverted in some water, the water does not rise up into the jar. Why?
2. Describe and explain your observations for step c of the Procedure. These questions should help you: Was all of the air involved in the chemical change that occurred when the candle burned? Is air composed of more than one gas? What gas do you think was used up in the chemical reaction?
3. What does the volume of water measured in step d indicate?
4. Does air contain enough carbon dioxide to make the limewater test work?
5. Does a burning candle produce carbon dioxide?

9.3 INVESTIGATION: Some Gases in Air: Further Studies

After Investigation 9.2 you probably made a conclusion stating what gas was used up by the burning candle and how much of that gas was in the air when you started the experiment. However, the burning candle produced a gas. This could have affected your results. This investigation uses the rusting of iron to remove the same gas that the burning candle removed. No gas is formed this time. Try this investigation and compare your results to those of Investigation 9.2.

Materials

steel wool (degreased)
wooden splints
2 test tubes (20 × 150 mm)
2 beakers (250 mL)
ruler

Procedure

a. Obtain a piece of degreased steel wool from your teacher. He/she has removed the grease using gasoline so the steel wool will rust more quickly.
b. Moisten the inside of one test tube with water. Moisten the steel wool and push it to the bottom of the test tube (Fig. 9-3). Invert this test tube in a beaker that is about ¼ full of water.

c. Remove air from the test tube by shaking the test tube and add water to the beaker until the heights of the water in the test tube and beaker are equal.

d. Set up a control using the second test tube and beaker. It should be the same as the first set-up, except that it does not contain the steel wool.

e. Measure the length of the air column in each test tube. Record your results in the first line of a table like Table 63.

f. Place the materials where they will not be disturbed for 3 or 4 days. Note the water levels in the test tubes each day until no further changes occur.

g. At the end of the 3-4 day period, add water to each beaker until the heights of the water in the test tube and beaker are equal. Measure the length of the air column in each tube. Record the results in Table 63.

h. Record any other observations in the bottom line of Table 63.

i. Raise the test tube with the steel wool 1-2 cm in the water. Put your thumb over the mouth of the test tube while it is still in the water. Stand the test tube upright with your thumb still over the mouth. Insert a burning wooden splint into the tube. Note your results in Table 63.

j. Repeat step i using the control test tube. Note the results in Table 63.

TABLE 63 Steel Wool and Air

	Test tube with steel wool	**Control**
Initial height (cm)		
Final height (cm)		
Other observations		

Discussion

1. Explain the purpose of the control.
2. Calculate the percentage of air used up in each test tube. Use the following formula:

$$\text{Percentage of air used up} = \frac{\text{Initial height} - \text{Final height}}{\text{Initial height}} \times 100\%$$

3. What is the percentage of air remaining in each test tube?
4. Explain why the water rises in the test tube containing the steel wool. Why does it eventually stop rising?
5. How does the gas left in the test tube containing steel wool compare to the air that was initially in the test tube? What gas do you think was used up in rusting the iron? What gas do you think remains?

6. What do you think iron rust is made of?
7. Compare your answer for part 2 of this discussion with the answers of other students in your class. Calculate the average for the class. Make a conclusion based on this answer.

9.4 Composition and Importance of Air

You have discovered that air is a mixture of gases. This section tells you what those gases are and how much of each one is present in air. It also discusses the importance of these gases.

Composition of Air

The percentage composition of clean dry air is listed in Table 64. These are average values at sea level. The composition changes with altitude. It also varies from place to place over the surface of the earth. For example, the air over a large marsh contains greater than average amounts of methane because the decay in a marsh produces this gas. Air contains many more gases than the 13 listed in the table, but they are normally present in amounts even smaller than that of ozone. For example, volcanoes release sulfur dioxide and hydrogen sulfide. Decaying plant and animal material produces ammonia. Cars produce carbon monoxide. Industries emit hundreds of other gases.

TABLE 64 Composition of Clean Dry Air

Gas	Percent by volume
Nitrogen	78.09
Oxygen	20.94
Argon	0.93
Carbon dioxide	0.034
Neon	0.001 8
Helium	0.000 52
Methane	0.000 2
Krypton	0.000 1
Nitrous oxide	0.000 05
Hydrogen	0.000 05
Xenon	0.000 009
Nitrogen dioxide	0.000 002
Ozone	0.000 002

The most common gases in air are clearly nitrogen (78%) and oxygen (21%). They make up 99% of the volume of air. Most of the remaining 1% is argon. In Investigation 9.3 you used the rusting of iron to remove oxygen from air. The amount that the water rose up the test tube indicated the percent oxygen in air. Did you get a value close to 21%? The gas that remained in the test tube was mainly nitrogen. Thus your value should have been close to 78%.

Air is never completely dry, not even above deserts. It always contains some water vapour. The amount ranges from 0.01% over deserts to as much as 4% over the oceans. Nor is air ever completely clean. This was true even before people began putting pollutants into it. For example, dust has always been present in air. Dusty air is a suspension of tiny solid particles in the air. Some of these particles are non-living; others are living. Volcanic ash and dust from desert storms are examples of non-living dust that nature puts into the air. Smoke, soot, and ashes are examples of dust that people put into the air. Bacteria and pollen grains are examples of living dust particles.

Some people are allergic to dust. One person may be allergic to a certain type of pollen. Another may be allergic to a pollutant in the smoke of a factory or a cigarette. The dust particles irritate the lining of the bronchial tubes and lungs, causing excessive mucus production.

When dust particles and chemicals in the air combine with water vapour, a **smog** is formed. This often happens when factories emit smoke into foggy air on a calm day. Thousands of people in cities have died as the result of breathing air polluted with smog. Many of those who died were old people and many had respiratory problems caused by smoking. Their weakened lungs and bronchial tubes were unable to get rid of the excess mucus formed because of irritation caused by smog particles.

Importance of Air

The most important gas in air is **oxygen**. Both plants and animals (including us) require it for respiration. We also need it to burn the fuels that warm our homes and run our industries. Oxygen helps to decay organic matter such as dead leaves and sewage.

The most important role of **nitrogen** is to dilute the oxygen. Without this diluting effect, reactions involving oxygen would be much too fast. Respiration of living things would be so fast that they would quickly "burn out". The burning of fuels would be almost uncontrollable.

Another important use of nitrogen is in the making of proteins, a basic part of plant and animal tissues. All proteins have nitrogen atoms in them. Animals get their nitrogen atoms by eating other animals or plants. Plants get their nitrogen atoms by absorbing nitrates (a compound of nitrogen) from soil or water. Certain bacteria and algae help replace these nitrates in the soil or water by changing nitrogen gas from the air into nitrates. The nitrate in commercial fertilizer is made from atmospheric nitrogen or mined from deposits in the ground.

Since only 0.034% of the air is **carbon dioxide**, you may think that this gas is not very important. But this is not so. Carbon dioxide is used by green plants and algae to make carbohydrates such as starches and sugars. This is their way of getting food. Inside the cells of the plants and algae, carbon dioxide combines with water to form carbohydrates and oxygen. Light and chlorophyll are needed for this process which is called **photosynthesis**. Some of the carbohydrates are used by the plants and algae as food. These carbohydrates react with oxygen during **respiration** to produce energy for growth and other life processes. Animals get their food supply by eating plants, or animals that have eaten plants. Thus they, too, are dependent upon carbon dioxide in the air for food.

The following two equations summarize the processes of photosynthesis and respiration. Study them carefully. What relationship do you see between them?

Photosynthesis in Green Plants

$$\text{Carbon dioxide} + \text{Water} + \text{Light energy} \xrightarrow{\text{chlorophyll}} \text{Carbohydrates} + \text{Oxygen}$$

Respiration in Plants and Animals

$$\text{Carbohydrates} + \text{Oxygen} \longrightarrow \text{Energy} + \text{Water} + \text{Carbon Dioxide}$$

Carbon dioxide is also important because it forms a "heat blanket" around the earth. The carbon dioxide in the atmosphere acts much like the glass in a greenhouse. It reflects heat back to the earth, helping to keep the earth warm. If there were no carbon dioxide in the air, most of the earth's heat would radiate into outer space and the earth would become very cold. This property is called the **greenhouse effect**.

Oxygen, nitrogen, and carbon dioxide are the most important gases in air. The other gases, though less important, still fit into nature's scheme. For example, you may recall that ozone shields life on the earth from dangerous ultraviolet radiation.

In the following sections of this unit you study the properties and uses of oxygen and carbon dioxide.

Discussion

1. **a)** What percent of the air is made up of nitrogen, oxygen, and argon combined?
   b) Why are the last 5 gases in Table 64 often called "trace" gases?
2. The nitrogen and oxygen in air can be separated from one another by a process called the "fractional distillation of liquid air". When air is cooled to about $-200°C$ it turns into a liquid. Explain how the fractional distillation of liquid air can

separate the two gases. (The b.p. of nitrogen is −196°C and the b.p. of oxygen is −183°C.)

3. **a)** Which air would contain more water vapour, the air over a polar sea or the air over a tropical sea? Why?
   b) Distinguish between living and non-living dust.
4. **a)** What is smog?
   b) How is smog formed?
   c) Why do some people suffer more from smog than others?
5. **a)** Summarize the importance of oxygen in air.
   b) Summarize the importance of nitrogen in air.
6. **a)** What takes place during photosynthesis?
   b) What takes place during respiration?
   c) How are the processes of photosynthesis and respiration related?
7. **a)** What is the greenhouse effect?
   b) Some scientists fear that the carbon dioxide which we put into the air from car exhaust and industrial emissions might seriously change the earth's climate. How might this happen?

9.5 INVESTIGATION: Properties of Oxygen

In this investigation you collect a few containers of oxygen gas. You then study some of the physical and chemical properties of this important component of air.

Materials

cylinder of compressed oxygen
overflow tray
2 gas bottles, 2 glass plates
2 large test tubes (20 × 150 mm), 2 stoppers
charcoal, sulfur, magnesium, steel wool
deflagrating spoon, coat-hanger wire
candle, wooden splint
limewater
Bunsen burner

Procedure A Some Basic Properties

a. Your teacher has set up a station for the collection of oxygen gas by the downward displacement of water (Fig. 9-4). Go to the station and get 2 test tubes and 2 gas bottles full of oxygen gas. Cover the mouths of the gas bottles with glass plates. Put stoppers in the mouths of the test tubes.

b. Note the colour and odour of the gas in one test tube. Add 2-3 mL of limewater to this test tube and shake the contents vigorously for a few seconds. Describe the results.

Figure 9-4 The collection of oxygen gas by the downward displacement of water.

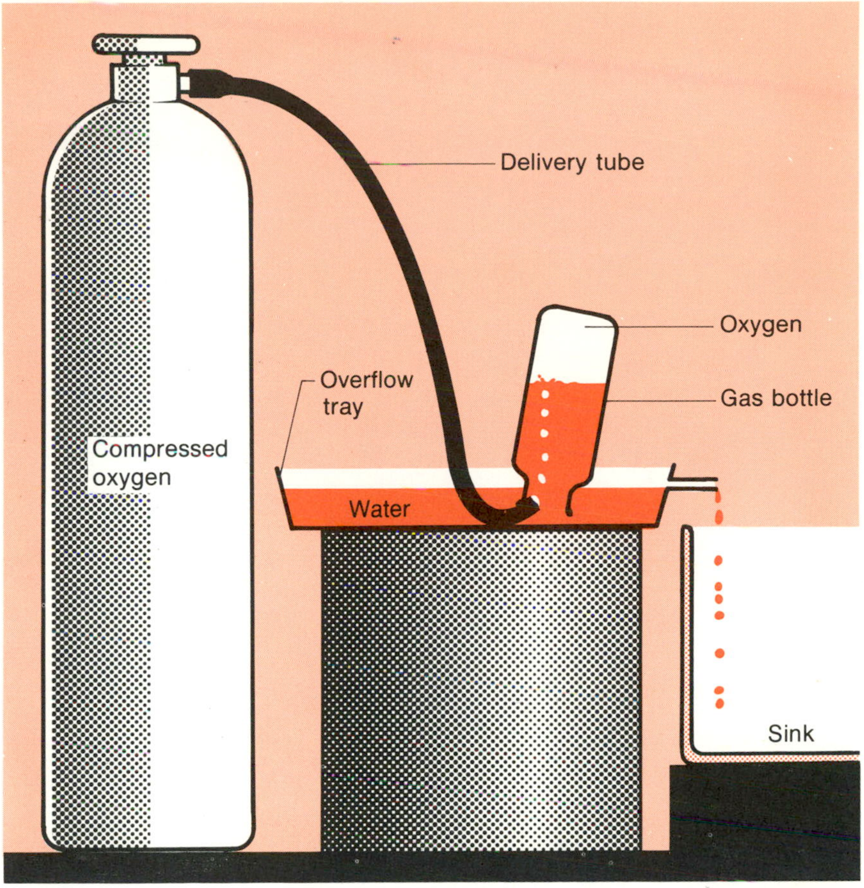

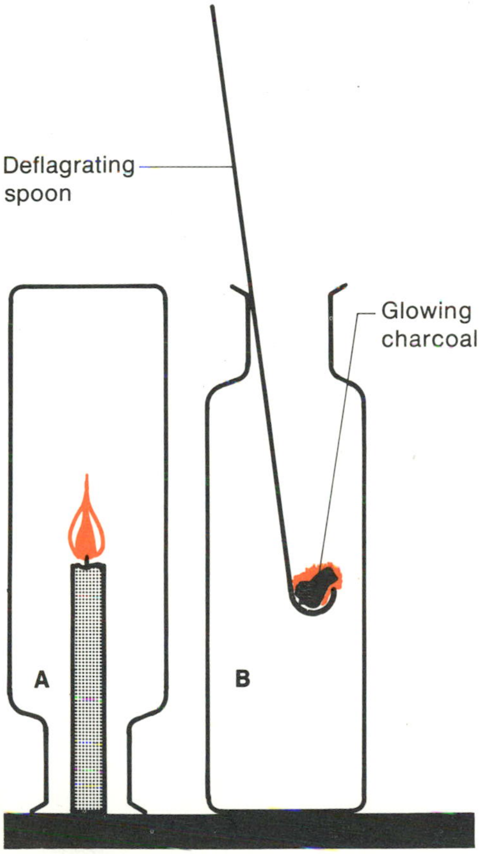

Figure 9-5 How does oxygen affect a burning candle?

c. Light the splint and blow out the flame. Make sure there is a glowing ember at the tip. Lower the glowing splint into the second test tube of oxygen. When no further changes occur, add 2-3 mL of limewater and shake the contents vigorously for a few seconds. Describe the results.

d. Light the candle. Invert one bottle of oxygen over the candle (Fig. 9-5,A). When the candle stops burning, add about 5 mL of limewater to the bottle and shake the contents vigorously for a few seconds. Describe the results.

e. Place a piece of charcoal in the deflagrating spoon (or in a loop at the end of the wire, if you have no spoon). Heat the charcoal until it glows brightly. Lower it into the second bottle of oxygen (Fig. 9-5,B). When no further changes occur, add about 5 mL of limewater to the bottle and shake the contents vigorously for a few seconds. Describe the results.

Procedure B Further Properties

CAUTION: Wear safety goggles during this investigation. Because some of these studies can be dangerous, your teacher may choose to demonstrate them.

a. Place a pea-sized piece of sulfur in the deflagrating spoon. Heat it gently until it melts and begins to burn. Lower it into a gas bottle of oxygen. Describe the results. **CAUTION:** Do not smell the gas. Perform this experiment only in a well-ventilated space such as a fume-hood.

b. Tie a 5 cm strip of magnesium ribbon to one end of the wire. Heat the free tip of the magnesium until it begins to burn. Quickly lower the burning magnesium into a bottle of oxygen. Do not touch the sides of the bottle; it will break.
CAUTION: Wear safety goggles. Do not stare directly at the bright flame. It could harm your eyes.

c. Fasten a wad of steel wool to a loop at the end of the wire. Heat the steel wool until it glows. Quickly lower it into a bottle of oxygen. Observe the precautions noted in step b.

Discussion A

1. List the physical properties of oxygen that you observed in this investigation. (The fact that oxygen can be collected by the downward displacement of water indicates one of these properties.)
2. What was the reason for adding limewater to oxygen gas (step b of Procedure A)?
3. An **organic** substance is one that is (or was at one time) part of a plant or animal. All organic substances contain the element carbon. The charcoal used in this investigation is almost pure carbon. The candle and wooden splint are also organic substances. What compound is formed when organic substances are burned in pure oxygen? Explain why you made this conclusion.
4. Do you think wool is an organic substance? Why?
5. Are Dacron and Orlon organic substances? Why?
6. Name one substance that you think will be formed when gasoline is burned in a car engine. Explain why you named that substance.

Discussion B

1. Compare the burning of sulfur in air and in oxygen. Explain the difference. The burning of sulfur produces a gas called sulfur dioxide. Write a word equation for the burning of sulfur.
2. Compare the burning of magnesium in air and in oxygen. Explain the difference. The white product formed is magnesium oxide. Write a word equation for the burning of magnesium.
3. Describe and explain the effect of oxygen on hot steel wool (iron).

9.6 Oxidation

The process by which oxygen combines with another substance is called **oxidation.** In the previous investigation you saw several examples of **rapid oxidation**, or **combustion.** When rapid oxidation occurs, you always notice heat and light. But sometimes oxygen reacts so slowly with another substance that no flame is produced. In this case the process is called **slow oxidation.** The rusting of iron and the decay of wood are examples of slow oxidation. Although no heat is noticed during slow oxidation, it is still produced. In fact, slow oxidation of 1 g of magnesium produces the same amount of heat as rapid oxidation of 1 g of magnesium. You simply do not notice it because it is given off over a long period of time. Slow oxidation also occurs in the cells of your body. Oxygen unites with food (carbohydrates) in the cells to produce heat and carbon dioxide.

The compounds that are formed when oxygen unites with other elements are called **oxides**. You have already met carbon dioxide, sulfur dioxide, magnesium oxide, and iron oxide. Very high temperatures are often needed to "excite" atoms to the point where they will start combining with one another. You discovered this in Investigation 9.5. However, once the atoms start combining, they give off heat. This heat is more than enough to keep the reaction going. Once you started the various substances burning, you did not have to keep heating them. We make use of this fact in many ways. You have to heat the charcoal in a barbecue to get it burning. But, once it begins combining with oxygen, it produces enough heat to keep the reaction going and to cook a few hamburgers as well. The gasoline in a car engine must be "sparked" to get it burning. But, once oxidation starts, enough heat is produced to burn the gasoline and make the engine go.

Even a rocket engine uses the process of oxidation. Since there is no oxygen in outer space, the rocket must carry oxygen (or some other oxidant) as well as fuel. The oxygen is mixed with the fuel and ignited. The hot oxides produced are directed out one end of the rocket to propel it.

Discussion

1. **a)** Define the terms oxidation and oxide.
   b) Describe the differences between slow and rapid oxidation.
2. **a)** List 3 examples of slow oxidation.
   b) List 3 examples of rapid oxidation.

9.7 INVESTIGATION: Combustion of Fuels

Acetylene gas is a fuel. It is combined with oxygen in welding torches to produce a very hot flame. It is an organic compound, thus it contains carbon. It should, therefore, produce carbon dioxide when it burns in pure oxygen. When this happens, we say that **complete combustion** has occurred. If insufficient oxygen is present for complete combustion, products other than carbon dioxide form. We call this **incomplete combustion**.

The purpose of this investigation is to compare the products of complete and incomplete combustion of acetylene. You then use the results to discuss combustion in automobile engines and household furnaces.

Materials

400 mL beaker
4 test tubes (20 × 150 mm)
a few lumps of calcium carbide
forceps
wooden splint
Bunsen burner

CAUTION: Wear safety goggles during this investigation.

Procedure

a. Calcium carbide reacts with water to form acetylene. Collect a test tube full of acetylene as follows: Invert a test tube full of water into a beaker that is half full of water. Using forceps, drop a small lump of calcium carbide into the water. Place the inverted test tube over the calcium carbide (Fig. 9-6). Allow the acetylene gas to completely fill the test tube. Lift the test tube from the water and place it mouth down on the desk top.

b. Repeat step a three more times but collect less acetylene each time. Collect one half of a test tube of acetylene in one tube, one third in another, and one twelfth in the last. Lift each tube from the water, allowing air to replace the water that runs out. Put your thumb over the mouth of each tube and shake it vigorously for 20-30 s to mix the acetylene and air. Then place the test tube mouth down on the desk top.

c. Hold each test tube in a horizontal position and bring a burning wooden splint to its mouth. After each reaction is complete, record all observations. Then add 3-4 mL of limewater to each test tube and shake it vigorously for several seconds.

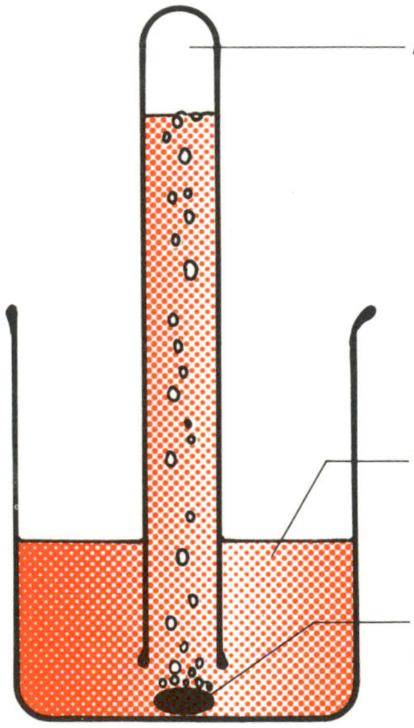

Figure 9-6 The preparation of acetylene gas.

Discussion

1. Which test tube contained the most oxygen? the least?
2. In which test tube did complete combustion occur? What product was formed in this case?

3. In which test tubes did incomplete combustion occur? What products form during incomplete combustion?
4. Where was the flame located in the first test tube? Why?
5. If a car engine is not properly tuned, incomplete combustion occurs. What product will be formed in the engine? How might it affect the engine?
6. If the burner of a gas or oil furnace in a home is not serviced properly, incomplete combustion of the fuel will occur. Describe the likely results.
7. Another product of incomplete combustion of fuels is carbon monoxide. This deadly gas has caused the death of many people. Explain why a car engine that is idling produces carbon monoxide.
8. Combustion of organic substances usually produces a compound that is well known to all of us. See if you can figure out what it is by trying this simple experiment at home: Hold a tall dry glass upside down in one hand. With the other hand, hold the flame of a burning candle just inside the mouth of the glass. Watch closely for any changes. What do you think the compound is? How could you prove that you are correct? What evidence have you noticed which indicated that the combustion of gasoline produces the same product?

9.8 INVESTIGATION: Spontaneous Combustion and Dust Explosions (*Teacher Demonstration*)

This investigation demonstrates two interesting but dangerous types of rapid oxidation. Countless fires and explosions are caused every year by people who do not understand how to prevent these types of oxidation.

Materials

sodium peroxide
absorbent cotton
dust explosion apparatus (Fig. 9-8)
lycopodium powder or dry corn starch
eyedropper
asbestos mat
Bunsen burner

CAUTION: Wear safety goggles during this demonstration.

Procedure A Spontaneous Combustion

a. Make a "bird's nest" out of the absorbent cotton and place it on the asbestos mat.
b. Place about 5 cm^3 of granular sodium peroxide in the "bird's nest".

c. Add a few drops of water to the sodium peroxide as shown in Figure 9-7. Stand back in case spattering occurs.

Procedure B Dust Explosions

a. Prepare the apparatus as shown in Figure 9-8. Make sure the funnel is completely full of lycopodium powder. The top of the funnel should be opposite the candle flame.
b. Light the candle and press the lid firmly in place.
c. Blow into the rubber tubing with a short quick puff.
d. Repeat the entire experiment without using the lid.
e. Place a pile of lycopodium powder on the asbestos mat. Try to light it with the Bunsen burner.

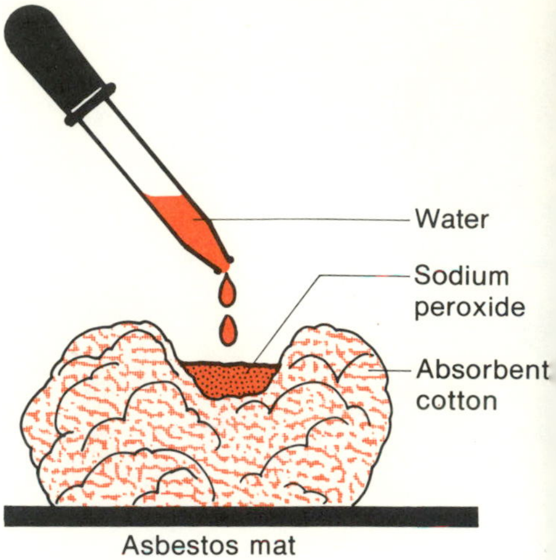

Figure 9-7 Spontaneous combustion of absorbent cotton.

Discussion A

1. Use a dictionary to find the meaning of "spontaneous". Why is this reaction called "spontaneous combustion"?
2. Sodium peroxide reacts with water to produce heat and oxygen. Knowing this, explain the results of the demonstration. (Hint: What actually burned? Why did it burn?)
3. Four conditions must be met before spontaneous combustion can occur.
a) a reaction must take place that produces heat
b) a substance must be present that has a low kindling temperature; that is, it catches fire at a low temperature
c) the substance must be enclosed or insulated so that heat builds up in it instead of escaping
d) an oxidizer such as oxygen must be present
Explain how the demonstration meets these conditions.
4. Rags soaked in an oil-base paint or linseed oil will often undergo spontaneous combustion if they are thrown in a heap in the corner of the basement. Oils are organic compounds that oxidize slowly. Explain how the four conditions for spontaneous combustion are met. How could spontaneous combustion be prevented in this case?
5. Poorly cured or moist hay will often undergo spontaneous combustion if it is packed tightly in a barn. Bacteria feed on sugars in the stems and leaves of the moist hay. This is a form of slow oxidation. Explain how the four conditions for spontaneous combustion are met. How could spontaneous combustion be prevented in this case?

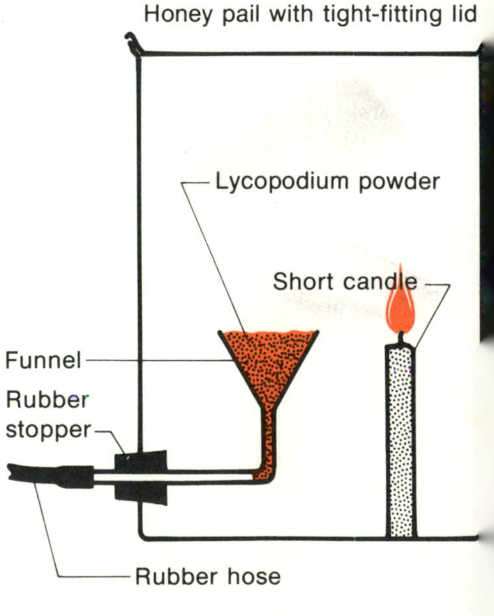

Figure 9-8 Demonstration of a dust explosion.

Discussion B

1. Lycopodium powder is the spores of a club moss. In other words, lycopodium powder is part of a plant and thus is living. What atom will be in the molecules of this powder? What compound will be formed when the powder undergoes complete combustion?

2. Was there any evidence of incomplete combustion when you observed step d of Procedure B? Explain.
3. Describe and explain the difference in the rates of oxidation between steps d and e of Procedure B.
4. Describe and explain the results of step c of Procedure B.
5. Serious fires have been caused by people who have thrown a full dust bag from a vacuum cleaner into a fire that is burning in a fireplace. Explain what happens.

9.9 INVESTIGATION: What Causes Rusting?

In Investigation 9.3 you discovered that the oxygen in air unites with iron (steel wool) to form iron oxide, or rust. But you did not prove that the oxygen alone causes the rusting. You probably know that wet iron rusts quickly. A garden shovel left outdoors on a rainy night will be rusty in the morning. Also, you have likely heard that salt makes the iron parts of cars rust quickly.

What causes rusting? Is it the oxygen, water, or salt? Or is it a combination of two or all of these?

Materials

9 test tubes (20 × 150 mm)
13 nails (4-5 cm long)
sandpaper and grease remover
(gasoline or dichloromethane will do)
silica gel (or anhydrous calcium chloride)
absorbent cotton
motor oil
common salt
magnesium ribbon, aluminum foil, zinc foil, tin foil,
copper foil (or wire)

Procedure

a. Clean the nails with sandpaper to remove any rust or protective coating. Then wash them in the degreaser to remove any grease that may have been added to prevent rusting.
CAUTION: The degreaser should be used in a fume-hood, away from flames, and with rubber gloves.
b. Number the test tubes from 1 to 9.
c. Place 2 nails in test tube 1. Half cover them with distilled water. These nails are exposed to both air and water. This is your control.

d. Make sure test tube 2 is completely dry. Place 2 nails in it. Then add a few pieces of silica gel. Stuff a plug of absorbent cotton into the mouth of the test tube. The silica gel absorbs water from the air. Thus the nails are exposed to air but not water.

e. Boil some water in test tube 3 for several minutes. Boiling expels dissolved air from the water. Place 2 nails in the water. They must be completely covered by the water. Then add 1-2 mL of motor oil. The oil floats on the water, keeping air out. These nails are exposed to water but not air.

f. Place 2 nails in test tube 4. Half cover them with water that contains dissolved salt. These nails are exposed to air, water, and salt.

g. Wrap a piece of magnesium ribbon tightly around part of a nail. Place the nail in test tube 5. Add water until the nail is almost covered.

h. Repeat step f for each of test tubes 6 to 9, using the following in place of the magnesium ribbon:
test tube 6—aluminum foil
test tube 7—zinc foil
test tube 8—tin foil
test tube 9—copper foil (or wire)

i. Let the test tubes stand undisturbed for at least 2 or 3 days. Examine the nails for rust.

Discussion

1. In which of test tubes 2 to 4 did more rusting occur than in the control (test tube 1)? In which of these test tubes did less rusting occur than in the control? What do you think causes rusting?

2. Compare the amount of rusting in each of test tubes 5 to 9 with that in the control. Which metals are best at preventing rusting of iron?

3. List 5 things you could do to slow down or stop the rusting of iron.

9.10 Preventing Corrosion

Corrosion is the name given to the slow oxidation of a metal. You have studied the corrosion of iron; it is called **rusting**. Many other metals also corrode. For example, if a piece of aluminum is polished until it shines, it soon corrodes if it is exposed to the air. The aluminum reacts with oxygen in the air to form aluminum oxide. This oxide appears as a dull thin film on the aluminum.

You have probably seen this film on aluminum. But you probably have not seen aluminum that has been badly damaged by corrosion. This is because the film of oxide that forms on aluminum completely covers the surface of the aluminum. Now oxygen can no longer come in contact with the aluminum. Therefore, once a complete film of oxide has formed, the aluminum no longer corrodes.

Many metals behave in a manner similar to aluminum. In fact, some behave exactly like aluminum. They react with oxygen in the air to form a thin protective layer of oxide. Others react with carbon dioxide in the air to form a thin protective coating of carbonate. Still others form a coating that is a mixture of oxide and carbonate. Magnesium, zinc and copper are examples of common metals that form protective coatings.

Unfortunately, the coating of iron oxide that forms on iron does not protect the iron from further oxidation. Iron oxide is very porous and flaky. It allows air to come in contact with the metal underneath. To make matters worse, iron oxide **catalyzes** or speeds up the rusting of iron. Therefore, once iron begins rusting, it continues to corrode at a rapid rate until it has been badly damaged or completely destroyed (Fig. 9-9).

Corrosion is responsible for billions of dollars of damage annually in North America. Much of this corrosion occurs on cars. Car engines and transmissions are now being built that will last for more than 200 000 km of driving. But, long before that distance has been reached, the car bodies have usually corroded. In addition to the expense caused by such corrosion, many dangers are involved. If the brake lines corrode, they may burst when the brakes are applied. If the exhaust system or floor corrodes, dangerous carbon monoxide may enter the car.

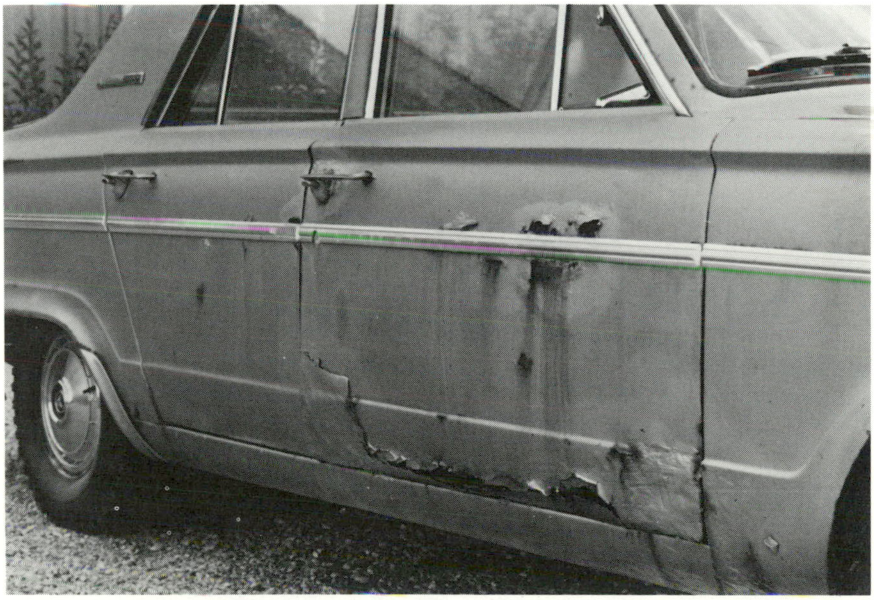

Figure 9-9 This corrosion could have been prevented or at least reduced. How?

Because of the cost and dangers involved in the corrosion of iron, many steps are taken to prevent or slow down this process. The main methods are discussed here.

Use of a Protective Coating

One way to prevent corrosion of iron is to cover the iron with a substance that keeps air away from the iron. A film of grease or oil will do this. Carpenters and gardeners often coat their tools with a layer of grease or oil if they are not going to use them for a long period of time.

A coat or two of paint will also serve as a protective layer. The iron must be free of rust before the paint is applied. Otherwise the rust will blister through the paint. Generally a primer coat is put on the iron before the final paint is applied. The primer is made of a substance that clings well to iron and covers it completely.

Another way to protect iron from corrosion is to cover it with a layer of metal that does not corrode easily. Zinc is often used for this purpose. The iron is covered with a thin coating of zinc by dipping the iron in molten zinc. This process is called **galvanizing**. Barn roofs and eavestroughs are often made of galvanized iron; rust-prone parts of cars like the rocker panels, fender wells, and exhaust systems may also be galvanized.

Sometimes the coating of metal is applied to the iron by **electroplating**. This process uses electricity to deposit a layer of nickel or chromium on the iron. The bumpers and hub-caps of some automobiles are electroplated with chromium. Toasters and irons are usually electroplated with nickel.

Rust-proofing, undercoating, and plastic liners are used by automobile manufacturers to lessen corrosion. In all three cases, a special film is applied to the iron that keeps air away from it.

Corrosion of iron is also reduced by forming an **alloy** of the iron with metals like nickel and chromium. The alloy is formed by mixing the metals while they are molten. The mixture solidifies into a homogeneous solid that resists corrosion. An alloy of iron, nickel, and chromium is called **stainless steel.**

Use of Sacrificial Corrosion

You do not know enough chemistry to completely understand this method. However, since it is widely used, it is briefly discussed here.

In this method another metal is "sacrificed" to save the iron from corrosion. You discovered in the preceding investigation that a piece of magnesium, aluminum, or zinc, when attached to iron, lessens the corrosion of the iron. These metals are "sacrificed" or corroded instead of the iron. Oil pipelines are made of iron. To prevent them from corroding, blocks of magnesium are attached to the pipeline every 0.5 km or so. Of course, every so often the magnesium blocks have to be replaced. But this is easier and cheaper than replacing the whole pipeline! The iron hulls of ships are often protected from corrosion by placing blocks of zinc in contact with them. The zinc blocks corrode in place of the iron.

Preventing Oxidation with Antioxidants

Oxygen attacks many substances besides metals. Foods and automobile antifreeze are examples. In these cases we do not call the process corrosion; it is simply called oxidation. It is prevented or lessened by the use of **antioxidants**. An antioxidant is a substance that unites with oxygen more easily than the substance it is protecting. In foods the antioxidant is called a **preservative**. If some oxygen gets into a jar of food, it oxidizes the preservative instead of the food.

Automobile antifreeze also contains an antioxidant. When oxygen gets into the radiator of a car, it oxidizes some of the antifreeze. This forms an acid which can corrode the radiator of the car. However, if an antioxidant is present, the oxygen unites with it instead of the antifreeze. Car manufacturers recommend changing the antifreeze every two years. This is because the antioxidant is usually used up in that time.

Discussion

1. **a)** What is corrosion?
   b) What is rusting?
2. **a)** Explain how metals such as magnesium, aluminum, zinc, and copper have a built-in resistance to corrosion.
   b) Explain why iron corrodes more easily than most metals.
3. List and describe the types of coatings that are used to protect iron.
4. Describe the use of sacrificial corrosion to protect iron.
5. **a)** What is an alloy?
   b) What is stainless steel? Name 2 items that are made of stainless steel. Explain why they were made of stainless steel.
6. The exhaust of a car contains water vapour that was produced by the burning of the gasoline. If the muffler of the car is cold, some of the water condenses in it. If the muffler is hot, less water condenses. Will mufflers last longer in cars that are used for short trips or in cars that are used for long trips? Why?
7. Look around your home for examples of corrosion. List as many examples as you can. Explain how each of these could have been prevented.

9.11 INVESTIGATION: Properties of Carbon Dioxide

In this investigation you collect a few containers of carbon dioxide gas. You then study some of the physical and chemical properties of this important component of air.

Materials

cylinder of compressed carbon dioxide
overflow tray
2 gas bottles, 2 glass plates
3 test tubes (20 × 150 mm), 2 stoppers
2 small beakers (150 mL)
candle, wooden splint
limewater
Bunsen burner
acid-base indicator

Procedure

a. Your teacher has set up a station for the collection of carbon dioxide gas by the downward displacement of water. (See Figure 9-4.) Go to the station and get 2 gas bottles and 3 test tubes full of carbon dioxide gas. Cover the mouths of the gas bottles with glass plates. Put stoppers in the mouths of the test tubes.

b. Note the colour and odour of the gas in one test tube. Add 2-3 mL of limewater to this test tube and shake the contents vigorously for a few seconds. Describe the results.

c. Light the wooden splint and while it is still burning lower it into a bottle of carbon dioxide gas. Describe the results.

d. Place a small piece of burning candle in the bottom of a beaker. "Pour" a bottle of the gas into the beaker (Fig. 9-10). Note the results.

e. Remove the stopper from a test tube of the gas and place your thumb over the mouth of the test tube. Invert the test tube in a beaker that is about two-thirds full of water. Shake the test tube gently to splash water up into the gas. Note any evidence of dissolving.

f. Add 2-3 mL of distilled water to the last test tube of gas. Replace the stopper and shake the contents vigorously for several seconds. Remove the stopper and add a few drops of acid-base indicator. (Your teacher will explain how the indicator works.)

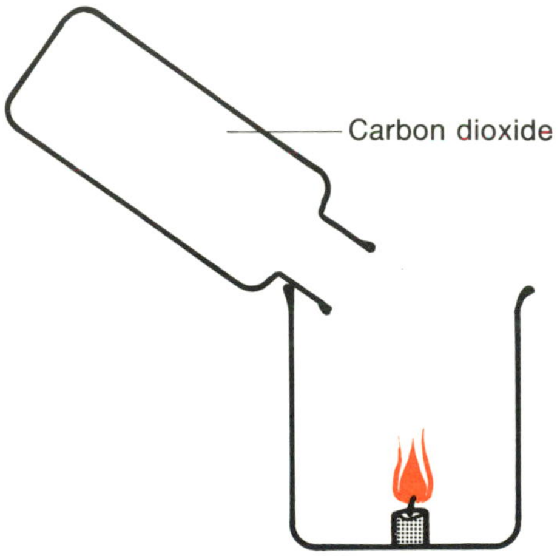

Figure 9-10 Is carbon dioxide more dense or less dense than air?

Discussion

1. Describe the colour and odour of carbon dioxide.

2. What do you conclude from the results of part c of the procedure? Suggest a use for carbon dioxide gas.

3. Is carbon dioxide gas more dense or less dense than air? Explain why you made this conclusion.
4. Does carbon dioxide dissolve in water? Does carbon dioxide form an acid in water? Explain why you made your conclusions.

9.12 INVESTIGATION: Uses of Carbon Dioxide

This section consists of several mini-investigations and readings that will help you understand the uses of carbon dioxide. Try to relate the uses to the properties that you studied in the preceding section. You will have to design your own procedure for some of these investigations.

Materials

soda water	Bunsen burner
candle, wooden splint	gas bottle
limewater	test tube (20 × 150 mm)
acid-base indicator	

baking soda, baking powder, vinegar
gas-generating apparatus (Fig. 9-11)

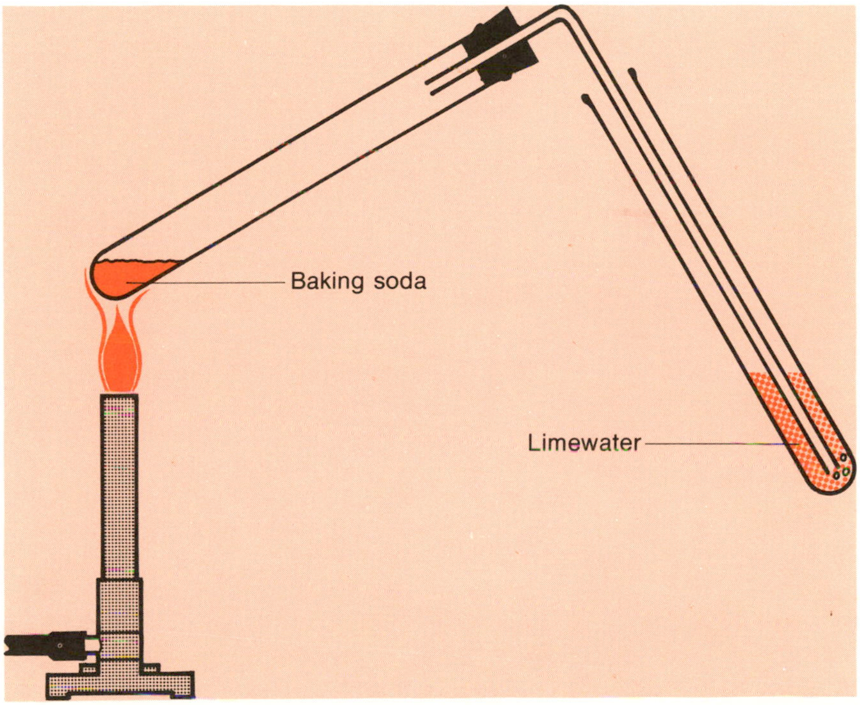

Figure 9-11 Gas – generating apparatus for producing carbon dioxide from baking soda.

USE 1 Preparation of Carbonated Beverages

You learned in the preceding investigation that carbon dioxide dissolves in water to form an acid. The acid is called carbonic acid. If carbon dioxide is forced into water under pressure, a concentrated solution of carbonic acid is formed. It is called carbonated water or soda water. It gives soft drinks their bubbly nature and tangy taste.

Obtain a sample of carbonated water from your teacher. Prove that it is an acid. Now prove that the gas in it is carbon dioxide. To do this, you will first have to get some of the gas out of the water. Then test this gas to see if its properties are the same as those of the carbon dioxide you studied in Investigation 9.11.

A

USE 2 Baking

A second use of carbon dioxide is in the leavening of the dough in cakes, bread, and biscuits. An ingredient that releases carbon dioxide is added to the dough. The bubbles of carbon dioxide puff up (leaven) the dough. The heat of the oven then cooks the dough so it stays in the risen position. You have probably noticed the holes that the carbon dioxide made in the dough of baked goods such as bread.

One way that carbon dioxide can be released in dough is to mix some baking soda (sodium bicarbonate) into the dough. When the dough is heated, the baking soda releases carbon dioxide. You can demonstrate how heat releases carbon dioxide from baking soda by using the apparatus shown in Figure 9-11. **CAUTION:** Take the limewater away from the delivery tube before you stop heating. If you don't, the limewater may back up into the hot test tube and break it.

Carbon dioxide can also be released from baking soda by an acid. Try this: Place a pinch of baking soda in a test tube. Add 2-3 mL of vinegar. Sour milk is often mixed with baking soda and dough to make biscuits. Sour milk contains lactic acid. It reacts with the baking soda to produce carbon dioxide, causing the biscuits to rise.

Baking powder is made of baking soda and a solid acid. When water is added, the solid acid dissolves. It then attacks the baking soda to form carbon dioxide. Place a pinch of baking powder in a test tube and add 2-3 mL of water.

Finally, yeast is used as the source of carbon dioxide during the baking of bread. Yeast is a living organism. It feeds on sugar in the dough. Like most living organisms, it gives off carbon dioxide as it respires. Obtain some powdered yeast from your teacher. Use the apparatus in Figure 9-11 to show it produces carbon dioxide. You will not need the Bunsen burner. Add a pinch of the yeast to 3 or 4 mL of sugar solution. You will have to

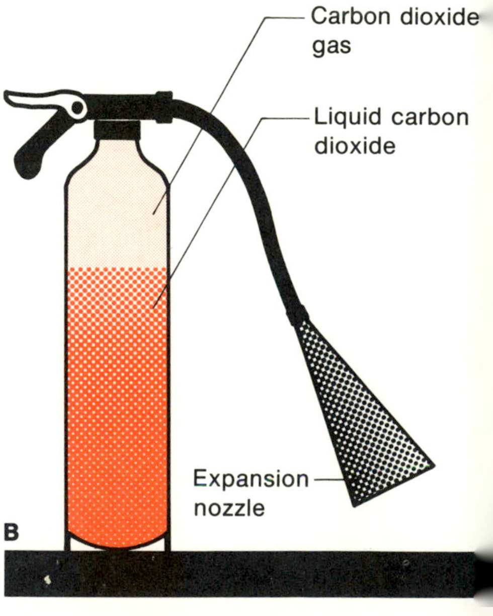

B

Figure 9-12 A liquid carbon dioxide extinguisher: A; exterior view; B, contents.

check the apparatus 2 or 3 h later, since it takes a while for the yeast to produce a noticeable amount of carbon dioxide.

USE 3 Dry Ice

Dry ice is solid or frozen carbon dioxide. It is very cold; it sublimes from the solid state to the gaseous state at $-78.5°C$.

In Unit 4 you learned that cooling takes place when molecules move further apart. Thus, to make dry ice, gaseous carbon dioxide is compressed and then allowed to expand rapidly through a valve. This cools the gas so much that it comes out of the valve as a snow-like solid. This solid is then compressed into blocks and sold as dry ice. Ice cream trucks and trucks that transport frozen foods often use dry ice. One reason is that it is very cold. A second reason is that, as it absorbs heat, it changes directly to carbon dioxide gas (sublimes) and there is no liquid to clean up. Ice is not as cold and it melts to give liquid water.

USE 4 Fire Extinguishers

As you have discovered, carbon dioxide has three properties that make it useful for putting out fires. It does not burn; it does not support combustion of other substances; it is more dense than air. All of the fire extinguishers discussed here release carbon dioxide gas. Each one produces it in a different way.

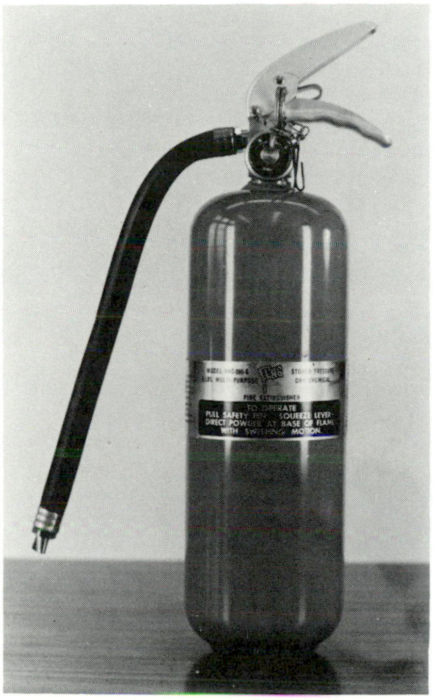

Figure 9-13 A dry chemical fire extinguisher.

A. Liquid Carbon Dioxide Extinguisher

In this type of fire extinguisher, liquid carbon dioxide is stored under pressure in a steel container (Fig. 9-12). When the trigger is pulled, carbon dioxide gas escapes into the nozzle. As the gas expands in the nozzle, it cools so much that it sublimes into solid carbon dioxide. It looks like snow as it leaves the nozzle. The low temperature of this "dry ice" cools the burning substance below its kindling temperature. Also, since carbon dioxide is more dense than air, it tends to stay where it is directed, smothering the fire.

B. Dry Chemical Fire Extinguisher

This is the most common household type of fire extinguisher (Fig. 9-13). Within its steel container is a powdery mixture of solids and a compressed gas. Many solids may be present, but one of them is usually sodium bicarbonate. The compressed gas may be carbon dioxide, nitrogen, or air.

You learned in the discussion of baking that heat releases carbon dioxide from sodium bicarbonate. When the trigger is pulled, the compressed gas blasts the powdery sodium bicarbonate on the flame. The heat of the flame decomposes the sodium bicarbonate, forming carbon dioxide which smothers the flame.

Discussion

1. Describe how you proved that carbonated water is an acid.
2. Describe how you proved that the gas in carbonated water is carbon dioxide.
3. Describe two methods that can be used to release carbon dioxide from sodium bicarbonate (baking soda). Explain how each of these is used in baking.
4. What is baking powder? What advantage does it have over baking soda?
5. Explain why yeast can be used as a source of carbon dioxide in baking.
6. What is dry ice? What advantages does it have over ice as a coolant?
7. What properties of carbon dioxide make it useful for extinguishing fires?
8. Explain how a liquid carbon dioxide extinguisher works.
9. Explain how a dry chemical fire extinguisher works.

9.13 Some Common Air Pollutants

Air pollution is defined as *the presence in the atmosphere of substances that are harmful to living things or their habitats.* Some common pollutants are particulate matter (tiny solid particles), sulfur dioxide, carbon monoxide, carbon dioxide, nitrogen oxides, hydrocarbons, ozone, and lead. Many of these are natural components of the atmosphere. For example, particulate matter in the form of pollen grains and dust from desert storms is a normal part of the atmosphere. Sulfur dioxide from volcanoes is also a natural part of air. Carbon dioxide is given off by the respiration of living things.

Until recently nature was able to absorb and recycle these substances so that they did no harm and were not called pollutants. Then our industries, homes, and cars began producing these substances faster than nature could absorb them. Some of the substances accumulated in high enough concentrations to harm living things. Now they are called pollutants. Here is an example to illustrate this point. Volcanoes give off sulfur dioxide. If this sulfur dioxide drifts over the ocean and gradually dissolves in the water, it does no harm to living things or their habitats. Therefore it is not called a pollutant. However, if a coal-burning power plant emits sulfur dioxide that drifts over farmland, beans that are being grown for human food will be damaged. The sulfur dioxide has harmed living things—the beans and us. Therefore it is called a pollutant.

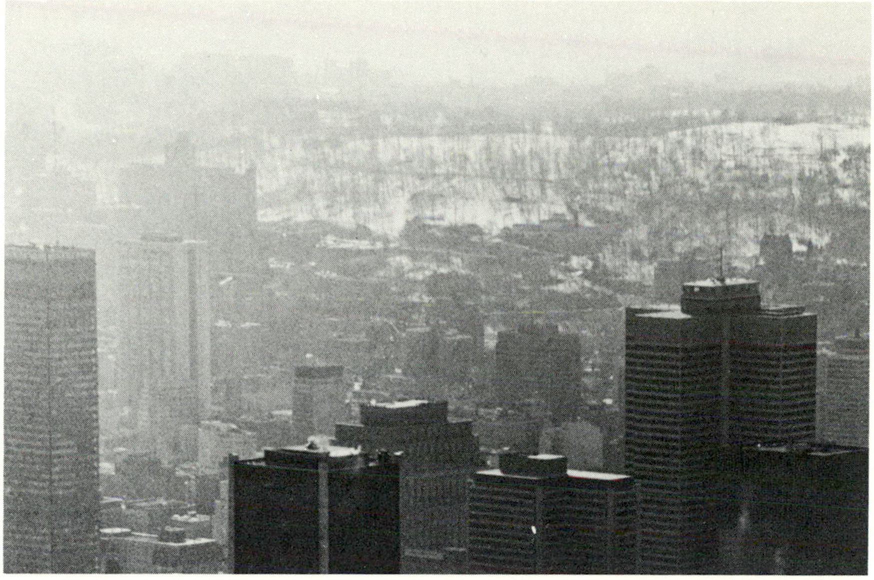

National Film Board of Canada

Figure 9-14 What are some of the substances causing this haze?

Particulate Matter

If you live in a large city, you probably inhale about 70 000 solid particles every time you breathe. If you live in the country, you inhale only 40 000 particles in each breath. Some of these particles are put into the air by natural sources such as volcanoes, forest fires, and dust storms. Most of them are put into the air by human activity—industries, cars, and homes. Most of these airborne particles are very small. They range in size from 10^{-5} cm to 10^{-3} cm in diameter. They may be made of soot, fly ash, grease, oil, or metal.

We are all familiar with the effects of particulate matter. These particles settle on our clothing, and invade our noses and throats. The smallest particles may even reach the deepest sections of our lungs. Soot is mainly carbon. Recent studies have shown that finely divided carbon can cause cancer. Thus tiny soot particles could be responsible for some cases of lung cancer.

Particles in the air are largely responsible for the haze that surrounds most cities (Fig. 9-14). This haze is gradually spreading into the countryside. Scientists fear that the haze may gradually lower the average temperature of the earth. The tiny particles tend to reflect incoming sunlight. If this is true, the earth could be in the next ice age sooner than expected.

Sulfur Dioxide

Sulfur dioxide has a sharp choking odour. You may have smelled it after a wooden match was struck. It forms sulfuric acid shortly after it comes in contact with water. This is why sulfur dioxide is so dangerous to people and other living things. It reacts with moisture in the lungs and other tissues to produce sulfuric acid.

Most of the sulfur dioxide in the earth's atmosphere is from natural sources. Decay of organic matter and volcanoes account for almost 80% of what is in the air. About 16% comes from the burning of coal and oil that contain sulfur; 4% comes from smelters and refineries. You might think that human sources are not very important. But remember that humans release most of their portion in cities. Thus the concentration in some cities becomes high enough to be a serious health hazard.

It takes very little sulfur dioxide to affect people. A sulfur dioxide concentration of 6 µL/L paralyzes and damages the bronchial tubes and lungs. Even at levels as low as 0.20 µL/L deaths in cities rise sharply. Asthma sufferers and heavy smokers are most affected.

Carbon Monoxide

This gas is colourless and odourless. About 50% of the air pollution in North America is carbon monoxide. Nearly all of it is man-made; few natural sources exist. Most of it comes from the incomplete combustion of organic compounds like gasoline and fuel oil. Automobiles produce about 80% of the earth's carbon monoxide; power plants, homes, incinerators, and industries produce most of the remainder.

Carbon monoxide is very dangerous. Just 10 µL/L in the air makes most people sick. About 1 300 µL/L is fatal. Levels of 120 µL/L impair driving ability. This level is often reached in tunnels, parking garages, and downtown streets. Traffic jams often cause levels as high as 400 µL/L. Many people who complain of headaches due to the tension of heavy traffic are probably suffering from carbon monoxide poisoning.

The smoking of just one cigarette in a poorly ventilated car can increase the carbon monoxide concentration to a level that affects both the driver and the passengers, including non-smokers. Imagine, then, the effects when 2 or 3 people are smoking in a car in heavy downtown traffic.

Carbon monoxide does its damage by slowing down the rate of transfer of oxygen from the lungs to the rest of the body. Hemoglobin in the red blood cells carries oxygen from the lungs throughout the body. Carbon monoxide attacks the hemoglobin and prevents it from carrying oxygen. A person who is suffering from carbon monoxide poisoning is suffocating for lack of oxygen. After a person has breathed air with 80 µL/L of carbon monoxide for 8 h, his/her hemoglobin can only carry 85% as much oxygen as it did in unpolluted air. This causes the same effect as losing about 1 L of blood.

People who are overweight and people with cardiovascular and respiratory problems are most affected by carbon monoxide.

Such people need all of the oxygen they can get. Heavy smokers are also among those who suffer first when air gets polluted with carbon monoxide. This is because the carbon monoxide in their smoke has already tied up 5-10% of their hemoglobin.

Carbon Dioxide

This gas is a natural part of the atmosphere. Respiration of living things constantly adds it to the air. Carbon dioxide is also an important part of the atmosphere. Without it, photosynthesis would cease and all life on earth would eventually die. Why, then, do we call it a pollutant? The burning of fossil fuels (coal, oil, and gasoline) has been putting carbon dioxide into the air faster than nature's cycles can take it away. About 6 000 000 t of carbon dioxide are accumulating in the atmosphere each year.

Carbon dioxide is not directly harmful to humans. You can breathe high concentrations without suffering any ill effects. Scientists are mainly concerned with the "greenhouse effect" that the extra carbon dioxide can cause(see Section 9.4, page 227).

Nitrogen Oxides

Traces of the oxides of nitrogen are natural components of air. But industrialization has increased the concentrations of these gases, particularly in cities. Automobile engines, household furnaces, and power plants put millions of tonnes of these oxides into the air every year.

The nitrogen oxides react with water to form acids. One of these acids is nitric acid which is one of the most corrosive acids known. When the nitrogen oxides contact the moisture in your nose, bronchial tubes, lungs, or eyes, acids are formed. An irritating sensation results. High concentrations of these gases are fatal. Long exposure to low concentrations can cause cancer. Because these oxides are so dangerous and because they come mainly from cars, the government has forced car makers to modify the engines of cars to lower the concentration of these gases.

Studies have shown that the smoke curling from a cigarette has about 160 times the concentration of nitrogen oxides thought to be safe for breathing.

Hydrocarbons

Hydrocarbons are molecules that are made of only two kinds of atoms—hydrogen and carbon. Since they contain carbon, they are organic compounds. Most of the hydrocarbons in the air come from natural sources. Decay of organic matter produces the hydrocarbon methane; forests and other vegetation produce countless other hydrocarbons.

Man-made emissions make up only 15% of the hydrocarbons in the air. Again, this does not seem to be much. But remember that nearly all of it is emitted in cities. Car exhausts give off about 200 kinds of hydrocarbons and are responsible for almost 50% of the man-made emissions. Industrial plants, aircraft, agricultural operations, and spilling of gasoline at service stations contribute most of the remainder.

Many hydrocarbons can cause cancer. For example, benzopyrene is a hydrocarbon that is emitted by cars and coal-burning furnaces. It has been proven to cause lung cancer. Benzopyrene is one of the hydrocarbons that is present in cigarette smoke. It is not surprising that almost all of the people who die of lung cancer were heavy smokers.

Hydrocarbons unite with nitrogen oxides, particularly in cities with sunny climates, to produce **photochemical smog**. Los Angeles is famous for the damaging and irritating brown haze that blankets the city much of the time. But even cities in more temperate climates have occasional episodes of photochemical smog. A sure way to avoid it is to stop putting hydrocarbons and nitrogen oxides into the air.

Ozone

We have already discussed the importance of the ozone layer in the atmosphere in Section 9.1, page 221. It is a necessary and natural part of the air. But, once again, industrial processes are producing ozone in certain areas faster than nature can absorb it.

Ozone is a strong oxidizing agent. It attacks or "burns" other substances. (An ozone molecule is made up of 3 oxygen atoms.) At very low concentrations it causes eye irritation, coughing, and chest pains. Only 1 μL/L breathed in during an 8 h period every day for a year produces bronchitis and other respiratory diseases. More important is its effect on plants. Very low concentrations attack and damage the leaves. Farmers can no longer grow beans in parts of southwestern Ontario because of the ozone from Detroit's factories many kilometres away. Ozone also damages rubber and most textiles.

Lead

Lead, too, is a natural component of the air. But humans are the greatest emitters of lead into the air. Sources include coal burning, pesticide spraying, and the burning of waste. But the most important source is the use of leaded gasoline.

Lead poisoning causes headaches, dizziness, insomnia, anemia, and general weakness. It eventually produces stupor, coma, and then death. Lead is a cumulative poison. Every bit you inhale stays with you, gradually building up to dangerous levels.

Government legislation is gradually eliminating leaded gasoline. Yet many people continue to buy it, even though they have a car that will run on non-leaded gasoline, just because it is cheaper. How often is pollution caused by people who are more concerned with money than they are with the welfare of others and themselves?

Discussion

1. **a)** Define air pollution.
   b) Explain why methane (natural gas) produced by a marsh would not likely be air pollution, whereas methane emitted by an oil refinery would be air pollution.
2. **a)** What is particulate matter?
   b) List the main natural sources of particulate matter.
   c) List the main human sources of particulate matter.
   d) Describe 2 hazards created by particulate matter.
3. **a)** Why is sulfur dioxide dangerous to humans?
   b) List the main natural and human sources of this gas.
   c) Describe the effects of sulfur dioxide on people.
4. **a)** What are the main sources of carbon monoxide?
   b) Describe the effects of carbon monoxide on humans.
5. **a)** Explain why carbon dioxide is considered a natural part of the atmosphere.
   b) What has destroyed nature's carbon dioxide balance?
6. **a)** What are the main sources of nitrogen oxides?
   b) How do these oxides affect people?
7. **a)** What are hydrocarbons?
   b) What are the natural sources of hydrocarbons?
   c) What are the man-made sources of hydrocarbons?
   d) What dangers do hydrocarbons present to people?
8. **a)** Why is ozone a natural and important part of the atmosphere?
   b) Describe the effects of man-made concentrations of ozone.
9. **a)** What are the man-made sources of lead?
   b) What are the symptoms and effects of lead poisoning?
   c) Do you think lead concentrations will get higher or lower in the future? Why?
10. This unit has mentioned that both carbon dioxide and particulate matter could affect the climate of the earth. Summarize the effect that each could have. What is your prediction for the future climate of the earth?
11. **a)** Read the definition of air pollution again. Is tobacco smoke in a room a form of air pollution? Defend your answer.
    b) A recent study concluded that a non-smoker who spent 4 or 5 h at a cocktail party where half the people were smoking would inhale more organic matter than a person would

inhale in one month from the air of New York City. Another study concluded that a non-smoker who spends 1 h in a smoke-filled room can inhale as much of a certain cancer-causing substance as a smoker who smokes 35 filter-tipped cigarettes a day. List some of the substances in that smoke and describe their possible effects on humans. On the basis of this and other information, your class may wish to conduct a debate on whether or not smoking should be banned from public places like restaurants and theatres.

9.14 INVESTIGATION: Examining Particulate Matter

The source and composition of a particle can often be determined by its size, shape, colour, and texture. For example, metal shavings and sand grains are easy to identify. In this investigation you collect samples of airborne particulate matter from several sites. You then use a microscope to examine them.

Materials

compound microscope (also a stereo microscope, if available)
several glass slides
petroleum jelly

Procedure

a. Coat the glass slide with a thin film of petroleum jelly or any other transparent sticky material.
b. Place one slide at each of several sites. Select sites that you suspect may have high concentrations of certain particulates. Some examples are construction sites, underground parking lots, factory areas, congested streets, and a smoky room. The slides should be placed where wind cannot blow settled dust onto them. You only want to collect airborne dust. Leave the slides exposed until you can see evidence of particulate matter. One day will often do.
c. If possible, examine the particulate matter on each slide with a stereo microscope. It is useful for studying the largest particles.
d. Examine the particulate matter on each slide with a compound microscope. Use oblique incident light as shown in Figure 9-15. Do not illuminate the slide from below. This method is the best for viewing particles larger than 10^{-3} cm in diameter.
e. Transmitted light is best for viewing the smallest particles. Illuminate the slide from below in the usual fashion. If the film

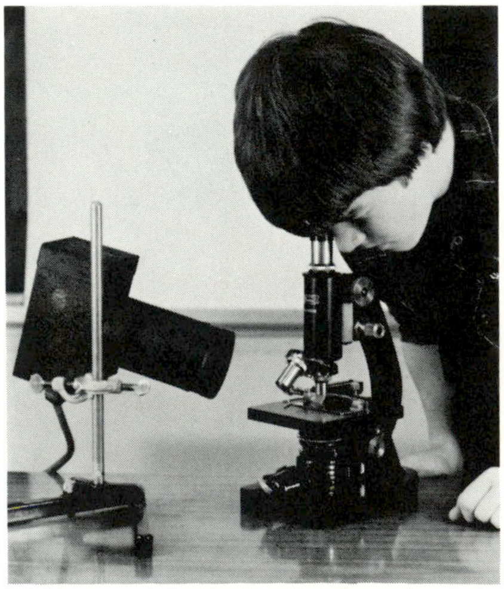

Figure 9-15 Use of oblique incident light in the examination of particulate matter. Note the position of the light relative to the microscope stage.

of petroleum jelly is not too thick, you may be able to study the smallest particles.

Discussion

1. Sketch and describe each different type of particle. If possible, name the material of which each particle is made.
2. Suggest the most likely source of each type of particle.

Highlights

The atmosphere consists of several layers, each of which has its own characteristics. From the earth outward the layers are the troposphere, stratosphere, mesosphere, thermosphere, and exosphere. The ozone layer is in the stratosphere and the ionosphere is in the thermosphere.

Air is a mixture of several gases, but is mainly nitrogen (78%) and oxygen (21%). The oxygen in the air causes oxidation of many substances. Rapid oxidation releases heat and light; slow oxidation releases heat only. The rusting of iron is an example of slow oxidation.

Organic substances such as fuels produce carbon dioxide and water when complete combustion occurs. When incomplete combustion occurs, carbon monoxide is also formed.

The corrosion of iron is called rusting. This costly process is controlled through galvanizing, electroplating, alloy formation, and sacrificial corrosion.

Carbon dioxide is more dense than air and does not support combustion. Therefore it is useful in fire extinguishers. This gas also plays a role in carbonated beverages and baking. In its solid form (dry ice), it is used as a coolant.

Air pollutants are substances in the atmosphere that are harmful to living things or their habitats. On a global scale, the most common air pollutants are particulate matter, sulfur dioxide, carbon monoxide, carbon dioxide, nitrogen oxides, ozone, and lead. From the viewpoint of human health, tobacco smoke is the most hazardous air pollutant.

UNIT

10 Forces in Nature

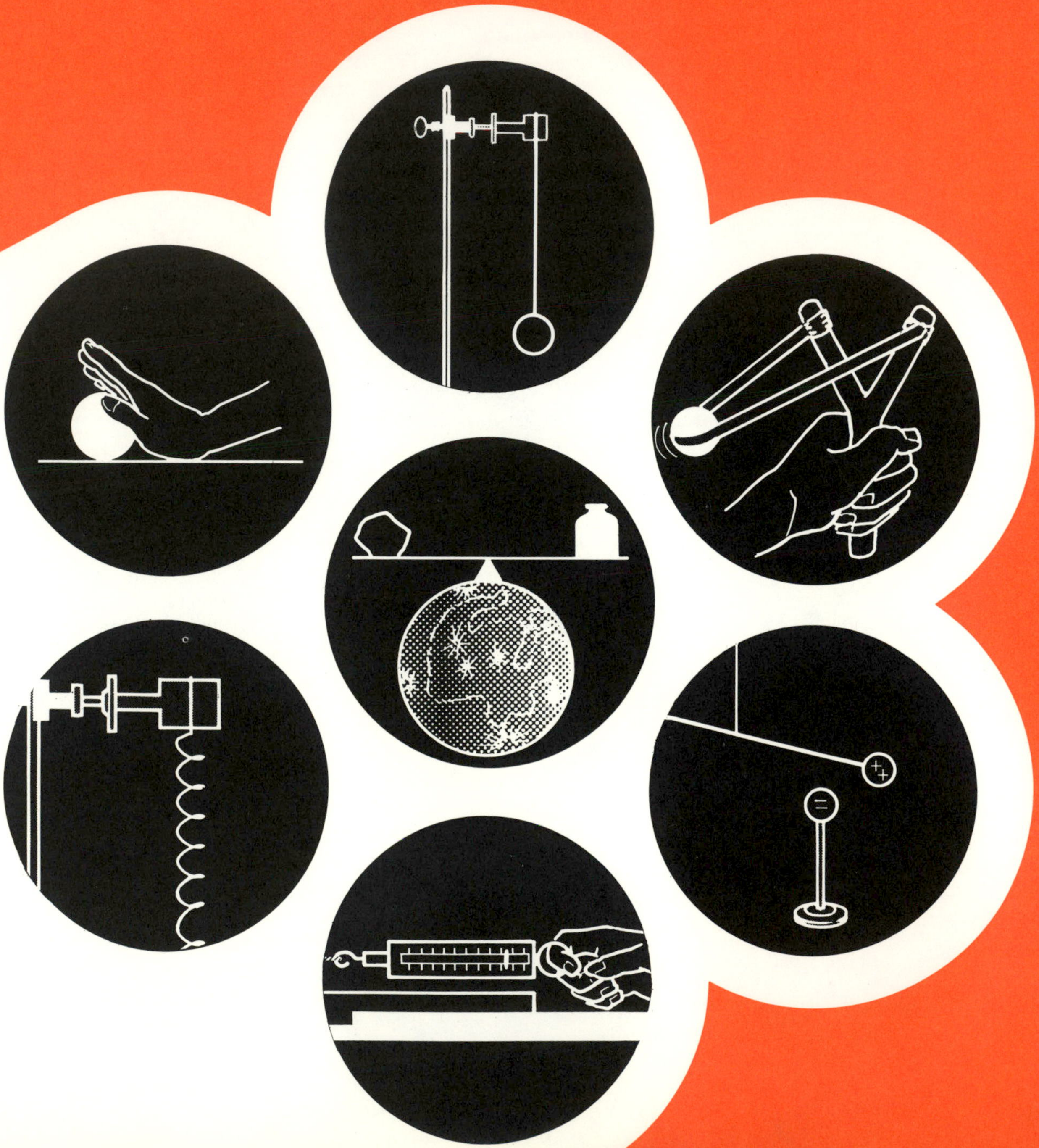

In Units 3 and 4 you learned that matter is made of small particles. These particles are attracted to one another by forces that exist between them. What is a force? How can we tell when a force is present? How are forces measured? What kinds of forces exist in nature? You find answers to these questions in this unit.

10.1 Force: A Description

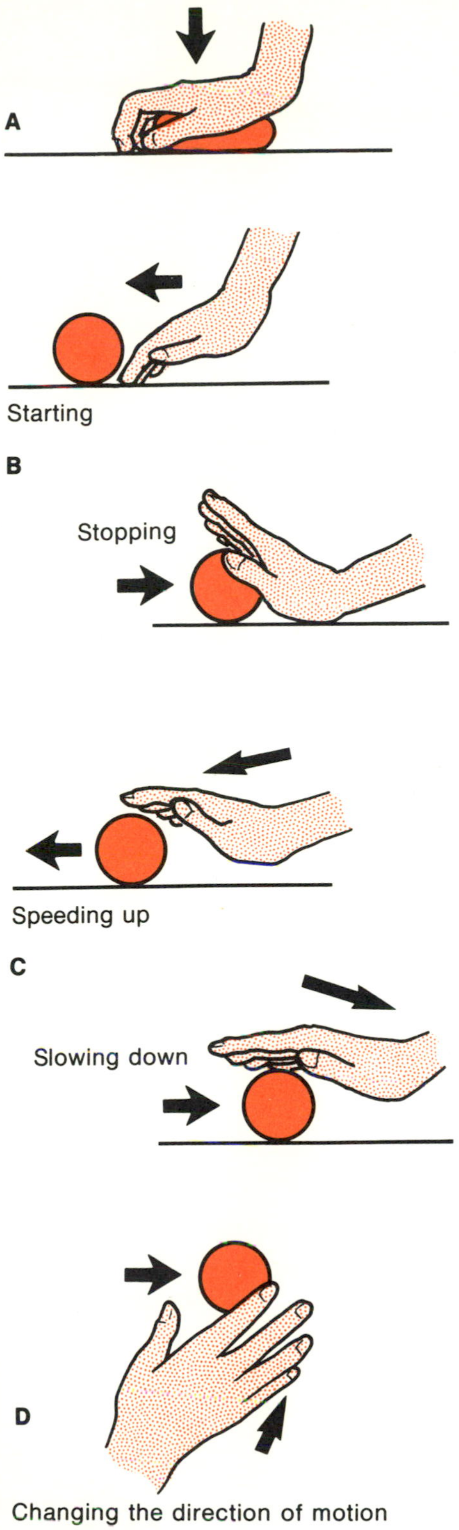

Figure 10-1 The effects of a force on a body.

A force is a push or a pull. Although this description is correct, it does not tell us very much about a force. We can get a better understanding of what a force is by studying the effects of a push or a pull on a body.

Consider a soft rubber ball that is lying on a level, hard table. When you place the palm of your hand on the top of the ball and push downward, the ball changes shape. It flattens vertically and lengthens horizontally. But the same size force applied down on the surface of the table does not change its shape. A much larger force is needed. *A force can change the shape of a body* (Fig. 10-1, A).

If, instead of pushing down on the stationary ball, you give the ball a quick horizontal push, the ball starts to move along the table. When the moving ball is given a quick push of the same size and for the same time in a direction opposite to its motion, the ball stops. However, if we replace the small rubber ball with a very large cube of iron and apply a small force, the cube does not move. The cube is too massive. Also, the force of friction opposing the motion is too large to overcome. *A force can start an object moving or can stop it* (Fig. 10-1, B).

If you give a continuous push to a ball at rest, it not only starts moving, but continues to speed up. This increase in speed continues for as long as the force is applied. A similar push in a direction opposite to the motion decreases the speed of the ball. *A force can change the speed of a body* (Fig. 10-1, C).

There is one other situation to consider. If you give the ball a sideways push as it is moving along the table, the direction of motion of the ball changes. Instead of speeding up or slowing down as before, the ball begins to move in a new direction. *A force can change the direction of motion of a body*(Fig. 10-1, D).

The examples illustrate that a force has a tendency to change the shape and/or the motion of a body.

Look at the examples again. They show two other properties common to all forces. Each force has a *definite strength*, or as scientists say, a **magnitude**. The magnitude is described qualitatively using adjectives such as huge, large, small, or minute. A large force is needed to push the iron cube along the table. A small force moves the ball. The magnitude of a force is expressed quantitatively by using a numeral and a unit to replace the adjectives. Thus a scientist may speak of a force of ten newtons. Each force also has a *definite direction*. Adjectives such as left, right, up, down, and compass directions such as north, south, southeast, and north-west are used. In our earlier examples, for you to lift the ball up off the table your hand had to exert a force on the ball in an upward direction. *All forces have a definite magnitude and a definite direction*.

There are many types of forces in nature. To distinguish among them, an adjective is used that describes the source of the force. The force applied to the ball in our example came from the person's muscles. This force is called a *muscular force*. In Unit 6 a magnet was used to separate iron from sulfur in a mechanical mixture. The force arising from the magnet is called a *magnetic force*. The forces holding atoms of carbon and oxygen together to form molecules of carbon dioxide are *electric forces*. It is this same electric force arising from a recently used comb that attracts paper to the comb. A wire or string stretches and is under tension when supporting a mass. The force exerted by a wire or string under tension is called a *tensile force*. Other kinds of forces are *gravitational forces* and *elastic forces*. These forces are also named by using an adjective to describe the source. Figure 10-2 shows some types of forces.

The effects of a force on a body and the source of the force give a more complete definition of a force. *A* **force** *is defined as a push or a pull of a definite magnitude and direction. A force is exerted by one body on a second body. A force can produce a change in the shape or a change in the motion of the second body.*

Figure 10-2 Types of forces.

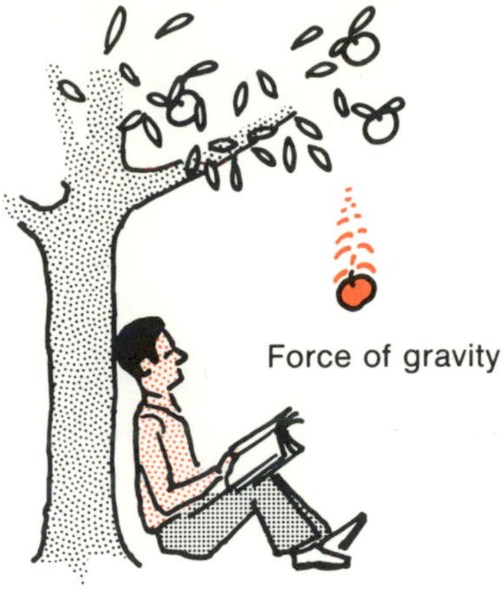

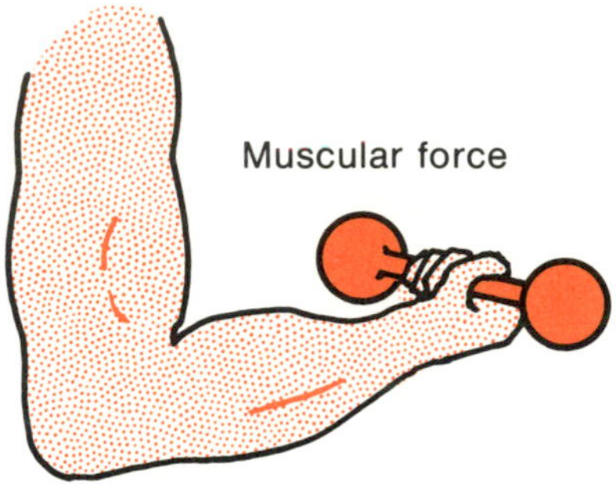

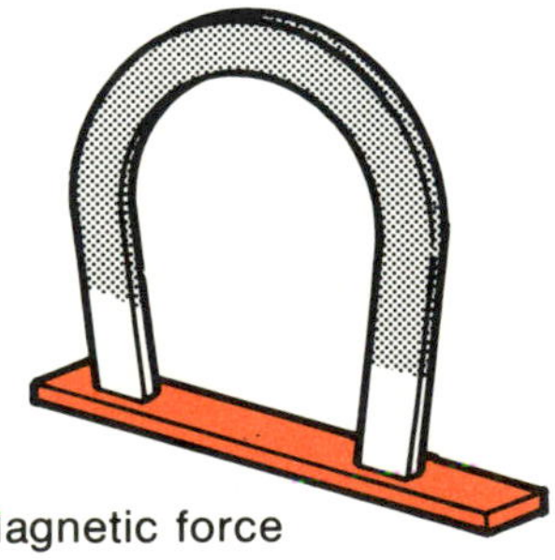

Discussion

1. Define a force.
2. What effects can a force have on an elastic object?
3. How does a physical quantity like force differ from a physical quantity like density?
4. What is the usual method for naming a force? Give 5 examples of forces named this way.
5. In each of the following, what is the name of the force causing the change in motion or shape of the object? a) a book falls to the floor; b) a person climbs a flight of stairs; c) an electromagnet lifts a bale of scrap; d) a pile driver drives a steel beam into the ground; e) a balloon rubbed on a sweater sticks to the wall.
6. Name the forces that are acting on a mountain climber climbing up a rope from one ledge to another.

Figure 10-2 Types of forces.

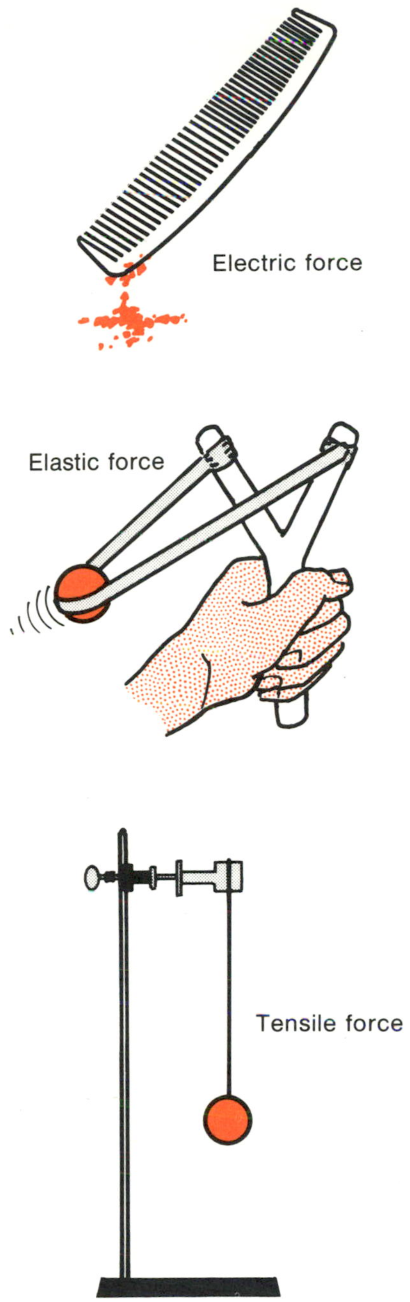

10.2 Vector and Scalar Quantities

Scalar Quantity

In Unit 2 you studied the physical quantities of length, area, volume, and mass. These quantities are all examples of scalar quantities. *A* **scalar quantity** *is any physical quantity which can be described by a single numeral and the correct unit.* For example, the height of a man could be expressed as 180 cm. The numeral 180 gives the height of the man only if the correct unit cm is included. The unit mm with the numeral 180 would not be correct in this example since no man on earth is that short.

Density is another example of a scalar quantity. The density of nickel is expressed in Table 25, page 58, using the numeral 8 900 and the unit kg/m^3. There are many examples of scalar quantities in nature. All scalar quantities can be described by a single numeral and a correct unit.

Vector Quantity

Force is not a scalar quantity. It cannot be described completely using only a numeral and a unit. A *direction* must be included with the numeral and unit. Force is an example of a vector quantity. *A* **vector quantity** *is any quantity which can be described completely only by stating both a magnitude (a numeral and a unit) and a direction.* A push of 30 units of force east on a toy car is an example of a vector quantity. The magnitude of the force is 30 units of force and the direction is east. (You will find out what the unit of force is in Section 10.5.) A vector quantity is represented on diagrams by a line with an arrowhead on one end. Such a line is called a **vector**. The length of the vector is drawn to scale to represent the magnitude of the vector quantity. The direction of the arrowhead shows the direction of the vector quantity. The vector drawn to represent a push of 30 units of force east acting on the toy car is shown in Figure 10-3. Each 1 cm of the directed line segment represents 10 units

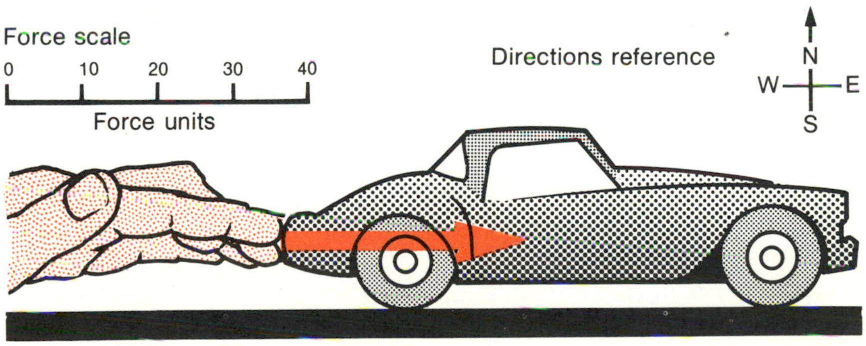

Figure 10-3 A force vector with scale indicated.

of force. The directed line segment to represent 30 force units is 3 cm long. The arrowhead points east, showing that the force is east. The scale is shown above the diagram. This scale enables anyone to interpret the vector to determine the magnitude of the vector quantity. The force vector must be drawn from the point of application of the force. There are many kinds of vector quantities, but force is the first one you have met in this text.

Discussion

1. Compare a scalar quantity and a vector quantity. How are they the same? How are they different?
2. What is a vector? When is a vector used?
3. Make a table with these headings: Scalar quantity; Numeral; and Unit. List 5 different scalar quantities in the left column. Give a specific numerical example of each by completing the table.
4. Name a vector quantity other than force. Why is it a vector quantity?
5. Draw vectors to represent the following forces. Use a scale of 1 cm to represent each 2 units of force. Consider up the page to be north, down—south, to the right—east, and to the left—west.
   a) 4 force units N
   b) 8 force units E
   c) 6 force units SW
   d) 9 force units NE
6. Draw the side view of a rectangular tool box. Show a handle on top. Draw a force vector to show a muscular force of 10 force units being applied upward on the handle. Show a force of gravity of 10 force units pulling downward. Indicate the scale used.

10.3 INVESTIGATION: Sensing a Force and Measuring Changes in Shape

Before we can describe any physical quantity we need to know two things. First we must know how to determine whether or not the quantity is present. Then we must have a method for measuring how much of the quantity is present.

We know how to detect the presence of a force. A force is being exerted on a body if it changes in shape and/or motion. An elastic material such as a rubber band stretches as it is pulled. The larger the pull, the larger the stretch. A ball speeds up as it is pushed horizontally on a table. The larger the push, the larger the increase in speed in a given time. But how can the magnitude of the force be measured? A rubber band or spring changes length when a force is applied. Perhaps a "stretchy" material can be used to measure the magnitude of a force.

In this investigation you study qualitatively the effect of the magnitude of a force on the stretch of a rubber band.

Materials

a rubber band about 5-10 cm long
ruler
pen or pencil

Procedure

a. Prepare a data table for each partner with the headings "Length rubber band is stretched" and "Estimated pull".
b. Divide responsibilities. One partner becomes the experimenter, and the other the recorder.
c. Place the ruler horizontally on the desk surface in front of the experimenter.
d. Hold a pencil (or pen) in the left hand. Loop a rubber band around the pencil and the index finger of the right hand. Hold the left end of the rubber band stationary on the desk with the pencil. Tighten the band with the index finger to take out the slack. Do not stretch the band.
e. Line up the index finger of the right hand with the zero mark on the ruler as shown in Figure 10-4.
f. Stretch the rubber band by moving the index finger from the 0 cm to the 6 cm mark. Stop. Call the force the index finger is exerting "the unit force". Record the length that the rubber band is stretched and "the unit force" in the data table.
g. Now move the index finger to the 12 cm mark. Stop. Have the experimenter estimate the size of the force being exerted on the rubber band by the index finger in terms of "the unit force". Is it bigger? If so, how many "unit forces" is it? Record the length the rubber band has been stretched from 0 and the new force in terms of unit forces.
h. Continue stretching the rubber band in increases of 6 cm until a maximum stretch of 30 cm is reached. The experimenter must estimate how the applied force compares to the unit force in each case. Record this estimate in the table.
i. Return the index finger of the right hand to line up with the zero mark on the ruler. The experimenter must now close his/her eyes. Without looking, stretch the elastic band by moving the index finger until you sense that a force of "2 unit forces" is being applied to the elastic band. Record the stretch corresponding to this estimated force in the table. Repeat this procedure for a force of "4 unit forces".
j. Reverse roles and repeat the experiment.
k. Compare results.

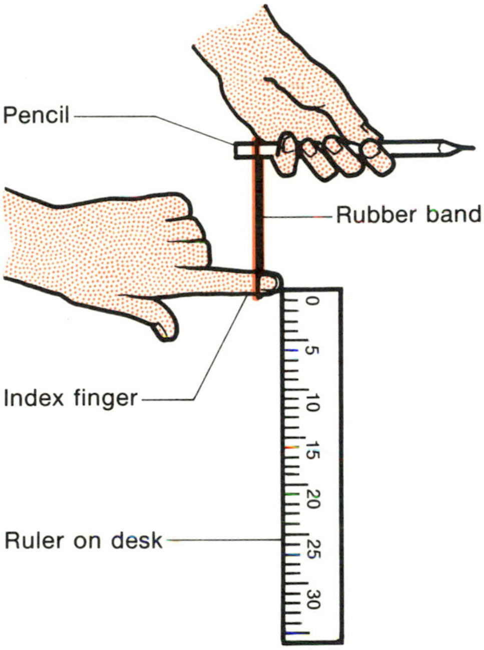

Figure 10-4 Aligning the rubber band.

Discussion

1. How much force did it seem to take to move the index finger from the 6 cm mark to the 12 cm mark? from the 12 cm mark to the 18 cm mark?
2. Compare the estimated force each partner recorded for each length the rubber band was stretched.
3. Compare the stretches of the rubber band each partner recorded when estimated forces of 2 units and 4 units were applied with the eyes closed.
4. How effective is the sense of touch for estimating a force?
5. How effective is the sense of touch for applying a force?

10.4 Mass and the Force of Gravity

The results of Investigation 10.3 showed that the sense of touch is not very accurate. It can be used to detect the presence of a force, but it is not suitable for measuring it. Nor can it be used to apply a specific force. However, if we use a set of identical forces to stretch the rubber band, we can mark fixed points called gradations on the scale. The result is a calibrated force measurer that can be used to measure unknown forces.

Suppose the rubber band used in Investigation 10.3 is hung from a support as shown in Figure 10-5. If a chunk of matter having a mass of 1 kg is hung from the rubber band, the rubber band stretches. This indicates that the 1 kg mass is applying a force to the rubber band. When this 1 kg mass is removed and another 1 kg mass is attached, the rubber band stretches the same amount. Identical masses can be used to apply identical forces to a rubber band in the laboratory. Where do these forces come from?

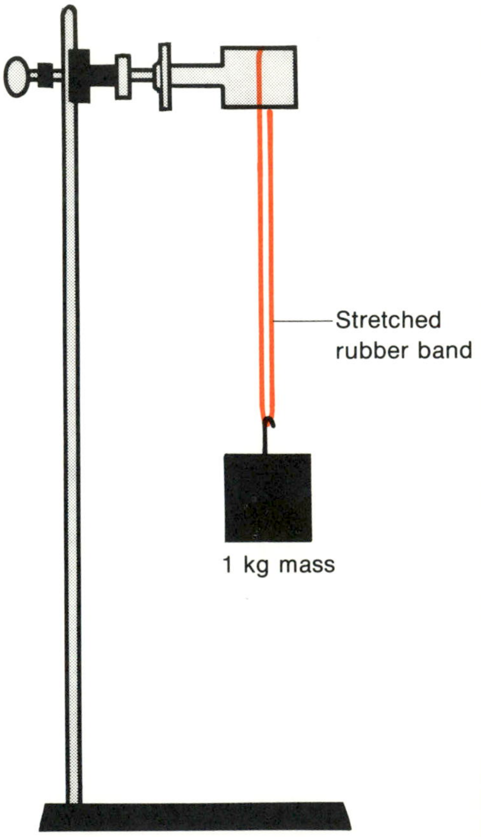

Figure 10-5 A rubber band stretched by using the force of gravity on a 1 kg mass.

Mass

You found out in Unit 2 that two objects have the same amount of matter in them if they have the same mass. **Mass** *is defined as a measure of the amount of material in an object*. The mass of an object can be measured using an equal arm balance. The unknown mass is placed on one pan and standard masses are added to the other pan until balance is achieved. The pull of gravity acting on the unknown mass is balanced against the pull of gravity acting on the standard mass. The equal arm balance and standard masses can be used to measure the mass of an object as long as a pull of gravity exists. When the mass of an object is measured using an equal arm balance, the same result is obtained whether the measurement is made on the earth or on the moon. The force of gravity on the unknown mass is different on the earth and on the moon. But the force of gravity on the standard masses on the other side of the balance changes in exactly the same way (Fig. 10-6). The mass of an object never changes unless, of course, some of the matter in the object is removed.

Force of Gravity

The **force of gravity** *is defined as the force of attraction between two masses in the universe.* Sir Isaac Newton, a 17th-century physicist, published the generalization which describes the force of gravity. It is called the **Principle of Universal Gravitation**. It states that *every mass in the universe exerts a force of gravitational attraction on every other mass. This force of gravity between any two masses varies directly as the product of the masses and inversely as the square of the distance between their centres.* The formula which summarizes this principle is $F = \frac{Gm_1m_2}{d^2}$. F is the force of gravity, G is a constant, m_1 and m_2 are the masses, and d is the distance between the centres of the masses. This principle is illustrated in Figure 10-7 using vectors.

Figure 10-6 Measuring mass on earth and on the moon. The same masses balance the unknown mass on the moon. But, on the moon, both the standard masses and the unknown mass experience a force of gravity about 1/6 that on earth.

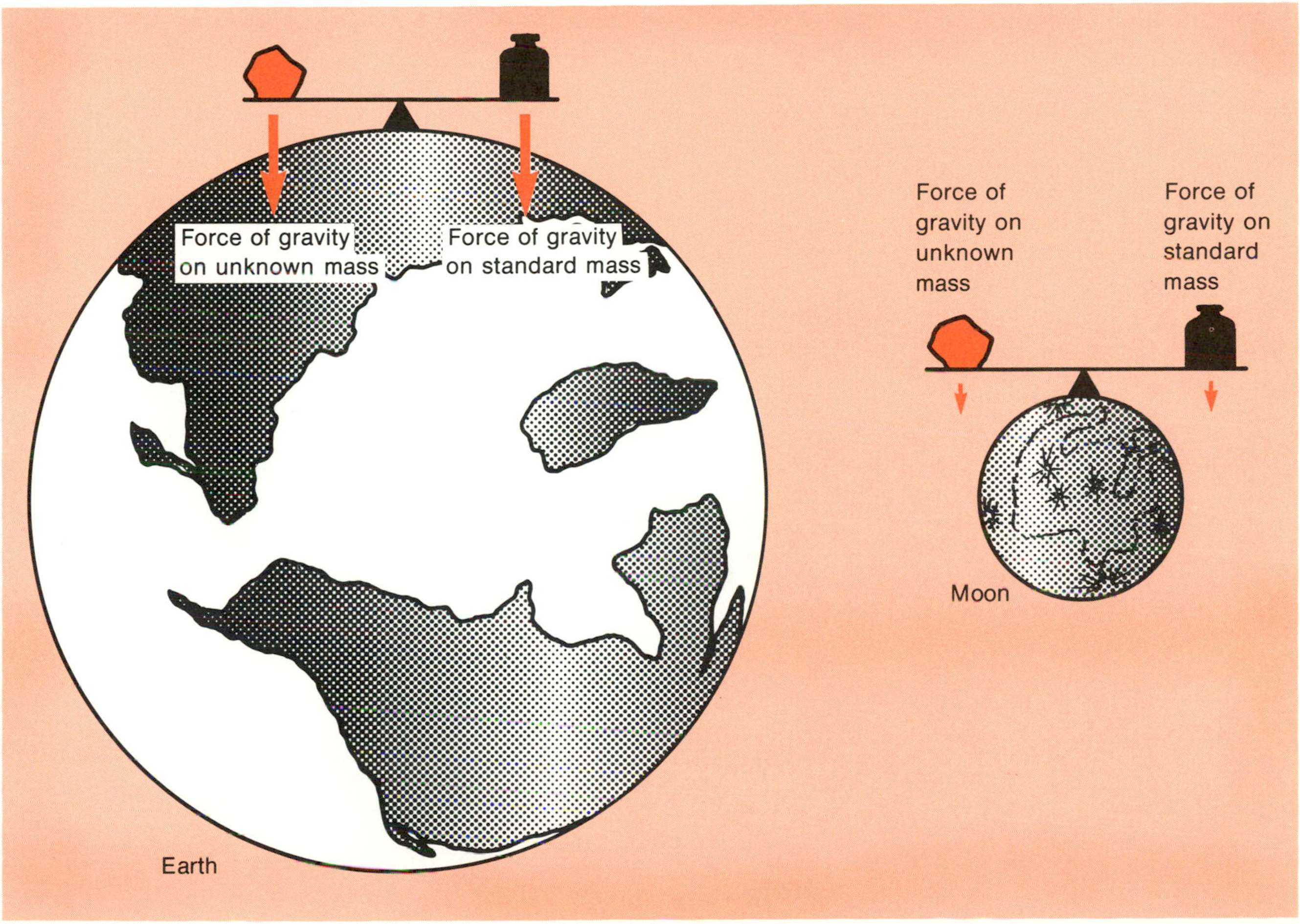

The force of gravity acting on an object can change. The force of gravity between an object on the earth and the earth itself depends on the mass of the earth and the mass of the object. The greater the mass of the object, the greater the force of gravitational attraction toward the earth. The force of gravity acting on a 2 kg mass measured in the lab is 2 times as great as the force of gravity measured for a 1 kg mass.

The force of gravity on an object depends on how far the object is from the centre of the earth. The greater the distance between the centre of the earth and the object, the less the force of gravitational attraction. The force of gravity acting on a 1 kg mass at sea level is greater than at the top of Mt. Everest. Since the radius of the earth is greater at the equator than at the poles, the force of gravity on a 1 kg mass is less at the equator. There is about 0.5% difference—not much to worry about. However, when the force is measured at the surface of the earth and then 1 earth radius above the earth, the force of gravitational attraction is one quarter as great—a significant difference. This was clearly evident during several U.S. space missions as the astronauts were able to move around easily in their space capsules. An astronaut being attracted to the earth with a force of 800 force units is only attracted toward the earth with 200 force units at this new distance. Imagine the high-jump record that could be set out in space!

Figure 10-7 The principle of universal gravitation. Notice how the masses and the distance between the centres of the masses affect the force of gravity.

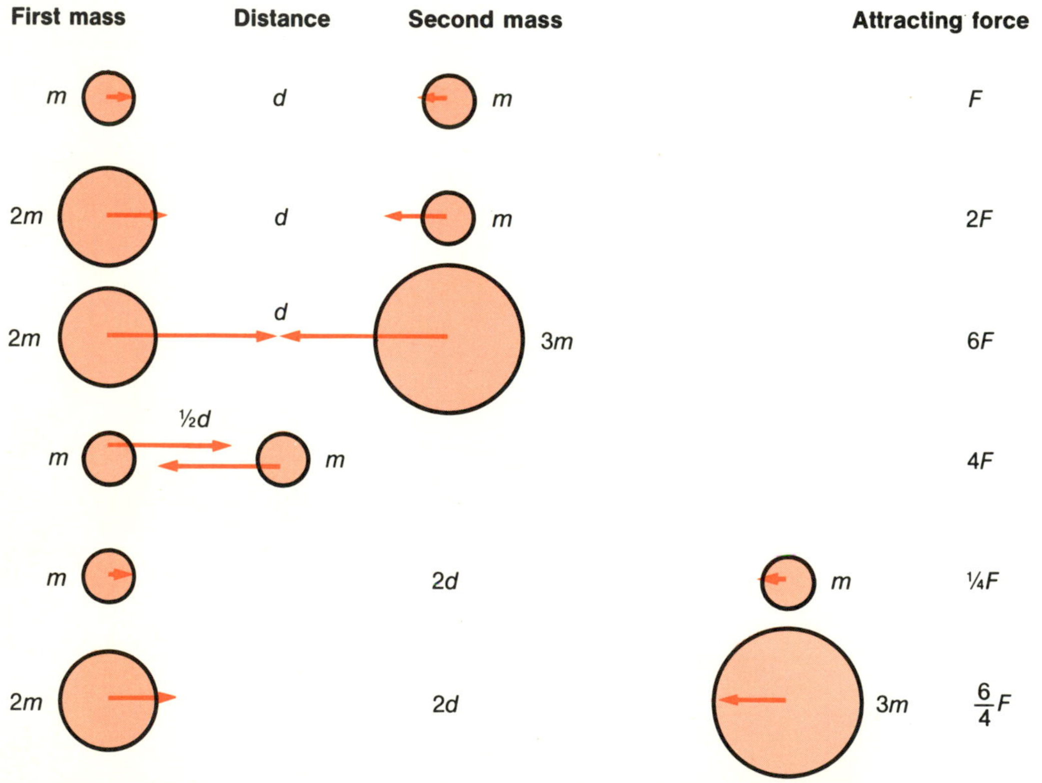

The force of gravity between an object and a celestial body such as the moon or Mars also depends on the mass of the celestial body. The greater the mass of the celestial body on which the force of gravity is being measured, the greater the force of attraction. The moon attracts a 1 kg mass with a force about 1/6 the force of the earth. Jupiter attracts it with a force about 2.6 times as great. In these cases the force of attraction is different from that on earth because of both the mass of the celestial body and its radius.

The force of gravity is measured in the laboratory using a spring scale. The spring scale is calibrated by stretching the spring using known forces. The forces which correspond to these stretches are marked on a scale. Once the spring scale is calibrated in this way, it can be used any place in the universe to measure forces within the range of the instrument.

Discussion

1. Look up the words "detect" and "measure" in a dictionary. Use each in a sentence to show that you understand the difference between detecting and measuring a force.
2. Define mass. What is the SI base unit for mass?
3. What instrument is used to measure mass?
4. Why is a spring scale not suitable for measuring mass in a spaceship?
5. Describe a method for measuring the mass of an object at a point in space where there is no force of gravity. (Hint: You have an equal arm balance, a set of standard masses, and a string.)
6. Define force of gravity.
7. Where would it be easier to set a new high-jump record, at the equator or at one of the poles? On the earth or on the moon?
8. Make a table like Table 65. Compare mass and the force of gravity by completing the table.

TABLE 65 A Comparison of Mass and the Force of Gravity

Feature compared	Mass	Force of gravity
Definition		
Measuring instrument		
How the quantity is measured		
Factor(s) it depends on		
Variability		

9. State Newton's Principle of Universal Gravitation. Then explain it using an example.

10. If we compare the earth and the moon we get these ratios:

$$\frac{\text{Radius of earth}}{\text{Radius of moon}} = \frac{3.7}{1.0} \qquad \frac{\text{Mass of earth}}{\text{Mass of moon}} = \frac{81}{1.0}$$

Use these ratios to show that the force of gravity on a 1 kg mass on the moon should be about one sixth the force of gravity on a 1 kg mass on the earth.

10.5 The SI Unit of Force

The Newton

The SI unit of force is the newton. The symbol for the newton is N. The newton is properly defined in terms of the effect a force has on the motion of a body of a definite mass. However, for our purposes, *one* **newton** *is roughly the force exerted by the earth on a* 0.1 kg *mass*. As a result, *a* 1 kg *mass experiences a force of gravity on earth of about* 10 N.

The newton is a rather small force. It is roughly the force exerted upward by your hand when supporting a medium-sized apple or a D cell for a transistor radio.

The champion weight lifter at the Summer Olympics in Montreal in 1976 lifted a mass of 255 kg. Thus he exerted an average muscular force of about 2 550 N on the bar bells. The cable of a crane exerts a tensile force of about one meganewton (1 MN) to hoist a large locomotive.

Derived Units

The newton is a derived unit. A derived unit is a unit that consists of two or more base units. There are two kinds of derived units: those with compound names, and those with special names.

A derived unit with a compound name is one that contains two or more units in its name. An example is the kilogram per cubic metre (kg/m^3)—a unit for density you studied in Section 3.2.

The newton is our first example of a derived unit with a special name. The derived unit of force expressed in terms of base units is the kilogram metre per second squared ($kg{\cdot}m/s^2$). This is given the special name "newton" to make it simpler and to honour Sir Isaac Newton. One newton expressed in base units is equal to one kilogram metre per second squared ($1\ N = 1\ kg{\cdot}m/s^2$).

Calculating the Force of Gravity of the Earth on a Mass

The earth exerts a force of gravity of about 10 N on a mass of 1.0 kg. To determine the force of gravity on any mass on earth, first express its mass in kilograms. Then multiply the mass by the force of gravity-mass ratio of 10 N/kg. The force of gravity (***F***) in newtons, the mass (***m***) in kilograms, and the force of gravity-mass ratio (***g***) are related in the following way: ***F*** = ***m g***.

Sample Problem

Calculate the force of gravity of the earth on a 320 g mass.

Solution

$$m = 320\text{ g} = 0.32\text{ kg}$$
$$g = 10\text{ N/kg}$$
$$F = mg$$
$$= 0.32\,\cancel{\text{kg}} \times 10\frac{\text{N}}{\cancel{\text{kg}}}$$
$$= 3.2\text{ N}$$

Discussion

1. **a)** What is a newton?
b) Define a derived unit.
c) Express the newton in base units.
d) Why was the derived unit for force replaced by the special name newton?

2. **a)** What is the force of gravity-mass ratio on earth?
b) What would it be on the moon?

3. What is the approximate force of gravity in newtons acting on each of the following masses on earth:
a) 1 kg **c)** 0.5 kg **e)** 50 g
b) 4 kg **d)** 100 g

4. Make a table with these headings: Name of item; Mass of item (kg); Force of gravity on item (N). Find out the mass of 5 common household items. (Mass is usually printed on the label.) Make sure you lift them when looking at the label to get a feeling for the pull of gravity in newtons. Complete the table.

5. Find out your mass in kilograms. What is the force of gravity acting on you on earth? What would it be on the moon? What would your mass be on the moon?

6. Determine the mass the earth attracts with each of the following forces:
a) 4 N **c)** 300 N **e)** 10 kN
b) 20 N **d)** 1.5 kN **f)** 8 MN

7. Draw the side view of a suitcase. The suitcase has a mass of 15 kg. Draw a force vector to show the force of gravity acting on the suitcase. Indicate the scale used. Where do you think the force vector should be drawn from? Draw a force vector exerted upward by the floor.

10.6 INVESTIGATION: Calibrating a Spring Scale Force Measurer

A stiff spring will not be extended as much for a given force as a weak spring would be. By choosing suitable coil springs, it is possible to construct spring scales which have a large range, but are not very sensitive to small changes in force. We can also construct spring scales that have a small range and are very sensitive to small changes in force. Commercial spring scales are enclosed in a protective case with a scale permanently marked on the case. They have a zeroing feature in case the spring becomes permanently stretched.

In this investigation you calibrate a force measurer for measuring forces in newtons. You then use it to measure the force of gravity on some common objects. Next you plot a graph of your results. Finally you find out the relationship between the force applied to an elastic spring and the extension produced.

Materials

one piece of graph paper per person

spring
set of hooked masses
ring clamp
ring stand
C-clamp
metre stick
adjustable clamp
roll of masking tape
golf ball
hockey puck
book

Procedure

a. Make a table like Table 66.
b. Record the number or symbol of the spring you are using.
c. Attach a strip of masking tape to the entire length of the metre stick beside the scale. Do not cover the scale.
d. Clamp the ring stand to the edge of the table using a C-clamp.
e. Attach an adjustable clamp to the ring stand. Face it out from the table. Clamp a metre stick in an upright position with the 0 end up.
f. Attach a ring clamp near the top of the ring stand. Face it out.
g. Suspend the metal spring over the edge of the table from the ring clamp. Adjust the metre stick so that the 0 is opposite the

free end of the spring. Make a mark on the masking tape next to the 0 of the metre stick and label it 0 N. The final set up is shown in Figure 10-8.

h. Hang a 100 g mass on the lower end of the spring. Lower it gently until the spring supports the mass. The spring is now being pulled by the force of gravity on a 100 g mass. This is about 1 N. Record in the table the extension of the spring, the mass supported, and the force pulling on the spring. Make a mark on the masking tape opposite the end of the spring. Label this 1 N.

i. Add more mass to the free end of the spring in steps of 200 g up to a maximum of 1 300 g. Each time the mass is added, record the total mass supported by the spring, the total force exerted on the spring, and the extension of the spring in centimetres. Also mark on the masking tape opposite the end of the spring the force being supported at that extension.

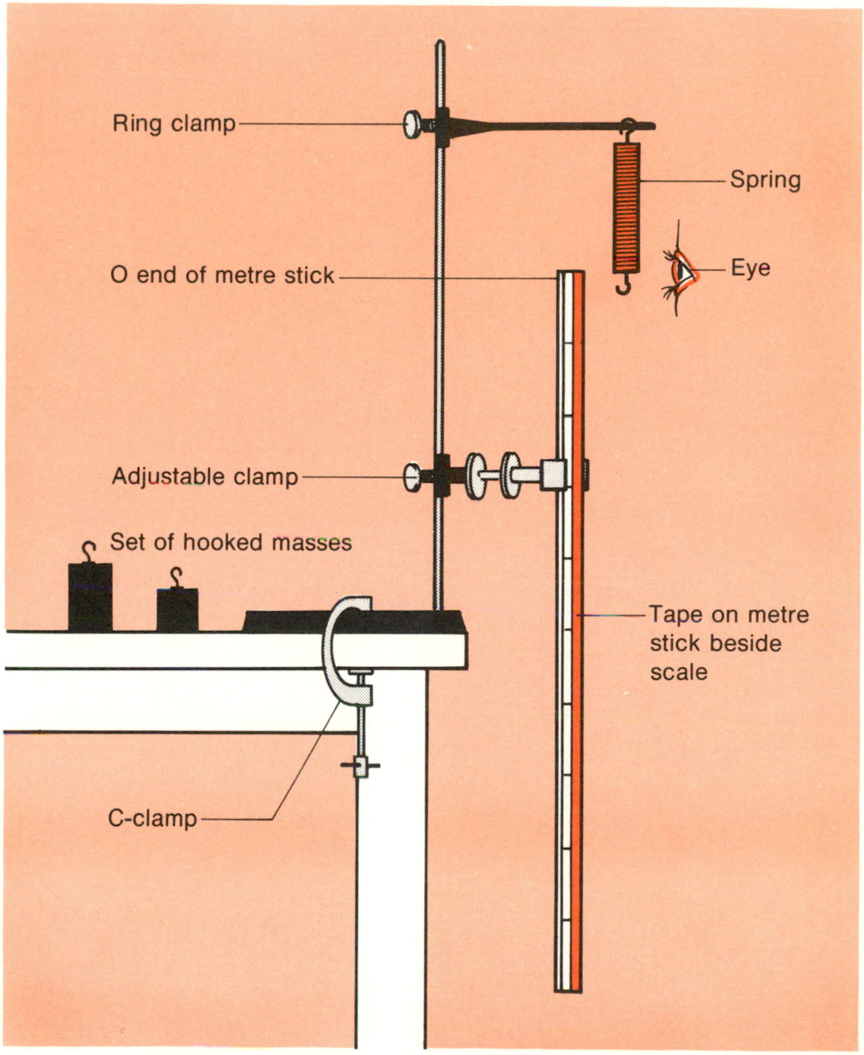

Figure 10-8 Calibrating a spring-scale force measurer.

j. Carefully remove the masses from the end of the spring. Attach the golf ball using a small length of masking tape. Record the extension of the spring. Place a mark on the masking tape opposite the end of the spring. Label it gb for golf ball.

k. Repeat step j for the hockey puck (hp) and for the book (b).

l. Estimate the force of gravity acting on the golf ball, the hockey puck, and the book. You can do this more accurately by dividing the interval corresponding to each newton on the tape into 10 equal parts.

m. Plot a graph with the force applied to the spring on the vertical axis and the extension on the horizontal axis. Convert extensions in centimetres to metres before doing this.

TABLE 66 Calibrating a Force Measurer
Spring # or symbol ________

Attached mass (g)	Force (N)	Extension of spring (cm)	Extension of spring (m)
100			
300			
500			
700			
900			
1 100			
1 300			
Golf ball			
Hockey puck			
Textbook			

Discussion

1. Compare the increase in stretch for each increase of 1 N force applied. Was this increase in stretch the same at the beginning of the experiment as at the end? Is there a pattern in the behaviour of the spring?
2. Can we assume that an elastic spring increases in length by the same amount for each 1 N of force applied?
3. What was the force of gravity acting on the golf ball? the hockey puck? the textbook? How do your values compare with those of other groups? Should they be the same?
4. Describe the shape of the graph obtained. Did it go through the origin? What relationship exists between the extension of the elastic spring and the applied force?
5. From your graph determine the extension (in metres) that would be produced by a force of 4 N. What is this procedure called?

6. From your graph determine the extension (in metres) that would be produced by a force of 14 N. What is this procedure called?
7. Compare the extensions you obtained corresponding to the forces of 4 N and 14 N with those of other groups. Are they the same? Should they be the same?

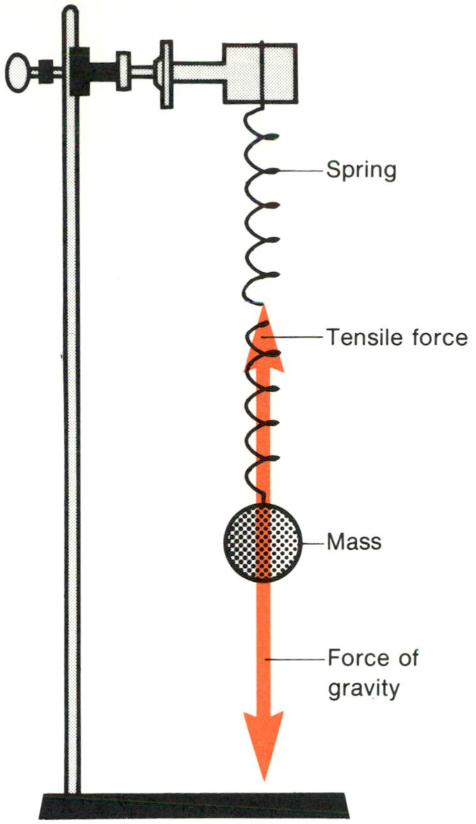

Figure 10-9 A spring supporting a mass. Note the magnitude and direction of the forces.

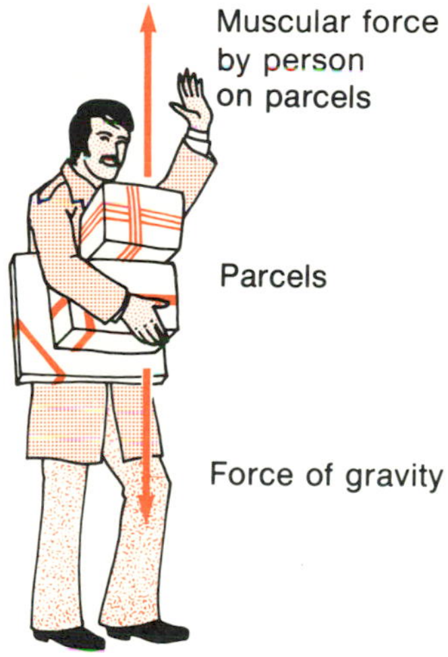

Figure 10-10 A person holding several parcels. Note the magnitude and direction of the forces.

10.7 Balanced and Unbalanced Forces and Newton's First Law

A mass suspended from a spring does not appear to have a force acting on it. The shape of the mass is not distorted, nor is its motion changing. However, there are two forces acting on it. They are the tension in the spring (the tensile force) and the force of gravity. These are shown using vectors in Figure 10-9.

Balanced Forces

The force of gravity exerted by the earth on the mass acts vertically downward. The tensile force exerted by the spring acts vertically upward. The two forces are equal in magnitude but opposite in direction. When two or more forces act on a body, but the body does not change in shape or in motion, we say the forces are balanced. The single force that can replace all the forces and give the same effect is called the **resultant or net force**. The resultant or net force when all the forces are balanced is zero.

The spring scale force measurer functions by balancing forces. The force to be measured is balanced against the tensile force exerted by the spring. Then the magnitude of the unknown force is read off the calibrated scale.

A person holding up several parcels is another example of balanced forces (Fig. 10-10). He applies a muscular force upward on the parcels. The earth exerts a downward force. The muscular force is balanced by the force of gravity. The resultant force is zero. Whenever an object is at rest, the forces acting on the object are balanced.

Consider the forces acting on a car moving at a constant speed of 50 km/h along a straight level road (Fig. 10-11). The vertical forces acting on the car are the force of gravity acting down and the force of the road acting up. Since the car pushes on the road and the road reacts by pushing back on the car, the force of the road is called the **reaction force**. Since these two forces are balanced, there is no motion of the car in the vertical direction. There are two horizontal forces acting on the car. The motor exerts a forward force through the wheels in contact with

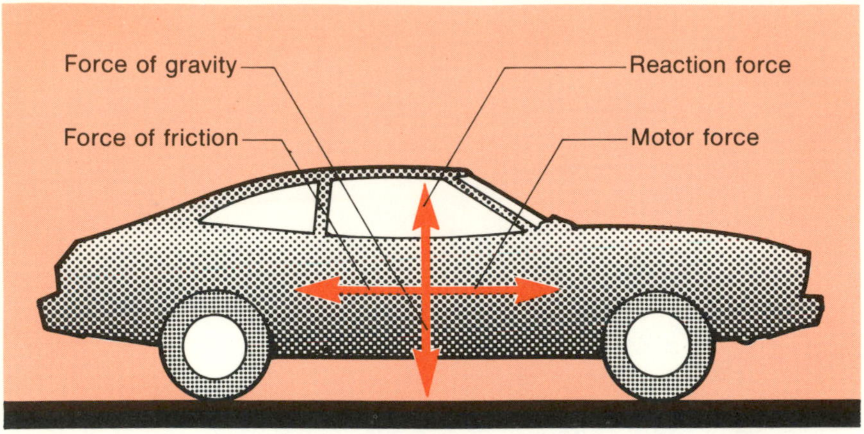

Figure 10-11 Forces acting on a car moving at constant speed in a straight line.

the road. The road and the air exert a frictional force backward on the automobile. Since there is no change in motion of the car (the car is not speeding up or slowing down), the forward and the backward forces must be balanced. Whenever an object moves at a constant speed in a constant direction, the forces acting on the object are balanced. The resultant force is 0. This suggests that a moving body with no forces acting on it will travel at a constant speed forever. A meteorite hurtling through space far away from any other matter is such an example.

Unbalanced Forces

Suppose the driver of a car moving at a constant speed presses on the accelerator. The forward force exerted by the motor through the wheels in contact with the road becomes greater than the force of friction. The car increases in speed (or accelerates) from say 50 km/h to 60 km/h. During the time of the acceleration the horizontal forces acting on the car are unbalanced.

If the driver eases up on the accelerator, the forward force becomes less than the force of friction and the car slows down. An unbalanced force can cause the speed of a body to increase or decrease. The speed increases in the direction of the unbalanced force.

An unbalanced force can also cause a change in the direction of motion of a body. When a driver approaches a curve in the road, he turns the wheels in a direction to go around the curve. The road then exerts a sideways force on the car that changes its direction of motion. If the road is icy, the sideways force may not be large enough to cause the correct change in direction. The car may skid off the road. An unbalanced force can cause a change in the direction of motion of a body.

If a driver approaching a curve takes his foot off the accelerator and also turns the wheel, two unbalanced forces come into play. The unbalanced force backward on the car slows it down, and the unbalanced force sideways on the car causes it to change direction. Whenever an object changes its speed or direction of motion, unbalanced forces are acting on the object. Whenever an unbalanced force acts, the resultant or net force on a body is not zero.

The great English physicist, Sir Isaac Newton (1642-1727), summarized these ideas concerning motion and force. He stated: *A body continues in its state of rest or constant speed in a straight line unless an external unbalanced force acts on it*. This is sometimes referred to as **Newton's Law of Inertia**, or **Newton's First Law**.

Discussion

1. What effects do each of the following forces have on a body: balanced forces and unbalanced forces?

2. What is a net or resultant force?

3. Name and describe the forces acting on a motorcycle travelling at a constant speed in a straight line.

4. Summarize how a spring scale force measurer uses the principle of balanced forces.

5. **a)** Look up the definition of inertia in a dictionary. Use an example to describe what is meant by the inertia of an object.

b) State Newton's First Law.

c) Explain the relationship between Newton's First Law and inertia. Use an example.

6. Try the following activities and describe how they illustrate Newton's First Law:

a) Place a small beaker or glass on a table. Cut a card slightly bigger than the top of the beaker. Cover the open end of the beaker with the card. Place a coin on the centre of the card. Flick the card away quickly with your index fingers.

b) Take care to catch the mass in this activity. Suspend a large mass from a support by a piece of cotton thread. Attach another piece of thread to the bottom of the mass. Attach a wooden rod or stick to the lower thread. Grasp the rod and pull down with a rapid jerk. Repeat the investigation with one difference. Pull down slowly on the wooden rod. Be careful!

c) Remove the end from an open box. Place a ball in the centre of the box. Move the box along the table with the open end forward. Stop the box suddenly.

d) Drive a large nail into a wooden block using a hammer with a massive head. Repeat with a hammer having a head with a smaller mass.

e) Pile two wooden blocks on top of one another. Give the bottom block a sharp horizontal tap with a hammer. Be careful!

7. A small car has a mass of 900 kg. It is travelling along a straight level road at a constant speed of 40 km/h. The force of friction acting on the car is 2 000 N. Draw a sketch of the car. Draw force vectors to show the forces acting on the car. Use a scale of 1 cm to represent 2 000 N.

10.8 INVESTIGATION: The Force of Friction

The **force of friction** *is the force that opposes motion whenever one surface either moves or tends to move past another.* It is difficult to slide a large object along a floor. A force is also needed to keep it moving. An unbalanced force is needed to start the object moving. A spring scale used to apply a horizontal force to a block of wood lying on a table does not move the block until the force is greater than the force of friction. Since the force applied by the spring increases from zero to the value needed to start the motion, the force of friction opposing the motion must also increase. The maximum value of the force of friction just before the object starts to move is known as the force of **limiting static friction**. A force is also needed to keep the block moving at a constant speed. A block moving at a constant speed has balanced forces acting on it. The force of friction which opposes the motion when an object is moving is called the force of **kinetic friction**.

The force of friction exists between the surfaces of all states of matter. But it is easier to study friction between solid surfaces.

In this investigation you compare the size of the two kinds of forces, the forces of limiting static friction and kinetic friction. You also study the effect the following factors have on the force of friction: the surface area in contact and the forces pushing the surfaces together. You use a calibrated newton spring scale to measure these forces.

Materials

newton spring scale
smoothly sanded wooden block
string
smoothly sanded wooden board
sheet of graph paper per person
set of masses

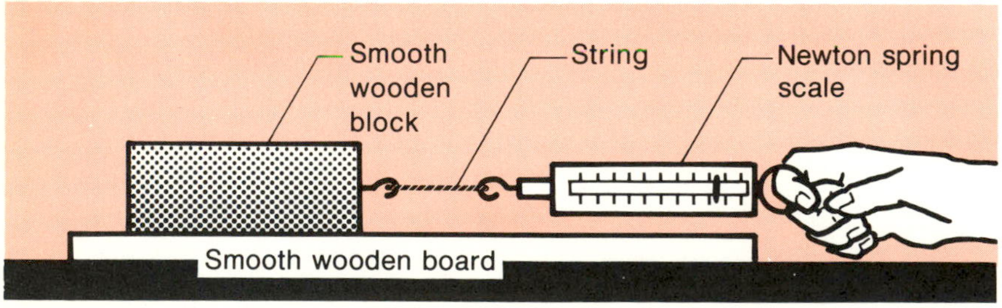

Figure 10-12 The force of friction between a wooden block and a wooden board.

Procedure A Comparing Limiting Static and Kinetic Friction

a. Attach a piece of string to the wooden block. Attach the newton spring scale to the string.
b. Place the wooden block on top of the wooden board on a level surface as shown in Figure 10-12. Be sure the largest surface is down.
c. Pull slowly on the spring scale. Record the force in newtons at the instant when the wooden block begins to move. This force is equal to the force of limiting static friction between the wooden block and the wooden board.
d. Repeat step c three times.
e. Calculate the average value for the force of limiting static friction.
f. Pull slowly on the spring scale so that the wooden block slides at a constant speed along the board. Read the force in newtons needed to keep the object moving at constant speed. This force is equal to the force of kinetic friction between the wooden block and the wooden board. Record this reading.
g. Repeat step f three times. Try to keep the object moving at the same constant speed each time.
h. Calculate the average value for the force of kinetic friction.

Procedure B Friction and the Surface Areas in Contact

Turn the wooden block so that one of its smaller sides is down. Measure the force of kinetic friction again. Take 3 readings and average them.

Procedure C Friction and the Force Pushing Surfaces Together

a. Make a table like Table 67.
b. Hang the wooden block from the spring scale and measure the force of gravity in newtons acting on it. Enter this value in the second column of the table. This force pushes the surfaces together. It is called the normal force since it acts perpendicular to the surfaces.
c. Place the wooden block on the wooden board as shown in Figure 10-12. Be sure the larger surface is down.
d. Measure the force of kinetic friction in newtons between the wooden block and the wooden board. Record this force in the table.
e. Place a 200 g mass on the upper surface of the wooden block. The normal force pushing the surfaces together is now

the sum of the force of gravity on the block and on the 200 g mass. Enter this force in the second column of the table.

f. Measure the force of kinetic friction for the block and its load.

g. Add more mass to the wooden block in steps of 200 g up to a maximum of 1 000 g. Each time the mass is added, record the total force pushing the surfaces together, and measure and record the force of kinetic friction.

h. Plot a graph with the force of kinetic friction on the vertical axis and the force pushing the surfaces together on the horizontal axis.

TABLE 67 Studying Friction

Mass of wooden block + load	Force pushing surfaces together (N)	Force of kinetic friction (N)
Block		
Block + 200 g		
Block + 400 g		
Block + 600 g		
Block + 800 g		
Block + 1 000 g		

Discussion A

1. Compare the average value for the force of limiting static obtained in step e with the average value for the force of kinetic friction obtained in step h. Make a generalization based on this comparison.

Discussion B

1. Compare the force of kinetic friction for the largest surface area (from Procedure A) to the force of kinetic friction for the smallest surface area (from Procedure B).
2. Does the force of friction between the wooden block and the board depend on the area of the surfaces in contact? Should it? Why?

Discussion C

1. What do you think the force of kinetic friction between the wooden block and the wooden board would be if there was no force of gravity acting on the wooden block?
2. What happened to the size of the force of kinetic friction as the force pushing the two surfaces together increased?

3. Describe the shape of the graph obtained. Did it go through the origin? What relationship exists between the force of kinetic friction and the force pushing the surfaces together?
4. Compare your results with those of other groups.

10.9 Friction between Solid Surfaces—Coefficients of Friction

Friction is present whenever the surface of one object moves or tends to move over another surface it touches. Friction is caused by the irregularities in the surfaces in contact. Friction can be useful. It holds a nail in a board; it makes it possible for us to walk; it stops moving vehicles when the brakes are applied. But friction also poses many problems. It causes machinery to heat up and wear out and it consumes valuable fuel.

The Nature of Matter and Friction

Scientists believe that friction is caused by the projections on the surfaces of touching objects and by the electrical force between the particles in the surfaces. Friction is a retarding force that acts parallel to the surfaces in contact. Friction acts opposite to the direction of motion of the object.

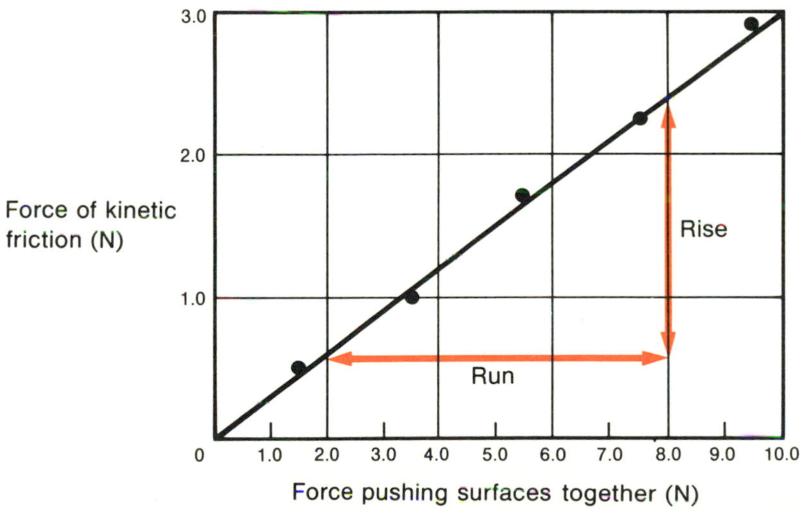

Figure 10-13 Coefficient of kinetic friction for oak on oak.

Friction differs for different substances. The force of friction of copper on copper is larger than that of steel on steel. It changes as the surfaces are made smoother. Friction usually decreases as the surfaces are polished. However, if the surfaces are made very smooth, the friction between them actually increases.

The sizes of the forces of both limiting static friction and kinetic friction depend very little on the areas of the surfaces in contact. It takes the same force to start a rectangular block moving or to keep it moving whether it is placed on its largest or smallest surface. Limiting static friction is usually larger than kinetic friction. It takes a larger force to start a block sliding than to keep it sliding.

Both limiting static friction and kinetic friction are directly proportional to the force pushing the surfaces together. That is, when the force of gravity on an object doubles, the force of friction doubles. When the force of gravity is halved, the force of friction halves. The force of friction on an incline is less than on the level. This is because the force pushing surfaces together is not as large.

The Coefficients of Friction

A graph of the force of kinetic friction against the force pushing two oak surfaces together is shown in Figure 10-13. Let's calculate the slope of the line. The formula for slope is:

$$\text{Slope} = \frac{\text{Rise}}{\text{Run}} = \frac{2.4\ \text{N} - 0.6\ \text{N}}{8.0\ \text{N} - 2.0\ \text{N}} = \frac{1.8\ \text{N}}{6.0\ \text{N}} = 0.3$$

Notice in this case that the slope has no units. The slope of the graph gives the **coefficient of kinetic friction** of the two surfaces in contact.

The force of friction is given the symbol $\boldsymbol{F}$. The force pressing the surfaces together, called the normal force, is given the symbol $\boldsymbol{n}$. The coefficient is given the symbol μ. The formula for the coefficient of kinetic friction is $\mu_k = \frac{F}{n}$ Each set of surfaces has two coefficients, a coefficient of kinetic friction and a coefficient of limiting static friction. The coefficients depend on the material and the roughness of the surfaces as shown in Table 68.

For automobile tires the coefficient of kinetic friction (wheel skidding) decreases with an increase in car speed. This means that a car with brakes locked skids further in slowing down from 50 km/h to 40 km/h than it does in slowing down from 20 km/h to 10 km/h. The force of friction between the tires and the road is smaller at higher speeds.

Discussion

1. Define the force of friction.
2. Use the particle theory and the knowledge that surfaces have projections to explain the following:

a) Why is the magnitude of the force of limiting static friction usually greater than the force of kinetic friction?
b) Why do the forces of limiting static friction and kinetic friction increase with an increase in the force pushing the surfaces together?
c) Why does the area of the surfaces in contact not affect the forces of limiting static friction or kinetic friction?

3. Describe how the coefficient of kinetic friction can be determined for greased steel sliding on greased steel.

4. Should a driver in a drag race spin the car tires during the takeoff? Why?

5. An inexperienced driver stuck in the snow tends to spin the car tires to gain more traction. What advice would you give to the driver? Why?

6. Why is traction improved if a bag of sand is placed in the trunk of the car?

7. Design and carry out an investigation to measure the coefficient of kinetic friction for a rubber tire on asphalt. Compare the value you get for asphalt with the value recorded in Table 68 for concrete. Pay careful attention to control of variables.

8. Why is it wise to pay attention to the road sign which states: "Drive carefully—pavement slippery when wet"?

9. Manufacturers of skis recommend different waxes for different snow and temperature conditions.
a) Design an experiment to test their recommendations. Pay careful attention to the control of variables.
b) If conditions permit, try your experiment.

TABLE 68 Typical Values of Coefficient of Static Friction μ_s and of Kinetic Friction μ_k

Materials	μ_s	μ_k
Copper on copper	1.6	1.0
Steel on steel	0.15	0.09
Oak on oak	0.5	0.3
Rubber tire on dry concrete	1.0	0.7
Rubber tire on wet concrete	0.7	0.5
Teflon on teflon	0.04	0.04
Waxed hickory skis on dry snow	0.06	0.04
Waxed hickory skis on wet snow	0.20	0.14

10.10 INVESTIGATION: The Electric Force

The force which has the greatest effect on our lives is the force of gravity. We never escape it. Even when astronauts are far out in space, they are attracted to celestial bodies, to one another, and to the space capsule.

The electric force also influences man greatly. His understanding of it has made possible electric power and improved transportation, communication, and health. In this investigation you study the characteristics of the electric force.

Materials

ring stand
adjustable clamp
support rod
silk or nylon thread
acetate strip and dry cotton cloth
glass rod and dry silk cloth
vinylite strip and dry wool cloth
ebonite rod and dry cat's fur
masking tape

Procedure

a. Construct a table with two headings: Procedure and Results.
b. Clamp the support rod horizontally to the ring stand about 30 cm above the surface of the table.
c. Attach a short length of thread to the end of the acetate strip using masking tape. Suspend the acetate strip from one end of the rod so that it sways freely as shown in Figure 10-14.

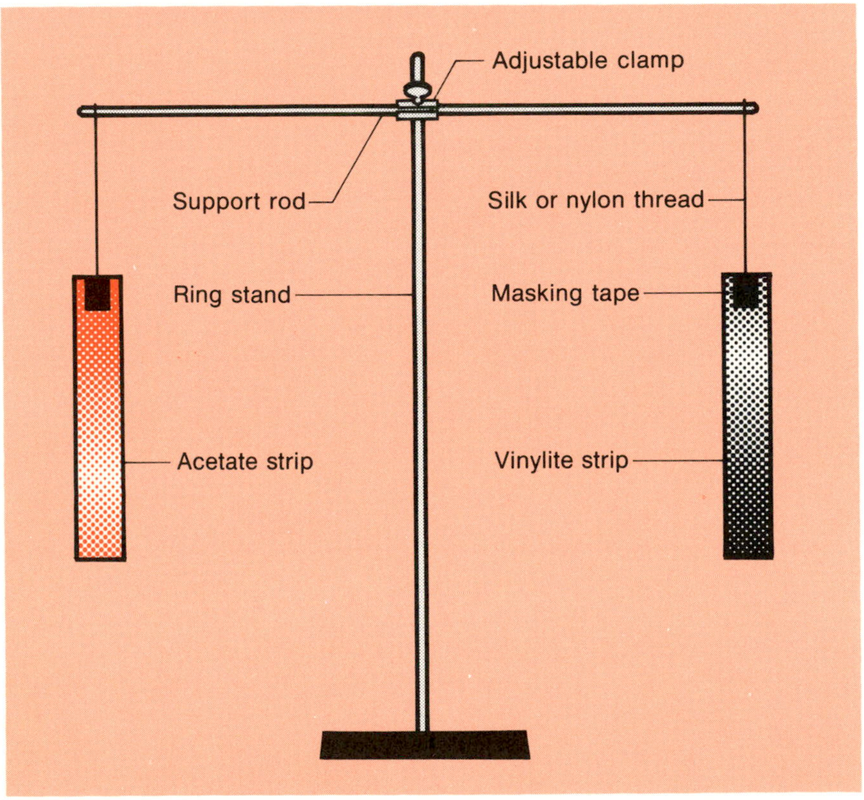

Figure 10-14 The electric force.

d. Repeat step c for the vinylite strip, but suspend it from the other end of the rod.
e. Rub the acetate strip rapidly with the cotton cloth to give it an electric charge. Do not touch the part of the surface that has been charged.
f. Rub the vinylite strip with the woollen cloth to give it an electric charge.
g. Rub a second acetate strip with the cotton cloth to give it a charge.
h. Bring it close to but not touching each of the suspended and charged strips. Record what happens.
i. Rub a second vinylite strip with the wool cloth to give it a charge and repeat step h. Record what happens.
j. Hold the charged acetate and vinylite strips in one hand so that they touch one another over their entire lengths. Bring the pair close to but not touching each of the suspended charged strips. Record what happens.
k. Rub an ebonite rod with cat's fur and repeat step h.
l. Rub a glass rod with silk and repeat step h.
m. Design and carry out a qualitative investigation to study the effect that changing the distance between two charged objects has on the magnitude of the electric force between them. Record your results.

Discussion

1. The force of gravity is an example of an action-at-a-distance force. Two masses do not need to be in contact for a force of gravity to exist between them. Is the electric force an action-at-a-distance force? Support your answer using data from the investigation.
2. The force of gravity always causes one mass to attract another. Is this the case for the electric force? Support your answer using data from the investigation.
3. What happens when two objects having the same kind of charge are brought close to one another? Write a statement summarizing the results.
4. What happens when two objects having a different kind of charge are brought close to one another? Write a statement summarizing the results.
5. What evidence supports the hypothesis that a glass rod rubbed with silk has the same kind of charge as an acetate strip rubbed with cotton?
6. What evidence supports the hypothesis that an ebonite rod rubbed with cat's fur has the same kind of charge as a vinylite strip rubbed with wool?
7. What happens to the size of the electric force as two charged objects are brought closer together? further apart? How does

the electric force vary with the distance between the charged objects?

8. Two charged objects are attracted to one another by electric forces. Compare the magnitude and direction of the two forces.

10.11 Properties of the Electric Force

A Way of Classifying Forces

Forces were named in Section 10.1 from the sources that cause them. The force exerted by a muscle is called a muscular force. The force exerted by a magnet is called a magnetic force. Forces can also be classified into two groups depending on whether or not the source of the force is in contact with the object on which the force acts. Both a muscular force and an elastic force are examples of **contact forces**. A person exerting a muscular force on an object must be in contact with the object to exert the force. However, the electric force can be exerted by one charged object on another without the objects being in contact. Both the force of gravity and the electric force are examples of **action-at-a-distance forces**.

The Law of Electric Charges

Two objects with the same kind of charge repel one another. If two acetate strips rubbed with cotton are given the same charge, they repel one another. A glass rod rubbed with silk repels an acetate strip rubbed with cotton. Therefore the glass rod must have the same charge as the acetate strip. Two vinylite strips rubbed with wool repel one another. An ebonite rod rubbed with cat's fur also repels a vinylite strip. The ebonite strip and the vinylite strip must have the same kind of charge.

Two objects having the opposite kind of charge attract one another. Charged acetate strips and vinylite strips are attracted to one another. A charged glass rod attracts a charged vinylite strip. A charged ebonite rod attracts a charged acetate strip.

Any two different materials when rubbed together become electrically charged. Two centuries of search has uncovered only two kinds of electric charge. The two kinds of charge are called positive and negative, or + and −. A **positive** charge is defined as the kind of electric charge that a glass rod gets when it is rubbed with silk. An acetate strip rubbed with cotton also gets a positive electric charge. A **negative** charge is defined as the kind of electric charge that an ebonite rod gets when it is rubbed with cat's fur. A vinylite strip with wool also gets a negative electric charge.

From Investigation 10.11 we know that charged objects exert electric forces on one another as follows: Positively charged objects repel one another; negatively charged objects repel one another; a positively charged object and a negatively charged object attract one another. These statements are summarized in the **Law of Electric Charges** which states that *like charges repel and unlike charges attract*.

The Electric Force and Distance

The electric force between two charged objects decreases rapidly as the distance between the charges increases (Fig. 10-15). The further away a charged acetate strip is from a charged vinylite strip, the less the force of attraction. The nearer a charged acetate strip is brought to another charged acetate strip, the greater the force of repulsion.

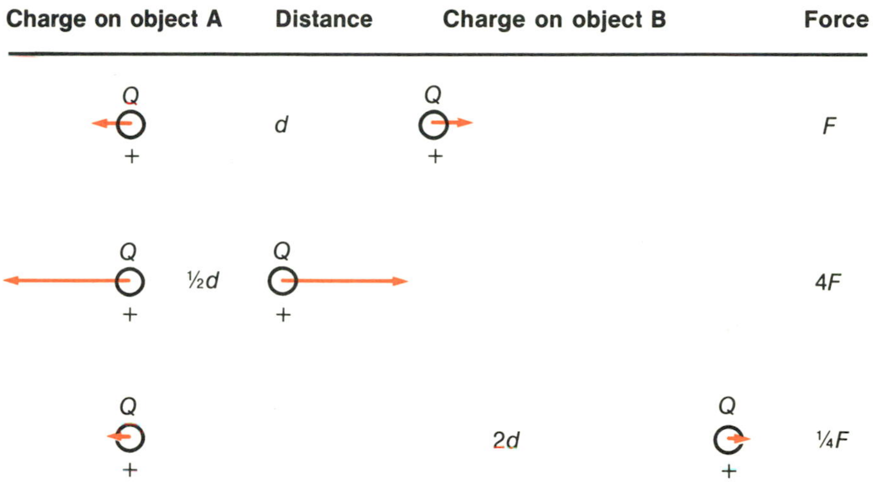

Figure 10-15 The electric force and distance. Since both object A and object B have a positive charge, the electric force is repulsive.

The French physicist, Charles Augustin Coulomb (1736–1806), performed experiments in the late 18th century to determine how the electric force depended on distance. He used the sensitive torsion balance shown in Figure 10-16 to measure the forces between charged objects as the distance between them was varied.

A torsion balance consists of a long, light, horizontal rod suspended by a fine vertical wire. Two small spheres which can be charged are mounted at opposite ends of the rod. A small force applied at right angles to the end of the rod causes the rod to turn. This twists the fine wire. The greater the force, the greater the twist given to the wire. When the horizontal rod comes to rest, any force acting to twist the wire is balanced by the elastic force in the wire. The force causing a given amount of

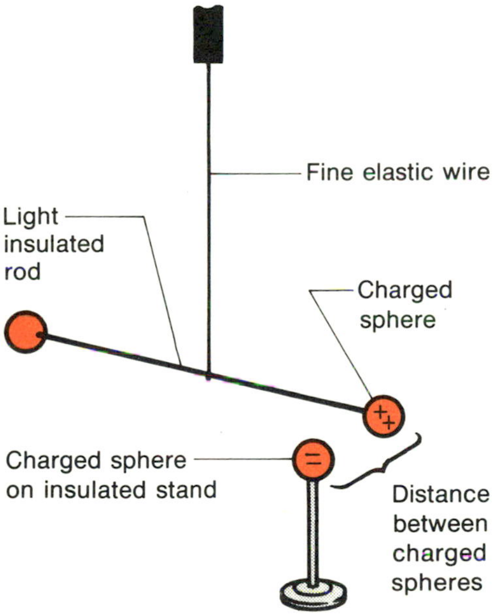

Figure 10-16 A schematic diagram of Coulomb's torsion balance.

twist can be determined. When a charged sphere is brought close to one of the suspended spheres, the electric force between them causes the horizontal rod to turn. The rod finally comes to rest with the two charged spheres separated by a certain distance. Both the separation and the electric force corresponding to this distance can be determined. The distance can be varied and the electric force which corresponds to each new distance can be determined.

Coulomb found, using this procedure, that *the force between two small charged objects varies inversely as the square of the distance between them*. If the distance between the two charged objects is doubled, the force between them decreases to 1/4 its original value. If the distance between the two charged objects is decreased to one quarter of the original distance, the force increases 16-fold.

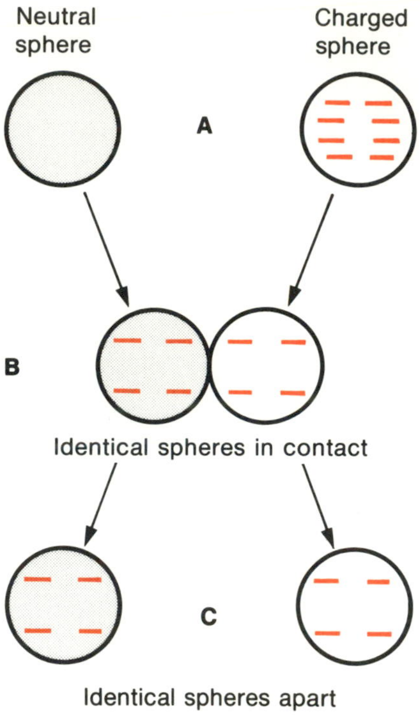

Figure 10-17 A charged sphere sharing its charge with an identical neutral sphere: A, a charged sphere and an identical neutral sphere; B, the charge being shared between the two identical spheres; C, the identical spheres apart, each with half the original charge.

The Electric Force and Quantity of Charge

Coulomb also did experiments with the torsion balance to find out how the electric force depended on the quantity of charge on the bodies. He found that a sphere could share its charge with an identical neutral sphere. Consequently, a given charge could be split into two, four, or more equal parts. Figure 10-17 shows how this is done. Part A shows a sphere with 8 negative charges and an identical neutral sphere. Part B shows the identical spheres in contact. The charge spreads itself uniformly over the two spheres. Part C shows the two spheres apart. Each sphere now has half of the original charge.

He found that the smaller the charge on a sphere, the smaller the electric force. In fact, the *electric force between two spheres*

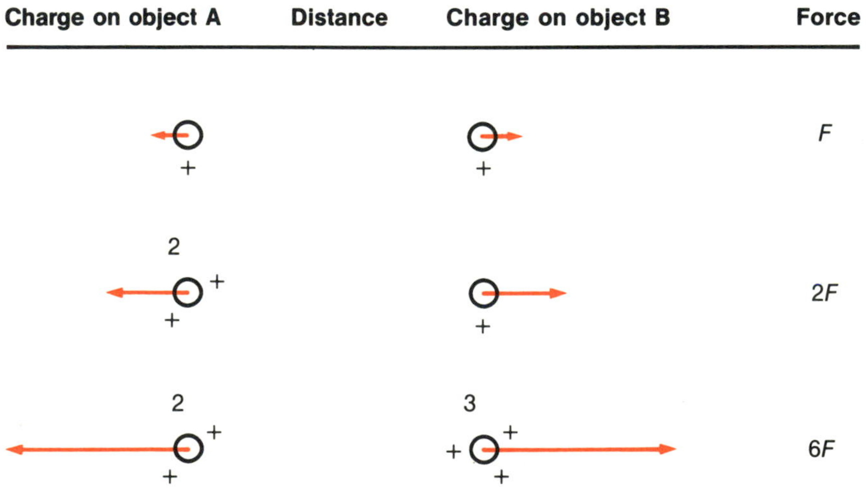

Figure 10-18 The electric force and quantity of charge.

was found to vary directly as the product of the charges on the spheres. If both spheres are given double the charge, the electric force becomes four times as great. If one of the charges is quartered and the other is made four times as large, the force stays the same (Fig. 10-18).

Coulomb's Law

The generalization describing the electrical force between charged objects is called **Coulomb's Law**. It states that *the electrical force between two charged objects varies directly as the product of the charges and inversely as the square of the distance between their centres*. The formula which summarizes this principle is $F = k\frac{q_1 q_2}{d^2}$. F is the electrical force, k is a constant, q_1 and q_2 are the charges, and d is the distance between the charges.

A Comparison of the Electric Force and the Force of Gravity

The electric force and the force of gravity are very similar. Both vary inversely with the distance between the objects. Both vary directly as the product of the quantity of mass or charge of the object. Both are action-at-a-distance forces. We have no explanation for these similarities.

There are two important differences between the electric force and the force of gravity. First, the force of gravity is always an attractive force. However, the electric force can be either attractive or repulsive, depending on whether the charges on the objects are opposite or of the same kind. Second, the electric force is much stronger than the force of gravity. This accounts for the ability of the particles making up a steel wire to attract one another with electric forces large enough to easily support a mass of 1 kg against the force of gravity of the whole earth.

The Electric Force and Matter

You found out in earlier units that all substances, solid, liquid, and gaseous, are made up of molecules. All molecules are composed of one or more particles called atoms. Scientists now know that each atom is made up of many smaller particles each with its own properties. Two of these particles are the **electron**, with an electric charge of negative one, and the **proton**, with an electric charge of positive one. Since the charge on the electron is equal in magnitude but opposite in kind to the charge on the proton, the atom is neutral.

The force that holds atoms together to make molecules is the electric force between protons and electrons. A molecule of hydrogen is made up of two atoms of hydrogen held together by the electric force. Two atoms of hydrogen and one atom of oxygen are held together by the electric force to make a molecule of water.

The electric force also acts between the molecules of matter. The electric force can hold molecules of water together in a set position to form solid water or ice. When heat energy is added to these molecules they eventually break away from their fixed positions and begin to slide or roll freely over each other. When this happens the solid becomes a liquid. The electric force between the molecules of the liquid is weaker because the molecules are further apart. But the force is still strong enough to prevent the molecules in a liquid from completely filling the container. The molecules near the surface experience a downward force made up of both the electric force and the force of gravity. However, when a liquid changes to a gas, the force of attraction between the molecules becomes very small. As a result, the molecules move about and occupy the entire container.

The attractive electric force has been used to explain all of the examples so far. Consider the collision of the molecules of a gas such as hydrogen or helium. If the electric force of repulsion did not exist, these molecules would collapse. However, when the molecules approach within a certain distance of one another, the electric force of repulsion causes them to change direction and move away from one another. It is also the electric force of repulsion between the molecules making up the walls of a balloon and the molecules of gas in the balloon which keeps a balloon inflated.

Discussion

1. Summarize the properties of the force of gravity and the electric force. How are they similar? How are they different?
2. What is an action-at-a-distance force? How does it differ from a contact force? Give two examples of each kind of force.
3. Define a negative charge. Define a positive charge. Give an example of each.
4. If an ebonite rod rubbed with cat's fur is brought near a suspended glass rod rubbed with silk, what will happen? Why?
5. An instrument called a pith ball electroscope is constructed by suspending a small sphere made of pith from a support using a silk or nylon thread. The pith is covered with aluminum foil. When a charged object, such as an ebonite rod or a glass rod, is brought close to but not touching a neutral pith ball, the ball is attracted toward the charged object. Write an explanation for this phenomenon.

6. If two small charged objects each experience an electrical force of repulsion of 0.08 N when they are 10 cm apart, what will be the force when they are: a) 100 cm apart, b) 25 cm apart, c) 5 cm apart, d) 2.5 cm apart?
7. Two small objects, A and B, are placed a set distance apart. When a charge of +6 units is placed on A and a charge of −8 units is placed on B, the force between them is 0.02 N.
   a) Is the electric force an attractive or repulsive force?
   b) If the charge on A is changed to +12 units and that on B stays the same at −8 units, what will the electric force be?
   c) If the charge on A is changed to +18 units and the charge on B becomes −2 units, what will the electric force be?
8. State Coulomb's Law.
9. When two neutral different substances are rubbed together they become charged. The electrons lost by one substance, giving it a positive charge, are donated to the second substance. Scientists have constructed an **electrostatic series** which shows the charge the substances receive when rubbed together. A material obtains a positive charge when rubbed with a substance following it in the series, and a negative charge when rubbed with a substance preceding it. The following is part of an electrostatic series: glass, wool, fur, silk, paraffin wax, ebonite, copper.
   a) Design and carry out an experiment to determine where acetate, vinylite, cotton, and the material making up a comb should be placed in the series.
   b) Which pairs of substances should receive a greater charge when rubbed together: glass rubbed with silk, or glass rubbed with fur? Why?
   c) How will the charges on the two rubbed substances in each pair compare in magnitude?
10. Predict what will be observed if a charged ebonite or glass rod is brought close to a fine, continuous stream of water from a water tap. Design an experiment to test your prediction. Write an explanation for what is observed using the idea of an electric force.

10.12 INVESTIGATION: The Electric Force and Matter

The electric force can be used to explain many phenomena we see in nature. In this investigation you will study some phenomena. You should be able to explain your observations using the electric force.

Materials

eyedropper
glass capillary tubing
10 mL graduated cylinder
plastic straw
wire loops
pins
clean sheet of acetate
clean sheet of wax paper
clean sheet of aluminum foil
200 mL beaker
50 mL beakers
spool of thread
paper clips
hot and cold water
soapy water
salad oil
ethyl alcohol
glycerine
detergent
soap solution

Procedure A Water Droplets

a. Fill a clean eyedropper with water at room temperature.
b. Squeeze the eyedropper to cause a drop of water to form on the end of the eyedropper. Allow it to fall into a beaker. Note and sketch the shape of the drop just before it falls.
c. Form other drops on the end of the eyedropper. Compare the size of the drops just before they fall.
d. Slowly allow drops of water to form and drop into the 10 mL graduated cylinder until 50 drops have been collected. Record the volume of 50 drops of water.
e. Repeat step d using hot water at a temperature of about 80°C.
f. Repeat step d using soapy water.
g. Clean the eyedropper and graduated cylinder after use.

Procedure B Liquid Droplets on Surfaces

a. Place a clean sheet of acetate horizontally on the table. Place one drop of each of the following liquids on the surface of the acetate at different locations: water, soapy water, salad oil, rubbing alcohol, glycerine. Hold the end of the eyedropper about 0.5 cm from the surface of the acetate when releasing the drops.
b. Note the size and shape of the drops as viewed from above and from the side.
c. Sketch the side view of each drop.
d. Repeat steps a, b, and c using clean aluminum foil.
e. Repeat steps a, b, and c using clean wax paper.
f. Place 2 drops of water close to one another on the surface of the wax paper. Use a pin to move the drops close to each other. Record what happens.

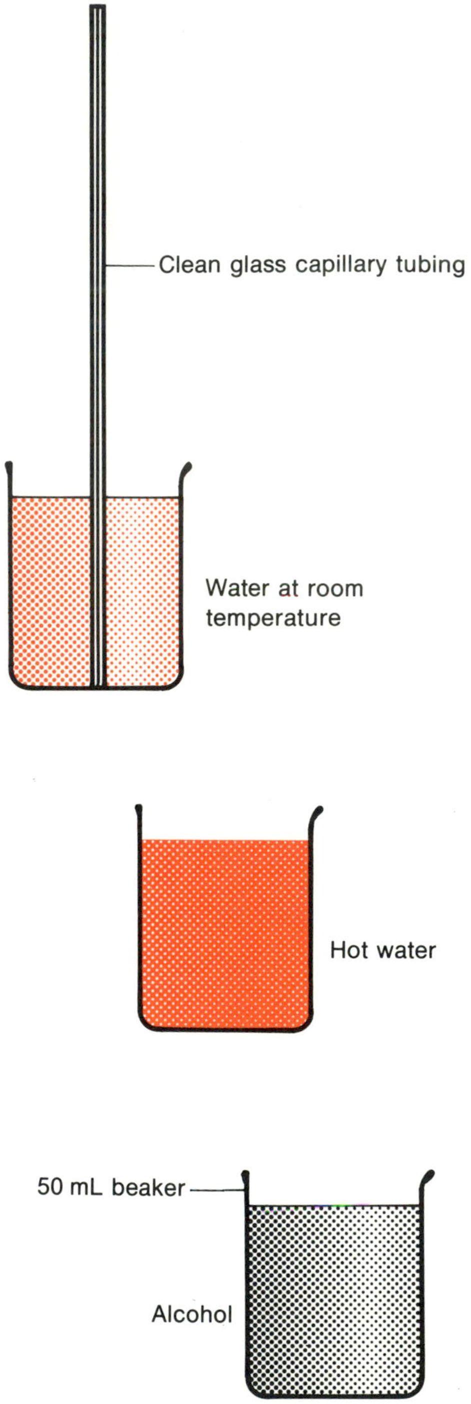

Figure 10-19 The behaviour of liquids in a capillary tube.

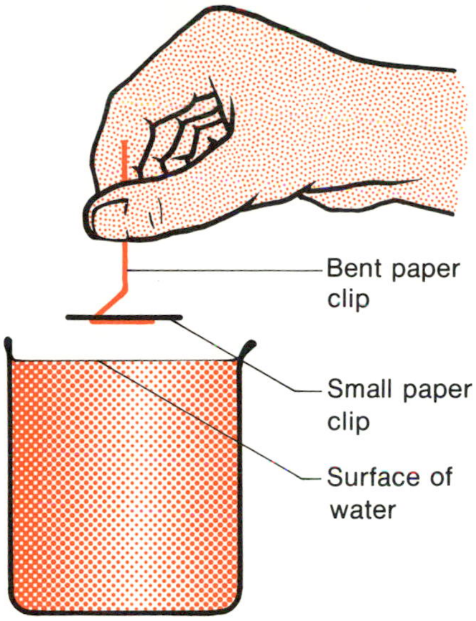

Figure 10-20 A paper clip being lowered onto the surface of water.

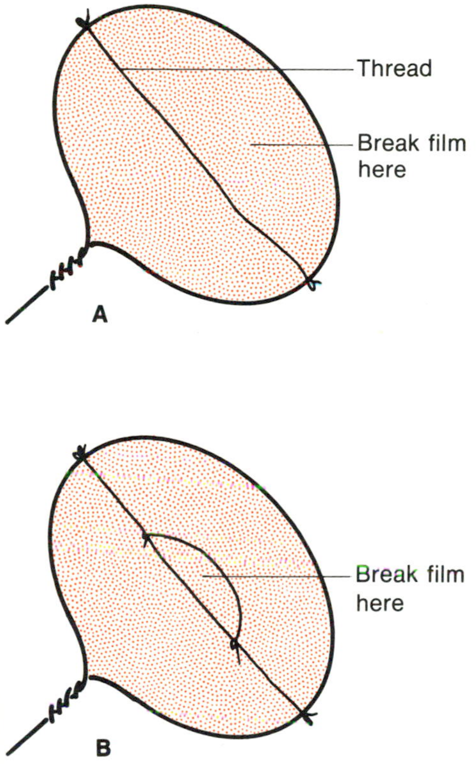

Figure 10-21 A soap film on a wire loop: A, soap film supported by a single thread; B, soap film supported by a thread loop.

Procedure C Capillary Action

a. Fill a 50 mL beaker with water at room temperature.
b. Hold a length of clean glass capillary tubing vertically in the water as shown in Figure 10-19. Leave about 10 cm of tubing sticking out of the water. Observe the behaviour of the water inside the glass tubing. This behaviour is called **capillary action**. Measure and record the final height the water reaches.
c. Repeat steps a and b using hot water at a temperature of about 80°C.
d. Repeat steps a and b using alcohol at room temperature.

Procedure D Surface Tension

a. Fill a 200 mL beaker about ¾ full of water.
b. Use a bent paper clip to carefully place another paper clip on the surface of the water as shown in Figure 10-20. Carefully observe the surface of the water where it touches the paper clip.
c. Add 3 or 4 drops of detergent to the water several centimetres away from the paper clip. Watch carefully and record your observations.

Procedure E Soap Films

a. Lower the end of a plastic straw into the soap solution. Lift it out and blow gently through the upper end. Sketch the shape of the bubble formed at the bottom.
b. Make a wire loop having a diameter of about 5 cm and a handle.
c. Tie a thread loosely from one side of the loop across the diameter to the other side as shown in Figure 10-21, A.
d. Submerge the wire loop with the thread in the soap solution. Remove the loop from the solution and rotate the loop back and forth. Observe the movement of the thread within the soap film.
e. Use a pin to break the soap film on one side of the thread. Sketch the resulting shape of the soap film and thread.
f. Tie a second thread at two places to the first thread to form a thread loop as shown in Figure 10-21, B. Repeat step d. Use the pin to break the soap film inside the thread loop. Sketch the results.

Discussion

1. Describe the shape and size of drops of water released from an eyedropper. Compare the volumes of a drop of cold water and a drop of hot water. Describe the effect of temperature on

the electric force. Compare the volume of a drop of cold water with the volume of a drop of soapy water. Use the electric force to account for both observations.

2. Compare the shapes of the following drops on the acetate sheet: water, soapy water, salad oil, alcohol, and glycerine. Which substance has the strongest electric force between like molecules? Why did you pick that substance?
3. Compare the shapes of water drops placed on the acetate sheet, the aluminum foil, and the wax paper. Account for any differences by describing the electric force between like molecules and between unlike molecules.
4. Use the electric force to account for the behaviour of two drops of water on wax paper when they are brought close together.
5. Compare the height to which cold water, hot water, and alcohol rise in clean glass capillary tubing. Account for capillarity (the rising of the water) by discussing the electric force between the molecules of the liquid and the molecules of the glass.
6. The surface of any liquid seems to have a skin. It appears that the surface is under a constant elastic stretch. This effect is called **surface tension**. Use the electric force to account for the existence of surface tension. Compare the surface tension of water with the surface tension of soapy water.
7. Explain the movement of the thread suspended in the soap film on a wire loop before and after breaking the film. Use the electric force to explain the shape the film takes. Make a generalization about the surface area of soap films based on the activities with the soap film.

Highlights

A force is a push or a pull of a definite magnitude and direction. Force is a vector quantity. It is described completely only when both a magnitude and a direction are stated. Vector quantities are represented on diagrams by a scaled, directed line segment called a vector.

A force can produce a change in the shape of an object. The lengthening or shortening of a spring is used to measure a force. The unit of force is the newton (N). 1 N is about equal to the force of gravity the earth exerts on a mass of 0.1 kg. The force of gravity to mass ratio on earth is about 10 N/kg.

Balanced forces do not change the motion of an object. Newton's Law of Inertia summarizes the behaviour of objects with balanced forces. An unbalanced force can change the speed or direction of motion of an object.

Two important forces in nature are the force of gravity and the electric force. The characteristics of these forces are summarized in the Principle of Universal Gravitation, the Law of Electric Charges and Coulomb's Law. These forces have a number of similarities. They are action-at-a-distance forces, and their magnitudes vary inversely as the square of the distance between the objects. Both the force of gravity and the electric force vary directly as the product of the mass or charge on the objects. There are differences. The force of gravity is always attractive, but the electric force can be either attractive or repulsive depending on whether the charges are opposite or alike.

Friction is another important force. Friction is present whenever one surface moves or tends to move over another. The force of friction varies for different surfaces because of the nature of their projections and the different electrical forces between particles. The force of friction is directly proportional to the force pushing surfaces together. It is not affected by the area of the surfaces in contact.

UNIT

11 Application of Forces

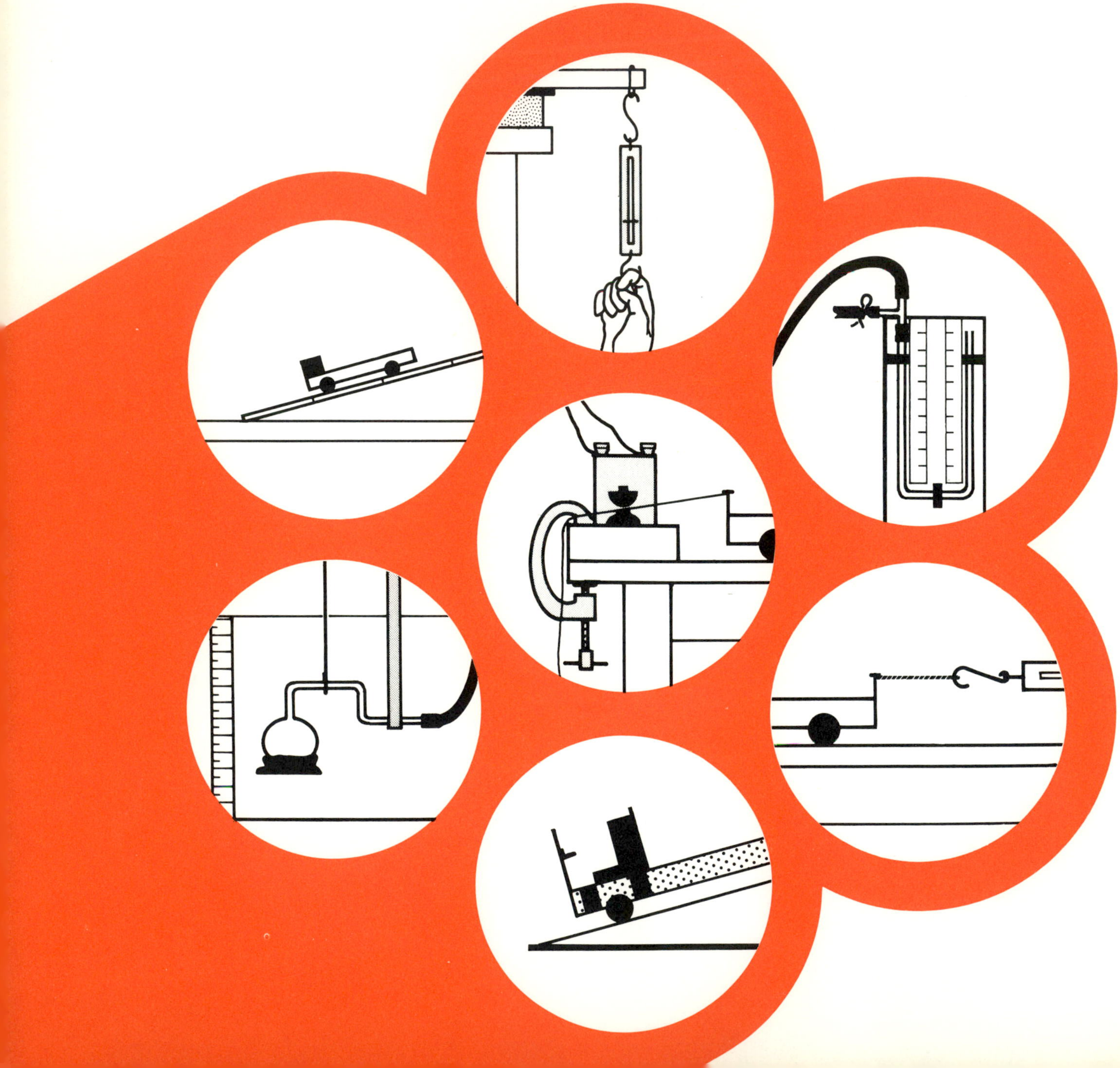

In Unit 10 you learned what a force is. You also studied several types of forces. In this unit you use your knowledge of forces to explain several important physical quantities in which forces are involved. Among these are speed and pressure.

11.1 Uniform Motion, Non-uniform Motion, and Forces

Motion

If the distance between two objects changes, one or both of the objects is in motion. A marble rolling down an incline moves away from its starting point. The distance between two cars moving in the same direction along a highway changes when one car travels at a greater speed than the other.

When the line joining two objects changes in direction, one or both of the objects is in motion. The swinging bob on the end of a pendulum moves with reference to the support. The length of the string does not change, but its direction from the support does.

We know that *one object is in motion with reference to a second object if the straight line between them changes in length or direction*.

Uniform Motion

An object is in uniform motion if it travels at a constant speed in a straight line. Newton's first law states that an object continues in uniform motion as long as no unbalanced force is applied. In real life a force must be applied to maintain a constant speed. This force is used to balance the force of friction.

Speed

Speed is a measure of the motion of an object. **Speed** *is defined as the distance travelled per unit time*. An object which covers 10 m in 1 s has an average speed of 10 m/s.

Constant Speed

An object moves at constant speed if it covers equal intervals of distance in equal intervals of time. Consider the following example. A car travels at a constant speed along a road. The driver records the distance travelled from the start at various times during the trip. The data are recorded in Table 69.

TABLE 69 Distance-Time Record for a Car Travelling at Constant Speed

Distance (km)	Time (h)
0	0
20	0.25
40	0.50
80	1.0
120	1.5
160	2.0
240	3.0

The car covers 40 km during each 0.50 h interval or 80 km during each 1.0 h interval. The car has a constant speed of 80 km/h. Figure 11-1 shows the distance-time graph of the motion.

A distance-time graph for an object moving at constant speed is a straight line passing through the origin. Let's calculate the slope of the line.

$$\text{Slope} = \frac{\text{Rise}}{\text{Run}} = \frac{220\ \text{km} - 60\ \text{km}}{2.75\ \text{h} - 0.75\ \text{h}} = \frac{160\ \text{km}}{2.0\ \text{h}} = 80\ \text{km/h}$$

The slope of a distance-time graph for an object moving at constant speed gives the magnitude of the constant speed.

The word equation for the slope of a distance-time graph is:

$$\text{constant speed} = \frac{\text{total distance travelled}}{\text{total time taken}}$$

Distance is given the symbol $\boldsymbol{d}$, time the symbol $\boldsymbol{t}$, and speed the symbol $\boldsymbol{v}$. Therefore the formula is $\boldsymbol{v} = \frac{\boldsymbol{d}}{\boldsymbol{t}}$.

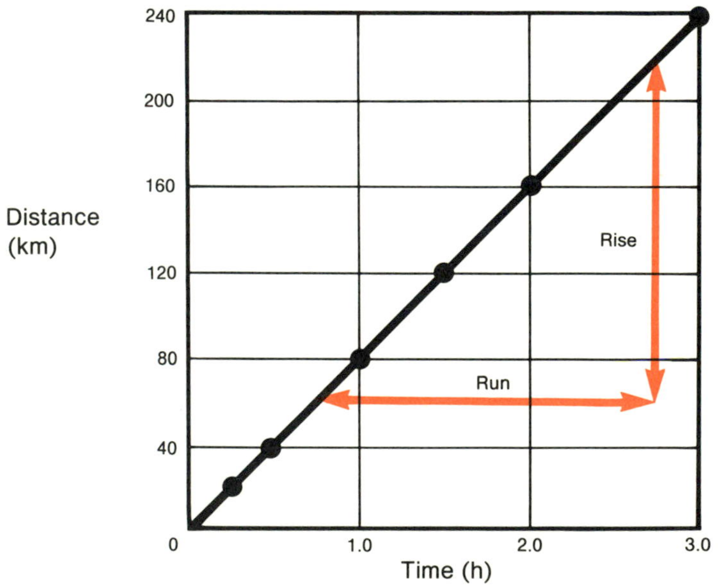

Figure 11-1 Distance-time graph of constant speed.

Average Speed

The speedometer of a car travelling along the highway at a constant speed of 80 km/h reads 80 km/h at all times. This kind of motion is difficult to maintain for long periods of time. Cars stop to be refuelled. Traffic jams decrease the safe driving speed. A car which travels 100 km during the first hour but only 60 km during the second hour travels a total distance of 160 km in a total time of 2 h. Its average speed is calculated as follows:

$$\text{average speed} = \frac{\text{total distance travelled}}{\text{total time taken}} = \frac{160\ \text{km}}{2\ \text{h}} = 80\ \text{km/h}$$

The average speed for the trip is the same as the constant speed the car would have to maintain to cover the same distance in the same time. The formula $v = \frac{d}{t}$ is used to calculate both the average and constant speed. Notice that speed is a scalar quantity since no direction is stated.

Non-uniform Motion

An object has non-uniform motion if its speed is changing or if its direction is changing, or both. An unbalanced force produces a change in the speed of an object. The unbalanced force of gravity increases the speed of a ball as it rolls down a hill. The unbalanced force of friction decreases the speed of the ball on the level and finally stops it. Non-uniform motion is called accelerated motion.

Acceleration

Acceleration is a quantity which describes how fast the speed of an object changes. **Acceleration** *is defined as the rate of change of speed with time.* An unbalanced force causes an object to accelerate. If the unbalanced force is constant in magnitude and direction, the acceleration is constant in magnitude and direction. Acceleration is a vector quantity.

Discussion

1. How can we tell if one object is in motion with reference to a second object? Give a specific example.

2. **a)** What is uniform motion?
b) Describe the forces acting on an object that has uniform motion.
c) What is non-uniform motion?
d) Describe the forces acting on an object that is accelerating.

3. **a)** Define speed.
b) Compare constant speed and average speed.

c) Sketch and describe the distance-time graph of an object that has constant speed.

d) What physical quantity does the slope of a distance-time graph represent?

4. A car is being driven between two towns 80 km apart. The driver records distance travelled every 20 min as a check on the uniformity of his driving. He obtained these data:

Distance (km)	0	16	32	48	64	80
Time (min)	0	20	40	60	80	100

a) Plot a distance-time graph from the data.

b) Using slopes, determine the average speed from the graph during the first 40 min and then during the last 40 min. Compare the values.

5. The formula for finding speed is $v = \frac{d}{t}$. Rearrange the formula to solve for d. Rearrange the formula to solve for t.

6. An object travelling at a constant speed travels 800 m in a time of 40 s. What is the speed of the object?

7. A car travels 80 km in the first hour of travel, 40 km in the second hour, and 130 km in the third hour. What was the average speed of the car for the trip?

8. A driver can travel at an average speed of 25 m/s on a level highway between two towns. It takes the driver 2 h to travel between the two towns. How far are the towns apart in kilometres?

9. A shell from a certain gun has an average speed of 960 m/s. How many kilometres does it travel in 5.0 s?

10. A pitcher throws a baseball with an average speed of 90 km/h. Calculate the time in seconds for the ball to travel from the pitcher's mound to the plate (a distance of 18.5 m).

11. One light year is defined as the distance that light travels in one year. Calculate the number of kilometres in a light year (speed of light = 3.00×10^8 m/s).

12. A runner on a 20 km run travels at an average speed of 20 km/h for the first 10 km and an average speed of 14 km/h for the last 10 km. Calculate his average speed for the run.

11.2 INVESTIGATION: The Speed of a Cart

An object travels at a constant speed in a straight line when no unbalanced force acts on it. In this investigation you apply a constant force to a cart in order to balance the forces of friction. Then you collect distance and time data. Next you plot a graph of distance against time for the motion. You finally calculate the average speed of the cart in centimetres per second.

Materials

roll of ticker tape	roll of masking tape
ticker tape timer	dynamics cart
power supply	newton spring scale
C-clamp	length of string
stop watch	ruler
sheet of graph paper per person	

Procedure A Preparation

a. Make a table like Table 70.
b. Set up the apparatus as shown in Figure 11-2.

Procedure B Finding the Period of the Timer

a. Attach the ticker tape timer to the end of the lab table using a C-clamp.
b. Draw two lines across a 1 m length of ticker tape, one near each end. Thread the tape through the timer. Let most of it hang down from the table.
c. Connect the timer to the power supply. Close the switch.
d. Pull the ticker tape through the timer. As the first mark passes beneath the vibrator, start the stop watch. As the second mark reaches the vibrator, stop the stop watch.
e. Try several runs to make sure that the vibrator is marking the ticker tape. Practise pulling the ticker tape so that the marks are well spaced for counting. Record the time elapsed in seconds for the ticker tape to travel the distance between the two lines on the ticker tape.
f. Count the number of tick marks between the two lines on the ticker tape. Divide the time elapsed by the number of tick intervals to calculate the elapsed time between ticks in seconds. This is the period of the timer. Record this above the table.

Figure 11-2 Measuring the speed of a cart.

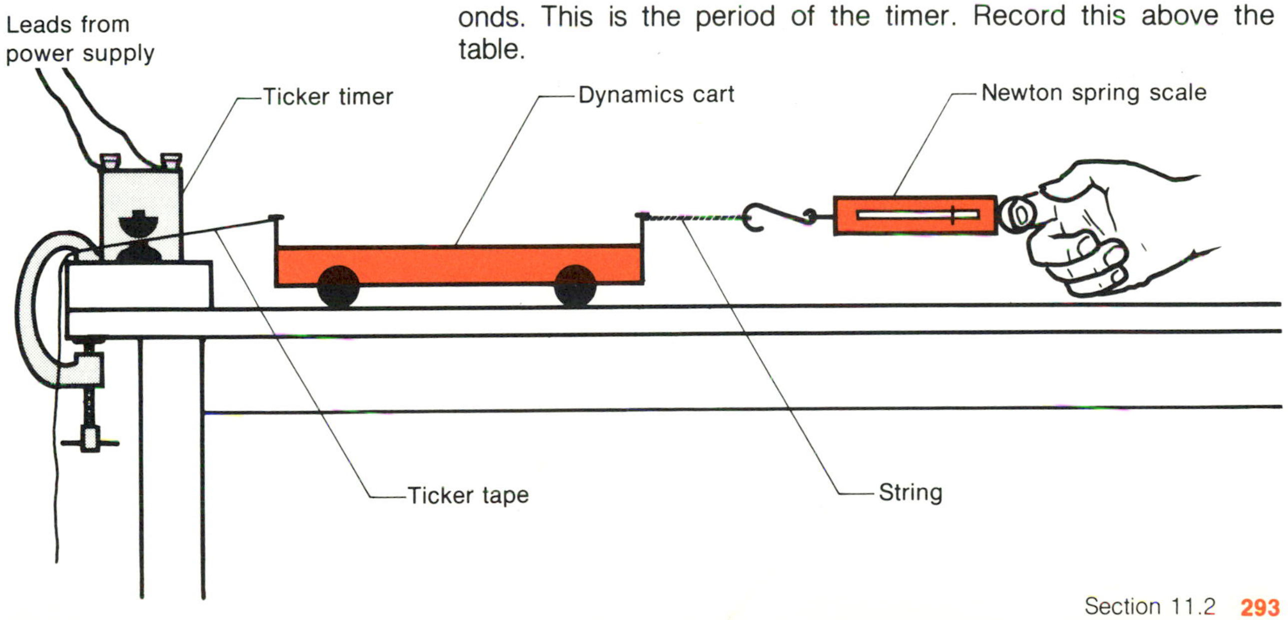

Procedure C Making a Ticker Tape Record of the Motion of the Cart

a. Attach the newton spring scale to the dynamics cart using the string. Apply sufficient force to move the cart at constant speed along the table. Practice this several times until you are able to maintain a constant speed. Record the force needed. This force is equal to the forces of friction on the cart.

b. Back the cart up to the ticker timer. Use the masking tape to attach about 1.5 m of ticker tape to the cart. Thread this tape through the timer. Tighten the tape between the cart and the timer.

c. Start the ticker tape timer and apply the force needed to move the cart at constant speed along the table. Repeat this until each person has a ticker tape of the object's motion to analyze.

Procedure D Analyzing the Motion of the Cart

a. Label the first mark ''starting point: $t = 0$''. Mark the tape into 10 tick intervals called ''tocks''. Number these 1, 2, 3, etc. Convert the tocks to seconds by multiplying the period (in seconds per tick interval) obtained in Procedure B by 10 tick intervals, 20 tick intervals, 30 tick intervals, etc. Record these times in the second column of the table.

b. Measure the distance in centimetres from each tenth tick (tock) to the starting point ($t = 0$). Record this in the distance column.

c. Plot a graph with distance from the start (centimetres) on the vertical axis and time (seconds) on the horizontal axis.

d. Calculate the slope of the graph.

TABLE 70 Distance-Time Data

Period of the ticker timer_____. Force of rolling friction_____.

Time (tocks)	Time (s)	Distance from the start (cm)
0	0	0
1		
2		
3		
4		
5		
6		
7		
8		
9		
10		

Discussion

1. What is the period of the timer in seconds per tick interval?
2. Describe the graph of distance against time for the cart's motion. Does it pass through the origin? Is the line straight? Is it horizontal or sloped?
3. Compare the distance travelled in equal intervals of time. Was the cart moving at a constant speed? How can you tell from the graph?
4. What was the average speed of the cart in centimetres per second? Compare the average speed of your cart with the average speed of other carts in the room. Account for any differences.
5. What was the magnitude of the forces of friction for your cart in newtons? How do you think this value would compare to the value for sliding friction for a cart without wheels?

11.3 INVESTIGATION: Non-uniform Motion and Seat Belts

Newton's First Law and unbalanced forces can be used to explain the motion of a passenger involved in a car crash. When a moving car collides with a large stationary rigid object such as a tree, the car slows down rapidly. But the passenger continues to move forward for a fraction of a second. Then an unbalanced force stops the passenger. This large unbalanced force should be applied by a seat-belt. However, in the case of an unbelted passenger, the dashboard and the windshield exert the force.

When a stopped car is hit from behind by a moving car, the person in the stopped car can receive a painful whiplash. The seat of the car exerts an unbalanced force forward on the lower part of the body. The head tends to remain at rest until it is pulled forward by the neck. However, when a head rest is present, it exerts the unbalanced force.

In this investigation you study the motion of a passenger involved in a collision. Plasticene represents the passenger. A cart takes the place of the car. You observe the motion of the passenger with and without seatbelts. You then apply Newton's First Law and your understanding of forces to account for the motion.

Materials

two dynamics carts
plasticene "passenger"
an obstacle (e.g. a book)
a 1.5 m ramp
front seat assembly
a method of inclining the ramp (e.g. a number of books)
ruler
shoelace
masking tape

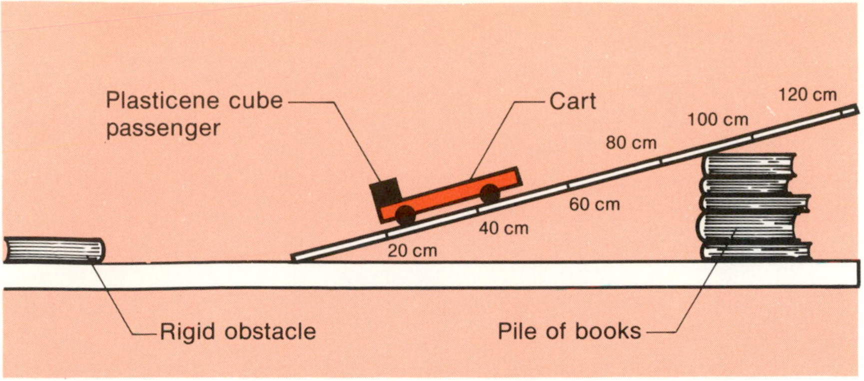

Figure 11-3 Non-uniform motion and seat belts.

Procedure A The Effect of Sudden Stops without Seatbelts

a. Mark the ramp into 20 cm intervals. Raise one end of the ramp about 30 cm. Position an obstacle such as a book about 30 cm from the bottom of the ramp. Hold the obstacle stationary.

b. Make a plasticene cube "passenger" with sides about 2 cm long. Place the passenger on the front of a dynamics cart. Place the front of the dynamics cart at the 20 cm mark on the ramp as shown in Figure 11-3.

c. Release the cart. Observe the motion of the passenger during and after the collision. Measure the distance the passenger moves from the collision point to where it stops. Repeat this step several times and average the distance.

d. Release the cart from several different distances up the incline to vary the speed. Observe the motion of the passenger and measure the distance as in step c. Repeat this procedure several times for each height to average the distances.

Procedure B The Effect of Sudden Starts without Seatbelts

a. Replace the obstacle in Procedure A with a cart. Position the back of this "target cart" on the level about 30 cm from the bottom of the ramp. Place the plasticene cube passenger at the front of the target cart.

b. Release another cart from the 20 cm mark on the ramp to collide with the back of the target cart. Observe the motion of the passenger during and after the collision. Measure the distance the passenger moves along the cart. Repeat this step several times and average the distance.

c. Repeat step b several times. Release the incoming cart from various distances up the ramp to vary the speed.

Procedure C Sudden Stops—Passenger in a Front Seat Assembly

a. Place the obstacle about 30 cm from the bottom of the ramp. Hold it stationary.

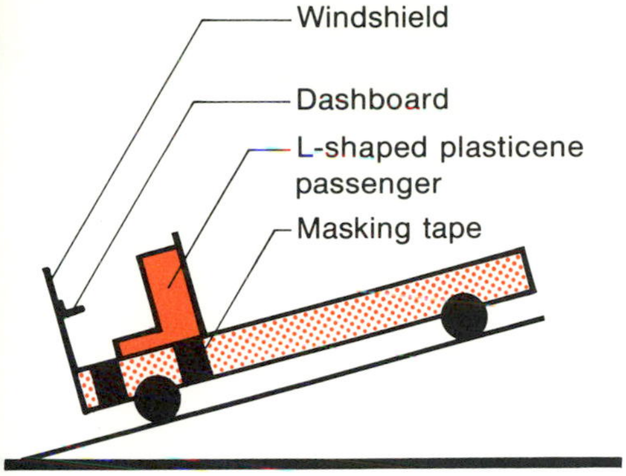

Figure 11-4 Passenger in a front-seat assembly without seat belts.

b. Use masking tape to attach the front seat assembly to a dynamics cart. Position the sharp dashboard toward the front of the cart.
c. Shape the plasticene to make an L-shaped passenger as shown in Figure 11-4.
d. Place the "passenger" in the seat without a seatbelt. Release the cart from about halfway up the ramp. Observe the motion of the passenger during and after the collision. Examine the passenger for damage caused by the collision.
e. Repeat step c. This time use the shoelace as a seatbelt to fasten the passenger in the seat. Try a lap belt and then a lap-and-shoulder belt.

Discussion A

1. Describe the motion of the passenger during and after the front-end collision.
2. How did the speed just before the collision change as the cart was released from further up the ramp?
3. How did the distance the passenger rolled after the collision change as the cart was released from further up the ramp?
4. Describe the motion of an unbelted passenger in a car which collides with a stationary obstacle.
5. Explain the observations using Newton's First Law and forces.
6. Seatbelts prevent a passenger from being thrown from the car. Why is it usually more dangerous to be thrown from the car than to remain in it?

Discussion B

1. The plasticene cube in this case represents the passenger's head. Describe the motion of the passenger's head during and after the rear-end collision.
2. What happens to the distance the passenger's head moves as the speed of the incoming cart increases? How does the force the neck exerts on the head change as the speed of the incoming cart increases?
3. Compare the motion of the passenger in a rear-end collision and in a front-end collision.
4. Explain the observations using Newton's First Law and forces.
5. Why should cars be equipped with head rests? Why should head rests be adjusted to the proper height?

Discussion C

1. Compare the injury the passenger suffered with and without seatbelts.
2. Which is safer, a lap belt or a lap-and-shoulder belt? Why?
3. Explain the observations using Newton's First Law and forces.

4. Why should cars be equipped with both seatbelts and head rests?
5. Properly adjusted seatbelts prevent a passenger's head from hitting the dashboard or windshield. How much space should exist between a seatbelt and a person's chest?
6. Why should dashboards be padded?
7. Discuss the merits of air bags which inflate in front of the driver at the instant of collision.

11.4 Pressure: A Force Applied to an Area

A force can change the speed of an object. It can also change the shape of an object. For example, in Investigation 11.3 the dashboard exerted a force on the "passenger" which slowed it down and changed its shape.

A force applied downward on a soft surface compresses it. If you were given the choice of having your foot stepped on by a man or a woman, you would probably choose the woman. The earth attracts a 50 kg woman with a force of about 500 N. It attracts a 80 kg man with a force of about 800 N. The woman will apply less force to your foot. However, if the woman is wearing shoes with narrow heels (a small surface area), you might be wiser to choose the man. The area over which the 800 N force is spread is greater. The surface of your foot may not compress as much and you may feel less pain. Both the force and the area over which the force is applied affect the amount that a body is compressed.

Let us estimate the force exerted on a unit area of your foot by the man and the woman. We assume that each person steps on your foot with one heel.

Man

The man is wearing a shoe with a heel 6 cm square. The area of the heel in square metres is $0.06\ \text{m} \times 0.06\ \text{m} = 0.003\,6\ \text{m}^2$. The force the man exerts on a unit area of your foot is the force of gravity pulling down on the man divided by the area of the heel.

$0.003\,6\ \text{m}^2$ of your foot supports 800 N

$$1\ \text{m}^2 \text{ of your foot supports } \frac{800\ \text{N}}{0.003\,6\ \cancel{\text{m}^2}} \times 1.0\ \cancel{\text{m}^2} = 222\,000\ \text{N} = 222\ \text{kN}$$

Woman

If the woman is wearing a shoe with a heel 3 cm square, the area of the heel in square metres is $0.03\ \text{m} \times 0.03\ \text{m} = 0.000\,9\ \text{m}^2$.

$0.000\,9\ m^2$ of your foot supports 500 N

$$1\ m^2 \text{ of your foot supports } \frac{500\ N}{0.000\,9\ \cancel{m^2}} \times 1.0\ \cancel{m^2} = 556\,000\ N = 556\ kN$$

The force exerted by the woman on a unit area of your foot is 2.5 times as great as the force exerted by the man on the same unit area.

Pressure

Pressure *is defined as the force acting on a unit area of a surface at right angles to the surface.* Pressure is a vector quantity since it has both magnitude and direction. Pressure is given the symbol $\boldsymbol{p}$, force the symbol $\boldsymbol{F}$, and the area the symbol $\boldsymbol{A}$. Therefore, the formula for pressure is $\boldsymbol{p} = \frac{\boldsymbol{F}}{\boldsymbol{A}}$

Units of Pressure

Since pressure is force per unit area, the derived unit of pressure is the newton per square metre (N/m^2). The newton per square metre was named the **pascal** (**Pa**) after the French scientist Blaise Pascal (1623–1662).

one pascal = one newton per square metre ($1\ Pa = 1\ N/m^2$)

The pascal is a relatively small pressure. A dollar bill lying flat on a table exerts a pressure of about 1 Pa.

A more useful multiple of the pascal is the **kilopascal** (**kPa**).

one kilopascal = one thousand pascals (1 kPa = 1 000 Pa)

Standard atmospheric pressure is 101.325 kPa. The difference between standard atmospheric pressure and the air pressure required in a standard bicycle tire (the tire-gauge reading) is about 400 kPa. Thus the total (absolute) pressure in this tire is 501.325 kPa. The woman in our earlier example would exert a pressure of about 556 kPa on your foot. Thus your foot would experience an absolute pressure of about 657 kPa.

The megapascal (MPa) is another useful multiple of the pascal. The yield strength of structural steel is about 300 000 000 Pa or about 300 MPa.

Sample Problem

A block of nickel 15 cm long, 20 cm wide, and 10 cm high rests on a horizontal table. The density of nickel is 8 900 kg/m^3. Calculate a) the mass of the block; b) the force of gravity acting on the block; c) the pressure exerted by the nickel block on the table (excluding air pressure).

Solution

Part a)

$l = 15\text{ cm} = 0.15\text{ m}$

$w = 20\text{ cm} = 0.20\text{ m}$

$h = 10\text{ cm} = 0.10\text{ m}$

$D = 8\,900\text{ kg/m}^3$

$V = l \times w \times h$

$= 0.15\text{ m} \times 0.20\text{ m} \times 0.10\text{ m}$

$= 0.003\text{ m}^3$

$D = \frac{m}{V}$ *or*

$m = D \times V$

$= 8\,900\,\frac{\text{kg}}{\cancel{\text{m}^3}} \times 0.003\,\cancel{\text{m}^3}$

$= 26.7\text{ kg}$

Part b)

$m = 26.7\text{ kg}$

$g = 10\text{ N/kg}$

$F = mg$

$= 26.7\,\cancel{\text{kg}} \times \frac{10\text{ N}}{\cancel{\text{kg}}}$

$= 267\text{ N}$

Part c)

$A = l \times w$

$= 0.15\text{ m} \times 0.20\text{ m} = 0.03\text{ m}^2$

Thus the area of the block in contact with the table is 0.03 m^2.

$p = \frac{F}{A}$

Thus $p = \frac{267\text{ N down}}{0.03\text{m}^2} = 8\,900\text{ N/m}^2\text{ down} = 8\,900\text{ Pa down}$

$= 8.9\text{ kPa down.}$

Discussion

1. **a)** Define pressure.
b) Explain how force affects the pressure on a surface.
c) Explain how the area on which the force acts affects the pressure on a surface.

2. What is a pascal in base units?

3. Calculate the pressure in each case. Express the answer in an appropriate unit.

	Force	Area	Pressure
a)	2 N down	1 m^2	
b)	1 N down	0.004 m^2	
c)	800 N down	0.4 m^2	
d)	7 000 000 N down	100 cm^2	

4. Explain why shoes with pointed heels can cause damage to tile floors.
5. If the pressure exerted by an object on a table is 6.25 kPa and the force of gravity on the object is 10 N, what is the area of the base of the object?
6. If an object exerts a pressure of 4 kPa on a surface, and if the area of the base is 49 cm^2, what is the force of gravity on the object?
7. Calculate the pressure in kilopascals you exert on a surface if you stand on the heel of one shoe.
8. **a)** What information would you need to find the pressure exerted by the atmosphere on the surface of the earth?
   b) How would you use this information to find atmospheric pressure?
9. Explain why a "strong man" at a carnival can allow a person to break a large flat stone placed on his chest by hitting it with a hammer. Use pressure and Newton's First Law in constructing your answer.

11.5 INVESTIGATION: Pressure and Stress on Polyurethane Foam

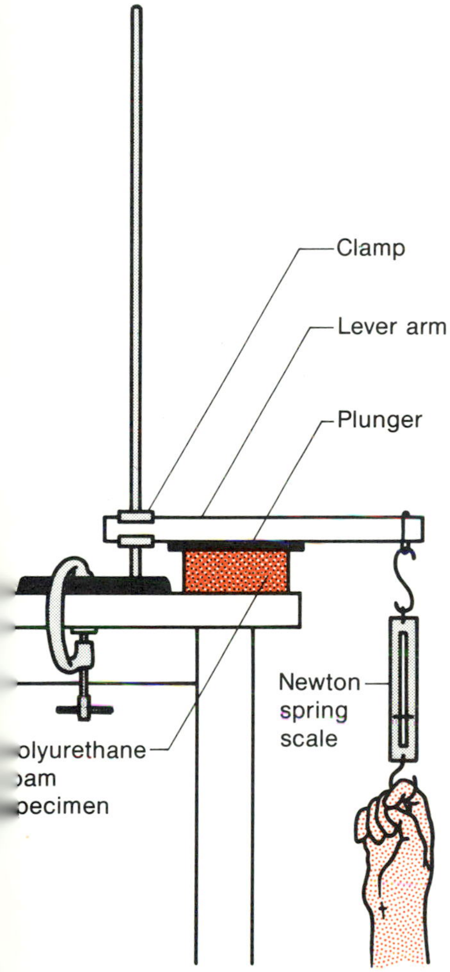

Figure 11-5 The compressibility of polyurethane foam.

When a force is exerted on an object which is not free to move, the object experiences a pressure and is put under stress. Pressure results in a stress. The magnitude of the stress is measured in the same way as pressure is measured—by the ratio of the force to the area over which the force acts. Stress can cause either compression or expansion, depending on the direction of the force. In this investigation you measure the pressure needed to compress polyurethane foam a certain amount. Therefore you measure the stress it experiences.

Materials

5 samples of polyurethane foam about 5 cm thick of increasing surface area
newton scale force measurer
ring stand
clamp
nail
C-clamp
lever with plunger
piece of string

Procedure

a. Make a table like Table 71.
b. Assemble the apparatus as shown in Figure 11-5.
c. Attach the lever arm and plunger assembly to the ring stand. Position the ring stand to leave about 5 cm of the lever arm

extending past the edge of the table. Clamp the ring stand in position.

d. Adjust the height of the plunger above the table to compress the foam to one half its usual thickness when the plunger is level. Record this compression in Table 71.

e. Measure the dimensions of the upper surface of each sample of polyurethane. Record these in the table.

f. Place a sample under the centre of the plunger.

g. Use the string to attach the newton spring scale to the lever arm. Apply the force to compress the sample evenly to the desired amount. Record this value in the table. Repeat this procedure for all samples. Make sure the centres of the samples are all placed at the same point beneath the plunger.

h. Calculate the force applied by the plunger to each sample. Record these in the table. The force applied by the plunger is greater than the force applied by the newton spring scale. If the sample is positioned halfway between the ring stand and the spring scale, the force is twice as much as the spring scale reading. Ask your teacher for the conversion factor for the set-up you are using.

i. Plot a graph with the force in newtons applied to the foam on the vertical axis and the area in square metres on the horizontal axis. Calculate the slope.

TABLE 71 Force-Area Data for the Compression of Polyurethane Foam

Amount polyurethane foam is compressed = ________ cm

Force applied to end of lever arm (N)	Force applied to foam block (N)	Length of foam block (m)	Width of foam block (m)	Surface area of foam block (m^2)

Discussion

1. How does surface area affect the force required to produce a given compression?
2. Describe the graph. Does it go through the origin? Is it a straight line graph?
3. What is the magnitude of the slope of the graph in newtons per square metre? in pascals? in kilopascals? The slope repre-

sents the stress the polyurethane foam experiences at a given compression.

4. Compare the slope you obtained with the values obtained by other groups in the class. Are they the same? Should they be the same?
5. Design and carry out an experiment to compare the stress at a given compression of different kinds of polyurethane foam pads.
6. Design an experiment to determine the pressure needed to compress the padding of a car dashboard by 2 mm. You may wish to try this experiment on the family car.
7. If you were building a bridge on concrete poles (piers), why would it be necessary to investigate the stress behaviour of concrete?

11.6 INVESTIGATION: Pressure in Liquids

Oceans and large lakes are the last remaining frontiers on earth. Scientists are designing and building equipment to explore their depths, but the pressure at these depths hinders exploration.

In this investigation you find out how the pressure on an object in a liquid changes with its depth below the surface and with the density of the liquid. You determine the direction pressure acts in water. Also you determine the effect the size of the body of water has on the pressure at a given depth. Your teacher may choose to demonstrate this study.

Materials

U-tube manometer containing coloured water
battery jar about the same height as the fish tank but with a smaller volume
flexible rubber tubing that fits the manometer, T-tube, and thistle tube
bent thistle tube and supports
piece of balloon rubber (diaphragm)
masking tape
rubber band
sheet of graph paper per person
fish tank
T-tube
pinch clamp
ditto fluid

Procedure A Assembling the Apparatus

a. Make a table like Table 72.
b. Fill the fish tank to within 5 cm of the top. Fill the battery jar to the same height.

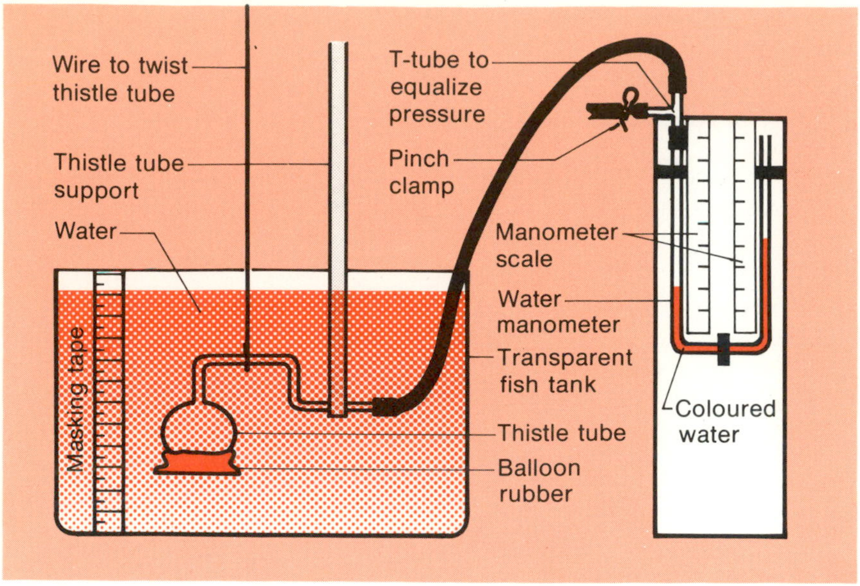

Figure 11-6 Pressure in water.

c. Mark two strips of masking tape at 1 cm intervals. Stick them to the outside of the fish tank and the battery jar as shown in Figure 11-6.
d. Stretch the balloon rubber (diaphragm) loosely over the end of the thistle tube. Secure it in this position with the rubber band.
e. Connect the thistle tube, rubber tubing, T-tube, and manometer as shown in Figure 11-6.
f. Hold the thistle tube above the water. Open the pinch clamp. Allow the water levels in the two arms of the manometer to reach the same height. Close the pinch clamp.

Procedure B Relationship between Pressure and Direction at Constant Depth

a. Lower the thistle tube to a depth of 15 cm in the fish tank.
b. Point the diaphragm sideways. The centre of the diaphragm should be at the depth of 15 cm. Read the difference in height of the water in the two arms of the manometer. Record this as the manometer reading at a depth of 15 cm.
c. Point the diaphragm down and then upward at exactly the depth of 15 cm. Record the manometer readings in the table.

Procedure C Relationship between Pressure and Depth of Water

a. Replace the thistle tube with a straight piece of glass tubing.
b. Point the open end of the glass tubing down.
c. Lower the open end of the glass tubing into the water, stopping at intervals of 1 cm.

d. Measure the difference in height between the arms of the manometer at each depth. Record these data in the table.
e. Plot a graph with difference in height of the arms of the manometer on the vertical axis and water depth on the horizontal axis.

Procedure D Relationship between Pressure, Depth, and Size of Body of Water

a. Remove the glass tubing from the water.
b. Repeat the steps outlined in Procedure C using the battery jar. Plot the data for the battery jar on the same set of axes as for the fish tank.

Procedure E Relationship between Pressure and Depth of Alcohol

a. Remove the glass tubing from the water.
b. Fill the battery jar with alcohol (ditto fluid).
c. Repeat the steps outlined in Procedure C. Plot the data for the manometer submerged in the alcohol on the same set of axes as before. Use a marker of a different colour to draw the line.

TABLE 72 Data for Pressure in Water and Alcohol

Procedure B		Procedure C		Procedure D		Procedure E
Direction diaphragm points	Manometer reading at 15 cm depth	Depth (cm)	Manometer reading in water (cm)	Depth (cm)	Manometer reading in water (cm)	Manometer reading in alcohol (cm)
Sideways		1				
		2				
Down		3				
		4				
Up		5				

Discussion

1. Compare the readings of a manometer pointed in different directions at the same depth. What is the generalization for the relationship between pressure and direction at constant depth?
2. Compare the direction of the pressure exerted by a liquid with the direction of the pressure exerted by a solid.
3. What happens to the pressure on an object submerged in a liquid as the object goes deeper in the liquid?
4. Describe the graph of manometer readings against depth for the fish tank. Does it go through the origin? Is it a straight line? Is it horizontal or sloped?

5. Describe the graph of manometer readings against depth for the battery jar.
6. Compare the volumes of water in the fish tank and in the battery jar.
7. What effect does the volume of water have on the pressure at a given depth?
8. Compare the density of water and alcohol. (See Table 27, page 63.)
9. What effect does the density of a liquid have on the pressure exerted on an object at a given depth?

11.7 Pressure in Fluids

Pressure and Direction

The results of Investigation 11.6 indicate that the pressure in a liquid is exerted equally in all directions. At a given depth, the pressure up, the pressure down, and the pressure sideways are equal. This is true in any fluid whether it is a gas or a liquid. The air pressure on the surface of your body is about 100 kPa at every point, and is directed at right angles to the surface of your body.

Pressure, Depth, and Density

The pressure on an object submerged in a fluid is dependent on two factors: the depth of the object below the surface of the fluid, and the density of the fluid. The greater the depth, the greater the pressure. The greater the density, the greater the pressure.

The force that is responsible for the pressure is caused by the force of gravity on the fluid vertically above the surface of the object. The force of gravity depends on the total mass of the fluid above the surface of the object. The total mass of this fluid depends on both the density and the volume of the fluid. The force exerted on a surface of an object at a given depth can be calculated if the area of the surface, the depth, and the density of the fluid are known. Since liquids are almost incompressible, the density of the liquid does not change significantly with depth.

Sample Problem Involving Pressure, Depth, and Density

Look at Figure 11-7. A surface about the size of the top of a person's head (20 cm × 20 cm) is shown submerged in water at a depth of 20 m. What is the pressure at this depth?

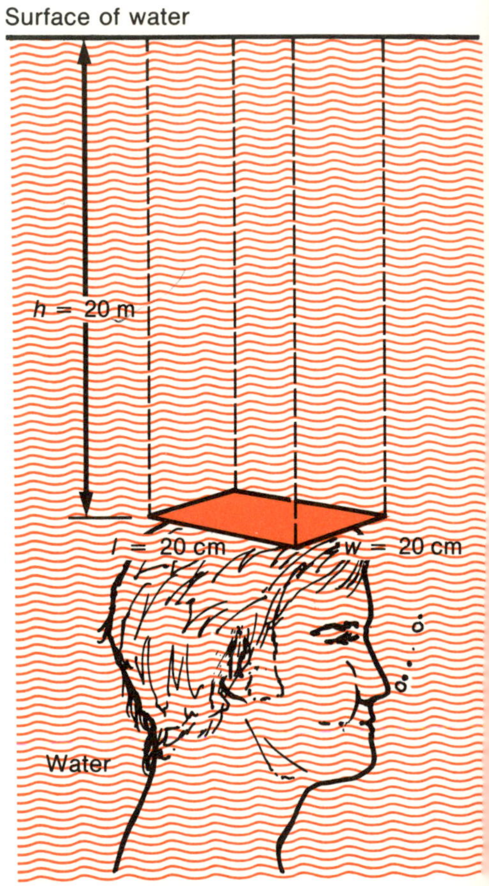

Figure 11-7 Pressure at a given depth in a fluid.

Solution

The force on this area is found by calculating the pull of gravity on the rectangular column of water supported by the area.

The **volume** of a regular object is $l \times w \times h$.
Thus the volume of water supported is $0.20\ m \times 0.20\ m \times 20\ m = 0.8\ m^3$.

The **area** of a rectangular surface is $l \times w$.
Thus the area supporting the volume of water is $0.20\ m \times 0.20\ m = 0.04\ m^2$.

The **mass of water** supported is $D \times V$
and the density of water is $1\,000\ kg/m^3$.

Thus the mass of water supported is $1\,000 \frac{kg}{\cancel{m^3}} \times 0.8\ \cancel{m^3} = 800\ kg$.

The **force of gravity** on a mass is $m\,g$.
and the force of gravity-mass ratio on earth is $10\ N/kg$.

Thus the force of gravity on the supported water is $800\ \cancel{kg} \times 10 \frac{N}{\cancel{kg}} = 8000\ N$.

Thus the force exerted on the submerged surface is 8 000 N.

The pressure is F/A.

Thus the pressure is $\frac{8\,000\ N}{0.04 m^2} = 200\,000\ Pa = 200\ kPa$.

The pressure exerted on the submerged surface by the water is about 200 kPa down.

But the air above the water exerts about another 100 kPa down. Thus a person submerged at a depth of 20 m has about 300 kPa exerted down on the top of the head.

Discussion

1. Define a fluid. Give two examples of fluids that are quite different in appearance.
2. Explain why the pressure in a fluid increases with depth.
3. **a)** A dam is constructed with the bottom thicker than the top. Why?
   b) The average thickness of a dam constructed to hold back the waters of a lake 12 km long is the same as for a lake 2 km long if the water is the same depth. Explain.
4. Otto von Guereck, a resident of Magdegurge, Germany, in 1654, took two tight-fitting hemispheres which were about a third of a metre in diameter and placed them together with their rims in contact. Air was withdrawn from the enclosure. It took four teams of horses attached to each hemisphere to pull them apart. Why? Show your calculations.
5. Try this. Fill a glass or bottle with water. Cover its mouth with a piece of cardboard. Invert the bottle and remove the hand supporting the cardboard. Describe and explain the results. (**CAUTION:** Do this experiment over a sink!)
6. Account for the shape of a submarine.

7. A Dutch boy is said to have saved his town by thrusting his finger into a hole in the dike holding back the North Sea. How could one finger hold back the pressure of the sea?
8. Doctors measure blood pressure on the arm at about heart level. Would they get the same pressure if they measured it on the leg? Explain.
9. **a)** Calculate the pressure of water at depths of 1 m, 10 m, and 1 km.
   b) Calculate the pressure in alcohol at depths of 1 m, 10 m, and 1 km.
10. Fresh water has a density of 1 000 kg/m^3. Sea water has a density of 1 030 kg/m^3. In which kind of water could a submarine go the deepest? Why?

Highlights

Our knowledge of forces enables us to explain motion and pressure.

An object in uniform motion travels at constant speed in a straight line. Speed is distance travelled per unit time. Both average and constant speed are calculated by dividing the total distance travelled by the total time taken. Some units of speed are the metre per second (m/s) and the kilometre per hour (km/h). Motion can be studied by drawing a distance-against-time graph. The distance-time graph for constant speed is a straight line with constant slope. An object in non-uniform motion changes its speed and/or direction. Non-uniform motion is called acceleration.

Pressure is determined by dividing the perpendicular force applied to a surface by the area of the surface. The unit of pressure is the newton per square metre (N/m^2) or the pascal (Pa). 1 Pa is exerted when a force of 1 N is applied to an area of 1 m^2. Standard atmospheric pressure is 101.325 kPa. Pressure in a liquid is exerted equally in all directions. Its magnitude depends on the density of the liquid and the depth of the object below the surface.

UNIT
12 Work, Basic Machines, and Power

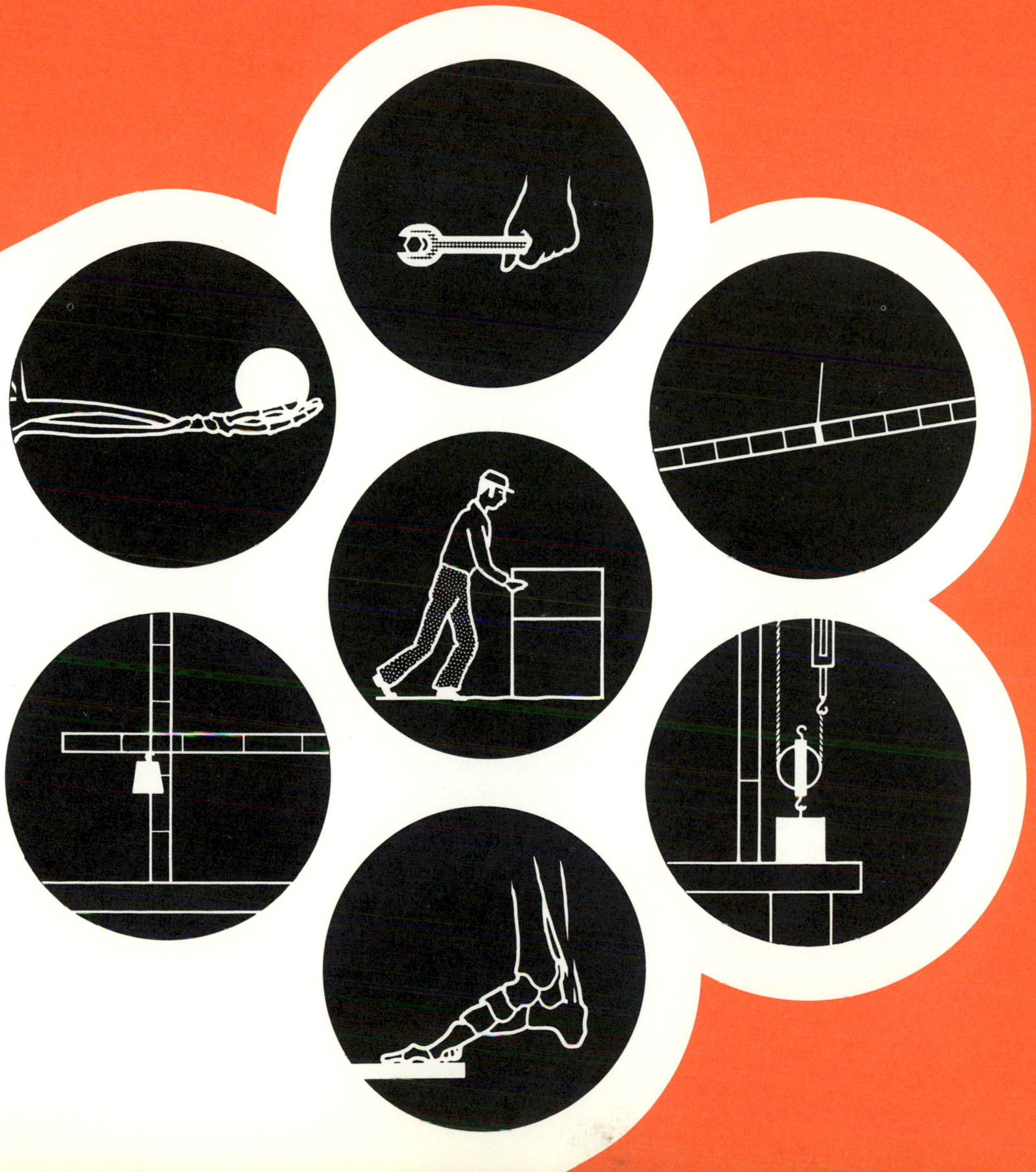

In this unit you continue your investigation of forces to discover how force and work are related. You also experiment with some simple machines. Finally you learn how to measure the power of a wide range of things, living and non-living.

12.1 A Description of Work

The Conditions of Work

The term "work" has different meanings for different people. A student doing homework calls it work. The same student probably calls a game of tennis play. However, a physical scientist would say the student is doing very little work during homework, but a great deal of work playing tennis.

In physical science the term work is restricted to activities involving both a force and a distance. Three conditions are necessary before work is done on an object: a force must be exerted on the object; the object must be moved by the force through a distance; the force and the distance moved must be in the same direction. Figure 12-1 shows three examples where work is being done on an object. Example A shows a person lifting a box from the floor to a table. The person exerts a force up. The box is moved a vertical distance. The force and the distance are in the same direction. Example B shows a person shoving the box along the floor. The person exerts a horizontal force forward to overcome the force of kinetic friction. The box moves through a distance in the direction of the force. Example C shows a rocket being launched into space. The burning of fuel provides the upward force on the rocket. The rocket moves in the direction of the force.

Figure 12-2 shows three examples where work is not being done. In Example A a person is holding a box stationary above the floor. A constant force up is exerted to balance the force of gravity. Since there is no motion in the direction of the force, no work is being done on the box. Example B shows the box being carried from one place to another. The person exerts a force up to support the box. The box is moved through a horizontal distance. But, since the force and the distance moved are at right angles, the person does no work on the box. Example C shows a space capsule coasting with its engines off far out in space where there is no force of gravity. Space can be considered a vacuum. Therefore the force of air resistance opposing the motion is negligible or almost non-existent. As a result, the space capsule moves through space at a constant speed. Since no force is being applied to the space capsule, no work is being done.

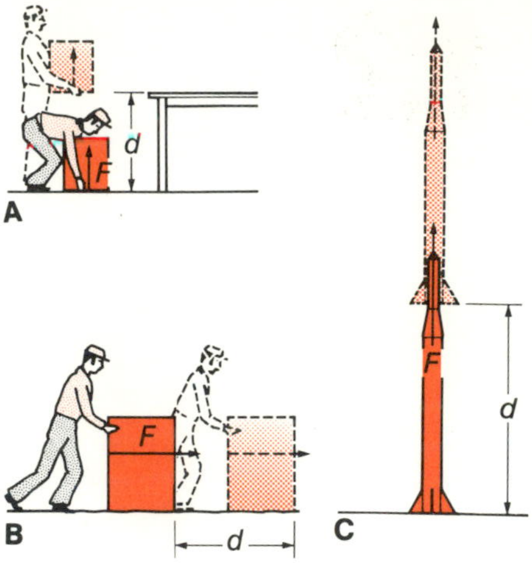

Figure 12-1 Work is being done in these examples.

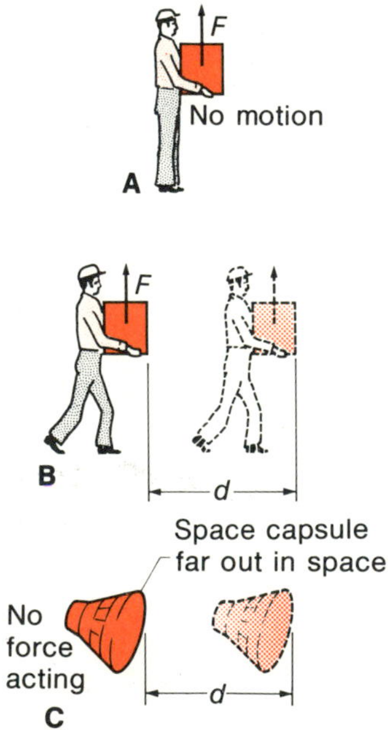

Figure 12-2 No work is being done in these examples.

For work to be done, a force must move an object through a distance in the direction of the force.

The Operational Definition of Work

An operational definition of work describes how work is measured.

Suppose a person uses a newton spring scale to measure the force applied to lift masses from one height to another. Figure 12-3 shows the person lifting an object with a mass of 1.0 kg to different heights. The force of gravity on a 1.0 kg mass is about 10 N. Therefore the person exerts a force of 10 N up to lift the mass. When the person lifts the mass 0.5 m, some work is done. When the mass is lifted 1.0 m, twice as much work is done. Three times as much work is done to lift the mass 1.5 m. In general, *when a constant force is applied to an object, the work done depends directly on the distance the object moves.*

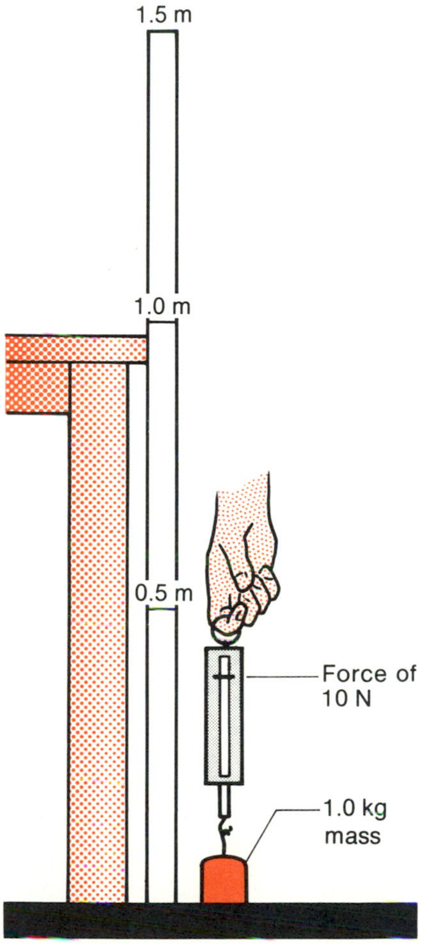

Figure 12-3 A constant force applied through different distances.

Figure 12-4 shows the person lifting loads having masses of 1.0 kg, 2.0 kg, and 3.0 kg respectively. The loads are lifted to the same height of 1.0 m. The person exerts a force of 10 N and does some work lifting the 1.0 kg mass. A force of 20 N is required to lift two 1.0 kg masses. When the two masses are lifted 1.0 m, twice as much work is done. A force of 30 N is used to lift three 1.0 kg masses. Three times as much work is done to lift this load to the height of 1.0 m. In general, *when objects are moved through a constant distance, the work done depends directly on the size of the force used.*

These examples show that the work done on an object depends directly on both the magnitude of the force applied and the distance the object moves.

The work done does not depend on time. A large force applied to an object for a short time might move the object a large distance. A smaller force applied for a longer time could move it the same distance. It is just the product of force and distance that is a measure of the work done.

The work done by a force is calculated by multiplying the force applied by the distance the object moves in the direction of the force.

Work = force × distance

If ***W*** stands for work, ***F*** for force, and ***d*** for distance, the equation for work is:

$$W = Fd$$

The SI Unit of Work

Since work is the product of force and distance, the SI derived unit of work is the newton metre (N·m). The newton metre has been given a special name in honour of the British physicist James Prescott Joule (1818–1889) who studied energy transformations. The newton metre (N·m) is called the joule (J). *One* **joule** *(J) is the work done when a force of one newton (1 N) is applied through a distance of one metre (1 m) in the direction of the force.*

1 J = 1 N·m

If one medium-sized apple is lifted through a vertical distance of 1 m, about 1 J of work has been done. The hydraulic elevator which lifts a car of mass 1 000 kg through a height of 2.0 m does about 20 kJ of work. A Russian weight lifting champion at the 1976 Summer Olympics in Montreal exerted an average force of about 2 550 N through a vertical distance of about 2.0 m and did 5 100 J of work on the barbells.

Discussion

1. Distinguish between the layman's and the scientist's use of the term "work".
2. **a)** What three conditions are necessary before work is done?
   b) Describe two common examples where work is being done.
   c) Under what conditions will an applied force perform no work on an object?
   d) A satellite orbits the earth because the earth exerts a force on it. Is the earth doing work on the satellite? Explain your answer.
3. **a)** Write the operational definition for work.
   b) Write the word equation and the symbol equation for work.
   c) What is the SI derived unit for work in terms of force and distance?
   d) What is the special name for the SI derived unit for work? What symbol is used for this unit?
4. Express the derived unit with a special name for work in terms of the base units of mass, length, and time.
5. **a)** Define the joule.
   b) How much work is done to lift one medium-sized apple to a height of 2.0 m?
   c) How much work is done to lift two medium-sized apples to a height of 1.0 m?
   d) How much work is done by a weight lifter holding a mass of 200 kg 2 m above the floor for 10 s?
   e) How much work was done in getting the 200 kg mass from the floor to the 2 m position?

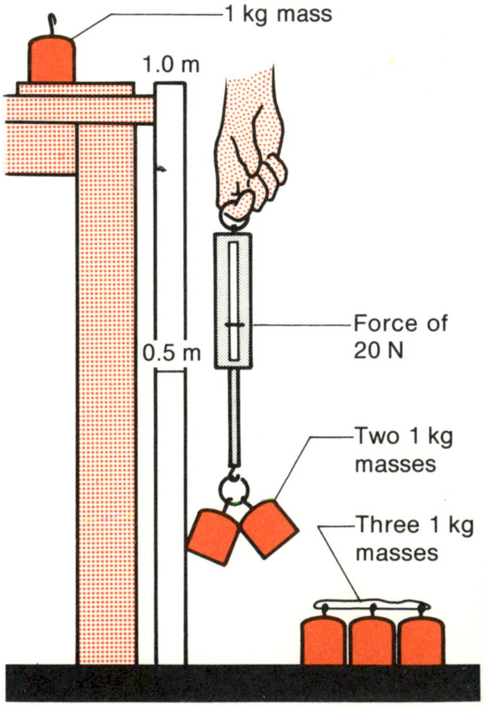

Figure 12-4 Different forces applied through a constant distance.

12.2 Problems Involving Work

Problems involving work are of three types. You may be given the force and distance and be asked to find work. You may be given the work and force and be asked to find the distance. You may be given the work and distance and be asked to find the force. The following are model solutions to these three types of problems.

Type 1 Finding Work, Given Force and Distance

Problem

An object has a force of 5.0 N applied to it and moves a distance of 4.0 m. Find the work done.

Solution

The formula for work is $W = Fd$
Thus $W = 5.0\ \text{N} \times 4.0\ \text{m} = 20\ \text{N·m} = 20\ \text{J}$.

Type 2 Finding Distance, Given Work and Force

Problem

20 J of work are done on an object when a constant force of 5.0 N is applied. Find the distance the object moves.

Solution

$W = Fd$

or $d = \frac{W}{F} = \frac{20\ \cancel{\text{N}}\cdot\text{m}}{5.0\ \cancel{\text{N}}} = 4.0\ \text{m}$

Type 3 Finding Force, Given Work and Distance

Problem

20 J of work are done by a constant force when it moves an object 4.0 m. Find the constant force on the object.

Solution

$W = Fd$

or $F = \frac{W}{d} = \frac{20\ \text{N}\cdot\cancel{\text{m}}}{4.0\ \cancel{\text{m}}} = 5.0\ \text{N}.$

Discussion

1. Calculate the work done in each of the following examples:
   a) Force = 4 N, distance 2 m
   b) Force = 120 N, distance 3.60 m
   c) Force = 20 kN, distance = 3.0 m
   d) Force = 8.0 N, distance = 150 cm
2. Calculate the force exerted in each of the following examples:
   a) distance = 20 m, work done = 500 J
   b) distance = 480 cm, work done = 288 J
   c) distance = 3.5 m, work done = 105 J
   d) distance = 4 km, work done = 20 kJ
3. Calculate the distance through which the force is exerted in the following examples:
   a) Force = 40 N, work done = 80 J
   b) Force = 2.2 N, work done = 660 J
   c) Force = 400 N, work done = 30 kJ
   d) Force = 2.5 kN, work done = 7.5 kJ
4. A force of 15 N is applied to a toy car and moves it 4.0 m. How much work is done?
5. A car travels along a level road a distance of 1.5 km. The forward force exerted by the road on the car is 200 N. Calculate the work done on the car in joules.
6. A force of 350 N is required to push a piano having a mass of 400 kg. How much work is done to push the piano 5.0 m across the floor?
7. A car accelerates from 0 to 100 km/h in 8.0 s. It covers a distance of 110 m in this time. The engine does 3.0×10^5 J of work to accelerate the car. What is the average force exerted by the engine?
8. Water flows over a waterfall at the rate of 70 kL/min. The waterfall is 30 m high. How much work is done by the force of gravity acting on the water in 1 s? (1 L of water has a mass of 1 kg.)
9. A 20 N force is applied to a body in its direction of motion. A 4 N force of kinetic friction acts opposite to the direction of motion. Calculate the work done by the applied force while the body moves 8.0 m.

12.3 INVESTIGATION: Using the Inclined Plane to Do Work

Man has devised many machines to help him work. One of the earliest machines was the inclined plane. The Egyptians may have built huge inclined planes to help them construct the pyramids.

In this investigation you study the characteristics of the inclined plane.

Materials

inclined plane
ring stand, support rod, and clamp
1.0 kg mass
newton spring scale
metre stick

Procedure

a. Construct a table like Table 73.
b. Lay the inclined plane flat on the table. Attach the newton spring scale to the mass. Pull the mass slowly along the horizontal plane at a constant speed. The force required to do this balances the force of kinetic friction. Record this force in Table 73.
c. Set up the inclined plane as shown in Figure 12-5. Incline the plane at an angle of about 30°.
d. Mark off a distance along the plane to a point x as shown. Record this distance (in metres) in the table.
e. Measure the vertical distance from the table top to the point x. Record this distance (in metres) in the table.
f. Use the spring scale to pull the mass at a constant speed up the inclined plane to the point x. Record the average force in newtons in the table.
g. Use the spring scale to lift the mass vertically to the point x. Record the force in newtons in the table.
h. Calculate the work done to move the mass to the point x using the inclined plane. Record this figure in the table.
i. Calculate the work done to move the mass to the point x using the inclined plane if the force of friction is ignored. (Assume the force of friction on the incline is about 0.75 times the force of friction on the level.) Record the work done in the table.
j. Calculate the work done to lift the mass vertically to the point x. Record this figure in the table.
k. Repeat steps b to j two more times. Use a different angle of the inclined plane each time.

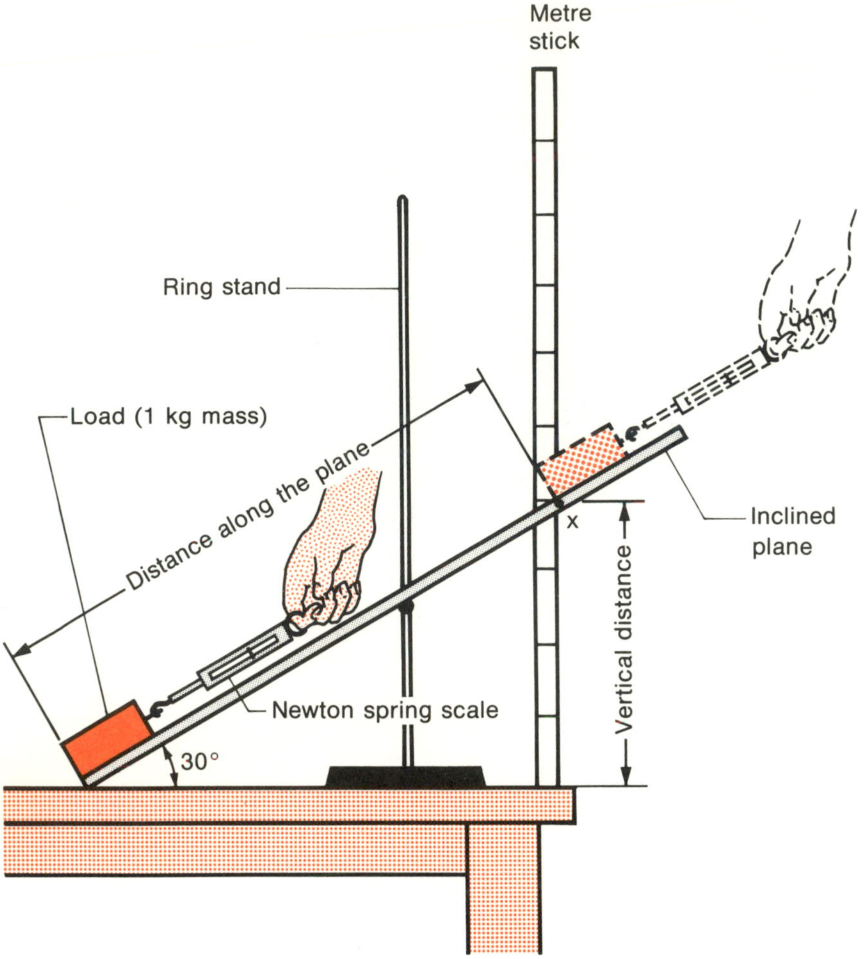

Figure 12-5 Work done using an inclined plane.

TABLE 73 The Inclined Plane and Work

	Moving mass on an inclined plane					Lifting mass vertically		
Trial	Force of friction (N)	Force (N)	Distance (m)	Work done (J)	Work done (No friction) (J)	Force (N)	Distance (m)	Work done (J)
1								
2								
3								

Discussion

1. Compare the force needed to lift the mass vertically to the point x with the force needed using the inclined plane.
2. How did the force change as the incline was made less steep?

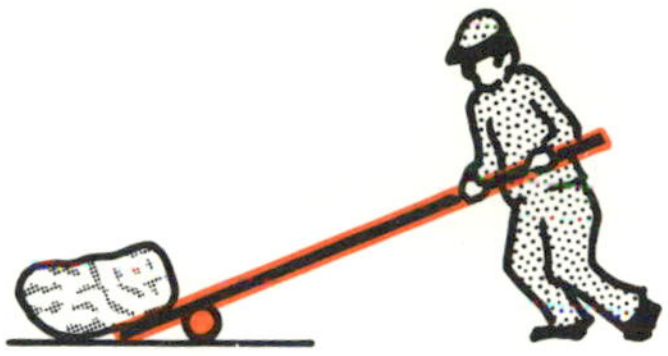

A Lever

B Inclined plane

C Pulley

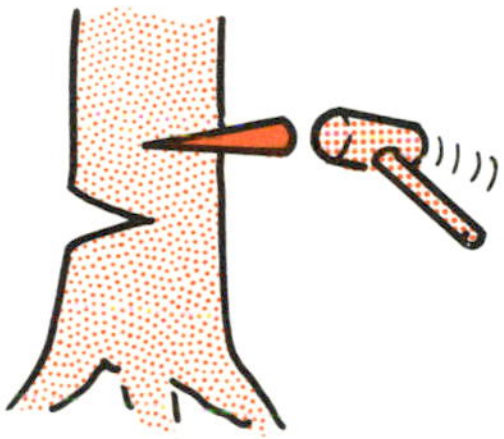

D Wedge

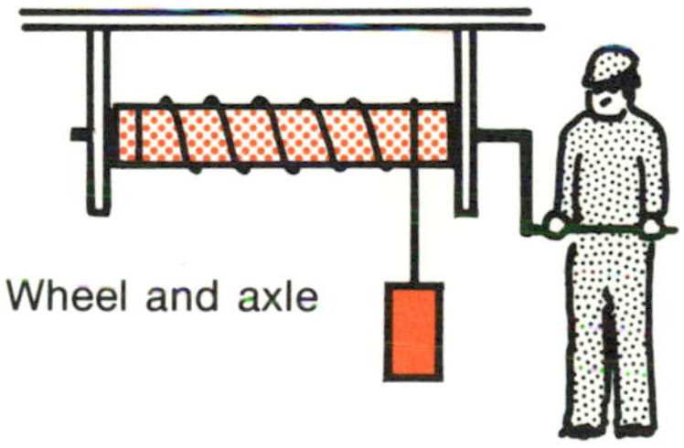

E Wheel and axle

F Jack screw

Figure 12-6 The six simple machines.

3. Compare the distance along the incline with the vertical distance to the point x.
4. How will the length of an inclined plane need to be changed if the force along the incline is to be decreased to raise an object to a constant height?
5. Compare the total work done using the inclined plane with the work to lift the mass vertically to the point x.
6. Compare the work using the inclined plane without friction with the work to lift the mass vertically to the point x.
7. A machine is a device that enables us to do work more conveniently than is otherwise possible. Use the results of this investigation to explain why people use inclined planes.

12.4 The Basic Machines and Work

The Basic Kinds of Machines

A machine is a device that enables us to do work more easily than is otherwise possible. There are two basic kinds of machines: the inclined plane and the lever. Four other machines gradually developed from these two. The wedge and the screw are modifications of the inclined plane. The pulley, the wheel, and the axle are adaptations of the lever. Figure 12-6 shows the six simple machines.

The Uses of Machines

Some machines change forces from one kind to another; a hydro-electric generator changes the force of falling water into an electric force. Other machines transfer forces from one place to another; the driveshaft of a car takes the force from the engine to the wheels. Some machines increase the size of a force; a car jack amplifies a muscular force to enable us to lift a car against the force of gravity. Others decrease the size of a force; the sprocket of a bicycle scales down the muscular force supplied to the pedals to a smaller force acting on the wheel in order to gain more speed. Some machines change the direction of a force; a single pulley mounted overhead changes a downward muscular force applied to a rope to an upward tensile force applied by the rope to a load.

The Inclined Plane as a Machine

Man usually finds it easier to exert a small force for a large distance rather than a large force for a small distance. The inclined plane is useful because less force is required to pull a load up an incline than to lift it vertically. As the incline is made less steep, the force required to move the load at constant speed decreases. However, the force on the incline has to be applied for a larger distance than the force to lift the load straight up.

Figure 12-7,A shows a baggage cart of mass 40 kg being pulled on the level. A force of 60 N is required to move it at constant speed. From our earlier study of motion we know that a second force of 60 N is acting back on the cart. This is the force of kinetic friction on the level.

Figure 12-7,B shows the baggage cart being pulled up the incline. The force of kinetic friction on the incline is 50 N. A force of 250 N is needed to balance the forces of kinetic friction and gravity acting backwards. However, if the baggage cart and its contents are to be lifted vertically, a force of about 400 N is needed. The inclined plane decreases the size of the force the man has to exert. Sometimes an inclined plane is essential if the load is heavier than a man can lift.

Mechanical Advantage

The mechanical advantage of a machine indicates how much easier a machine makes the work. **Mechanical advantage** *is the ratio of the force exerted by the machine to the force applied to the machine*. There are two kinds of mechanical advantage—ideal mechanical advantage and actual mechanical advantage.

Ideal Mechanical Advantage of the Inclined Plane. The force applied by the machine in Figure 12-7,B is a force of 400 N since this is the force needed to lift the object of mass 40 kg vertically to the top of the incline. If no force of friction exists, the man needs to apply a force of 200 N parallel to the incline to move the loaded luggage cart up the inclined plane. Therefore, the ideal mechanical advantage of this inclined plane is calculated as follows:

$$\text{Ideal mechanical advantage} = \frac{\text{Force applied by machine}}{\text{Force applied to machine without friction}} = \frac{400\ \text{N}}{200\ \text{N}} = \frac{2}{1}$$

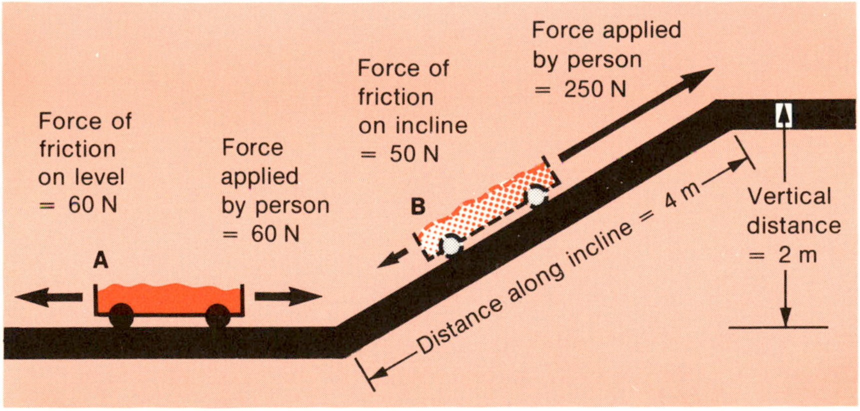

Figure 12-7 Using an inclined plane to lift a baggage cart (mass of cart = 40 kg).

From Investigation 12.3 you know that the force applied to the machine decreases as the ramp is made less steep. The ideal mechanical advantage of the inclined plane increases as the ramp is made less steep.

Actual Mechanical Advantage. The force applied by the machine to lift the object vertically in Figure 12-7,B is 400 N even when the force of friction is present. However, an actual force of 250 N parallel to the incline has to be applied to the machine to move the cart up the incline. 50 N of this force are used to balance the force of kinetic friction. The actual mechanical advantage of this inclined plane is:

$$\text{Actual mechanical advantage} = \frac{\text{Force applied by machine}}{\text{Force applied to machine with friction}} = \frac{400\text{ N}}{250\text{ N}} = \frac{1.6}{1}$$

The actual mechanical advantage of a machine decreases as the force of friction gets larger for the same degree of steepness.

Efficiency of Machines

The purpose of a machine is to enable man to do useful work. In Investigation 12.3 you found that the work done to lift the mass vertically to the point x is less than the work done to get it to point x using the incline. This is because some work is done to overcome the force of friction.

The ratio of the useful work done by a machine to the work done on the machine is a measure of its efficiency.

$$\textbf{Efficiency} = \frac{\textbf{useful work done by a machine}}{\textbf{work done on the machine}}$$

In Figure 12-7 the useful work is the work done to lift the luggage cart and its contents without friction to the top of the incline. This is calculated either by using the force required with the machine (ignoring friction) and the distance this force is applied, or the lifting force and the vertical distance.

Useful work = 200 N × 4 m = 800 N·m = 800 J or

Useful work = 400 N × 2 m = 800 N·m = 800 J

The actual work done on the machine is calculated using the force required with the machine and the distance the force is applied.

Work done on machine = 250 N × 4 m = 1 000 N·m = 1 000 J

$$\text{Efficiency} = \frac{800\text{ J}}{1\,000\text{ J}} = 0.80 \text{ or } 80\%$$

Efficiency is usually expressed as a percentage by multiplying the ratio by 100.

Whenever a machine is used, more work has to be done using the machine than without it. No machine or system is 100% efficient. Table 74 shows efficiencies for some systems. These efficiencies vary depending on the condition of the engine.

TABLE 74 Efficiency of Some Systems

System	% Efficiency
Diesel engine	50 – 55
Gasoline engine	30 – 35
Gasoline turbine	20 – 30
Jet engine	50 – 70
Steam engine	10 – 15
Steam turbine	35 – 40

Discussion

1. **a)** Name the two basic kinds of machines.
   b) Look at the diagram of the wedge and the screw in Figure 12.6. Summarize why they are considered to be modifications of the inclined plane.
2. List five ways machines change forces.
3. **a)** Define mechanical advantage.
   b) How is the ideal mechanical advantage of a machine calculated?
   c) How is the actual mechanical advantage of a machine calculated?
   d) How does the actual mechanical advantage of an inclined plane change as the plane is made steeper? as the force of friction increases?
4. **a)** How is the efficiency of a machine calculated?
   b) What happens to the efficiency of a machine as the force of friction increases?
5. A ramp is used to load crates onto a railway car. The length of the ramp is 6.0 m. The platform of the railway car is 1.5 m above the ground. Each crate has a mass of 450 kg. A force of 900 N is needed to overcome kinetic friction on the ramp. A force of 2 025 N applied parallel to the ramp moves the crate at constant speed up the ramp. Calculate:
   a) The ideal mechanical advantage of the ramp;
   b) The actual mechanical advantage of the ramp;
   c) The useful work done by the ramp;
   d) The actual work done on the ramp;
   e) The efficiency of the ramp.
6. In 1977 Volkswagen introduced a car called a "Rabbit" powered by a diesel engine rather than a gasoline engine. Why?

12.5 INVESTIGATION: Using the First-Class Lever to Do Work

A lever is a rigid bar that can rotate around a fixed point called a fulcrum. When a load is placed on a lever, a force (effort) applied at some other point moves the load. There are three classes of levers depending on the relative location of the load, the fulcrum and the applied force (effort).

In the first-class of lever, the fulcrum is between the load and the applied force (Fig. 12-8). Both forces act in the same direction on the lever. A pump handle is an example of a first-class lever. In this investigation you study the first-class lever to determine its mechanical advantage and efficiency.

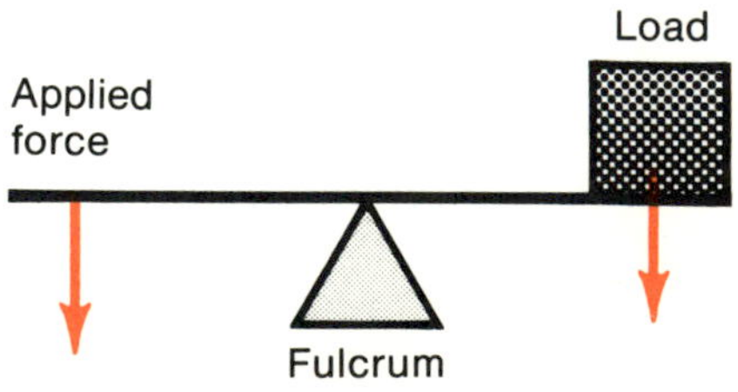

Figure 12-8 A first-class lever.

Materials

2 newton spring scales (range 10 N, 20 N)

3 metre sticks	C-clamp
knife-edge holder	0.5 kg mass
ring stand, clamp and rod	thread or unbent paper clip

Procedure

a. Copy Table 75 into your notebook.
b. Attach the knife-edge holder (fulcrum) at the 50 cm mark on the metre stick. Suspend the metre stick from the ring stand as shown in Figure 12-9.
c. Use a loop of thread or paper clip to suspend the load (500 g mass) from the 15 cm position on the metre stick.

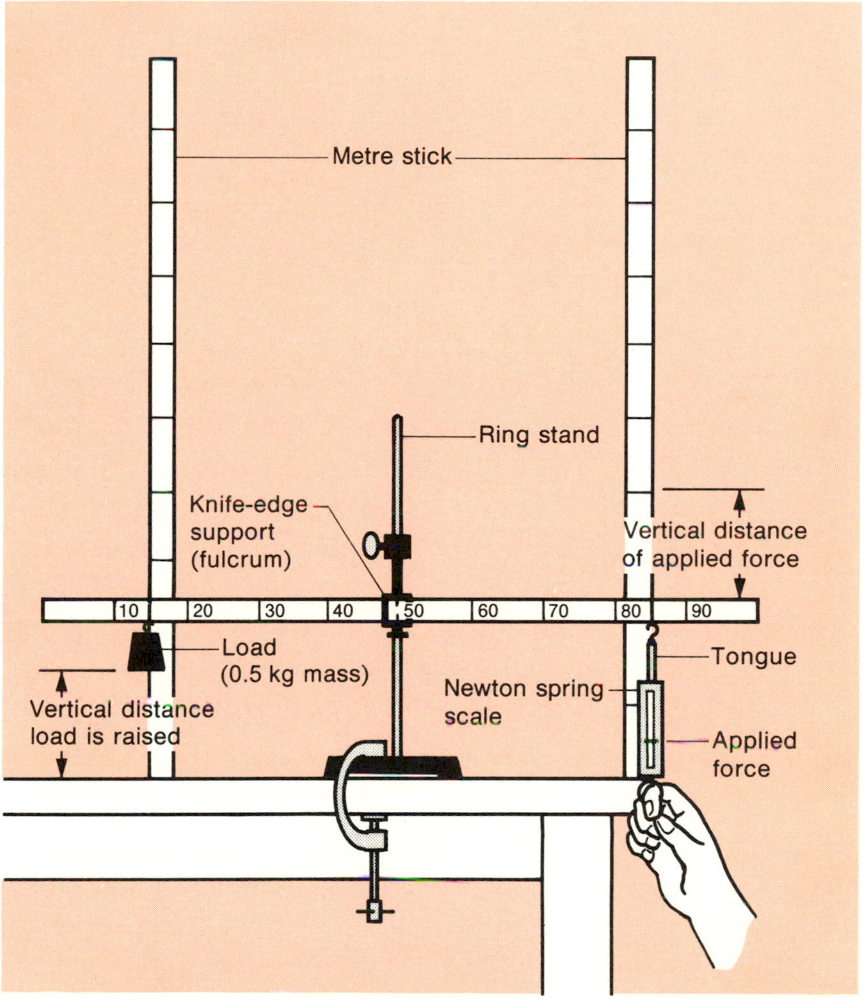

Figure 12-9 A first-class lever.

TABLE 75 The First-Class Lever

Distance from load to fulcrum (m)	Load 0.5 kg mass (N)	Vertical distance load is moved (m)	Work done by lever (J)	Applied force (N)	Vertical distance force is applied (m)	Work done on lever (J)	Mechanical advantage
0.35	5	0.05					
0.20	5	0.05					
0.50	5	0.05					

d. Use a loop of thread or paper clip to attach the newton spring scale at the 85 cm end of the metre stick.

e. Apply a downward force using the newton spring scale. Raise the load through a vertical distance of 5 cm. Measure the size of the applied force and its vertical distance. Record these in the table.

f. Move the load closer to the fulcrum (to the 30 cm mark on the metre stick). Repeat step e.

g. Move the load farther from the fulcrum (to the 0 cm mark on the metre stick). Repeat step e.

h. Calculate the mechanical advantage of the lever for the three positions of the fulcrum. Enter these in the table.

i. Calculate the work done by the lever and the work done on the lever for each case. Record these in the table.

Discussion

1. When is the actual mechanical advantage equal to 1? greater than 1? less than 1?
2. Compare the distance the load and applied force move for each of the above cases.
3. Compare the work done by the first-class lever with the work done on the lever.
4. What is the efficiency of this first-class lever?
5. What is the function of the knife-edge support? Why is a knife-edge used?
6. An equal arm balance and a pair of needle-nose pliers are first-class levers. Why?
7. Name three other first-class levers.
8. When the spring scale is used upside down as in this investigation, a correction factor should be added to the reading of the spring scale. This correction factor is two times the pull of gravity on the movable tongue. Why is this so?

12.6 INVESTIGATION: The Second-Class Lever

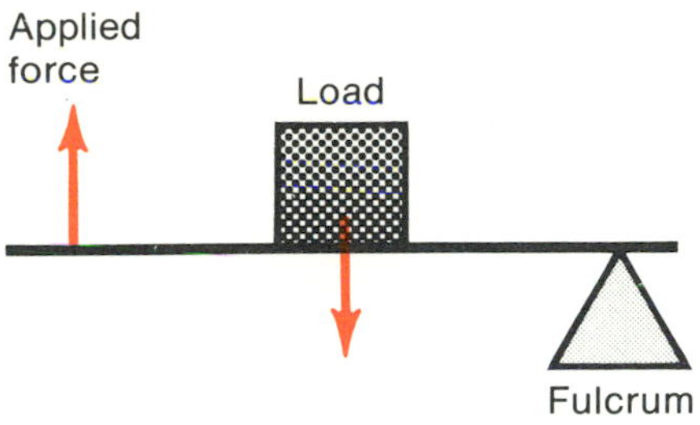

Figure 12-10 A second-class lever.

In a lever of the second-class, the load is between the fulcrum and the applied force (Fig. 12-10). The applied force and the load act in opposite directions on the lever. A wheelbarrow is a second-class lever. In this investigation you study the second-class lever to determine its mechanical advantage and efficiency.

Materials

As in Investigation 12.5

Procedure

a. Copy Table 76 into your notebook.
b. Attach the knife-edge holder (fulcrum) at the 50 cm mark on the metre stick. Suspend the metre stick from the ring stand as shown in Figure 12-11.
c. Use a loop of thread or paper clip to suspend the load (0.5 kg mass) from the 75 cm mark on the metre stick.
d. Use a loop of thread or paper clip to attach the newton spring scale at the 100 cm mark of the metre stick.
e. Apply an upward force using the newton spring scale. Raise the load through a vertical distance of 10 cm. Measure the size of the applied force and its vertical distance. Record these in the table.
f. Move the load closer to the fulcrum (to the 65 cm mark on the metre stick). Repeat step e.
g. Move the load farther from the fulcrum (to the 85 cm mark on the metre stick). Repeat step e.

TABLE 76 The Second-Class Lever

Distance from load to fulcrum (m)	Load 0.5 kg mass (N)	Vertical distance load is moved (m)	Work done by lever (J)	Applied force (N)	Vertical distance force is applied (m)	Work done on lever (J)	Mechanical advantage
0.25	5	0.10					
0.15	5	0.10					
0.35	5	0.10					

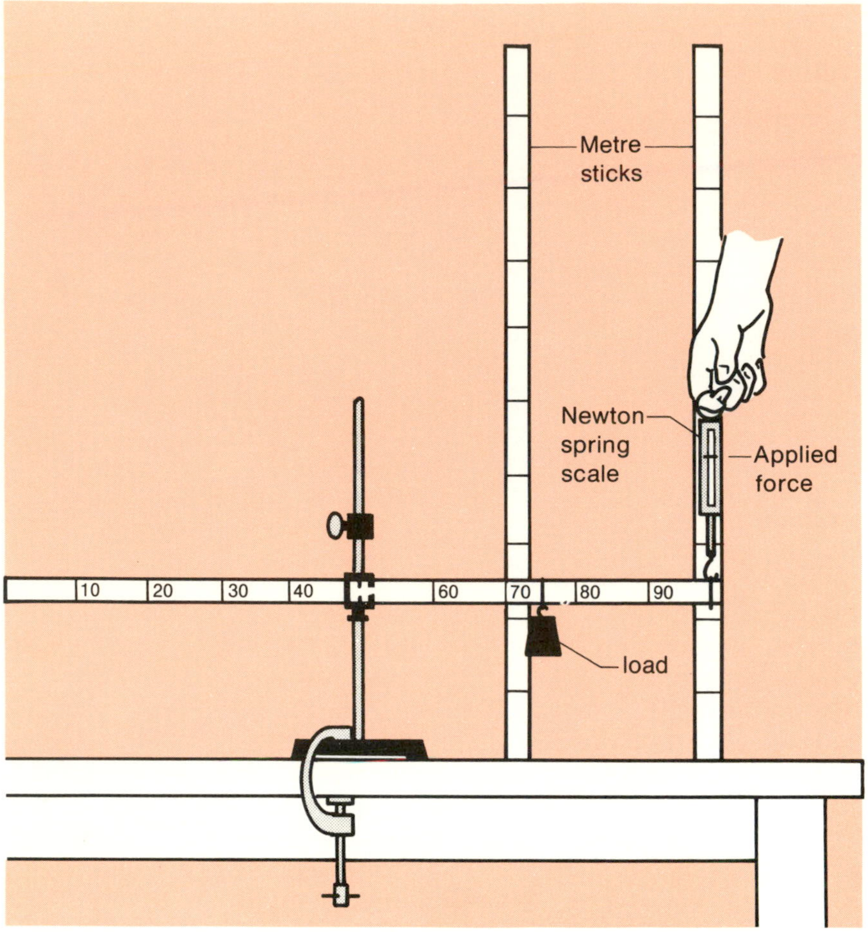

Figure 12-11 A second-class lever.

h. Calculate the mechanical advantage of the lever for the three positions of the load. Enter these in the table.

i. Calculate the work done by the lever and the work done on the lever for each case. Record these in the table.

Discussion

1. What conclusion do you make about the mechanical advantage of a second-class lever?
2. Compare the work done by the second-class lever with the work done on the lever.
3. What is the efficiency of this second-class lever?
4. If the knife-edge fulcrum is replaced by a flexible joint, what might happen to the efficiency?
5. Nutcrackers are second-class levers. Why?
6. Name three other second-class levers.
7. How do second-class levers make man's work easier?

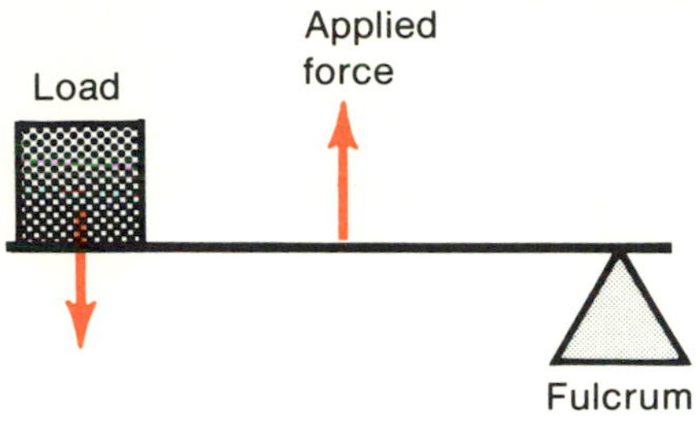

Figure 12-12 A third-class lever.

12.7 INVESTIGATION: The Third-Class Lever

In a lever of the third class, the applied force is between the load and the fulcrum (Fig. 12-12). The applied force and load act in opposite directions on the lever. Tweezers are third-class levers. In this investigation you study the third-class lever to determine its mechanical advantage and efficiency.

Materials

As in Investigation 12.5

Procedure

a. Copy Table 77 into your notebook.
b. Attach a knife-edge holder (fulcrum) upside down at the 50 cm mark on the metre stick. Support the metre stick from the ring stand as shown in Figure 12-13.

TABLE 77 The Third-Class Lever

Distance from force to fulcrum (m)	Load 0.5 kg mass (N)	Vertical distance load is moved (m)	Work done by lever (J)	Applied force (N)	Vertical distance force is applied (m)	Work done on lever (J)	Mechanical advantage
0.25	5	0.10					
0.15	5	0.10					
0.35	5	0.10					

c. Use a loop of thread or paper clip to suspend the load (0.5 kg mass) from the 100 cm mark on the metre stick.
d. Use a loop of thread or paper clip to attach the newton spring scale at the 75 cm mark of the metre stick.
e. Apply an upward force using the newton spring scale. Raise the load through a vertical distance of 10 cm. Measure the size of the applied force and its vertical distance. Record these in the table.
f. Move the applied force closer to the fulcrum (to the 65 cm mark on the metre stick). Repeat step e.
g. Move the applied force further from the fulcrum (to the 85 cm mark on the metre stick). Repeat step e.

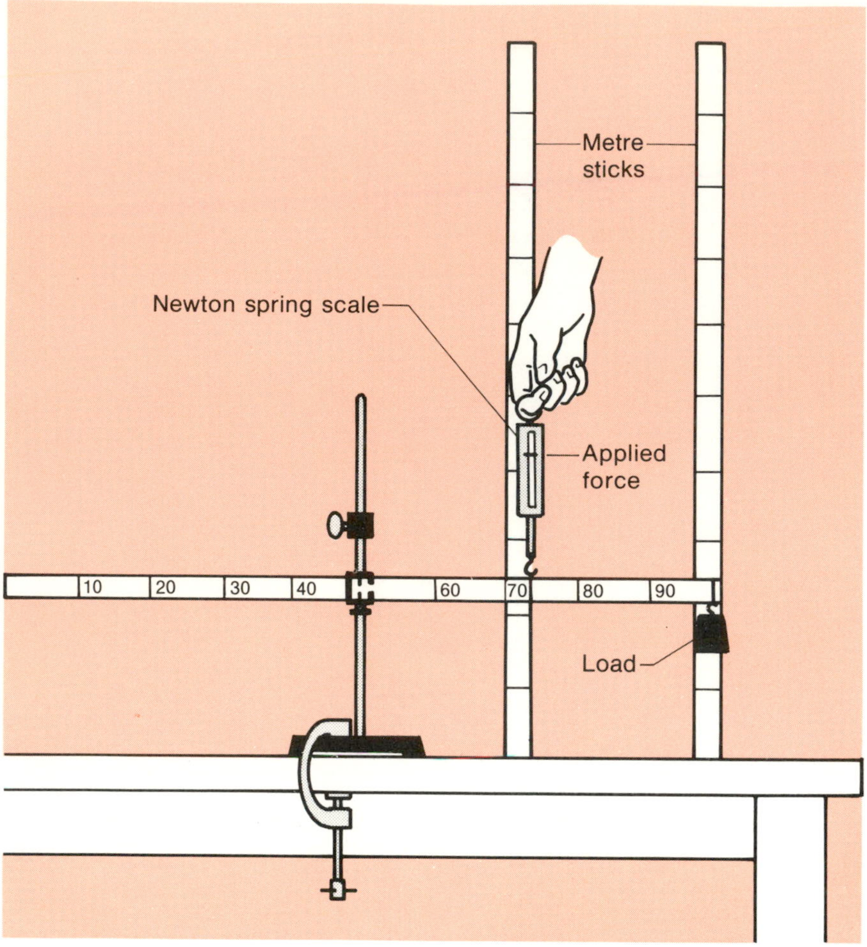

Figure 12-13 A third-class lever.

h. Calculate the mechanical advantage of the lever for the three positions of the applied force. Enter these figures in the table.
i. Calculate the work done by the lever and the work done on the lever for each case. Record these in the table.

Discussion

1. What conclusion do you make about the mechanical advantage of a third-class lever?
2. Compare the work done by the third-class lever with the work done on the lever.
3. What is the efficiency of this third-class lever?
4. If the knife-edge fulcrum is replaced by a flexible joint, what might happen to the efficiency?
5. The human arm is a third-class lever. Why?
6. Name three other third-class levers.
7. How do third-class levers make our work easier?

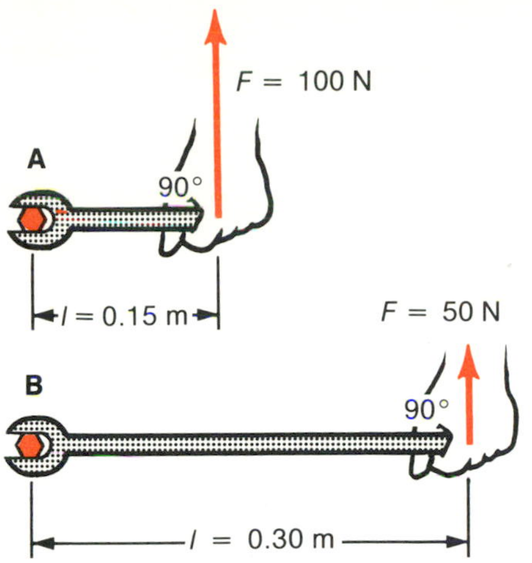

Figure 12-14 Using two different wrenches to produce the same torque on a nut.

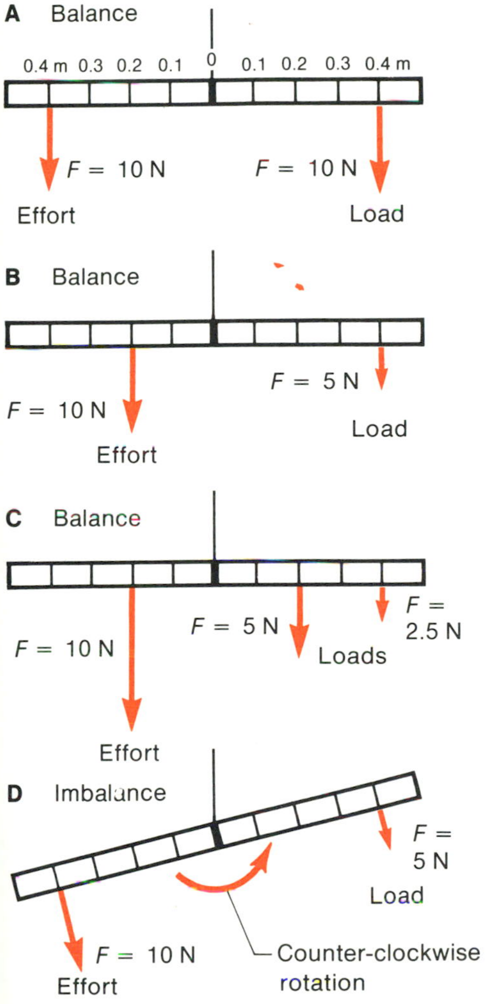

Figure 12-15 The law of the lever.

12.8 Torque and the Law of the Lever

Torque

Whenever a force acts perpendicular to a lever, the force tends to turn the lever about its fulcrum.

A mechanic applies a force perpendicular to a wrench to loosen or tighten a nut. This turning effect is called the **torque** (T).

The magnitude of the torque is calculated by multiplying the force applied by the perpendicular distance of the force from the fulcrum or turning point. Figure 12-14 shows the torque produced by a mechanic using two different wrenches to loosen a nut. Both wrenches are being used as second-class levers. However, a greater force must be applied with the shorter wrench to produce the same torque.

The formula for torque ($\boldsymbol{T}$) is $\boldsymbol{T} = \boldsymbol{F}\,\boldsymbol{l}$, where $\boldsymbol{F}$ is the applied force and $\boldsymbol{l}$ is the perpendicular distance from the fulcrum to the point of application of the force.

The shorter wrench (A), requires a force of 100 N at a distance of 0.15 m from the centre of the bolt to turn the nut. The torque applied is: $T = F\,l = 100\text{ N} \times 0.15\text{ m} = 15\text{ N·m}$. When the longer wrench (B) is used, a force of 50 N at a distance of 0.30 m from the centre of the bolt has the same effect. The torque applied is: $T = F\,l = 50\text{ N} \times 0.30\text{ m} = 15\text{ N·m}$. Although the same torque is applied, the mechanic is able to exert a smaller force to loosen the nut when the longer wrench is used.

The SI Unit of Torque

The unit of torque is the unit of force times the unit of distance. The unit for torque is the **newton metre (N·m)**. This appears to be the same unit as for work. But torque is not work. In the case of torque the force and its distance from the fulcrum must be at right angles to one another. Although this yields the unit newton metre (N·m) for torque, the unit joule (J) is never used since torque is not work. We say the torque is 15 N·m, but we never say it is 15 J.

The Law of the Lever

Whenever a lever is not rotating, the torque caused by the load is balanced by the torque caused by the effort. Figure 12-15 shows four cases where forces are being applied to a first-class lever. In

TABLE 78 Calculating Effort and Load Torques for a Lever

Case	Effort			Load			Result
	Force (N)	Perpendicular distance (m)	Torque N·m	Force (N)	Perpendicular distance (m)	Torque N·m	
A	10	0.4	4	10	0.4	4	Balance
B	10	0.2	2	5	0.4	2	Balance
C	10	0.2	2	5 2.5	0.2 0.4	1 1	Balance
D	10	0.4	4	5	0.4	2	Counter-clockwise rotation

cases A, B, and C, the lever is in balance. In case D, the lever rotates counter-clockwise. Table 78 shows the calculation of the torque caused by the load and effort for all four cases.

The fact that the effort torque is equal to the load torque when a lever is balanced is known as the **Law of the Lever**. This is a very useful law, since it enables us to determine either the magnitude or the position of the effort required to balance a given load.

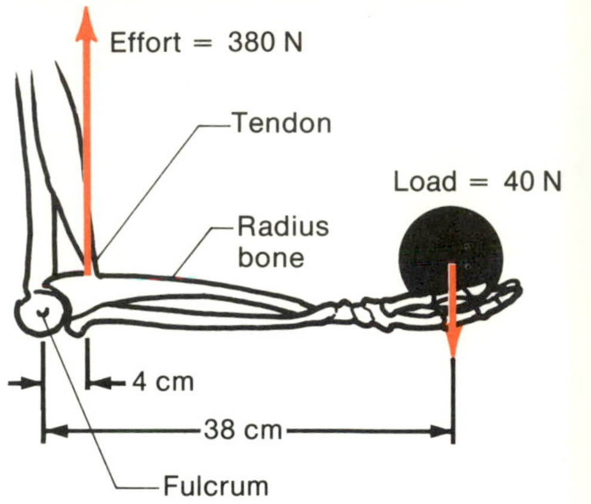

A The elbow is a third-class lever.

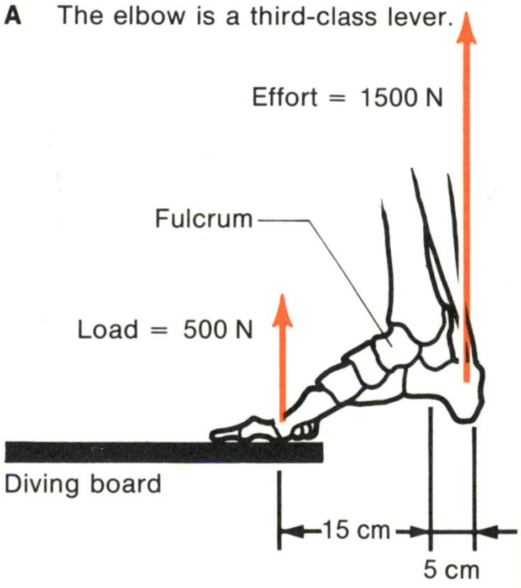

B The foot is a first-class lever.

Figure 12-16 Two levers in the skeleton of the human body (simplified).

Discussion

1. **a)** What effect does changing the mechanical advantage of a lever have on the force applied to the lever? on the distance the force must be applied to do a constant amount of work?
   b) What causes the efficiency of a machine to be less than 100%?
2. Determine the efficiency and mechanical advantage of each of the following levers:
   a) A load of mass 20 kg is lifted vertically through a height of 4.0 m by an applied force of 150 N exerted through a distance of 6.0 m.
   b) A force of 250 N downward on one end of a teeter-totter through a distance of 0.5 m exerts an upward force of 700 N on a man at the other end and raises him through a distance of 0.15 m.
3. Look up information on the wheel and axle. Summarize why they are considered to be adaptations of the lever.
4. Design and carry out an investigation to determine the efficiency and mechanical advantage of a car jack (be careful).
5. Can any machine be devised which has an efficiency greater than 100%? Why?

6. The skeleton of the human body is a system of levers. Figure 12-16 shows two such levers. The diagrams show that the force exerted by a muscle is much greater than the load it supports. This is true of all skeletal muscles.
   a) Is the mechanical advantage greater than, equal to, or less than 1 for the levers shown?
   b) Compare the distance a muscle must shorten or lengthen with the distance that the ends of the limbs move.
   c) The human skeleton is built for speed, not strength. Discuss.
7. Archimedes is supposed to have said of levers: "Give me a place to stand, and I will move the world." Which class of lever would be best for this purpose? Why?
8. What, if anything, prevents the design and use of an inclined plane with a mechanical advantage of 1 000 000?
9. Why does a road to the top of a mountain wind around the mountain?
10. Examine the data from Investigations 12.5, 12.6, and 12.7 for each class of lever and answer the following questions.
   a) For the load and effort compare the magnitude of the torque and the direction of the torque.
   b) Describe how the Law of the Lever can be used to determine either the magnitude or the position of the effort required to balance a given load.
   c) Torque has units of force × distance. Is torque work?
   d) What is the unit of torque?
   e) Suppose you are holding an object in your hand as shown in Figure 12-16. This load has a force of gravity acting on it of 35 N. Perform the necessary measurements on your arm to calculate the force the tendon is exerting on the radius bone to support this load.

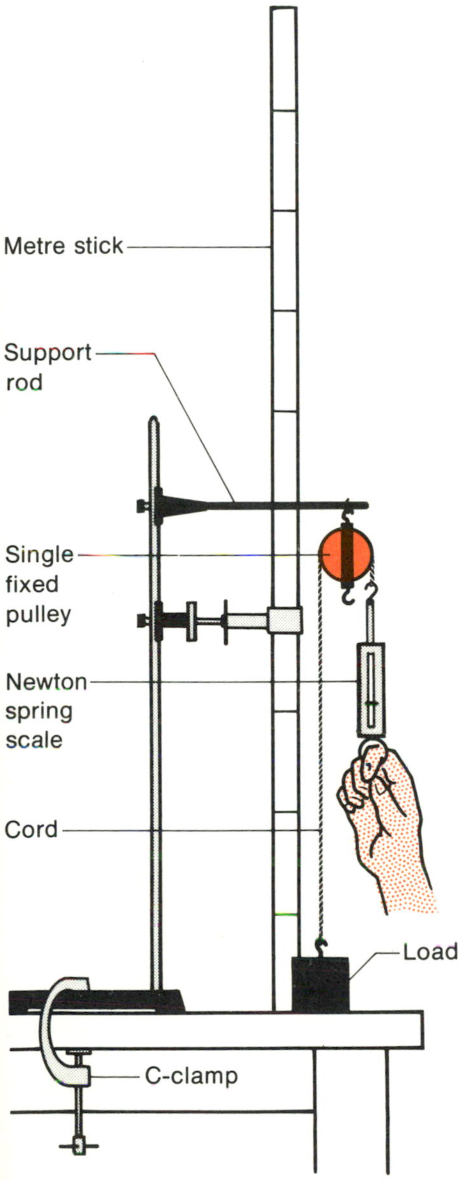

Figure 12-17 A single fixed pulley.

12.9 INVESTIGATION: Pulleys and Pulley Systems

A pulley is a modified lever. It consists of a wheel which turns on an axle. The axle and wheel are mounted on a frame. There are two kinds of pulleys: fixed pulleys and movable pulleys. Many different pulley systems can be made by connecting fixed pulleys and movable pulleys with a rope. A combination of a set of fixed pulleys called blocks with a set of movable pulleys called a tackle is called a block and tackle.

In this investigation you study a single fixed pulley, a single movable pulley, and a simple block and tackle. You measure the applied force, the load force, and the distance the forces move. Then you determine the mechanical advantage and the efficiency of each pulley system.

Materials

a newton spring scale
cord
ring stand, clamps, rod
C-clamp
several single pulleys
several multiple pulleys
metre stick
masses to act as loads

Procedure A   A Single Fixed Pulley

a. Copy Table 79 into your notebook.
b. Set up the ring stand, clamps, support rod, and pulley as in Figure 12-17.
c. Measure and record the force needed to lift the load at constant speed without the pulley. This is the load force.
d. Attach a cord to the load. Pass the cord through the single fixed pulley. Attach the free end of the cord to the newton spring scale.
e. Measure and record the force needed to raise the load using the pulley. This is the applied force.
f. Raise the load through a set vertical distance, say 0.25 m. Record the distance the applied force moves.
g. Calculate the actual mechanical advantage of the single fixed pulley.
h. Calculate and record the work done by the pulley and the work done on the pulley.
i. Calculate the efficiency of the single fixed pulley.

Procedure B   A Single Movable Pulley

a. Set up the apparatus as in Figure 12-18.
b. Measure and record the load force, the applied force, and the distance each force moves to raise the load a set vertical distance.
c. Calculate and record the actual mechanical advantage, the work done, and the efficiency of the single movable pulley.

Procedure C   A Simple Block and Tackle

a. Set up the simple block and tackle shown in Figure 12-19. To do this connect the single fixed pulley (tackle) and the single movable pulley (block) using the cord as shown.

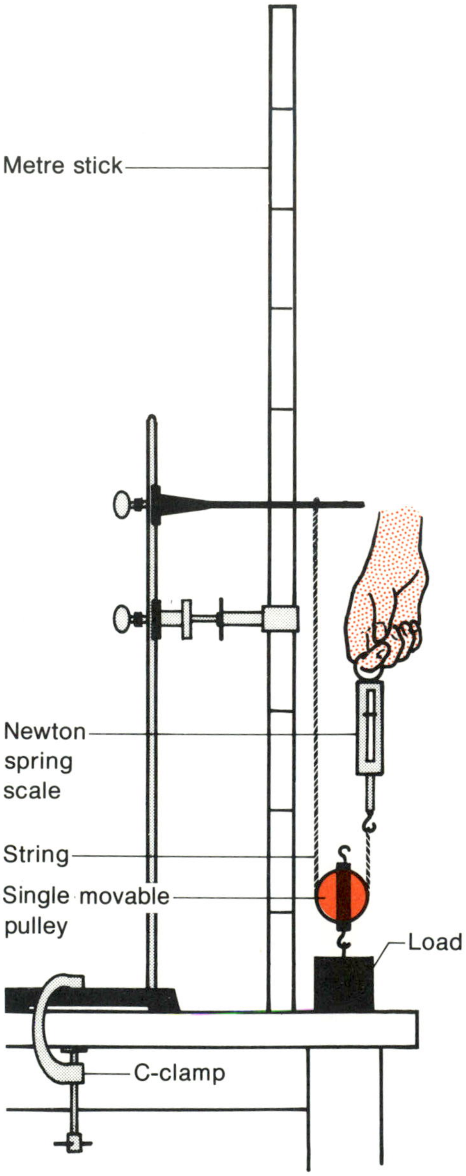

Figure 12-18   A single movable pulley.

b. Measure and record the load force, the applied force, and the distance each force moves to raise the load a set vertical distance.
c. Calculate and record the actual mechanical advantage, the work done, and the efficiency of the simple block and tackle.

Discussion A

1. Compare the magnitude and direction of the load force and the applied force.
2. What is the actual mechanical advantage of the single fixed pulley? the ideal mechanical advantage?
3. Compare the distance the applied force moves with the set vertical distance moved by the load force.
4. Compare the work done by the pulley with the work done on the pulley.
5. What is the efficiency of the single fixed pulley?
6. What useful purpose is served by a single fixed pulley? Give an example.
7. What class of lever does a single fixed pulley represent? Determine this by comparing the position of the fulcrum, the load force, and the applied force.
8. A slight error is introduced by using the newton spring scale upside down. Explain.

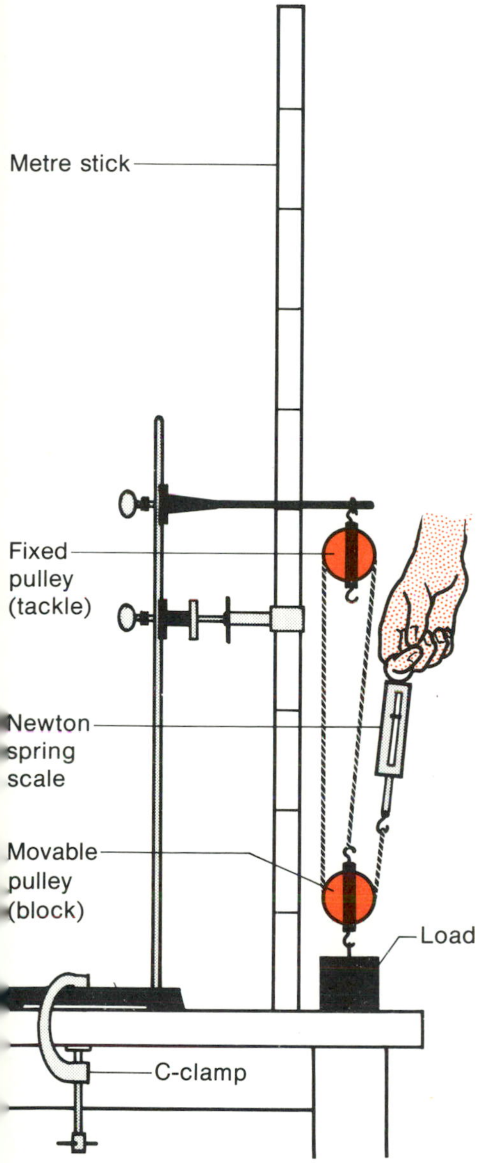

Figure 12-19 A simple block and tackle.

Discussion B

1. Compare the magnitude and direction of the load force and the applied force.
2. What is the actual mechanical advantage of the single movable pulley? the ideal mechanical advantage?
3. Compare the distance the applied force moves with the set vertical distance moved by the load force.
4. Compare the work done by the pulley with the work done on the pulley.
5. What is the efficiency of your single movable pulley?
6. What useful purpose is served by a single movable pulley? Give an example.
7. What class of lever does a single movable pulley represent?
8. Describe how a single fixed pulley and a single movable pulley can be connected with a cord to obtain the same mechanical advantage as a single movable pulley.

Discussion C

1. Compare the magnitude and direction of the load force and the applied force.

2. What is the actual mechanical advantage of the block and tackle? the ideal mechanical advantage?
3. What relationship exists between the ideal mechanical advantage and the number of cords supporting the movable pulley?
4. Compare the distance the applied force moves with the set vertical distance moved by the load force.
5. Compare the work done by the block and tackle with the work done on the block and tackle.
6. What is the efficiency of the block and tackle? Predict what will happen to the efficiency if more and more pulleys are connected.
7. What useful purpose is served by a block and tackle? Give some examples.
8. The mechanical advantage of a block and tackle is changed when more than one pulley is put on the block and on the tackle. These additional pulleys are usually mounted side by side on a common shaft. Use the multiple pulleys provided to design a block and tackle with a larger mechanical advantage. Predict the ideal mechanical advantage of the pulley system. Perform an investigation to test your prediction.

TABLE 79 Pulleys and Pulley Systems

Quantity	Single fixed pulley	Single movable pulley	Simple block and tackle
Load Force (N)			
Vertical distance load is moved (m)			
Work done by pulley (J)			
Applied force (N)			
Vertical distance applied force is moved (m)			
Work done on pulley (J)			
Mechanical advantage			
Efficiency			

12.10 Power

A Description of Power

The work done by a force exerted through a distance does not depend on the time taken. This is also true of the energy consumed. Power is the term used to describe the rate at which work is done, that is, the rate at which energy is consumed.

Power depends on the time taken to do a given amount of work. A person lifting 10 boxes from one height to another does the same amount of work whether the boxes are lifted in 5 min or 10 min. In both cases the same average force is applied on each box through the same vertical distance. However, the person's power is different since the time taken to do the work is different. The person lifting the boxes in half the time has twice the power.

Power also depends on the work done in a given amount of time. Suppose two people are working lifting the boxes. One lifts 10 boxes in 5 min while the other lifts 20 boxes in 5 min. The person doing twice the work in the same time has twice the power.

Two factors determine power: the amount of work done, and the time taken to do the work.

The Operational Definition of Power

Power *is defined as the rate of doing work, or the rate of using energy.* The operational definition for power describes how power is measured. Power is calculated by dividing the work done by the time taken. *Power* $= \frac{Work}{Time}$. If ***P*** stands for power, ***W*** for work, and ***t*** for time, the equation for power is $\boldsymbol{P} = \frac{\boldsymbol{W}}{\boldsymbol{t}}$.

The SI Unit of Power

Since power is work done divided by time taken, the SI derived unit of power is the joule per second (J/s). The joule per second has been given the special name watt in honor of James Watt (1736–1819), a Scottish engineer who made steam power practical in 1769. *One* **watt** (W) *is the power available when one joule* (J) *of work is done in one second.*

$1\ \text{W} = 1\ \text{J/s}$

Another unit of power in SI is the kilowatt.

$1\ \text{kW} = 1\,000\ \text{W} = 1\,000\ \text{J/s}$

Some Examples of Power

Man is capable of doing a large amount of work in a short time for brief intervals. The world record in a heavyweight class for weight lifting stood at 255 kg in 1976. The Russian weight lifter at the Olympics exerted an average force of about 2 550 N through a vertical distance of about 2.0 m and did 5 100 J of work. It took him about 5 s. His maximum power output was $\frac{5\,100\text{ J}}{5\text{ s}} = 1\,020\text{ W}$. Actually his power was larger since he lifted part of his own mass as well as the mass of the barbells. People are not capable of such a high power output for very long at one time. Our sustained power output is about 75 W. Table 80 shows some power data.

Discussion

1. **a)** Define power.
   b) What two factors determine the power of a machine or person?

2. **a)** Compare the work required to lift a 20 kg bag of potatoes a vertical distance of 1 m with the work required to lift a 10 kg bag 2 m.
   b) If both bags are lifted their distances in the same time, how does the power required in each case compare?
   c) If the bag having the lesser mass is moved its distance in half the time, compare the power required in each case.

3. A man has a mass of 100 kg.
   a) What average force in newtons would he have to exert to climb a staircase?
   b) How much work would he have to do to climb stairs between floors 5 m apart?
   c) If he climbs the stairs in 15.0 s, what power is he developing?

4. A crane lifts a 1 000 kg load through a distance of 30 m in 40 s. Calculate:
   a) the work done by the crane.
   b) the power of the crane in watts and kilowatts.

5. An automobile engine exerts a force of 4 000 N to accelerate a car over a distance of 100 m in 8.0 s. Calculate the power of the engine in watts and kilowatts.

6. Water flows over a waterfall at a rate of 7.0×10^4 kg/min. The waterfall is 30 m high. How much power does the falling water have in watts and kilowatts?

TABLE 80 Some Sample Powers

Biological system		Household appliances	
System	**Power (W)**	**Appliance**	**Power (W)**
Left ventricle	3	Clock	2
Humming bird flying	7	Radio	71
Human at rest	17	Television (black and white)	237
Dolpin swimming	210	Television (colour)	332
Human walking at 3.2 km/h	230	Washing machine	512
Horse working steadily	750	Refrigerator	615
Human running	1 000	Dishwasher	1 200
Top athlete	1 700	Oven (microwave)	1 500
		Oven (standard)	4 800
		Air conditioner	1 566
		Water heater	4 474
		Clothes dryer	4 856
Transportation system		**World system**	
System	**Power (kW)**	**System**	**Power (kW)**
Automobile (100 km/h)	29	Canadian consumption (1976)	3×10^7
Automobile (top acceleration)	100	World total consumption	6×10^9
Bus	150	Solar power involved in photosynthesis	4×10^{10}
Commuter train	3 000	Incoming solar power	1.73×10^{14}
Ocean liner	21 000		
707 jet	21 000		
Moon rocket	100 000		

7. Calculate the work done in each of the following cases:
a) power = 60 W, time = 2.0 s
b) power = 1 000 W, time = 4.0 min
c) power = 2.0 kW, time = 5.0 s
d) power = 4.0 kW, time = 8.0 h

8. Calculate the time required in each of the following cases:
a) work = 2 000 J, power = 500 W
b) work = 8 000 J, power = 4 kW
c) work = 8.0 MJ, power = 200 kW
d) work = 4.0×10^5 J, power = 1.5×10^2 kW

9. **a)** How many times more powerful is a moon rocket than a 707 Jet?
b) Compare the Canadian consumption of power in 1976 with the incoming solar power.

12.11 INVESTIGATION: Determining Your Power

Table 80 shows that a top athlete has a power of about 1 700 W. A human running has a power of about 1 000 W. In this investigation you determine your power in watts.

Materials

1 roll of string and a bob
1 metre stick
1 stop watch

Procedure

a. Attach a bob to the roll of string. Use the string, bob, and metre stick to measure the vertical height between two or more floors at a stairwell.
b. Determine your mass in kilograms.
c. Determine the fastest time you can climb between the floors.
CAUTION: Do not over-exert yourself!
d. Calculate your power in watts.

Discussion

1. What is your maximum power in watts?
2. Compare your power with the power of a human walking at 3.2 km/h and with the power of a human running.
3. Predict whether your right arm or left arm is more powerful. Design a procedure to find the power of each arm. Check with your teacher before you try the procedure. Was your prediction correct? One might predict that a right-handed person would have a more powerful right arm than left arm. Does this prediction agree with your results?

Highlights

Work is done on an object when a force is applied and the object moves in the direction of the force. Work is the product of the force and the distance. The unit of work is the newton metre (N·m) or the joule (J). 1 J is the work done when a force of 1 N is applied through a distance of 1 m.

Machines are devices to help man work. They are used to change the kind of force and its point of application, size, or direction. The six simple machines are the inclined plane, the wedge, the screw, the lever, the pulley, and the wheel and axle. The mechanical advantage of a machine is the ratio of the force exerted by the machine to the force applied to the machine. Mechanical advantage can be greater than, equal to, or less than 1, depending on the machine. The efficiency of a machine is the ratio of the useful work done by the machine to the work done on the machine. Because of friction, the efficiency of a machine is always less than 100%.

The turning effect of a force applied at right angles to a lever is called torque (T). The size of the torque is the product of the force and the perpendicular distance from the fulcrum. The unit of torque is the newton metre (N·m). Torque is not the same as work because of the difference in direction of the force and the distance. According to the Law of the Lever, when a lever is balanced, the effort torque is the same size as the load torque. They act to oppose one another.

Power is a measure of the rate of doing work. Power is determined by dividing the work done by the time taken to do the work. The unit of power is the joule per second (J/s) or the watt (W). 1 W is the power when 1 J of work is done in 1 s. A machine which exerts a force of 1 N through a distance of 1 m in a time of 1 s uses a power of 1 W.

UNIT

13 Heat Energy

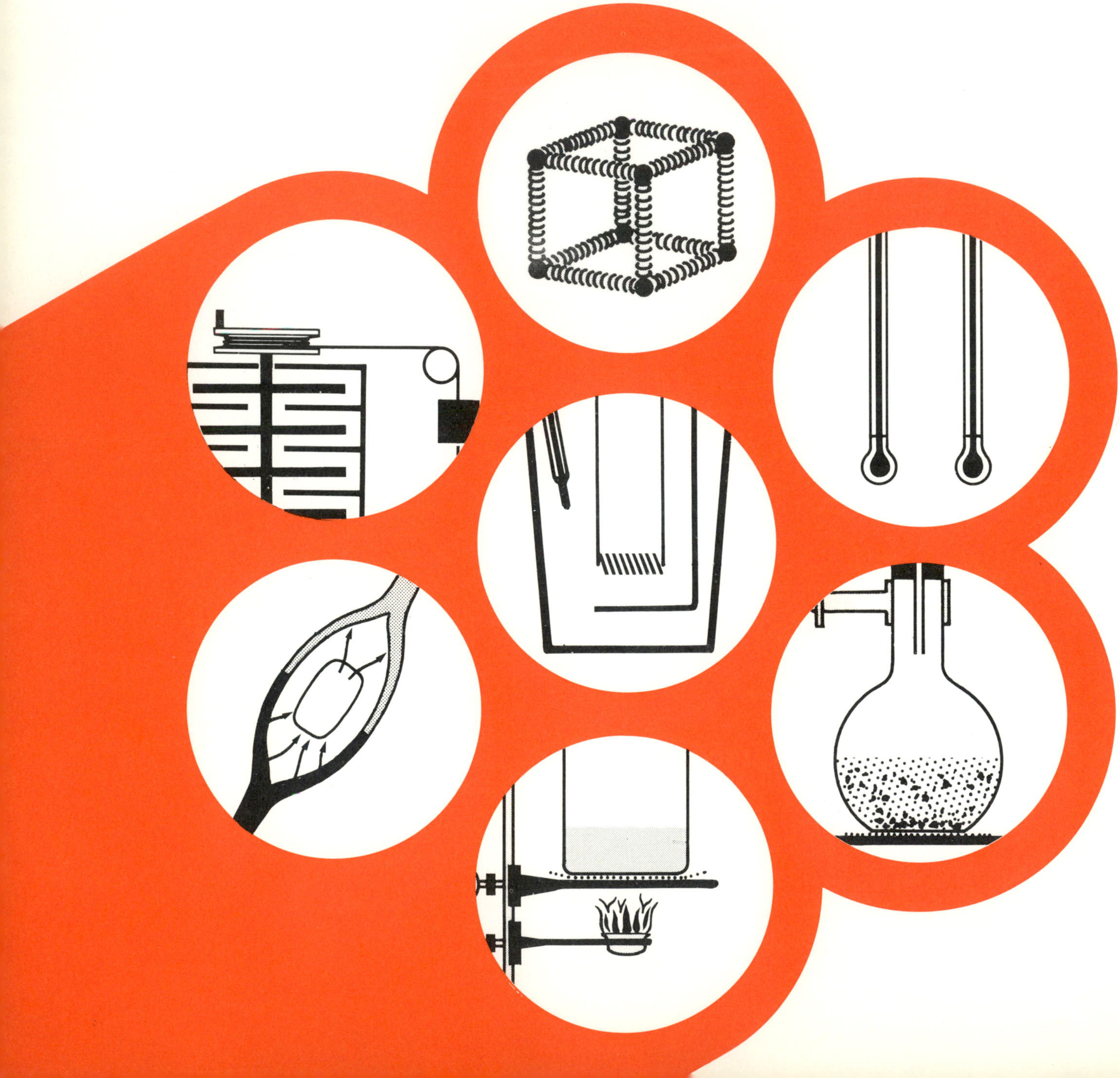

13.1 Measuring Temperature

The Senses and Temperature

You have probably used the sense of touch at one time or another to estimate temperature. Touch provides a rough estimate but it is not very reliable. Try this demonstration. Place one hand in hot water, and the other in cold water, then plunge both into lukewarm water. The hand from the hot water will feel cold; the hand from the cold water, warm. The nerves in each hand send a different message to the brain. Also, if you touch a piece of wood outdoors on a cold day with one hand, and a piece of iron with the other, the iron feels colder to the touch than the wood. Both objects have the same temperature, but our senses are fooled. Why?

A Description of Temperature

Heat flows from an object to its surroundings when its temperature is higher than that of its surroundings. Heat flows into an object when its temperature is lower than that of its surroundings. **Temperature** *is the property of an object that determines the direction of heat flow*. Although the sense of touch is unreliable for measuring temperature, it can detect the direction of heat flow. When heat flows away from the hand, the object it touches feels cold. The faster heat flows away, the colder the object feels. Iron conducts heat away from the hand faster than wood. Therefore, the iron feels colder than the wood.

Thermometers

Before scientists could study heat quantitatively, they had to invent an instrument to detect and measure temperatures accurately. You are no doubt familiar with this instrument—the thermometer.

There are many kinds of thermometers. Each kind of thermometer depends for its operation on a property of a substance that changes with an increase or decrease in temperature. For example, the hardness of some substances changes with temperature. Butter softens as the temperature increases. Ceramic materials of different composition soften at various known temperatures. Thus cones made of ceramic materials are used to estimate the temperature inside furnaces. The colour of metals depends on their temperature. The filament of a lamp is at a lower temperature when it glows red than when it glows white. The whiter a star is, the higher is its temperature. This temperature can be estimated by matching the star's colour with the colour of a filament which corresponds to a known temperature.

The volume of gases, liquids, and solids varies with temperature. This property is used in the design of most common thermometers. The first thermometer invented by Galileo in 1592 depended on the expansion of a gas for its operation. Fortunately, the volume changes of solids, liquids, and gases are linear over small ranges in temperature. That is, the volume of a substance changes the same amount for a certain change in temperature throughout the range in temperature. This enables thermometers to be constructed that have scales divided into equal-sized increments. Most liquid thermometers contain mercury or alcohol. Mercury has a higher boiling point than alcohol. Therefore it is used in thermometers designed to measure much higher temperatures than those measured by alcohol thermometers. Alcohol is used in thermometers designed to measure temperatures to below –100°C, since it has a lower freezing temperature than mercury.

Thermometers and Fixed Points

Until the mid-1600's, all thermometers had one major fault. They did not have a common scale. In order to have a common scale, thermometers must have the same two reference points. These reference points or standard temperatures are called **fixed points**. Once fixed points are identified, the interval between the fixed points is marked off into equal-sized divisions.

Common temperature reference points were difficult to find. Scientists looked for behaviours that always happened at the same temperature. The freezing point of water is constant. Therefore scientists chose it as one of the fixed points. Early thermometers used the temperature of the human body as the second fixed point. This is unsuitable because body temperature varies slightly from person to person. It also depends on the health of the individual. Scientists finally chose the boiling point of water at normal atmospheric pressure as the second fixed point.

Because scientists assigned different numbers to the two fixed points, three temperature scales evolved: the Celsius scale, the Kelvin scale, and the Fahrenheit scale.

The Celsius Temperature Scale

About twenty different temperature scales were proposed during the first half of the nineteenth century. The one in common use today is the Celsius scale. The Celsius scale was determined by setting the freezing point of water at 0°C and the boiling point at 100°C. The interval between these two fixed points was then divided into 100 equal intervals known as degrees Celsius. The Celsius scale extends below 0°C using negative numbers. It extends above 100°C without known limit. For example, deuterium nuclei must reach a temperature of about 10^8 °C before fusion will take place. The symbol °C is used to denote temperature intervals as well as specific temperatures. Thus room temperature is quoted as 20°C; a temperature rise of twenty degrees Celsius is also written as 20°C.

The Kelvin Temperature Scale

The SI base unit of temperature is the kelvin (K). The Kelvin scale (sometimes called the absolute scale) has one advantage over the Celsius scale. There are no negative readings on the Kelvin scale. The lowest possible temperature, –273.15°C, is called 0 K or absolute zero. The freezing point of water is 273.15 K and the boiling point is 373.15 K. These temperatures correspond to the fixed points of 0°C and 100°C. The degrees on the two scales are the same size: **1 K = 1°C.** Figure 13-1 shows the Celsius scale and the Kelvin scale and indicates some common temperatures.

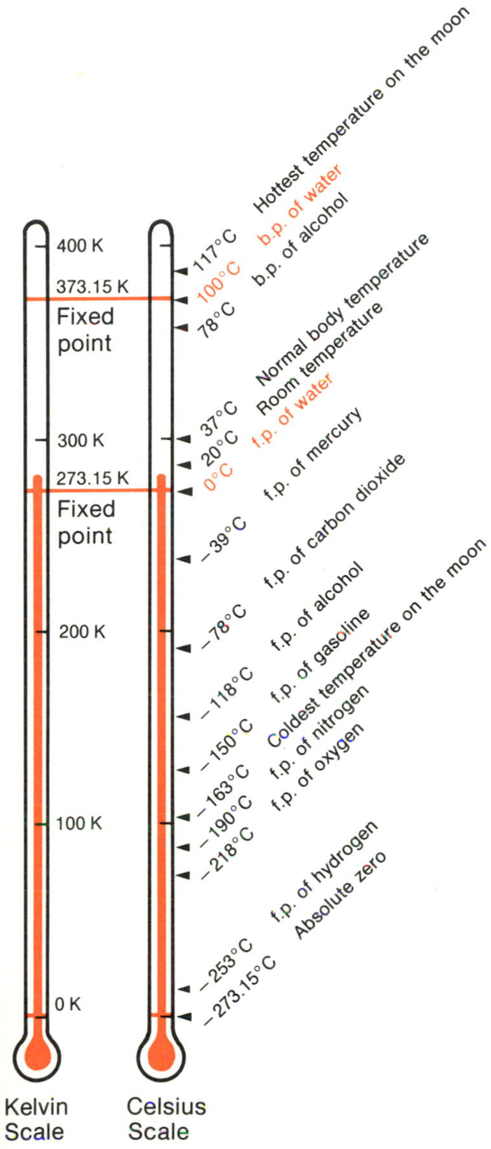

Figure 13-1 Some temperatures on the Kelvin and Celsius scales.

Discussion

1. **a)** Why is it unreliable to estimate the temperature of objects by touching them?
   b) A mother believes that her child has a high temperature. She feels the child's forehead. Her husband comes in from outside on a cold day. He feels the child's forehead. Which person will sense a higher temperature? Why?
2. **a)** Describe temperature.
   b) Two objects A and B are placed in contact. Object A has a higher temperature than object B. In which direction will heat flow? How will their temperatures change with time?
   c) Describe the direction of heat flow when you hold a piece of ice in your hand.
3. **a)** What characteristic must a property of an object have if it is to be used to measure temperature?
   b) Can the mass of an object be used as a measure of the temperature of the object? Explain.
4. **a)** What characteristic do fixed points on temperature scales have in common?
   b) One of the fixed points used in early thermometers was the temperature of the human body. Why is the temperature of the human body unsatisfactory as a fixed point on temperature scales?
5. **a)** How are Celsius and Kelvin temperature scales the same? How are they different?
   b) What advantage does the Kelvin scale have over the Celsius scale?

c) For what temperature range can each of mercury and alcohol be used in modern thermometers?

6. Make a liquid thermometer using a flask, a rubber stopper, and a length of capillary tubing. Find two fixed points and calibrate your thermometer. Use it to measure the temperature of the palms of your hands. (Hint: See Investigation 5.14, page 119.)

13.2 The Caloric Theory of Heat

Until the middle of the nineteenth century, there were two competing theories of heat. The caloric theory stated that heat is a material substance. The kinetic molecular theory stated that heat is related to the motion of the particles making up matter.

The Caloric Theory

The caloric theory of heat was proposed in 1779. It described heat as an invisible elastic fluid called **caloric**. Caloric is made up of particles which repel one another. These particles are attracted to particles of matter. Different kinds of matter attract caloric with different strengths. Caloric cannot be created or destroyed. Caloric can be either felt or stored. When caloric is stored, it is combined with particles of matter.

The caloric theory was able to explain a number of phenomena. A hot body contains more caloric than a cold body. When a body is heated, caloric is added to the spaces between the particles of matter. When a body is cooled, some of the caloric particles escape. It explained the direction of heat flow. Heat always flows from a warmer body to a colder body. When two bodies at different temperatures are placed in contact, particles of caloric flow from the hot body to the cold body. The repulsive forces between the particles of caloric cause this movement. The caloric theory accounted for the expansion of objects when heated. During heating, caloric is forced into the spaces between the particles of matter. The caloric particles push apart the particles making up the body. This results in an increase in volume. The caloric theory explained conduction. Heat moves through a material from the hot region to the cold region. Since particles of caloric are most numerous where the material is hottest, the particles repel one another and spread out until they achieve a uniform distribution. This results in the material's reaching the same temperature throughout. This theory explained the observation that different forms of matter contain different amounts of heat. Since different substances attract caloric with different strengths, different substances can absorb different quantities of heat. Believers in the caloric theory were even able to explain the heat generated by rubbing two surfaces together. The two objects simply lose some of their ability to hold caloric. The excess caloric is repelled to the outside and sensed as heat.

The Downfall of the Caloric Theory

An American amateur scientist, Benjamin Thomson (Count Rumford [1753-1814]), made a major contribution to the downfall of the caloric theory. He reasoned that, if caloric exists, hot objects should have more mass than cold objects. He tested this by using the most sensitive balances of the time to determine the mass of alcohol, gold, mercury, and water before and after heating. No increase in mass was detected even though gold was heated from the temperature of freezing water to a red hot glow. But this did not discourage believers in the caloric theory. They simply revised their model to say that the mass of the caloric is too small to detect.

Later Count Rumford supervised the boring of cannons for the Bavarian army. He used this opportunity to test the hypothesis that, when two objects are rubbed together to produce heat, they lose some of their caloric. To do this he added a measured mass of brass turnings heated by the boring process to a measured mass of cold water. The final temperature of the mixture was recorded. Then an equal mass of unbored brass was heated to the same initial temperature as the turnings. This was added to an identical mass of cold water at the same temperature as before. The water reached the same final temperature in both cases. Count Rumford also did an experiment using a blunt borer to show that an endless supply of heat is produced by rubbing two surfaces together.

His experiments led him to conclude that caloric did not exist. Its mass could not be detected. If caloric existed, the brass turnings should have lost some caloric during the boring process to produce the heat generated. As a result, the turnings should not have produced the same temperature changes in water as the same mass of unbored brass heated to the same initial temperature. Also, it should have been impossible to squeeze an inexhaustible supply of caloric from an object. He concluded that heat must come from the work done moving the surfaces past one another. He postulated that heat was a result of the vibration of the particles making up matter. However, he did not propose that heat and mechanical energy are equivalent or that energy is conserved during the boring process.

Sir Humphrey Davy, a famous lecturer in chemistry at the Royal Institute in London, England, contributed to the downfall of the caloric theory one year later in 1799. He rubbed two pieces of ice together in a vacuum at a temperature below the freezing point of water. Some of the ice melted. According to the caloric theory it is not possible to release caloric and, at the same time, attract additional caloric to melt the solid. Davy concluded that the heat required to melt the ice was created by motion. Friction causes the particles making up a body to vibrate. These vibrations are heat.

Discussion

1. **a)** What are the main points of the caloric theory of heat?
b) Use the caloric theory of heat to explain the following: the contraction of a metal as it cools; the constant temperature that accompanies a change in state.
c) What observations could the caloric theory not explain?

2. **a)** How did Count Rumford's experiments help to disprove the caloric theory?
b) What factors determine the amount of heat that can be produced by boring a cannon barrel?

3. Compare Davy's experiment using two chunks of ice with Rumford's experiment using two chunks of brass.

4. **a)** Why did it take so long to convince believers in the caloric theory that the theory was wrong?
b) Do you think it would take as long to change people's minds about a theory today? Discuss.

13.3 The Kinetic Molecular Theory of Heat

We now know that heat is a form of energy rather than a material substance called caloric. But what kind of energy? The modern theory of heat is based on the kinetic molecular (particle) model of matter. We learned in earlier units that all matter is made up of particles called molecules. The distance between molecules is large compared to their size. This distance is largest for gases, smaller for liquids, and least for solids. The molecules making up solids and liquids are held together by electric forces. The molecules attract one another at certain distances and repel one another when they are pushed closer together.

Thermal Energy

Figure 13-2 shows a cubic array of molecules making up a solid. The electric forces between the molecules are represented as elastic springs. Such springs do not exist; but they provide a simple and useful model. The coils of springs attract when stretched apart and repel when pushed together. Molecules behave the same way. We know that molecules are in constant

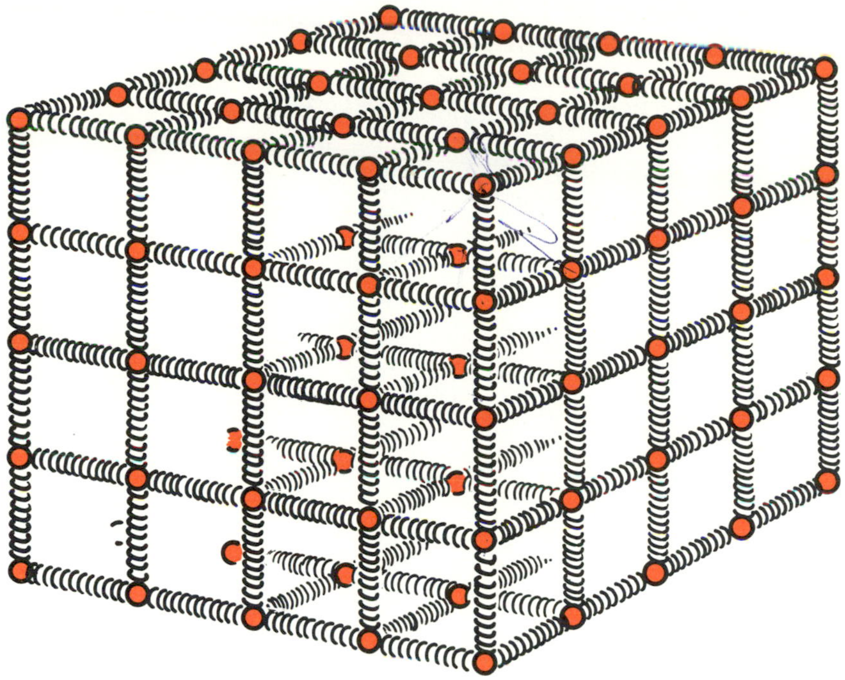

Figure 13-2 A model of a cubic array of atoms making up a crystalline solid. The springs represent electric forces.

motion. You observed the Brownian motion of particles in Sections 5.5 and 5.6. The molecules making up a solid vibrate back and forth about a fixed position much like a mass on the end of a spring pendulum. Molecules have two kinds of energy: elastic potential energy resulting from the electric force and kinetic energy due to their motion. Energy is continually being converted in a random manner between these two forms. However, the total energy of the moving molecules remains constant as long as no additional work is done. This energy is called thermal energy. **Thermal energy** *is the total elastic potential and kinetic energy an object has as a result of the random motion of its particles.*

The Difference between Thermal Energy and Heat

Suppose we add thermal energy to the cube using a hot flame. The molecules of gas in the hot flame are moving at high speed. They possess thermal energy. As they strike the cube, they transfer some of their energy of motion to the molecules making up the cube. The molecules vibrate faster. The thermal energy of the cube increases. The thermal energy transferred from the molecules of gas to the molecules making up the solid is called heat energy. **Heat energy** *is thermal energy being transferred from one material substance to another. It may be added to or removed from a given material..*

Heat travels naturally from a substance at a higher temperature into a substance at a lower temperature. The average kinetic energy of the particles in the hotter substance is greater than in the colder substance. With time, some of this kinetic energy is transferred to the slower particles through collisions.

But heat is not always transferred from a substance with more thermal energy into a substance with less thermal energy. For example, there is twice as much thermal energy in 100 g of boiling water as in 50 g of boiling water. Since both are at the same temperature, no heat is transferred. Heat can be transferred from a substance with less thermal energy to a substance with more thermal energy. A cup of hot coffee has much less thermal energy than a swimming pool. However, if the cup is partially submerged in the swimming pool, heat flows from the hotter cup to the cooler pool.

Increasing Thermal Energy without Transferring Heat

The thermal energy of an object can be increased without transferring heat to it. When a rubber band is stretched, the work done moves the molecules of rubber further apart. This increases the elastic potential energy of the rubber molecules. Some of this elastic potential energy changes to kinetic energy. The thermal energy of the rubber increases. When an iron anvil is struck with a hammer, the collision causes the molecules in both the hammer head and the anvil to vibrate faster than before. The kinetic energy of the molecules increases. As a result, the thermal energy of the substance increases. But heat has not been transferred in either example. In the first example, chemical potential energy stored in the food eaten by the person who is stretching the rubber band is converted into thermal energy. In the second example, kinetic energy of a moving object is converted into thermal energy.

Discussion

1. **a)** Define thermal energy.
b) Distinguish between thermal energy and heat.
c) What two kinds of energy make up thermal energy?

2. **a)** Summarize the main points of the kinetic molecular theory of heat.
b) Compare the total of the kinetic energy and the elastic potential energy of the particles in a hot substance and a cold substance.

3. Compare the amount of thermal energy in each of the following:
a) a soldering iron and a nail, both at 150°C;
b) a kettle of boiling water and a cup of boiling water;
c) a cup of coffee at 80°C and a cup of coffee at 60°C;
d) a 20 kg solid block of hydrogen at −260°C and 10 kg of hydrogen at 100°C.

4. Use the kinetic molecular theory of heat to explain
   a) the expansion of a metallic bridge in the summer;
   b) the constant temperature that accompanies a change of state.
5. **a)** Describe a way to increase the thermal energy of a substance without transferring heat to it.
   b) Explain how rubbing two rough surfaces together increases the thermal energy of the materials.

13.4 INVESTIGATION: Heat and Temperature Change

When heat energy is added to a substance, several changes can result: the temperature can change; the state can change; both the temperature and the state can change; or a chemical change can take place.

In this investigation you study the relationship between the heat energy added to a constant mass of liquid water and the temperature change of the water. You use electric energy to heat an immersion heater which in turn heats the water. The immersion heater has a known power. You calculate the heat energy added to the water in joules by multiplying the power in watts by the time in seconds. For example, an immersion heater with a power of 40 W produces 40 J of heat energy every second. Your teacher will tell you the power rating of your heater.

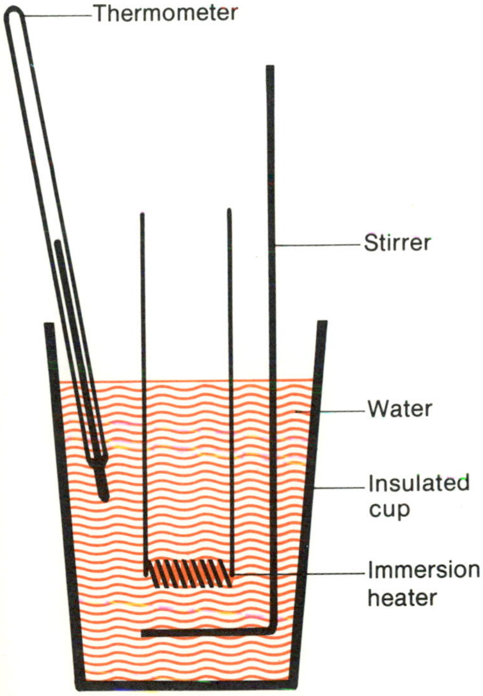

Figure 13-3 Apparatus for studying heat and temperature change.

Materials

insulated cup	stirring rod
immersion heater	graduated cylinder
thermometer	graph paper

Procedure

a. Copy Table 81 into your notebook.
b. Add 100 g of cold water to the insulated cup. To do this, measure out 100 mL of water using the graduated cylinder. Assume that 1 mL of water has a mass of 1 g.
c. Place the immersion heater in the water as shown in Figure 13-3. Be sure the heating element of the immersion heater is covered with water. Read and record the initial temperature of the water.
d. Turn on the heater. Stir the water with the stirring rod and record the temperature of the water every 60 s until a temperature of about 80°C is reached. Record the time and the temperature readings in the table.

e. Calculate the total heat energy added for each time interval. Enter this total in the table.
f. Plot a graph of heat energy added on the vertical axis against temperature on the horizontal axis.
g. Calculate the slope of the graph in joules per degree Celsius (J/°C) for a 100 g mass.
h. Obtain class results for the slope of the graph. Your teacher will help you.

TABLE 81 Heat Energy—Temperature Data for Heating Water

Time (s)	Heater power (W)	Total heat energy (J)	Temperature (°C)
0			
60			
120			
etc.			

Discussion

1. Describe the graph.
2. As the thermal energy of the water increases, what happens to the temperature?
3. Compare the heat added to the water during an interval of 120 s with the heat added during an interval of 240 s.
4. Compare the temperature change of the water during an interval of 120 s with the temperature change during an interval of 240 s.
5. As the heat energy doubles, what happens to the temperature change produced?
6. What relationship exists between the heat energy and the temperature change?
7. The slope of the graph has units of joules per degree Celsius (J/°C) for a mass of 100 g of water. Compare the value you obtain with other values in the class.
8. What is the significance of the slope?
9. Some heat energy is being lost to the container, the stirrer, and the surroundings. What effect does this have on the slope?

13.5 Thermal Energy and Temperature

Heat and Temperature Change

When heat is added to water at room temperature, the temperature of the water increases. The greater the heat transferred to the water, the larger the temperature change. Twice as much heat produces twice the temperature change. The heat transferred is directly proportional to the temperature change produced.

Thermal Energy and Temperature

The thermal energy of water increases as heat is added to it. Thermal energy and temperature are related, but they are different. Figure 13-4,A shows two cubes, one with eight times the mass of the other. Both cubes are at the same temperature. There are eight times as many molecules in the larger cube. Each molecule has some kinetic and elastic potential energy. The thermal energy of the cube is the total of the kinetic and potential energy for all the molecules. Since the larger sample has eight times as many molecules, it has eight times the thermal energy. Clearly, *the thermal energy of any substance is the total of the kinetic and potential energy for all the molecules in it*.

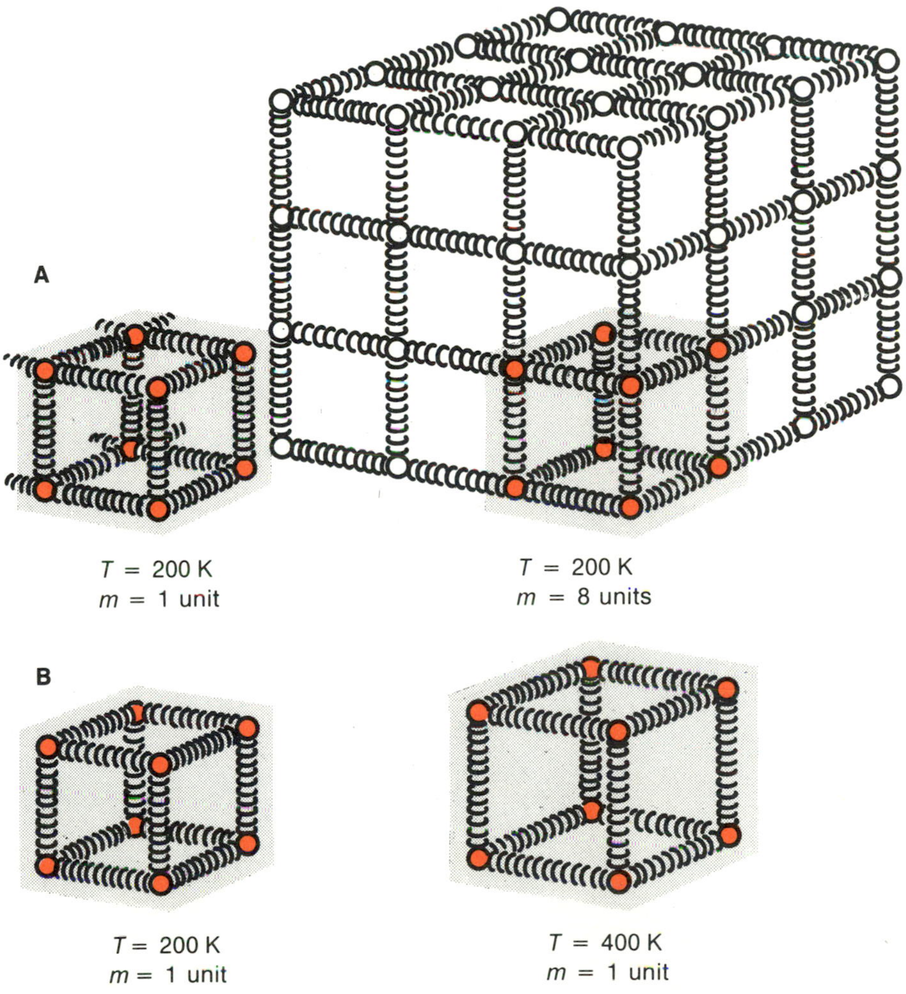

Figure 13-4 Thermal energy and temperature.

A thermometer measures only the average kinetic energy of the molecules near the bulb. Although each molecule transforms energy back and forth between elastic potential energy and kinetic energy, the average kinetic energy of each molecule in each cube is the same. *Temperature is a measure of the average kinetic energy of the molecules in a substance.* It is not a measure of the thermal energy. The greater the average kinetic energy of the molecules, the greater the temperature.

Figure 13-4,B shows two cubes with the same mass. One is at a temperature of 200 K. The other is at a temperature of 400 K. One has twice the temperature of the other. The two cubes have both a different thermal energy and a different temperature. Each molecule at the higher temperature has twice as much kinetic energy, on the average, as each cooler molecule. The total of the kinetic and potential energies for all the molecules is also larger for the hotter cube. Thus both the thermal energy and the temperature of the hotter cube are larger. Therefore, as the average kinetic energy of the molecules in a sample increases, so does the temperature. Thermal energy increases with temperature. If the temperature stays the same, then the thermal energy can increase with an increase in mass.

Discussion

1. **a)** Compare the heat transferred to a substance and the temperature change produced.
b) Explain this relationship in terms of the kinetic molecular theory of heat.

2. **a)** Distinguish between thermal energy and temperature.
b) Account for the temperature change of a piece of metal which results from a blow by a hammer.
c) Account for the temperature change produced in a rubber band as it is stretched.

3. **a)** How can a change in thermal energy take place without a change in the temperature of a substance?
b) When heat is added to a substance that is undergoing a change in state there is no temperature change until the change in state is complete. Why?

4. Two different samples of water are both at the boiling point at standard atmospheric pressure. They have the same temperature but one has more thermal energy than the other. Explain.

13.6 INVESTIGATION: Heat and Mass

We know from Investigation 13.4 that the heat added to a constant mass of water is directly proportional to the temperature change. A quantity of heat that produces a temperature change of 20°C is twice as large as a quantity of heat which produces a temperature change of 10°C. But if two different masses of water are heated through the same temperature change, how is the heat related to the mass of material?

In this investigation you study the relationship between the heat energy added and the mass of water.

Materials

insulated cup
immersion heater
thermometer
stirring rod
graduated cylinder
graph paper

Procedure

a. Copy Table 82 into your notebook.
b. Add 100 g of water to the insulated cup. Use the graduated cylinder and assume that 1 mL of water has a mass of 1 g.
c. Measure the initial temperature of the sample and record it in the table.
d. Place the immersion heater in the 100 g sample. Make sure it is covered by the liquid. Turn on the heater. Stir the water with the stirring rod and record the temperature of the water every 60 s until a temperature of about 80°C is reached. Record the time and the temperature readings in the table.

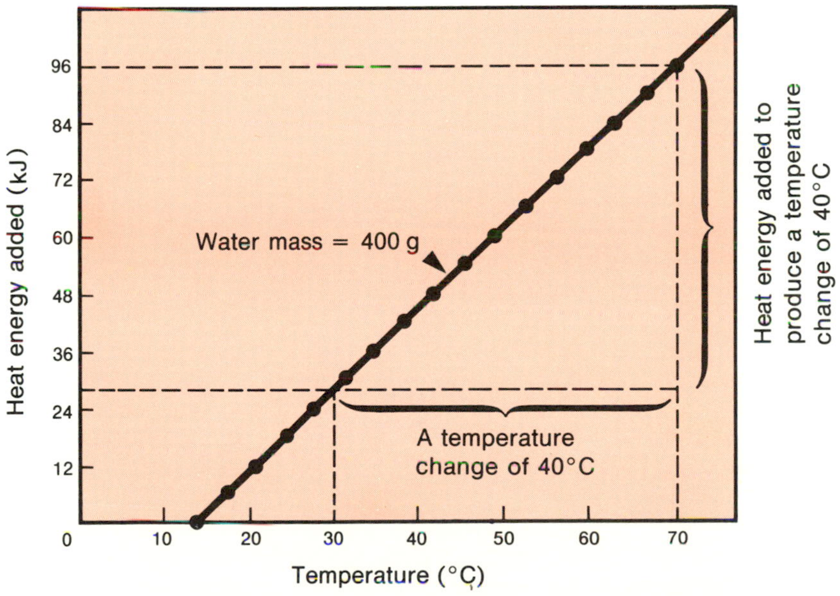

Figure 13-5 Graph of heat energy against temperature for a mass of 400 g.

TABLE 82 Heat and Mass				
Time (s)	Heater power (W)	Total heat energy (J)	Temperature (°C)	
			50 g of water	100 g of water
0 60 120 etc.				

e. Repeat steps b, c, and d using 50 g of water. Be sure to cool the immersion heater before placing it in the second sample.
f. Calculate the total heat energy added for each time interval. Enter this in the table.
g. Plot on the same graph paper the heat energy added against the temperature for each mass. Label each graph.
h. Calculate for each mass the heat energy needed to produce a temperature change of 40°C. To do this, draw two vertical dotted lines, one at a temperature of 30°C and the other at a temperature of 70°C as shown in Figure 13-5. Extend these up to cut the 50 g and 100 g mass graphs. Read from the vertical axis the heat energy in joules to produce a temperature change of 40°C in the two masses.
i. Obtain class results for the heat to change the temperature of a 50 g and a 100 g mass of water by 40°C. Your teacher will assist you.

Discussion

1. Compare the graphs of heat added against temperature for the two masses of water.
2. What is the heat needed to change the temperature of 50 g of water by 40°C; 100 g of water by 40°C?
3. Compare the heat needed to change the temperature of 50 g of water by 40°C with the heat needed to change the temperature of 100 g of water by 40°C.
4. What relationship exists between heat energy and mass for a constant temperature change?
5. List the sources of heat loss in the investigation.

13.7 INVESTIGATION: Heat and Different Substances

We know that the heat needed to change the temperature of a substance depends on two factors, temperature and mass. The larger the temperature change desired, the greater the quantity of heat required. The larger the mass of water heated, the greater the quantity of heat required. But if equal masses of two different substances are heated through the same temperature change, is the heat required the same or different? In this investigation you study the relationship between the heat energy added and the kind of substance.

Materials

insulated cup
immersion heater
thermometer
stirring rod
graduated cylinder
cold ethylene glycol
graph paper

Procedure

a. Copy Table 83 into your notebook.
b. Add 100 g of cold ethylene glycol to the insulated cup. Use the graduated cylinder. Since the density of ethylene glycol is 1 100 kg/m^3, 91 cm^3 have a mass of 100 g.
c. Measure the initial temperature of the sample. Record it in the table.
d. Place the immersion heater in the ethylene glycol. Make sure it is covered by the liquid. Turn on the heater. Stir the ethylene glycol and record the temperature every 60 s until a temperature of about 80°C is reached. Record the time and temperature readings in the table.
e. Repeat steps b, c, and d using 100 g of cold water. Be sure to cool the immersion heater before placing it in the water.
f. Calculate the total heat energy added for each time interval and each sample. Enter this total in the table.
g. Plot on the same graph paper the heat energy added against the temperature change for ethylene glycol and water. Label each graph.
h. Calculate the slope of a straight portion of each graph in joules per degree Celsius (J/°C) for 100 g of each substance.
i. Obtain class results for the slopes of the graphs. Your teacher will help you. Calculate the average value of the slope for the class.
j. Determine the joules per degree Celsius (J/°C) for a mass of 1 kg of water and for 1 kg of ethylene glycol. To do this divide the slope calculated in i by 0.1 kg.

Discussion

1. Compare the graphs of heat added against temperature change for the different substances.

TABLE 83 Heat and Different Substances

Time (s)	Heater power (W)	Total heat energy (J)	Temperature (°C)	
			100 g of ethylene glycol	100 g of water
0				
60				
120				
etc.				

2. Which substance requires the greater amount of heat to produce a temperature change of 40°C?
3. Which substance has the greater ability to store heat for a given temperature change?
4. What is the average value of the slope for each substance? This is called the **heat capacity** of the sample. It has units of joules per degree Celsius (J/°C).
5. What is the average value of the slope for a mass of 1 kg of each substance? This is called the **specific heat capacity** of the substance. It has units of joules per kilogram degree Celsius J/(kg·°C).
6. Compare the class values for the specific heat capacity of ethylene glycol and water.
7. The accepted value for the specific heat capacity of water is 4 200 J/(kg·°C). For ethylene glycol it is 2 200 J/(kg·°C). Compare the class values with the accepted values.
8. Account for any differences between the class values and the accepted values.
9. How could the experiment be modified to obtain more accurate values of specific heat capacity?

13.8 Measuring Quantities of Heat

Factors Affecting the Quantity of Heat

The expansion of the liquid in a thermometer provides a quick and accurate method of measuring temperature. But there is no instrument for directly measuring the quantity of thermal energy transferred from one body to another. Quantities of heat must be measured by the effects they produce on a substance. We have already seen that the heat added to a body during an increase in temperature depends on three factors: the mass of the substance; the temperature change produced; and the material making up the object. The quantity of heat transferred to an object varies directly with the mass of the object heated, the temperature change produced, and a property of the material called specific heat capacity.

The Heat Capacity of a Body

A lake stores more heat than a swimming pool. A car radiator filled with water stores more heat than the same radiator filled with ethylene glycol. Objects with a different ability to store heat have a different heat capacity. Objects with a small heat capacity warm rapidly because they absorb less heat energy for a given temperature change. They also cool more rapidly because they have less heat to give up. As a result, the heat capacity of a large body of water influences the climate in an area. The temperature near an ocean or a large lake changes more slowly than the temperature inland. Water has a moderating influence on climate.

Heat capacity *is the quantity of heat in joules needed to raise the temperature of an object by* 1 °C.

$$\textbf{Heat capacity} = \frac{\textbf{quantity of heat}}{\textbf{temperature change}}$$

In SI, the derived unit for heat capacity is the joule per degree Celsius (J/°C). Heat capacity is affected by the mass of the body. The greater the mass, the greater the heat capacity.

Specific Heat Capacity

The heat capacity of a large body such as a lake is difficult to measure directly. Heat capacity depends on the mass and the kind of material making up the object. The heat capacity of 2.0 kg of water is larger than that of 1.0 kg of water. The heat capacity of 1.0 kg of water is larger than that of 1.0 kg of ethylene glycol.

A better description of the heat capacities of different substances is obtained if equal masses are compared. Such a quantity is called specific heat capacity. **Specific heat capacity** *is the quantity of heat needed to raise the temperature of* 1 kg *of the substance through a change in temperature of* 1 K.

$$\textbf{Specific heat capacity} = \frac{\textbf{quantity of heat}}{\textbf{mass} \times \textbf{temperature change}}$$

Specific heat capacity is given the symbol $\boldsymbol{c}$, quantity of heat the symbol $\boldsymbol{Q}$, mass the symbol $\boldsymbol{m}$, and temperature change the symbol $\Delta \boldsymbol{t}$. Thus $\boldsymbol{c} = \frac{\boldsymbol{Q}}{\boldsymbol{m}\Delta \boldsymbol{t}}$

In SI, the derived unit for specific heat capacity is the joule per kilogram kelvin [J/(kg·K)]. Since a temperature interval of 1 K = 1 °C, an equivalent unit is the joule per kilogram degree Celsius [J/(kg·°C)]. Samples having a smaller mass than 1 kg are sometimes used in laboratory experiments. The derived unit for specific heat in this case is the joule per gram degree Celsius [J/(g·°C)]. The relationship between the two units is 1 000 J/(kg·°C) = 1 J/(g·°C).

Solving the specific heat capacity equation for heat, we obtain

$$\boldsymbol{Q = m \, \Delta t \, c}$$

Thus the quantity of heat needed to produce a certain change in temperature in a body is equal to the product of the mass of the object, its temperature change, and its specific heat. If m is in kilograms (kg), Δt in degrees Celsius (°C), and c in joules per kilogram degree Celsius [J/(kg·°C)], Q is expressed in joules. If m is in grams, and c in joules per gram degree Celsius, Q is expressed in joules.

Sample Problem

Methyl alcohol has a specific heat capacity of 2 500 J/(kg·°C). How much heat must be added to 500 g of the alcohol to heat it from 20°C to 60°C?

Solution

c = 2 500 J/(kg·°C)
m = 500 g = 0.5 kg
t_1 = 20°C
t_2 = 60°C
Δt = 40°C

$$\begin{aligned} Q &= m \times \Delta t \times c \\ &= 0.5\ \cancel{\text{kg}} \times 40\ \cancel{^\circ\text{C}} \times 2\,500\ \text{J}/(\cancel{\text{kg}}\cdot\cancel{^\circ\text{C}}) \\ &= 0.5 \times 40 \times 2\,500\ \text{J} \\ &= 50\,000\ \text{J} \\ &= 50\ \text{kJ} \end{aligned}$$

Specific Heat Capacity and the Size of Molecules

Aluminum has a specific heat capacity of 904 J/(kg·°C). Silver has a specific heat capacity of 236 J/(kg·°C). Thus 1 kg of aluminum can absorb about four times as much heat as 1 kg of silver for the same temperature change. But each silver atom has a mass about four times the mass of an aluminum atom. In general, the specific heat capacity is smaller for substances with more massive atoms. Why?

In 1 kg of the two metals there must be more than four times as many aluminum atoms as silver atoms. The difference in the size and number of atoms is represented in Figure 13-6. The 904 J of heat added to 1 kg of aluminum are divided among about four times as many aluminum atoms as the 236 J of heat added to 1 kg of silver. Each aluminum atom gets, on the average, as much energy from the 904 J of heat as every atom of silver receives from 236 J of heat. As a result, the temperature of 1 kg of aluminum increases by about the same amount as 1 kg of silver with the addition of four times the heat. Table 84 shows the specific heat capacities of some common substances. In general, the more massive the particles, the lower the specific heat capacity of the substance. However, there are exceptions.

TABLE 84 Specific Heat Capacities

Substance	Relative mass of particles	Specific heat at 25°C J/(kg·°C)	Substance	Relative mass of particles	Specific heat at 25°C J/(kg·°C)
Alcohol, methyl	32	2 500	Lead	207	130
Aluminum	27	900	Magnesium	24	980
Carbon (graphite)	12	710	Mercury	201	140
Copper	64	390	Nitrogen (s.p.)	28	1 000
Ethylene glycol		2 200	Oxygen (s.p.)	32	920
Gold	197	130	Paraffin oil		2 100
Helium (s.p.)	4	5 200	Silver	108	240
Hydrogen (s.p.)	2	14 300	Uranium	238	120
Ice at 0°C	18	2 100	Water	18	4 200
Iron	56	450	Water vapour (100°C)	18	2 000
			Zinc	65	390

NOTE: s.p. = standard pressure

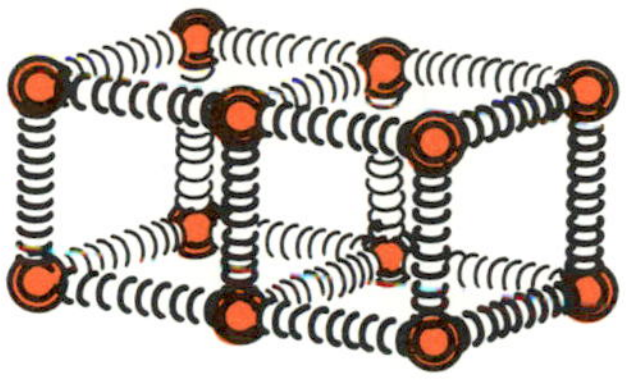

A unit mass of silver

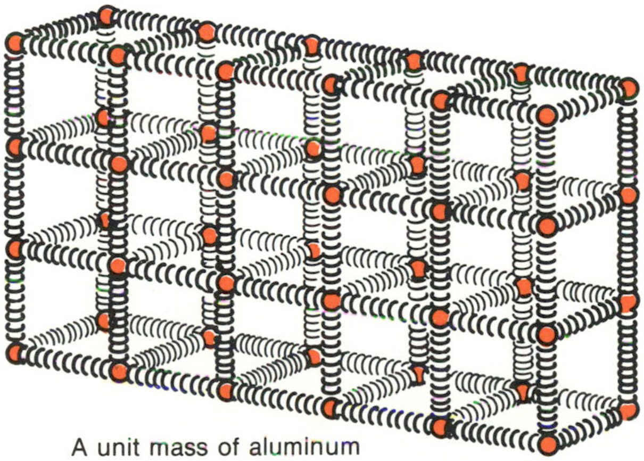

A unit mass of aluminum

Figure 13-6 Specific heat capacity and the size of atoms. A unit mass of aluminum has about four times as many atoms as the same mass of silver. The crystal structure has been simplified for comparison.

Discussion

1. **a)** What three factors affect the quantity of heat transferred when a substance changes in temperature?
   b) State the relationship between the quantity of heat transferred and each of the three factors.
2. **a)** Define the term "heat capacity".
   b) What is the derived unit of heat capacity?
   c) Explain why the climate near a large body of water is more moderate than the climate near a small body of water.
3. **a)** Define the term "specific heat capacity".
   b) How is the specific heat capacity of a substance determined?
   c) What is the SI derived unit for specific heat capacity?
   d) What is the derived unit for specific heat capacity that is usually used in laboratory work? Why?
4. **a)** The heat energy collected in a solar energy system is sometimes stored in liquids in underground tanks. Which liquid will hold more heat—ethylene glycol or water? Why?
   b) What advantage does a mixture of ethylene glycol and water have over pure water?
5. A copper atom has a mass about 2.4 times that of an aluminum atom.
   a) Predict how the specific heat capacity of aluminum will compare with that of copper. Why? Check your prediction using Table 84.
   b) Get the densities of copper and aluminum from Table 27, page 63. Compare their densities and specific heat capacities. What relationship exists between density and specific heat capacity? Why?
   c) Compare the specific heat capacities of ethylene glycol and paraffin oil. How should the masses of their molecules compare?
6. Calculate the amount of heat transferred in each of the following cases:
   a) 200 g of water warms from 15°C to 35°C;
   b) 0.40 kg of water warms from 30°C to 72°C;
   c) 75 g of water cools from 90°C to 52°C;
   d) 4.8 kg of ethylene glycol cools from 35°C to 20°C.
7. Determine the temperature change in each of the following cases:
   a) 200 g of water gains 8 400 J of heat;
   b) 15 kg of water loses 252 kJ of heat.
8. Determine the mass of water when 126 kJ of heat produce a temperature change of 30°C.
9. **a)** How much heat is required to raise the temperature of 40 g of aluminum from 20°C to 40°C?
   b) How much heat must be removed from 0.60 kg of iron to lower its temperature from 100°C to 50°C?

10. How much heat is lost by 40 g of ice in cooling from $-5.0°C$ to $-15°C$?
11. What temperature will the steam be if 40 kJ of heat are added to 1 kg of water vapour at 100°C?

13.9 INVESTIGATION: Heat Exchange in Mixtures

A transfer of heat occurs when two substances at different temperatures are mixed together. The warmer substance loses heat in cooling. The cooler substance gains heat in warming. In this investigation you study the heat transfer when different masses of water at different initial temperatures are mixed. You determine if the Law of Conservation of Energy holds for the transfer of heat.

Materials

Bunsen burner
ring stand, wire gauze, ring clamp
adjustable clamp
Erlenmeyer flask
100 mL graduated cylinder
several large insulated cups
thermometer

Procedure

a. Copy Table 85 into your notebook.
b. Set up the heating apparatus as shown in Figure 13-7.
c. Add 100 g of cold water to each of four insulated cups. Label the cups A, B, C, and D.
d. Record the initial temperature of the water in each cup just before the hot water is added.
e. Add 40 g of water to the Erlenmeyer flask. Bring it to a boil as rapidly as possible.
f. When the water begins to boil, remove it from the heat. Be careful! Record its temperature. Use the clamp as a handle and pour the 40 g of hot water into cup A.
g. Stir the mixture in cup A. Read and record the highest temperature reached.
h. Repeat steps e, f, and g, adding 80 g of hot water to cup B, 100 g to cup C, and 200 g to cup D.
i. Some hot water may have boiled away before it was added to the cold water. To see if this has happened, measure the total volume of the mixture using a graduated cylinder. Subtract from this total the volume of cold water initially in each cup. Enter the mass of the volume of hot water in the table.

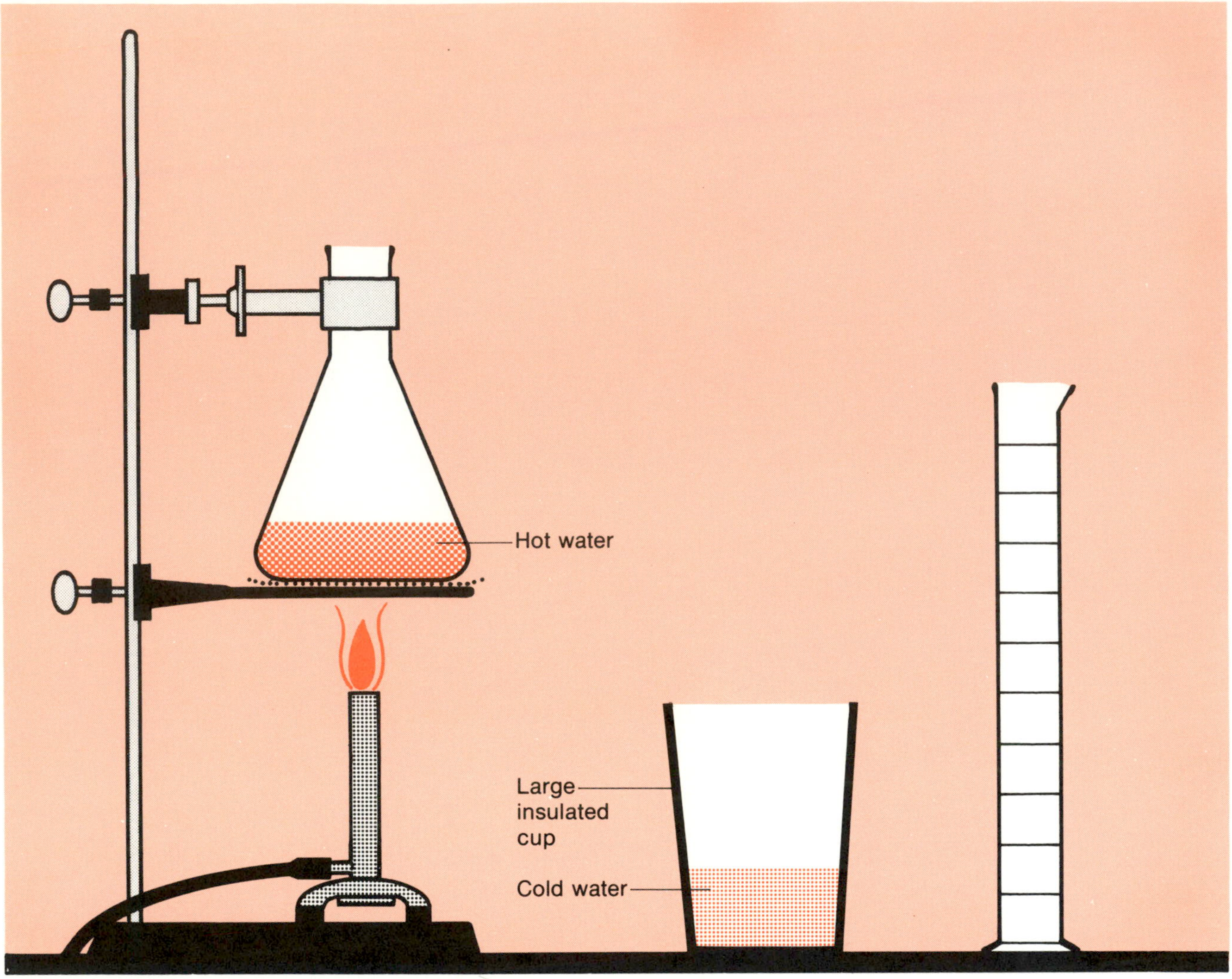

Figure 13-7 Apparatus for studying heat exchange in mixtures.

j.) Calculate the heat lost by the hot water in cooling and the heat gained by the cool water in warming for each mixture. Use the formula $Q = m \Delta t c$

Discussion

1. Compare the heat lost by the hot water with the heat gained by the cold water.
2. If the heat lost is not equal to the heat gained, which should be greater and why?
3. How could you modify the investigation to get more accurate results?
4. The Law of Conservation of Energy stated for heat exchange in mixtures is called the Principle of Heat Exchange. State the Principle of Heat Exchange.

TABLE 85 Heat Exchanged in Mixtures of Water

Mixture	Water	Mass (kg)	Temperature (°C)			Specific heat capacity J/(kg.°C)	Heat exchanged (J)
			Initial	Final	Change		
Cup A	Hot						
	Cold	0.100					
Cup B	Hot						
	Cold	0.100					
Cup C	Hot						
	Cold	0.100					
Cup D	Hot						
	Cold	0.100					

5. Use the Kinetic Molecular Theory of Heat to explain the temperature changes which occur when a hot liquid is added to a cold liquid.
6. A mixture is made by adding 120 g of water at a temperature of 80°C to 30 g of water at a temperature of 20°C. Calculate the final temperature of the mixture.
7. A mixture is made by adding 75 g of an unknown liquid at a temperature of 25°C to 60 g of water at a temperature of 90°C. The final temperature of the mixture is 65°C. Calculate the specific heat capacity of the liquid. Refer to Table 84 and identify the liquid. Can you be sure your identification is correct? Discuss.
8. You have used the Principle of Heat Exchange to determine the final temperature of a mixture and the specific heat capacity of one component. What other information can be determined for a mixture using this principle?

13.10 INVESTIGATION: The Specific Heat Capacity of Metals

Every substance has a definite specific heat capacity at a given temperature. This is used as a characteristic property together with density, melting point, and freezing point to identify substances. However, it is difficult to determine the specific heat capacity of a solid directly. The heat source and the thermometer cannot be easily immersed in the solid. As a result, the Principle of Heat Exchange is used to determine the specific heat capacity of solids. *The* **Principle of Heat Exchange** *states that if no heat* *is lost to the surroundings, the heat gained by the object at the lower temperature is equal to the heat lost by the object at the higher temperature.*

To determine the specific heat capacity of a solid, the solid is first heated to a known temperature in a liquid bath. Then the hot solid is immersed in a cold liquid. A liquid is used which does not react chemically with the solid. Heat is transferred until both components of the mixture reach the same final temperature. In this investigation you use the method of mixtures and the Principle of Heat Exchange to determine the specific heat capacity of some metals.

Materials

samples of several metals (aluminum, copper, iron, lead, zinc)
Bunsen burner
ring stand, ring clamp, wire gauze
200 mL beaker
100 mL graduated cylinder
insulated cup
thermometer
balance

Procedure

a. Copy Table 86 into your notebook.
b. Choose one sample of metal. Find the mass of the metal. Enter this in the table.
c. Attach a thread to the metal sample. Suspend the metal in water in a beaker. Do not let it touch the bottom. Heat the water to the boiling point. Allow it to boil for about five minutes so that the metal reaches the temperature of the water. Take the temperature of the water. Assume that this is the temperature of the metal. Enter this temperature in the table.
d. Add 100 g of cold water to an insulated cup. Measure and record the temperature of the cold water. Quickly remove the metal sample from the hot water bath. Shake it to remove excess water. Transfer the sample to the cold water in the cup.
e. Stir the mixture gently. Record the maximum temperature of the mixture.
f. Use the Principle of Heat Exchange to calculate the specific heat capacity of the metal.

TABLE 86 Determining the Specific Heat Capacity of a Metal

Component of mixture	Mass (kg)	Temperature (°C)			Specific heat capacity J/(kg.°C)	Heat exchanged (J)
		Initial	Final	Change		
Metal						
Water	0.100					

Discussion

1. Compare the value you obtain with other values obtained for the same metal by the class. Your teacher will help you.
2. Compare the class value for the specific heat capacity of the metal with the accepted value (Table 84).
3. Compare the specific heat capacities of the different metals used with their densities. Do this by arranging the metals in order of increasing specific heat capacity and then of density.
4. Account for any relationship between the density of a metal and its specific heat capacity in terms of the kinetic molecular model of heat.
5. List possible sources of error in the experiment. Indicate for each error whether it raises or lowers the value of specific heat capacity. Suggest modifications to the experiment to overcome the errors.
6. Design and carry out an experiment to determine the specific heat capacity of dry sand. Use your results to explain the temperature changes on a sandy beach on a sunny day. Also, explain why rocks and soil cool off and warm up so rapidly.

13.11 Heat Transfer during Changes of State

A definite amount of heat is required to change the temperature of 1 kg of a substance by 1 °C. This quantity is called the specific heat capacity of the substance. The specific heat capacity of a substance is different in each of the three states. For example, the specific heat capacity of ice is 2 100 J/(kg·°C), whereas that of liquid water is 4 200 J/(kg·°C) and of water vapour is 2 000 J/(kg·°C). This difference is due to the molecular arrangement in each of the three states.

Specific Latent Heat of Fusion

A definite amount of heat is also required to change the state of a substance. You know from your study of melting and boiling points that no temperature change takes place during a change of state. *The quantity of heat required to change* 1 kg *of a substance from the solid state to the liquid state without changing its temperature is called the* **specific latent heat of fusion** **(l_f)**. Since the temperature does not change, the average kinetic energy of the molecules does not change. The heat added to the solid is used to break up the orderly arrangement of the molecules. The molecules are moved further apart. As a result, the elastic potential energy of the molecules increases. The heat of fusion is used to increase the elastic potential energy of the molecules.

Specific Latent Heat of Vaporization

Heat is also added to a substance to change it from a liquid to a gas. *The quantity of heat required to change* 1 kg *of a substance from the liquid state to the gaseous state without a change in temperature is called the* **specific latent heat of vaporization** ***(l_v)***. During this change of state, there is a large increase in the distance between molecules. One millilitre of water expands to occupy about 1 700 mL when it becomes water vapour. The specific latent heat of vaporization is used to increase the elastic potential energy of the molecules. Molecules in the gaseous state have a much greater potential energy than those in the liquid state.

Units of Specific Latent Heat

Specific latent heat has units of energy per unit mass. If heat energy is represented by ***Q***, mass by ***m***, and latent heat by ***l***, the equation for latent heat is $l = \frac{Q}{m}$. Rearranging this becomes $Q = m\,l$. The SI derived unit for specific latent heat is the joule per kilogram (J/kg). However, since small masses are used in the laboratories, another acceptable unit is the joule per gram (J/g).

Discussion

1. Define specific latent heat of fusion (l_f) and specific latent heat of vaporization (l_v).
2. Compare specific heat capacity and specific latent heat.
3. Why is the temperature constant during a change of state?
4. What kind of energy changes take place in a substance when it changes state?
5. What is the SI unit for specific latent heat?
6. What is the relationship between joules per kilogram and joules per gram?
7. Look up the meaning of the word "latent" in a dictionary. Why is it used to describe the heat transferred during changes of state?

13.12 INVESTIGATION: Specific Latent Heat of Fusion of Ice

It requires more heat to melt 1 kg of ice than to change its temperature by 1 °C. In this investigation you use the method of mixtures and the Principle of Heat Exchange to determine the specific latent heat of fusion of ice (l_f).

Materials

Bunsen burner
ring stand, ring clamp, wire gauze
adjustable clamp
Erlenmeyer flask
100 mL graduated cylinder
two insulated cups
thermometer
balance
ice cubes

Procedure

a. Copy Table 87 into your notebook.
b. Set up the heating apparatus as shown in Figure 13-7 (p. 360).
c. Add 100 mL of water to the Erlenmeyer flask. Heat the water to a temperature of about 80°C.
d. Find the mass of an insulated cup.
e. Dry an ice cube and add it to the cup. Measure the mass of the cup plus ice cube.
f. Subtract d from e to obtain the mass of the ice cube.
g. Pour the hot water into the second cup. Read and record its temperature.
h. Add the hot water to the ice cube. Stir the mixture. Measure the final temperature of the water as soon as all the ice is melted.
i. Measure the mass of the cup plus contents. Subtract e from i to obtain the mass of the hot water.
j. Calculate the quantity of heat needed to melt 1 g of ice as follows:
1) Calculate the heat gained by the ice in melting. Use the formula $Q = m\,l_f$. The specific latent heat of fusion l_f is the unknown in the equation.
2) Calculate the heat gained by the melted ice water in warming. Use the formula $Q = m \Delta t\, c$. Remember that this water has an initial temperature of 0°C.

TABLE 87 Determining the Specific Latent Heat of Fusion of Ice

Components of mixture	Mass (kg)	Temperature (°C)			Specific heat J/(kg.°C)	Heat exchanged (J)
		Initial	Final	Change		
Hot water						
Melted ice						
Ice					Specific latent heat of fusion l_f (J/kg)	

3) Calculate the heat lost by the hot water in cooling. Use the formula $Q = m\,\Delta t\,c$

4) Set up the following relationship:

Heat gained by ice in melting	+	Heat gained by melted ice in warming	=	Heat lost by hot water in cooling

5) Solve for the specific latent heat of fusion of ice (l_f).

Discussion

1. How much heat is required to melt 1 kg of ice?
2. Compare your value with other values in the class.
3. The accepted value for the specific latent heat of fusion of ice is 336 kJ/kg. Compare the class value with the accepted value.
4. What was the temperature of the ice cube when the hot water was added? How do you know?
5. List sources of error in the experiment. Suggest changes to make these errors smaller.
6. Why does the temperature of an ice water mixture remain constant during melting and freezing?
7. Why is ice used as a refrigerant in coolers?
8. Explain how large containers of water placed in root cellars near vegetables prevent freezing.
9. Explain the use of smudge pots in orchards to prevent frost damage.
10. How much heat is given off when 200 g of water at 0°C freeze into ice at 0°C?
11. What mass of ice at 0°C can be changed to a liquid at 0°C by 0.8 kg of water at 100°C?
12. Calculate the quantity of heat required to melt 100 g of ice at an initial temperature of –20°C.
13. Calculate the quantity of heat needed to convert 50 g of ice initially at –10°C, into water at 80°C.

13.13 INVESTIGATION: Specific Latent Heat of Vaporization of Water

It requires more heat to vaporize 1 kg of water than to melt 1 kg of ice. In this investigation you use the method of mixtures and the Principle of Heat Exchange to determine the specific latent heat of vaporization of water (l_v).

Materials

Bunsen burner
ring stand, ring clamp, wire gauze

adjustable clamp	thermometer
steam generator	insulated cup
water trap	balance

Procedure

a. Make a table like Table 88.
b. Assemble the apparatus as shown in Figure 13-8.
c. Heat the steam generator until steam is coming from the trap.
d. Find the mass of the insulated cup.
e. Fill the cup two-thirds full with cold water. Find the mass of the cup and its contents.
f. Subtract c from d to find the mass of the cold water.
g. Record the temperature of the cold water.
h. Insert the outlet of the trap to a depth of about 2 cm below the surface of the cold water in the cup. Handle the trap carefully!

Figure 13-8 Apparatus for studying specific latent heat of vaporization of water.

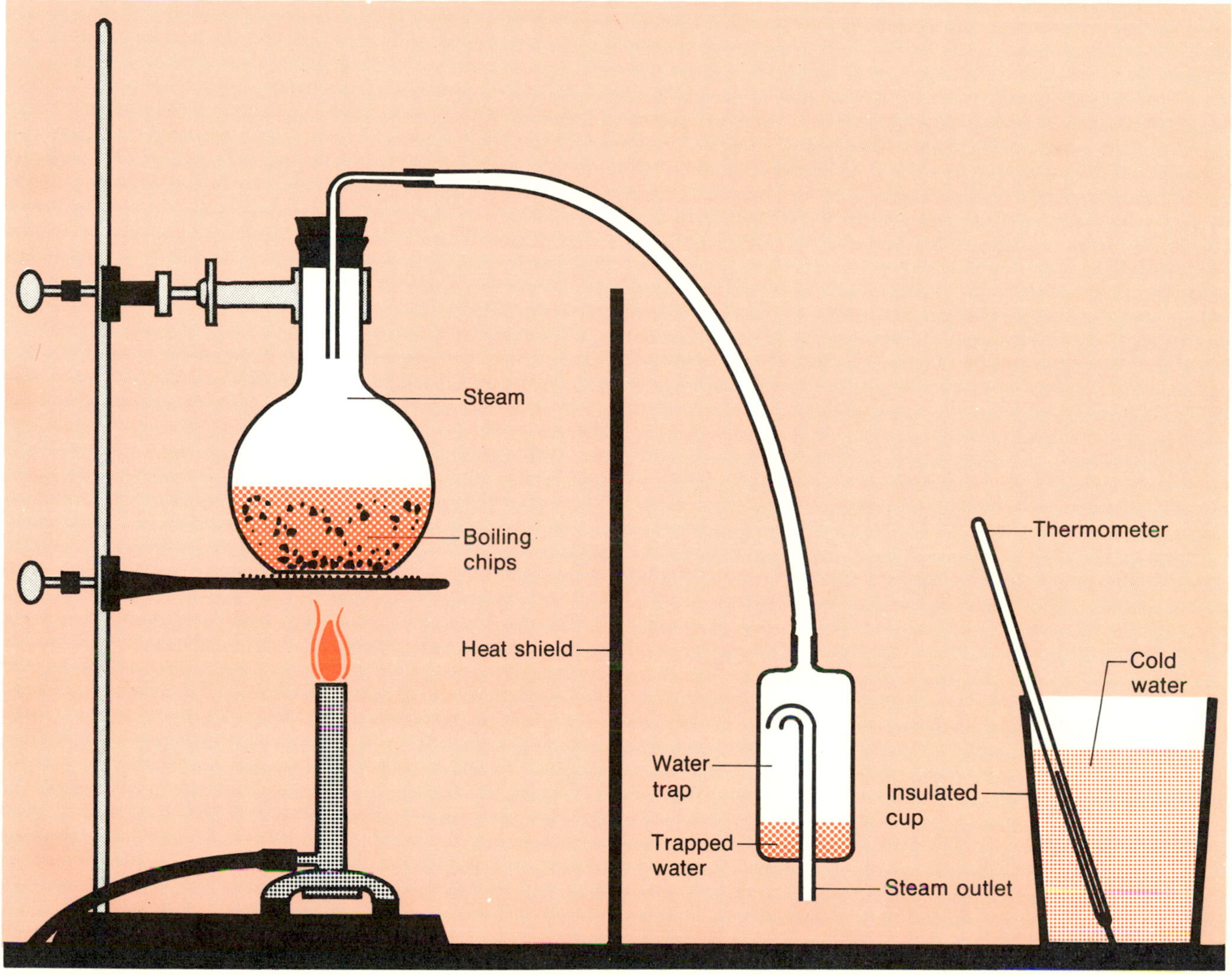

i. Stir the water and, when the temperature has risen about 40°C, quickly remove the trap. Then turn off the Bunsen burner. Continue stirring the water and record the maximum temperature reached.

j. Find the mass of the cup and its contents.

k. Subtract e from j to find the mass of the steam added to the water.

l. Calculate the quantity of heat given out when 1 kg of steam condenses as follows:

1) Calculate the heat lost by the steam in condensing. Use the formula $Q = m\,l_v$. The specific latent heat of vaporization is the unknown in the equation.

2) Calculate the heat lost by the condensed steam in cooling. Use the formula $Q = m\,\Delta t\,c$. Assume that this water has an initial temperature of 100°C.

3) Calculate the heat gained by the cold water in warming. Use the formula $Q = m\,\Delta t\,c$.

4) Set up the following relationship:

Heat lost by steam in condensing + Heat lost by condensed steam in cooling = Heat gained by cold water in warming

5) Solve for the specific latent heat of vaporization of water (l_v).

TABLE 88 Determining the Specific Latent Heat of Vaporization of Water

Components of mixture	Mass (kg)	Temperature (°C)			Specific heat capacity J/(kg. °C)	Heat exchanged (J)
		Initial	Final	Change		
Steam					Specific latent heat of vaporiz-ation (l_v) (J/kg)	
Water from steam						
Cold water						

Discussion

1. How much heat is given out when 1 kg of a steam condenses?
2. Compare your value with other values in the class.
3. The accepted value for the specific latent heat of vaporization is 2 268 kJ/kg. Compare the class value with the accepted value.
4. Why was the water trap used?

5. Why was cold water used?
6. List sources of error in the experiment. Suggest changes to make these sources of error smaller.
7. Why do you feel cool when you first step out of a swimming pool?
8. A burn from steam at 100°C is much more severe than a burn from hot water at 100°C. Why?
9. Some hot water radiators have a single pipe connected to them. The steam comes to the radiator and the water leaves the radiator in the same pipe. The incoming steam and the outgoing water are both at a temperature of 100°C. Where does the heat come from to warm the room?
10. Explain why rapid run-off of water from city streets increases the temperature in a city in the summer.
11. What role does the specific latent heat of vaporization play in the moderation of climate?
12. How much heat is needed to change 60 g of water at 100°C into steam at 100°C?
13. Calculate the minimum amount of heat needed to vaporize 4 kg of water at 80°C.
14. How much heat is released when 150 g of steam at 130°C is condensed and cooled to 30°C?
15. Calculate the quantity of heat needed to convert 80 g of ice at −40°C into steam at 160°C.

13.14 INVESTIGATION: The Specific Heat of Combustion of a Fuel

Fossil fuels such as coal, oil and gas contain chemical potential energy. We burn these fuels to convert the chemical potential energy into heat energy to warm homes and power cars. In this investigation you burn alcohol and use the heat energy to warm water. You determine the specific heat of combustion of the fuel in kilojoules per kilogram (kJ/kg).

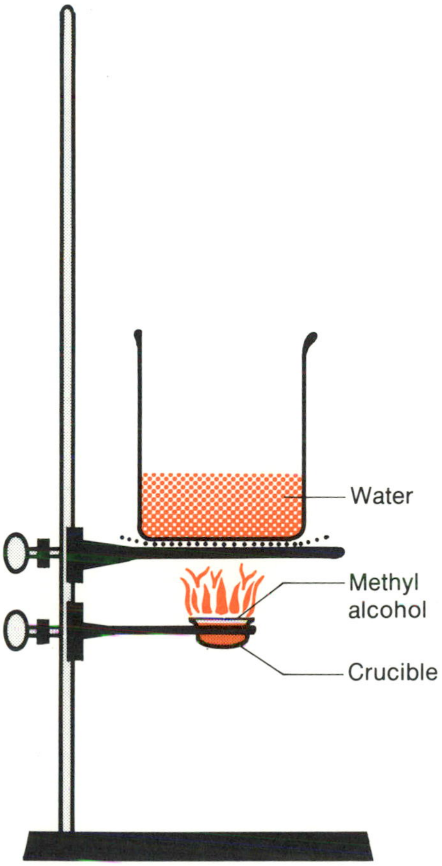

Figure 13-9 Apparatus for studying the specific heat of combustion of a fuel.

Materials

250 mL beaker
ring stand, ring clamp, wire gauze
small crucible
adjustable clamp
thermometer
100 mL graduated cylinder
10 mL graduated cylinder
methyl alcohol

Procedure

a. Set up the apparatus as shown in Figure 13-9.
b. Add 100 mL of cold water to the beaker.

c. Place 3 g of alcohol in the crucible. (Since the density of methyl alcohol is 729 kg/m^3, a volume of 3.8 mL has a mass of 3 g.) Place the crucible close to the bottom of the beaker as shown in Figure 13-9.
d. Measure and record the temperature of the water.
e. Ignite the alcohol with a match. Be careful! Stir the water gently until the alcohol has completely burned.
f. Record the highest temperature reached by the water.
g. Calculate the heat transferred to the water by the burning alcohol.
h. Calculate the specific heat of combustion of methyl alcohol in kilojoules per kilogram. Determine the class value.

Discussion

1. How many joules of heat energy did the water receive?
2. How many joules of heat energy did the alcohol transfer to the water?
3. What is the class value for the specific heat of combustion of methyl alcohol in kilojoules per kilogram?
4. The accepted value is 22 000 kJ/kg. Compare the class value with the accepted value.
5. Why is the class value lower than the accepted value? Discuss sources of error. Suggest ways to improve the experiment.

13.15 INVESTIGATION: The Specific Heat of Combustion of Foods

The foods we eat contain chemical potential energy. The body "burns" food to produce the different forms of energy needed by the body. Some of the chemical potential energy is converted into elastic potential energy in the muscles. Some is converted into kinetic energy as the arms, legs, and blood move. Much of the energy is converted into heat energy. This heat energy is used to replace the heat lost by the body and to maintain the normal body temperature of 37°C.

In this investigation you burn some foods. You use the heat produced to increase the temperature of water. You determine the specific heat of combustion of the foods in kilojoules per kilogram.

Materials

50 mL beaker
beaker tongs
two paper clips
balance
10 mL graduated cylinder

sheet of aluminum foil 10 cm x 15 cm
samples of food (walnut halves, peanuts, spoon-sized shredded wheat)

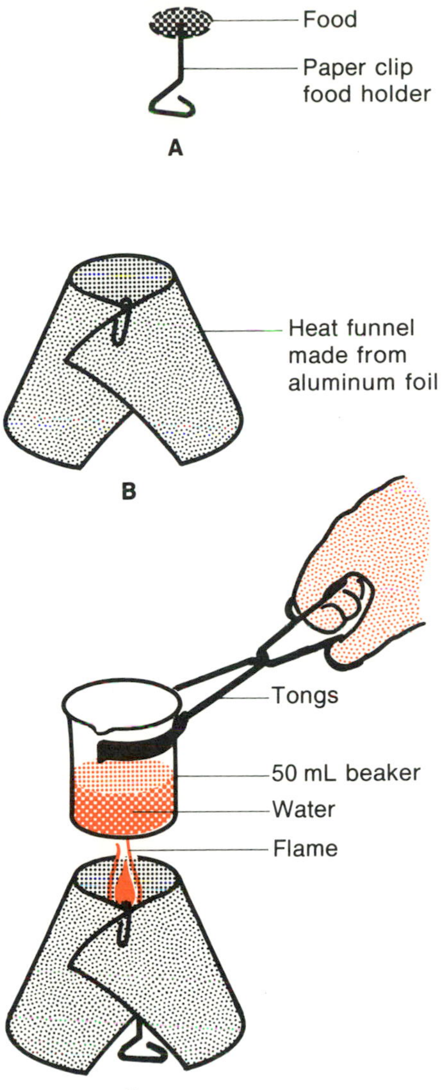

Figure 13-10 Apparatus for studying the specific heat of combustion of foods.

Procedure

a. Make a food holder out of a paper clip by bending it as shown in Figure 13-10, A.
b. Make a heat funnel by folding the sheet of aluminum foil to form a cone with an opening at both ends. Fasten the ends together with a paper clip as shown in Figure 13-10, B.
c. Find the mass of a piece of walnut in grams. Insert the end of the paper clip holder into the sample. Be careful not to puncture your hand or lose any of the food!
d. Place 30 mL of cold water in the 50 mL beaker. Record the temperature of the water.
e. Ignite the food with a match. Place the heat funnel over the food as shown in Figure 13-10, C. Hold the bottom of the beaker in the tip of the flame until the food is burned to an ash. Support the beaker using the tongs.
f. Record the highest temperature reached by the water.
g. Repeat steps c to f for a peanut and a spoon-sized shredded wheat.
h. Calculate the specific heat of combustion in kilojoules per kilogram for each food. Determine the class values.

Discussion

1. What is the class value for the specific heat of combustion of walnuts, peanuts, and spoon-sized shredded wheat?
2. The accepted values are: walnuts 27 300 kJ/kg; peanuts 24 600 kJ/kg; spoon-sized shredded wheat 14 900 kJ/kg. Compare the class values with the accepted values.
3. Why is the class value much lower than the accepted value? Discuss sources of error. Suggest ways to prevent heat loss and to improve the combustion of the food.

13.16 The Energy Content of Fuels and Foods

Specific Heat of Combustion

Organic compounds such as foods and fuels contain carbon and hydrogen. Organic compounds store energy in the form of chemical potential energy. When completely burned in oxygen, organic compounds produce carbon dioxide, water, and heat energy. *The heat energy produced when* 1 kg *of an organic substance is burned is called the* **specific heat of combustion**. The specific heat of combustion of substances is measured in kilojoules per kilogram. The complete combustion of 1 kg of crude oil produces 46 600 kJ of heat energy.

The Bomb Calorimeter

The values you obtained for the heats of combustion of foods in Investigation 13.15 were low. Much of the heat produced by burning the foods is lost to the surroundings. You observed a black, sooty film on the bottom and sides of the beaker. This indicates that the combustion of the food is not complete. Some chemical potential energy is not being completely converted into heat. A better supply of oxygen is needed to produce complete combustion.

Scientists use a special apparatus called a bomb calorimeter to obtain complete combustion and to prevent heat loss. This enables them to get accurate values for the heats of combustion of foods and fuels. A bomb calorimeter is shown in Figure 13-11. The sample is placed in a strong vessel. This vessel contains enough oxygen at high pressure for complete combustion. The vessel is immersed in a known mass of water at a known temperature. An electrically-heated wire ignites the sample. The temperature change of the water and the combustion chamber is recorded. The specific heat capacities of the water and the metal making up the combustion chamber are known. This and the mass and temperature change are used to calculate the heat of combustion of the sample.

Some heats of combustion for fuels and foods are given in Table 89. Heats of combustion of fuels are used by technicians and engineers to determine fuel requirements for homes and power plants. The heats of combustion of fats, carbohydrates, and proteins are used in planning diets and in nutrition studies.

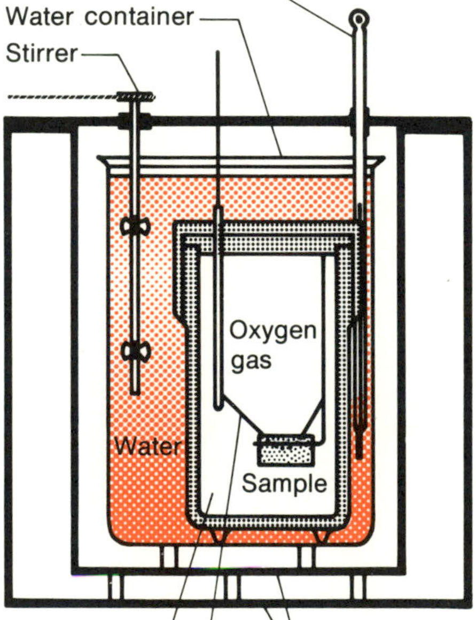

Figure 13-11 A bomb calorimeter for determining heat of combustion.

Discussion

1. **a)** How is a bomb calorimeter used to determine the heat of combustion of a substance?
   b) How is a bomb calorimeter designed to prevent heat loss and to obtain complete combustion?
2. **a)** Compare the heat of combustion of crude oil and coal. Why is crude oil an excellent fuel?
   b) Look at the heat of combustion of the foods in Table 89. Why should a person on a diet stay away from potato chips, fried foods, and nuts?
3. A 1.5 g sample of bituminous coal is placed in a bomb calorimeter. The iron combustion chamber has a mass of 60 g. 200 g of water at 10°C surround the combustion chamber. When the sample is burned in an excess of oxygen, the

TABLE 89 Specific Heats of Combustion

Fuel		Food	
Substance	Specific heat of combustion (kJ/kg)	Substance	Specific heat of combustion (kJ/kg)
Alcohol, ethyl	29 800	Bacon raw	27 900
methyl	22 300	Beef, T-bone Steak	16 700
Coal, bituminous	25 600-36 500	Bread, white	11 300
hard anthracite	29 400-35 300	whole wheat	10 200
coke	33 600	Carrot, raw	1 800
lignite	18 900-33 200	Cereal, oatmeal	16 400
Crude oil	46 600	shredded wheat	14 900
Gasoline	47 300	Cooking oil	20 800
Hydrogen	142 400	Eggs, fried	9 100
Methane	55 000	hard boiled	6 800
TNT	15 200	Peanuts, washed & salted	24 600
Wood, oak	19 300	Potato, boiled	3 200
pine	21 400	Potato chips	23 900
		Walnut	27 300

combustion chamber and water reach a temperature of 70°C. Calculate the heat of combustion of the coal.

4. A 2.00 g sample of potato chips is completely burned in a bomb calorimeter. The iron combustion chamber has a mass of 50 g. 100 g of water surround the chamber. The temperature change of the water and chamber is 90°C. Calculate the heat of combustion of the potato chips.

13.17 Heat and Energy Conservation

Early believers in the kinetic molecular theory of heat considered heat to be a form of energy. But the relationship between the mechanical forms of energy and heat energy was not known. Nor did they know that energy is conserved. Two men contributed to our understanding of thermal energy and energy conversions during the same period of history.

The Contribution of Julius Robert Mayer (1814–1878)

A young German doctor, Julius Robert Mayer, made observations which indicated that mechanical energy is converted to thermal energy, and that energy is conserved in the process.

Doctors at the time knew that venous blood is not as bright red as arterial blood (Fig. 13-12). Arterial blood carries food and oxygen to the cells of the body. Food is oxidized in the cells to produce energy. Some of this energy is used to do work. The rest is given off as heat. Oxygenated blood is brighter red than deoxygenated blood. The difference in colour between arterial and venous blood is related to the amount of oxygen removed from the blood to oxidize the food. The more food that is oxidized, the greater the amount of oxygen removed and the greater the colour difference.

Mayer, while a surgeon in the tropics, noticed that the venous blood of his patients was brighter red than the venous blood of patients in the colder German climate. He reasoned that the energy supplied by oxidizing food is related to the heat energy lost to the surroundings. Less heat is lost from the body to the surroundings in the tropics. Less food is oxidized to keep the temperature of the body constant. He was convinced that there must be a balance between the chemical energy consumed and the heat energy given off by the body. Mayer wondered if energy is conserved when other forms of energy are changed to heat. He did a number of qualitative experiments on energy conversions. For example, he dropped objects into water and noticed that the temperature of the water increased. He concluded that when energy is changed from other forms of energy to thermal energy it is neither created nor destroyed in the process. But because his experiments were qualitative in nature he had difficulty convincing others.

The Contribution of James Prescott Joule (1818–1898)

A wealthy English brewer became interested in the conversion between mechanical energy and thermal energy about the same time as Mayer did. He conducted precise quantitative experiments with various forms of energy to study the relationship between the energy consumed and the thermal energy produced. The apparatus he used to study the relationship between gravitational potential and thermal energy is shown in Figure 13-13. When the mass falls, it causes the paddles to stir the water in the insulated container. This increases the temperature of the water. Joule found that a certain change in gravitational potential energy produces the same quantity of thermal energy. He stirred materials other than water and the results were the same. His experiments showed that whenever a certain amount of various forms of energy is consumed the same amount of thermal energy is produced.

Because his experiments were more precise and quantitative than Mayer's, his results were accepted ahead of Mayer's. But both men finally received the highest honour bestowed by the Royal Society, the Copley medal, for their work.

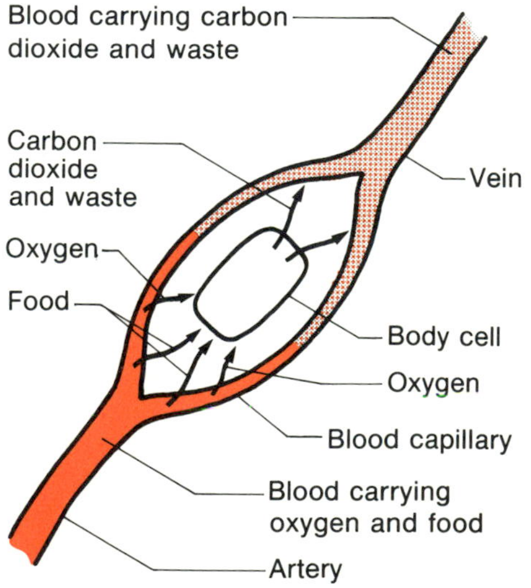

Figure 13-12 Materials leaving and entering capillaries.

The Calorie

Before scientists knew that heat is just another form of energy, they defined a unit of heat called the calorie (cal). No doubt you have heard of it. *A* **calorie** *is equal to* 4.2 J.

Now that we know that heat is just another form of energy, there is little sense in using two different units to measure the same thing. The calorie is an outdated unit. The SI unit of heat, the joule, replaces the calorie. Although the energy content of foods may still be quoted in some cookbooks in Calories (one Calorie = one thousand calories), joules should be used according to SI.

A 50 kg boy at the age of 13-15 needs about 13 000 kJ of energy daily for good nutrition. A 50 kg girl of the same age needs about 11 000 kJ of energy. Table 90 shows the joule content of a typical dinner.

TABLE 90 Joule Content of a Dinner

Food	Energy content (kJ)
85 g sirloin roast	1 071
1 medium sized baked potato	777
115 mL cooked carrots	84
115 mL cooked broccoli	84
57 mL white sauce	441
1 large white-flour roll	483
9 cm wedge of double-crusted apple pie	1 995
1 glass of whole milk	693
	Total 5 628

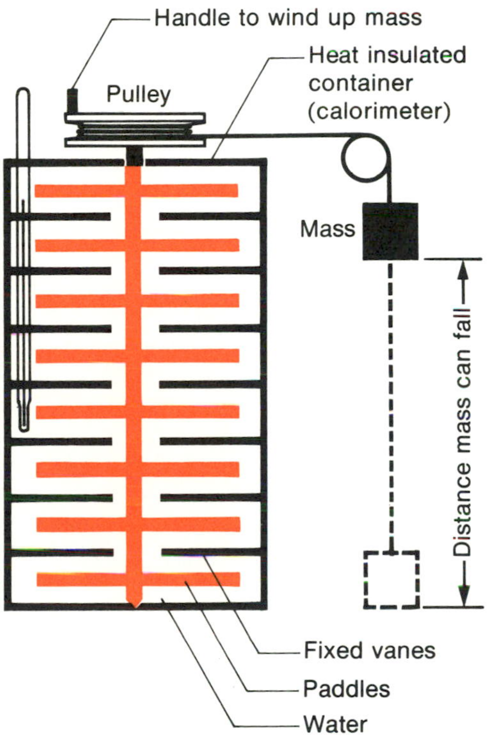

Figure 13-13 A simplified diagram of Joule's apparatus for observing the conversion of mechanical energy into thermal energy.

The Fate of All Forms of Energy

All forms of energy eventually become thermal energy as they are used. You converted electric energy and chemical potential energy into heat energy in earlier investigations. The heat energy became thermal energy in the environment. Joule and other scientists converted gravitational potential energy and kinetic energy into thermal energy. It is easy to convert various forms of energy into thermal energy. But it is difficult to convert thermal energy into other useful forms.

Suppose, for example, that a litre of gasoline is placed in the gas tank of a car. Where does the chemical energy in the gasoline go? It is first burned in the engine to produce heat. Some of the heat is carried to the environment by the car's cooling and exhaust systems. Some is converted into kinetic energy in the pistons and other moving parts of the car. This kinetic energy gradually becomes heat energy in the cylinders, tires, brakes, and air molecules. Eventually all of the chemical potential energy in the fuel is thermal energy. This thermal energy is so widely spread in the environment that it is impossible to get it back. Imagine the task of collecting all of the heat energy produced during a trip of 200 km!

All forms of energy, through use, eventually become thermal energy. The chemical energy in foods and fuels becomes thermal energy. The kinetic energy of the moon is gradually becoming thermal energy. The nuclear energy in uranium is becoming thermal energy. This thermal energy is so widely spaced that it is difficult to harness. This is one more reason for conserving the chemical energy the sun has given us. It makes sense to harness the energy in the tides, the sun, and the wind. We might as well use these forms of energy and save the chemical energy stored in fossil fuels for other uses. Besides, the energy from tides, the sun, and the wind is non-polluting.

The study of the various forms of energy and the change from one form to another has led man to an important law of science. This generalization, called the **Law of Conservation of Energy**, is stated as follows: *Energy cannot be created or destroyed; it can be changed from one form into another, but the total amount of energy in the universe is constant.*

Discussion

1. **a)** Why is the blood in arteries brighter red than the blood in veins?
b) Explain why the blood in the veins of a person living in the tropics is brighter red than the blood of a person living in a cold climate.
c) What conclusions did Mayer make based on his observations of the difference in colour of venous blood in the two climates?

2. **a)** Why were Joule's conclusions accepted more readily than Mayer's?
b) List the conversions in mechanical energy that take place after the mass in Figure 13-13 is released.
c) How many joules of mechanical energy are converted into thermal energy if a mass of 1 kg falls through a height of 1 m?
d) What conclusions did Joule make, based on his quantitative experiments converting different forms of energy to thermal energy?

3. **a)** Define the calorie.
b) Why is the calorie an outdated unit of energy?
c) What is the relationship between the calorie and the joule?
d) Explain why shaking water in an insulated bottle raises its temperature.

4. **a)** What eventually happens to all forms of energy as they are used?

b) A meteor enters the earth's atmosphere with a speed of 3 600 km/h. What happens to its kinetic energy?

c) Is it possible to cool the kitchen by leaving the refrigerator running with the door open? Discuss.

5. **a)** A 100 W immersion heater is placed in an insulated container holding 250 g of water at 10°C. What is the minimum length of time needed for the water to reach a temperature of 30°C?

b) What assumption did you make to do the problem? Is this a valid assumption?

6. The more the human body exercises, the more energy it uses. A person shovelling snow for an hour uses about 2 710 kJ of energy. The same person using a snow-blower uses 1 115 kJ of energy. But, in addition, the snowblower uses 0.2 kg of gasoline.

a) How much energy is consumed by the man and the machine when the snowblower is used? Assume that 1 kg of gasoline is equivalent to about 50 000 kJ of energy.

b) Trace the energy conversions that take place in both cases until the initial chemical energy becomes thermal energy.

c) Should a snowblower be used to remove snow from small sidewalks and driveways? Discuss.

7. **a)** One kilogram of fat is the equivalent of about 30 000 kJ of energy. If you take in 500 kJ more energy per day than you burn off through physical activity (about 20 g of potato chips), how many kilograms will you gain in 36 d? in 1 a? in 10 a?

b) Outline why it is important to count your joules daily.

8. Exercising "burns" off excess kilograms of fat. A sedentary male needs no more than 10 000 kJ of energy from food daily. An active farmer requires about 15 000 kJ. Women need 6 700 to 12 000 kJ depending on the activity. An exercise program is based on the energy used to do various exercises. Table 91 summarizes the energy cost of some activities.

TABLE 91 Energy Cost of Activities

Activity	Energy cost for a 70 kg individual (kJ/min)
Sitting in a chair	5.5
Walking at 6.6 km/h	21.2
Riding a bicycle	34.5
Swimming	47.0
Running	81.5

a) Assume that you are carrying 5 kg of excess fat. How many days of running for 20 min/d would it take to get rid of the excess fat? Assume that you don't go on a diet or change your eating habits in any way. Remember that using up 1 kg of fat requires about 30 000 kJ of energy.

b) Explain why both exercising and dieting are essential components of a program for losing excess mass.

Highlights

The Caloric Theory described heat as an invisible elastic fluid. This theory was discarded because it could not explain the results of experiments performed by Thomson and Davy. The Kinetic Molecular Theory is able to explain all heat phenomena. Thermal energy is the total of the elastic potential energy and the kinetic potential energy an object has because of the random motion of its molecules. Heat energy is thermal energy being transferred from one object to another. The quantity of heat energy transferred is equal to the product of the mass, the temperature change, and the specific heat capacity of the substance. Specific heat capacity is the quantity of heat needed to raise the temperature of 1 kg of a substance through a change in temperature of 1 K or 1 °C. Specific heat capacity is measured in joules per kilogram kelvin, J/(kg·K), or joules per kilogram degree Celsius, J/(kg·°C). In general the specific heat capacity is smaller for substances with more massive atoms.

Temperature is a measure of the average kinetic energy of molecules. According to the Principle of Heat Exchange, when two substances are mixed, the heat gained by the substance at the lower temperature is equal to the heat lost by the substance at the higher temperature.

The heat energy required to change the state of 1 kg of a substance without changing its temperature is called the specific latent heat. The unit of specific latent heat is the joule per kilogram (J/kg). The heat energy produced when 1 kg of an organic substance, such as food or a fuel, is burned is called the specific heat of combustion. The specific heat of combustion is measured in kilojoules per kilogram (kJ/kg).

Mechanical energy can be converted entirely into thermal energy. All forms of energy eventually become thermal energy. Unfortunately thermal energy cannot be converted completely into mechanical energy.

UNIT

14 Light Energy

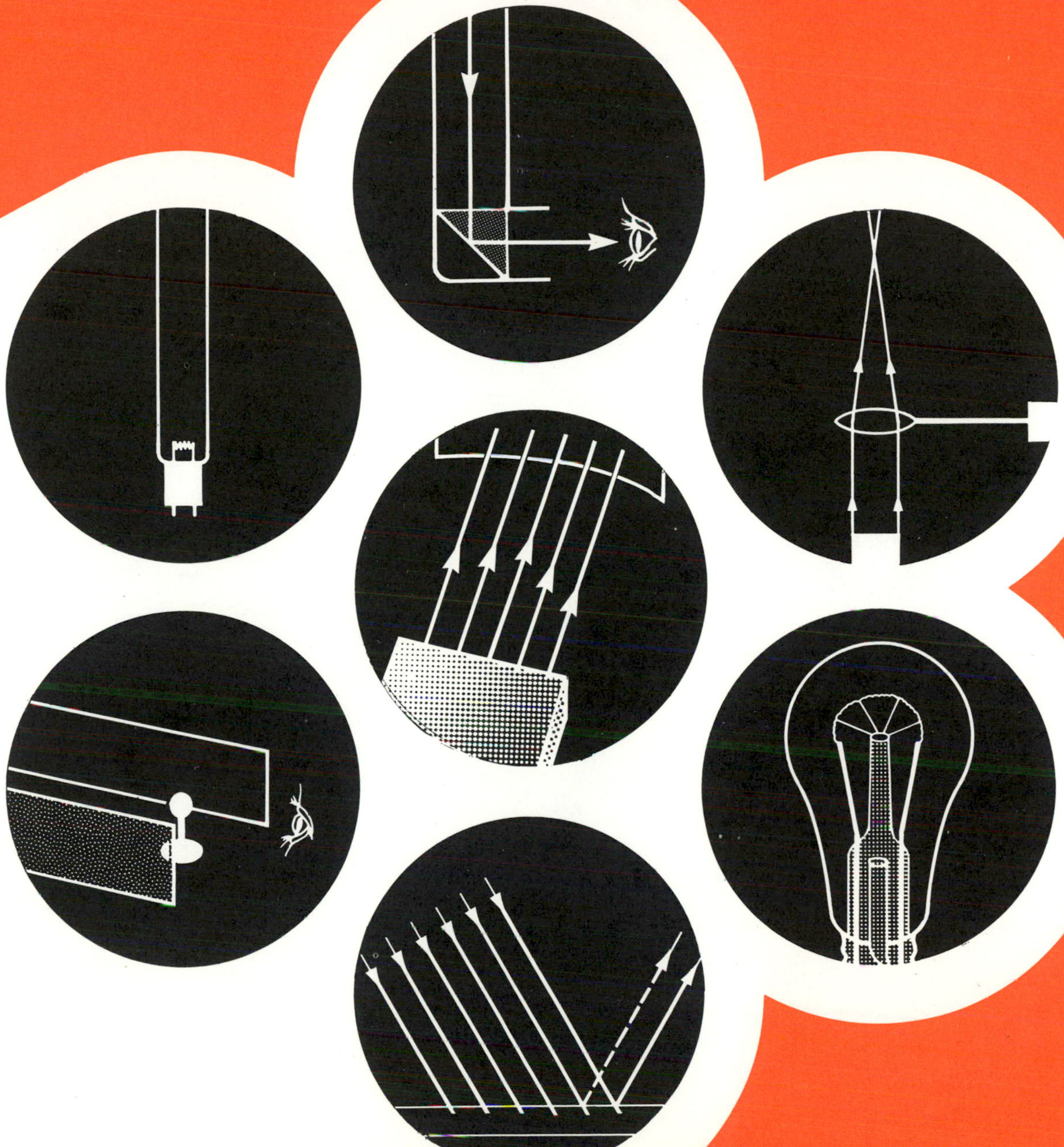

Whether you are looking through a telescope at a giant star or whether you are looking through a microscope at a tiny cell, what you see is the result of light coming from that object to your eye. What is light? How does it get from an object to your eye? What happens to the light after it enters your eye? How do the mirrors and lenses in telescopes and microscopes work? What gives an object its colour?

In this unit you seek answers to questions like these. However, do not expect to find exact answers to all such questions. Light has been studied by scientists for hundreds of years, yet an exact description of what light is has not yet been found.

A. Nature and Transmission

14.1 Sources of Light

There are two kinds of light sources, luminous and non-luminous. **Luminous sources** are those objects that actually produce the light that they give off. The sun, the stars, glowing lamps, and fireflies are examples. **Non-luminous sources** are those objects that can be seen only after light from a luminous source has been reflected from them. The moon, people, and books are examples. In a totally darkened room, non-luminous objects cannot be seen. For example, you cannot see another person sitting in a chair in a totally darkened room. However, if a lamp is turned on in the room, you can see the person. Light from the lamp, a luminous source, reflects off the person, a non-luminous source, making the person visible (Fig. 14-1).

Figure 14-1 The lamp is a luminous object. The person, book, and chair are non-luminous objects.

Whether an object is luminous or non-luminous depends on two factors. These are: the substance of which it is made and the conditions to which it is exposed. For example, a candle is non-luminous until it is lit; a lamp is non-luminous until it is turned on.

A non-luminous object often becomes luminous because of an energy conversion. That is, some other form of energy changes into light energy. Here are some examples of such energy conversions.

Chemical Potential Energy

A common source of light energy is chemical potential energy. When a fuel such as gasoline, fuel oil, or wood is ignited, light energy as well as heat energy are given off. The chemical potential energy stored within the molecules of the fuels is converted into light and heat energy as burning occurs.

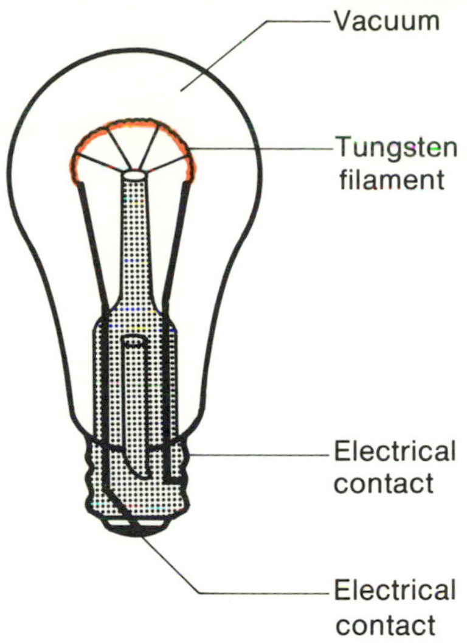

Figure 14-2 An incandescent lamp. Electrical energy flowing through the tungsten filament is changed into heat and light energy.

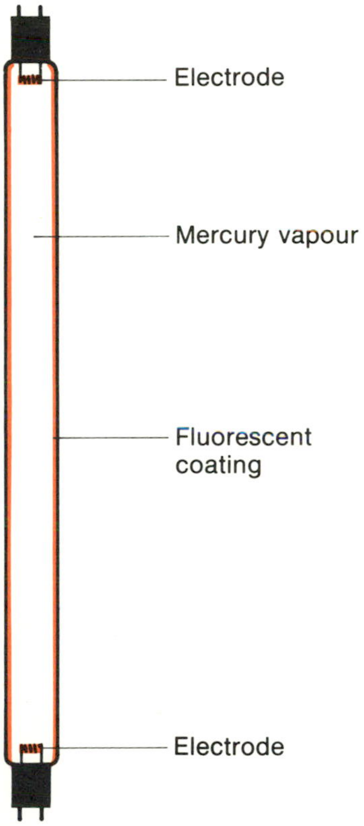

Figure 14-3 A fluorescent lamp is much more efficient than an incandescent lamp in converting electrical energy into light energy.

Heat Energy

When a piece of steel is heated with a welding torch to a temperature of about 800°C, it gives off red light. We say it is "red hot". If the steel is heated to about 2 500°C, it gives off white light. In this case we say it is "white hot". In both cases, some of the heat energy given to the steel by the welding torch is changed into light energy. When an object like the steel gives off light as a result of being heated, it is said to have been heated to **incandescence**.

Electrical Energy

An ordinary household "light bulb", properly called an **incandescent lamp**, converts electrical energy into light energy (Fig. 14-2). An electrical current is passed through a fine wire filament that is made of the metal, tungsten. Tungsten is used because it has a high melting point (3 400°C) and, therefore, will not melt easily when it gets hot. As the electrons in the electrical current move through the filament, some of their electrical energy is changed into heat energy. The heat energy warms the filament to incandescence, and light energy is given off. Unfortunately, incandescent lamps are energy wasters. Only about 5% of the electrical energy that enters an incandescent lamp is changed into light energy. The rest is given off as heat energy. You have probably noticed that an incandescent lamp becomes very hot when it is used.

A **fluorescent tube** (Fig. 14-3) is much more efficient than an incandescent lamp. It changes about 20% of the electrical energy into light energy. If you have ever felt a fluorescent tube that is operating, you will have noticed that very little heat is produced.

A fluorescent tube has an electrode at each end. When the tube is connected to an electrical circuit, electrons leave the electrodes. As the electrons move back and forth in the tube, some of them collide with mercury atoms that make up the mercury vapour in the tube. These collisions cause ultraviolet radiation to be formed. When the radiation strikes the fluorescent coating on the inside of the tube, light is given off.

The coloured signs that are used for advertising produce light in a manner similar to that of the fluorescent tube. The colour produced depends on the nature of the gas in the tube and on the nature of the fluorescent coating. For example, the signs that glow red, commonly called neon signs, are filled with a gas called neon. No fluorescent coating is used. When electrons strike the atoms of neon gas, some of their kinetic energy is changed into red light.

The picture tube of a television set directs a beam of electrons against a coating on the inside of the tube. Here, too, some of the kinetic energy of the electrons is changed into light energy.

A picture is formed due to a complex arrangement of electron beams and fluorescent materials.

Nuclear Potential Energy

Both **fission** (the splitting of atomic nuclei) and **fusion** (the joining of atomic nuclei) involve the changing of potential energy stored in the nuclei of atoms into other forms of energy. When a nuclear bomb is exploded, large quantities of light energy, as well as heat energy, are given off.

What is Light?

You have just seen how four different kinds of energy can be changed into light energy. The Law of Conservation of Energy suggests that the reverse might also be true; that is, light energy can be changed into other forms of energy. The solar cells which power many of the artificial satellites that now circle the earth change light energy into electrical energy. Many camera exposure meters also change light energy into electrical energy.

Since other forms of energy can be changed into light and light can be changed into other forms of energy, we can assume that light is a form of energy. The one thing that distinguishes light energy from all other forms of energy is the fact that it can be detected by the eye. Thus light can be defined in this way: **Light** *is that form of energy which the eye can detect*. This definition is called an operational definition, since it permits us to distinguish light from other forms of energy. It is not a true definition since it does not tell us what light really is.

Discussion

1. **a)** Distinguish between a luminous and a non-luminous light source. Give two examples of each source other than those mentioned in this section.
b) In outer space the stars and the sun can be seen at the same time. The sky around them is completely black. Account for these facts.

2. Give an example that is not described in this section which shows that chemical potential energy can be changed into light energy.

3. **a)** What is meant by the term "incandescence"?
b) Describe an event that you have seen which shows that heat energy can be changed into light energy.

4. **a)** Explain how an incandescent lamp changes electrical energy into light energy.
b) Explain how a fluorescent lamp changes electrical energy into light energy.

c) An incandescent lamp is much less efficient than a fluorescent lamp. In spite of this, most homes are lit by incandescent lamps. Why do you think this is so?

5. a) What is nuclear potential energy?

b) What evidence have you seen which shows that nuclear potential energy can be converted into light energy?

6. a) State an operational definition of light.

b) Why is this definition called "operational"?

14.2 Transmission of Light

One interesting characteristic of light is the fact that it can move from one place to another through a vacuum. Unlike sound and many other forms of energy, it does not require a material medium for its transmission. The best evidence for this is the observation that light travels from stars that are billions of kilometres away to the earth through the near-perfect vacuum of outer space. Also, if you look at a vacuum tube of a television set or radio, you can see the filament in the tube. Light must have left the filament and travelled through the vacuum of the tube to your eye or you would not be able to see the filament.

Types of Media

Although light does not require a material medium for its transmission, it can travel through many substances. Thus substances are often classified according to their ability to transmit light. Some substances, such as glass, air, and water, transmit light so well that you can see through them. They are called **transparent media**. Other substances, such as frosted glass, a clay suspension in water, and dense smoke in air, transmit some of the light that falls on them. But in doing so, they disperse or scatter the light so that you cannot see a clear image through them. Such substances are called **translucent media**. Still other substances, such as steel, wood, and black plastic, transmit no light at all. They are called **opaque media**.

Rectilinear Propagation of Light

"Rectilinear propagation" is the term used by scientists to describe the fact that light travels in straight lines. Even though you may feel that you do not know that light travels in straight lines, many of your actions reveal that you assume that it does. For example, if someone asks you to show them where a certain object is, you usually point to the object. By doing so, you are really just showing the person the direction from which light came to your eyes from the object. You have assumed that the object is in the direction from which the light came and that the light came straight to your eyes from the object. Here is another example: You do not expect to be able to see around a corner. However, if light travelled in a curved path, you should be able to see around a corner. Apparently the light travels in straight lines.

Figure 14-4 represents a projector emitting a beam of light into dusty air. You have likely seen this evidence that light travels in straight lines. The path followed by the light is represented by a straight line called a **ray** and also by a bundle of rays called a **beam**. If the projector is adjusted so that the rays come out of the projector parallel to each other, the resulting beam is called a **parallel beam** (Fig. 14-5). If a hand lens ("magnifying glass") is inserted into the parallel beam, the rays come together or converge. The resulting beam is called a **converging beam**. Once the rays have converged on a point, they spread out again or diverge. Now the beam is called a **diverging beam**. Review the meanings of the terms in Figure 14-5 as your teacher demonstrates this behaviour of light for you.

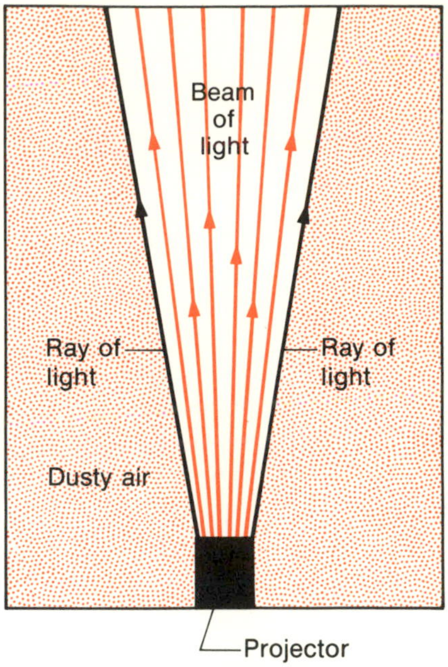

Figure 14-4 The beam of light coming from a projector shows that light travels in straight lines.

Formation of Shadows

If an opaque object such as a book is placed in front of a point source of light as shown in Figure 14-6, A, some of the diverging beam of light is blocked by the object. Only those rays beyond the ends of the object continue past the object. Thus a dark shadow is cast behind the object. This shadow region, called the **umbra**, receives no light from the source.

If an opaque object is placed in front of a large source of light, two shadow regions are formed as shown by the ray diagram in Figure 14-6, B. The dark central region which receives no light from the source is still called the umbra. Surrounding it is a region that is a partial shadow because it receives some light from the source. This region is called the **penumbra**.

During an **eclipse** of the moon (Fig. 14-7) both an umbra and a penumbra are formed. When the moon is totally within the umbra, as it is at A, it is said to be in **total eclipse**. When the moon is only partly within the umbra, as it is at B, it is said to be in **partial eclipse**.

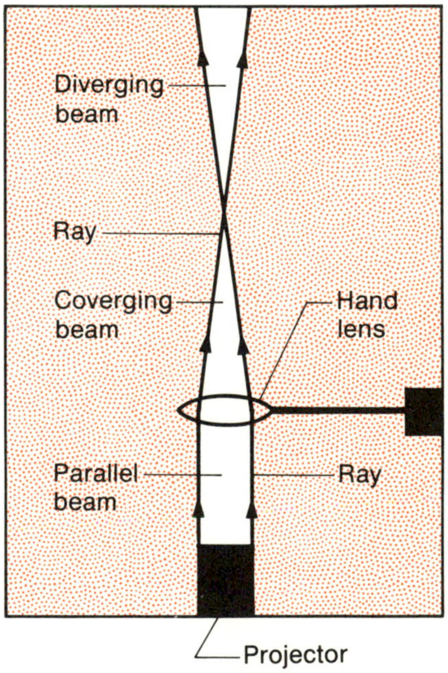

Figure 14-5 Chalk dust in the air in front of a projector shows what happens to a beam of light that is passed through a hand lens.

Discussion

1. a) Give an example, other than the ones discussed in this section, of the fact that light does not require a material medium for its transmission.

b) Describe any evidence you have which proves that sound requires a material medium for its transmission.

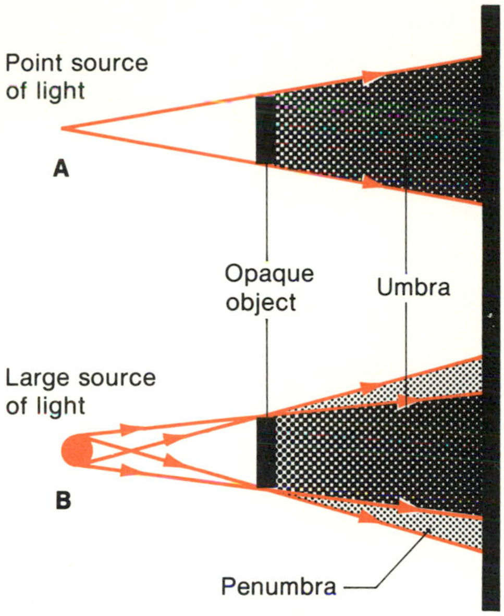

Figure 14-6 The shadows formed when a point source (A) and a large source (B) shine light on opaque objects of the same size.

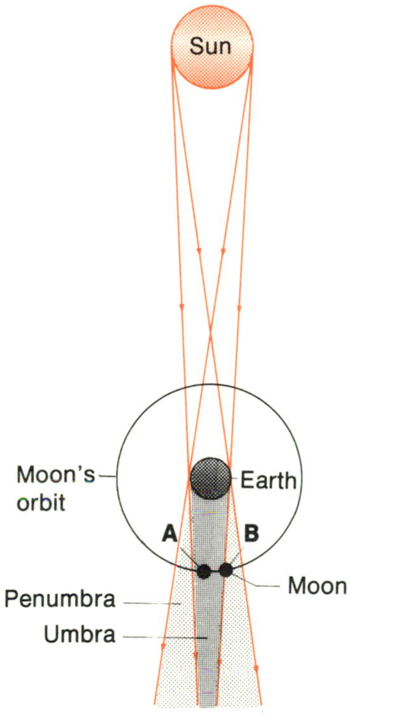

Figure 14-7 The moon is in total eclipse at A and in partial eclipse at B. (The diagram is not drawn to scale.)

2. a) Name and describe the 3 types of media that might be met during studies of light.
b) Give an example, other than the ones discussed in this section, of each type of medium.

3. a) What does the term "rectilinear propagation" mean?
b) What evidence do you have that light travels in straight lines?

4. Define these terms: ray, beam, parallel beam, converging beam, diverging beam.

5. a) Explain the terms umbra and penumbra.
b) Describe how a total and a partial eclipse of the moon occur.

6. Design a demonstration of a total and a partial eclipse of the moon. Try it to make sure it works. Write a description of your demonstration.

7. Consult an encyclopedia or other reference book to find out how a total and a partial eclipse of the sun occur. Prepare a report consisting of a diagram and a half-page description for submission to your teacher.

14.3 INVESTIGATION: Intensity and Distance from the Source

When you want better illumination of a book that you are reading, you move the book closer to the source of light. By so doing, you are assuming that the light intensity (brightness) increases as the distance from the light source decreases. One purpose of this investigation is to see if that assumption is true. A second purpose is to figure out how far light will travel from a source.

Materials

slide projector
light meter
metre stick

Procedure

a. Place a slide projector at the back of a completely darkened classroom. Project a beam of light toward the front of the room. Adjust the lens of the projector to form a beam that diverges as much as possible.
b. Stand in the beam of light and face the projector.
c. Use the light meter to measure the light intensity at a distance of 1 m from the projector lens.
d. Repeat step c for distances of 2 m, 3 m, 4 m, and 5 m. Record all of your results in tabular form.

e. Plot a graph with distance on the horizontal axis and light intensity on the vertical axis. Use the long side of the graph paper for the horizontal axis.

Discussion

1. Do your results support the assumption that was stated in the introduction to this investigation? Explain.
2. How far do you think light can travel before it can no longer be seen? Hint: Extrapolate your graph to find the distance at which the intensity would be zero.
3. Figure 14-8 shows light diverging from a point source and passing through a square hole in A. A screen is placed at B to catch the light. Note that B is twice as far from the source as A. Compare the area that is covered by light on B with the area of the hole in A. The screen is then moved to C which is three times as far from the source as A is. Compare the area on C that is covered by light with the area of the hole in A. If the screen were now moved to position D which is four times as far from the source as A is, how would the area covered by light compare with the area of the hole in A? What is the mathematical relationship between intensity and distance?

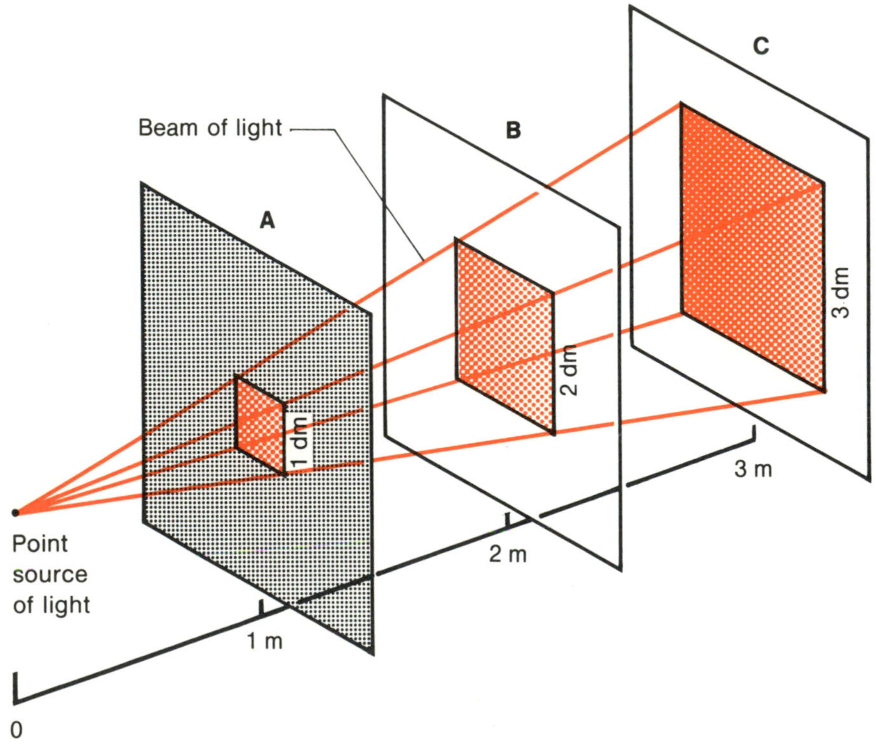

Figure 14-8 What is the mathematical relationship between intensity and distance?

14.4 INVESTIGATION: The Pinhole Camera

The pinhole camera is a simple application of the rectilinear propagation of light. It consists of an opaque box with a pinhole at one end and a translucent screen at the other end (Fig. 14-9). If the pinhole is pointed at a bright object, an image forms on the translucent screen.

In this investigation you use the assumption that light travels in straight lines to predict the nature of the image in a pinhole camera. You then check your predictions experimentally.

Three characteristics are commonly used to describe an image:

a. Size—Is the image *larger* or *smaller* than the object?
b. Attitude—Is the image *erect* or *inverted*?
c. Kind—Is the image *real* or *imaginary*? (A real image can be caught on a screen; an imaginary one cannot.)

Materials

pinhole camera or materials to make one (see Procedure A)
bright light source like a candle, lamp, or window

Procedure A Making a Pinhole Camera

If a pinhole camera is not available, make one as follows:

a. Obtain an opaque box such as a shoe box or a large tissue-dispensing box.
b. Remove a rectangular section from one end of the box. Replace it with a piece of oiled paper, waxed paper, or other translucent material.
c. Cut a hole 2-3 cm in diameter in the other end of the box. Cover this hole with aluminum foil. Make a pinhole in the centre of the foil with a fine pin.
d. Make sure all of the box except the pinhole and translucent screen is light tight.

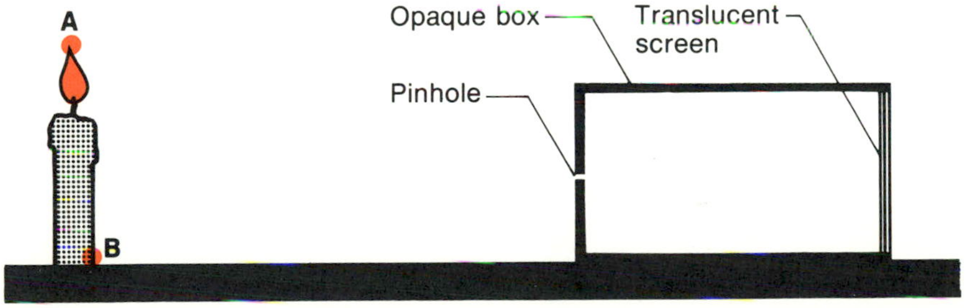

Figure 14-9 What are the characteristics of the image of the candle on the screen?

Procedure B Characteristics of the Image

a. Use the assumption that light travels in straight lines to predict the characteristics of the image of a candle placed in front of a pinhole camera as shown in Figure 14-9. You can do this as follows: Copy Figure 14-9 into your notebook. Assume that rays of light leave point A on the candle and travel in all directions. Draw the ray that will enter the box and strike the screen. Draw a similar ray from point B. Now write in your notebook your prediction of the characteristics of the image (size, attitude, and kind).

b. Place the pinhole camera 2-3 m from a lit candle or incandescent lamp. Note the characteristics of the image (size, attitude, and kind).

c. Gradually move the camera closer to the object. Note the characteristics of the image and how they change as the camera gets closer and closer to the object.

d. Move the camera further away from the object. Note the characteristics of the image and how they change as the camera gets further and further from the object.

Discussion

1. Compare your predictions regarding the characteristics of the image with your observations. Does this comparison support the assumption that light travels in straight lines? Explain your answer.
2. What happens to the size of the image when you change the distance from the camera to the object?
3. Under what circumstances is the image larger than the object?
4. Fairly good photographs of distant objects can be obtained by replacing the translucent screen with photographic film. However, the exposure time must be very long since the tiny pin hole lets in only a small amount of light. What do you think would happen to the image if the pin hole were enlarged? Explain your answer using a diagram similar to the one you drew in step a of Procedure B.
5. A regular camera has a large opening to let in more light than a pin hole. How is it equipped to overcome the problem that occurs when the opening is larger than a pin hole?

14.5 The Speed of Light

You learned in Unit 11 that speed $= \frac{\text{distance}}{\text{time}}$.

Therefore, to calculate the speed of light, one must find out how long it takes light to travel a certain distance. Galileo (1564–

1642) was the first person to try this. One night he stood on a hilltop with a lantern, and a second person with a lantern stood on a second hilltop about 2 km away. Galileo flashed his lantern at the other person. As soon as this person saw the flash, he sent a flash from his lantern back to Galileo. Galileo tried to measure the time required for the light to travel the 4 km. His attempt failed because the time required for light to travel 4 km is much less than the reaction time of the two men. Galileo repeated the experiment several times, using greater and greater distances between the two people, but he could not measure the short time needed by light to travel these distances. The speed of light has since been found to be about 3×10^8 m/s. In other words, light travels 3×10^8 m or 3×10^5 km in just 1 s. At this speed, light can travel from the sun to the earth, a distance of 148 808 000 km, in about 8 min. If it could be bent, it would travel around the earth over 7 times in just 1 s. Knowing this, it is not surprising that Galileo had problems!

Although Galileo's experiments were unsuccessful, they did reveal that the speed of light could be measured only if very long distances were used or if some means could be developed for measuring short time intervals. Roemer's method used long distances. Michelson's method used short time intervals.

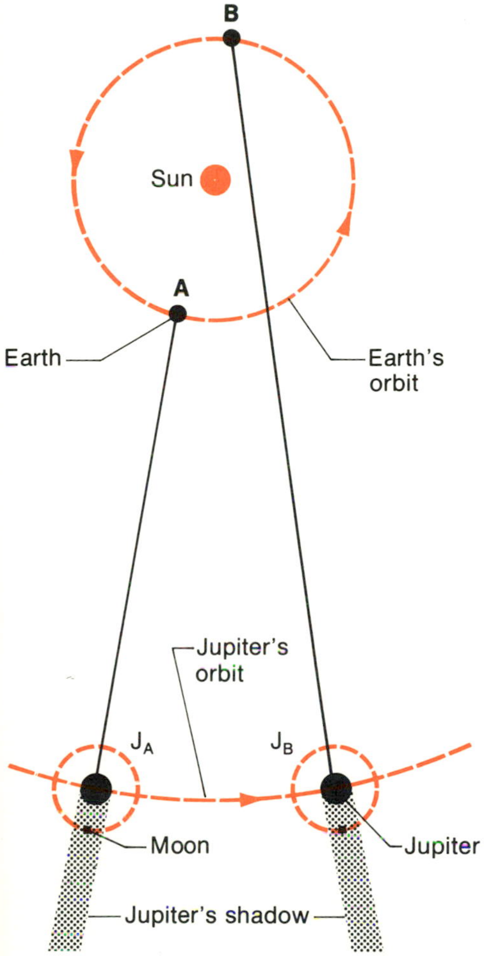

Figure 14-10 Roemer's method for finding the speed of light.

Roemer's Method

In 1676 Olaf Roemer measured the speed of light by a method that used long distances. One of Jupiter's moons takes 42 h to complete an orbit around the planet. Thus an eclipse of this moon takes place once every 42 h. Roemer measured the time between successive eclipses when the earth was at A (see Figure 14–10). Jupiter and its moon were at J_A. He continued to measure the time between successive eclipses for 6 months. The earth was now at B and Jupiter and its moon were at J_B.

Roemer discovered that, as the earth moved from A to B, the time interval between successive eclipses gradually became larger. Finally, when the earth was at B, the time interval was 1 300 s larger than it was when the earth was at A. Apparently it took light 1 300 s to go from A to B, the diameter of the earth's orbit. This distance was thought to be about 3×10^8 km. Thus the speed of light would have been $\frac{3 \times 10^8 \text{ km}}{1\,300 \text{ s}} = 2.3 \times 10^5$ km/s. Roemer's measurement of the time was not very accurate nor was the supposed diameter of the earth's orbit correct. Therefore his answer is not correct. However, he did prove for the first time that light required time to travel from one point to another.

More recent studies have shown that the diameter of the earth's orbit is 1.47×10^8 km and that the time required for light to cross the earth's orbit is 980 s. Thus the speed of light is

$$\frac{1.47 \times 10^8 \text{ km}}{980 \text{ s}} = 3.0 \times 10^5 \text{ km/s}.$$

Michelson's Method

Albert Michelson was the first person to measure the speed of light by developing a way of measuring short time intervals. His ingenious apparatus is shown in Figure 14-11. For his efforts, Michelson became, in 1907, the first American scientist to be given a Nobel prize.

Michelson mounted an octagonal (8-sided) mirror on a revolving axis. He set up this apparatus on Mt. Wilson in California. He then shone a bright beam of light on side A of the mirror. This beam was reflected to a curved mirror on Mt. San Antonio several kilometres away. The curved mirror reflected the beam back to side C of the octagonal mirror. When the octagonal mirror was exactly in the correct position, the light was reflected into a telescope where an observer could detect it. Michelson gradually increased the speed of the octagonal mirror until he saw flashes of light in the telescope. He now knew that, every time the mirror rotated ⅛ of a turn, the light travelled from the octagonal mirror to the curved mirror and back to the octagonal mirror. Thus the time required for light to travel to the curved mirror and back was ⅛ of the time required for the octagonal mirror to make one complete turn. Michelson then accurately measured both the distance between the mirrors and the speed of rotation of the octagonal mirror. From these two measurements, he calculated the speed of light.

Over the years Michelson improved this apparatus. By 1924 he was using a distance of 35 km between the mirrors. After his death, his co-workers continued his experiments. The last one involved measuring the speed of light in a vacuum. For this purpose the scientists used a vacuum tube 1.6 km long.

The best estimate of the speed of light in a vacuum is now $2.997\,928 \times 10^5$ km/s. In air at sea level, the speed is about 70 km/s slower. The speed in water is about 0.7 of the speed in a vacuum. In glass the speed is about 0.6 of the speed in a vacuum.

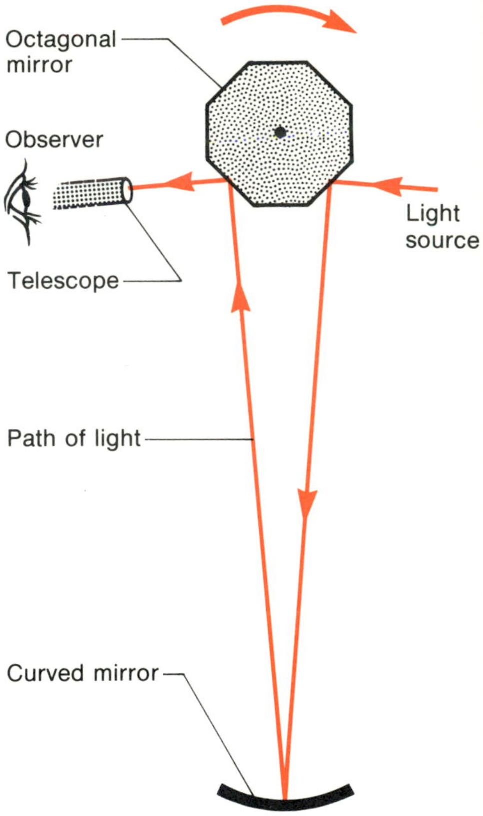

Figure 14-11 Michelson's apparatus for finding the speed of light.

The Light Year

Since the speed of light is such a large number, astronomers have used it to create a unit for measuring the large distances found between the stars and galaxies. This unit is called a light year. One **light year (ly)** is the distance that light travels in one year. 1 ly = 9 460.5 Tm, or about ten thousand billion kilometres.

The nearest star to us, α Centauri, is 4.3 ly away. This means that it takes light 4.3 a to travel from α Centauri to the earth, even though it travels at the incredible speed of 300 000 km/s. If α Centauri exploded today, we would not see the explosion for 4.3 a. A time interval of about 300 000 a would be required to make a round trip to α Centauri travelling at 30 000 km/s, which is more than the average speed of many rockets.

The brightest star in the sky, Sirius, is 8.7 ly away. The Great Spiral Nebula in Andromeda is 2 000 000 ly away. Ursa Major II, a galaxy that forms part of the "Big Dipper", is 2 425 000 000 ly away. Photographs have been taken of galaxies that are over 6 000 000 000 ly away.

Discussion

1. **a)** Describe Galileo's attempt to measure the speed of light.
   b) Why did this experiment fail?
2. Outline Roemer's method for measuring the speed of light.
3. Outline Michelson's method for measuring the speed of light.
4. **a)** State the speed of light in a vacuum in kilometres per second.
   b) Compare the speed of light in various transparent media. What do you think causes the differences in speed?
5. **a)** What is a light year?
   b) The star Sirius is 8.7 ly away. What does this mean?
   c) If Sirius vanished today, when would we first notice that this star was gone?
6. **a)** Imagine that an astronomer on a planet near Sirius has a telescope trained on the spot where you now are. What would the astronomer see?
   b) Would an astronomer in the Great Spiral Nebula in Andromeda see the same thing? Why?

14.6 Theories of Light

Light travels in straight lines. It travels at a very high speed but does not travel as fast in media such as air, glass, and water as it does in a vacuum. It will not travel at all through some media such as wood. Light can be reflected from mirrors and it can be refracted (bent) by lenses. What must light be made of in order to have such properties? This section takes a brief look at three theories that were developed to explain the properties of light. You can test the usefulness of these theories by using them to explain the results of the investigations you do in some of the following sections.

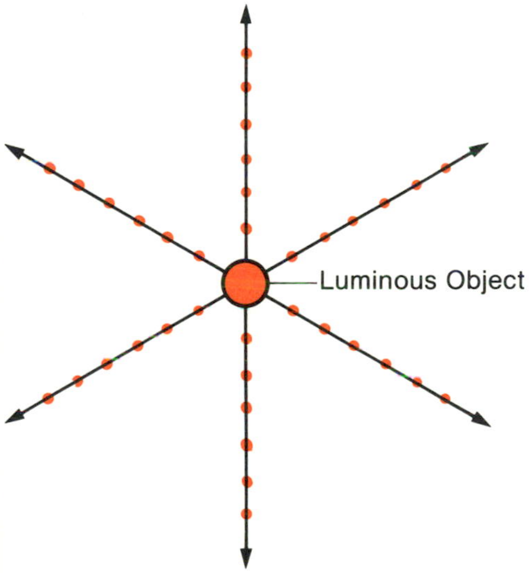

Figure 14-12 The particle theory states that light consists of tiny particles given off in all directions by a luminous object.

The Particle Theory

The particle theory of light says that light is made of a stream of tiny **particles** that are given off by a luminous object (Fig. 14-12). These particles are given off at high speed and in all directions.

This theory explains many of the properties of light. Sir Isaac Newton (1642–1728), the founder of this theory, used it to explain rectilinear propagation. He reasoned that gravity would have no noticeable effect on the path of particles of low mass and high speed. Thus particles emitted by a luminous object would travel in straight lines. Newton also used this theory to explain reflection of light from smooth surfaces. He suggested that the particles of light would bounce off the surface in much the same manner that a ball bounces off a wall.

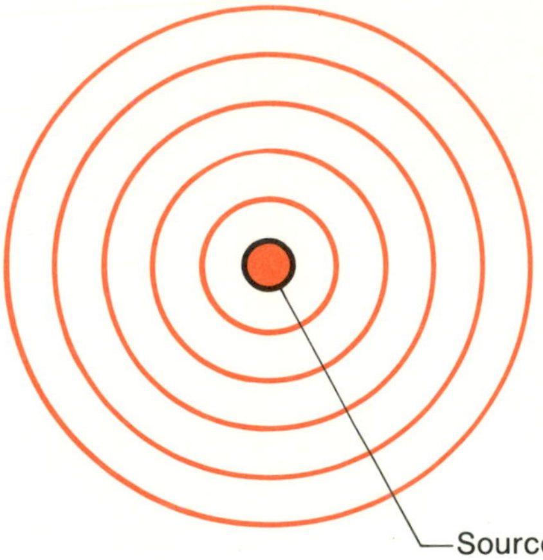

Figure 14-13 The wave theory states that light consists of waves that spread as concentric spheres from the source.

The Wave Theory

This theory was first proposed by Christian Huygens (1629–1695). It says that light consists of **waves** that spread out from the source in the form of concentric spheres much like the waves from a stone dropped into a lake (Fig. 14-13).

This theory also explains many of the properties of light. It has been used to explain reflection and refraction. In addition, it explains a phenomenon called interference which the particle theory cannot explain. By 1820 the particle theory had been laid aside in favour of the wave theory.

The Quantum Theory

In 1887 Heinrich Hertz discovered a phenomenon called the photoelectric effect. Like interference, this phenomenon is beyond your understanding at this time. However, you should note that the photoelectric effect could not be explained using the wave theory but could be explained using the particle theory. In 1901 Max Planck combined the wave theory and the particle theory and stated that light had a **dual nature**. He said that light consists of bundles of energy that can behave as either particles or waves. Planck called these bundles of light **photons** or **quanta**. This theory is widely accepted by scientists today.

Discussion

1. Compare the particle theory and the wave theory of light.

2. a) What is meant by the term "dual nature"?

b) What is a photon or quantum?

c) Why is the quantum theory more widely accepted than the particle theory or the wave theory?

d) Now that the quantum theory is widely accepted, is it proper to say that the other two theories are incorrect? Explain your answer.

B. Reflection of Light

Most objects that you can see are non-luminous. That is, you can see them because they reflect light to your eyes. Most non-luminous objects have rough surfaces. Such surfaces reflect light in a manner that reveals their shape, colour, and texture. Some non-luminous objects have smooth shiny surfaces that reflect light in such a way that images are formed. Such surfaces are called **mirrors**. The surfaces of polished metal and calm water are mirrors. But the mirror you know best is a piece of glass with a thin coating of silver on the back surface.

Reflecting surfaces are named according to their shapes. Those that are flat are called **plane**. The rest are **curved**. In the next few sections you study the properties of plane and curved surfaces.

14.7 INVESTIGATION: Reflection at Plane Surfaces

This investigation consists of two parts. You can do Procedure A at home. Its purpose is to develop the terms you need for the study of the reflection of light. It will also help you make a prediction of the First Law of Reflection. Procedure B will be done in class. Its purpose is to test your prediction of the Laws of Reflection.

Materials

plane mirror
flashlight
metre stick
powder
ray box plus accessories kit
thin protractor (preferably paper)

Procedure A The Terminology of Reflection

a. Place a plane mirror on the floor. Darken the room as much as possible. Shine a beam of light on the mirror at an oblique (slanting) angle as shown in Figure 14-14.

b. Blow some powder into the air above the mirror so you can see the beam of light and its reflection.

c. Hold a metre stick so it is perpendicular to the mirror at the point where the beam of light strikes the mirror. Compare the angle between the incoming beam and the metre stick with the angle between the reflected beam and the metre stick.

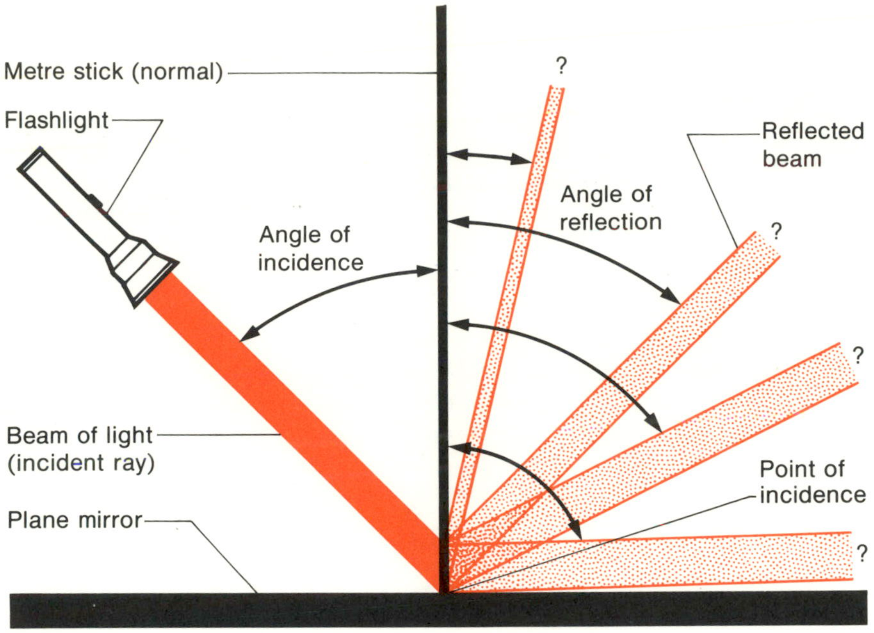

Figure 14-14 Reflection of light at a plane surface. Where will the reflected beam be?

d. Repeat this procedure using several different angles for the incoming beam.
e. Answer the questions in Discussion A before starting Procedure B.

Procedure B The Laws of Reflection

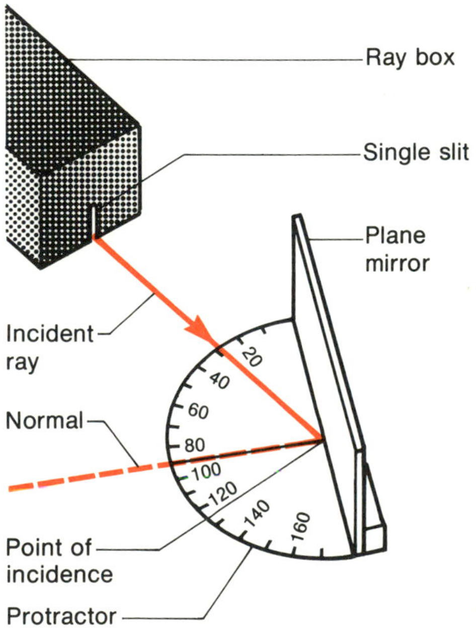

Figure 14-15 Using a ray box to study the Laws of Reflection for a plane mirror.

a. Stand a plane mirror vertically on a piece of paper on the desk. Place the protractor in front of the mirror as shown in Figure 14-15.
b. Direct a single ray of light from the ray box to the mirror so that the ray meets the mirror at the centre of the protractor scale. Draw the normal using a coloured pencil.
c. Record the angle of incidence and the angle of reflection in a table.
d. Move the ray box to get a different angle of incidence. Use the same point of incidence. Record the angle of incidence and the angle of reflection in your table.
e. Repeat step d for 3 more angles of incidence.
f. Replace the single slit of the ray box with a multiple slit. Adjust the ray box until the incident rays are parallel. Compare the angle of incidence with the angle of reflection for all of the rays.

Discussion A

The point at which a ray of light strikes a mirror is called the **point of incidence**. The ray from the source of light to the point of incidence is called the **incident ray**. The ray leaving the mirror is called the **reflected ray**. A line perpendicular to the mirror at the point of incidence is the **normal**. The **angle of incidence** is the angle between the incident ray and the normal; the **angle of reflection** is the angle between the reflected ray and the normal.

1. The First Law of Reflection compares the angle of incidence with the angle of reflection. Although you were not able to make accurate measurements in this investigation, you should be able to predict this law based on what you saw. What is your prediction?
2. A ball that bounces from a smooth surface obeys the First Law of Reflection provided it has no spin. Discuss the importance of this fact to a player of one of the following games: basketball, table tennis, tennis, billiards, handball.

Discussion B

1. State a generalization that compares the angle of incidence with the angle of reflection. This is the **First Law of Reflection**. Does it agree with your prediction in Part A of this investigation?
2. The **Second Law of Reflection** states that *the incident ray, reflected ray, and normal all lie in the same plane*. Describe any evidence you have from this investigation that supports this generalization. (Hint: Look up the meaning of the word "plane" if you do not know what it means.)
3. Do the Laws of Reflection hold when many rays strike the mirror at the same time?
4. Use one of the theories of light to explain reflection from a plane mirror and the Laws of Reflection.

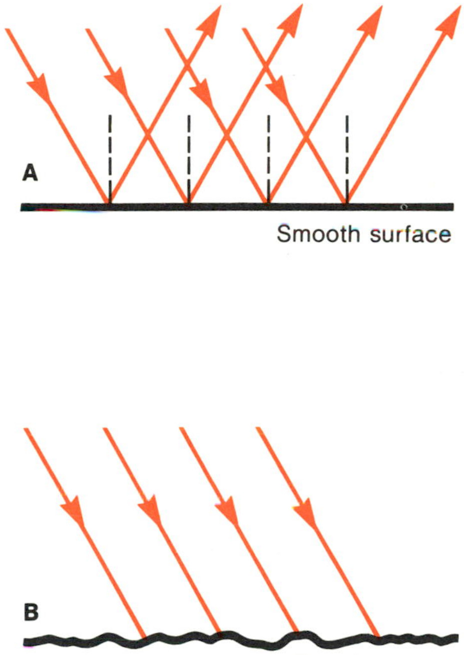

Figure 14-16 Comparing reflection from smooth and rough surfaces: A, regular reflection and B, diffuse reflection. Where will the reflected rays in B be located?

14.8 INVESTIGATION: Regular and Diffuse Reflection

In this investigation you compare reflection by smooth and rough surfaces.

You discovered in Investigation 14.7 that, when a parallel beam of light is shone on a smooth surface like a plane mirror, the light is reflected as a parallel beam. This behaviour is called **regular reflection** (Fig. 14-16,A). Regular reflection occurs because all incident rays in the parallel beam form equal angles of incidence with the smooth surface. As a result, the reflected rays all have equal angles of reflection. Therefore they leave the surface as a parallel beam. When the reflected beam meets a surface like a wall, it forms a spot about the size of the original light source.

Reflection from a rough surface is called **diffuse reflection**. Proceed with this investigation to find out how diffuse reflection differs from regular reflection.

Materials

ray box plus accessories kit
plane mirror
aluminum foil

Procedure

a. Draw a straight line across a sheet of paper. Stand a plane mirror on the line. Direct a single ray of light from the ray box to the mirror as you did in Investigation 14.7 (See Figure 14-15.)
b. Move the mirror back and forth along the line. Record your observations.
c. Repeat steps a and b using a piece of wrinkled aluminum foil instead of the plane mirror. The foil should have about the same dimensions as the mirror. Record your results.
d. Replace the single slit by a multiple slit. Adjust the ray box until the incident rays are parallel. Repeat steps a, b, and c. Record your results.
e. Look at your own image in both of the reflecting surfaces used in this investigation. Note any similarities and differences.

Discussion

1. Complete Figure 14-16, B in your notebook. Find the position of each reflected ray by first drawing a normal at each point of incidence. Then draw a reflected ray at each point of incidence by making the angle of reflection equal to the angle of incidence. Look up the meaning of the term "diffuse". Use your diagram to explain the term.
2. Account for the difference in the behaviour of the reflected rays when the mirror and then the aluminum foil were moved back and forth.
3. Use your completed Figure 14-16, B to explain the differences between your images in the two reflecting surfaces.
4. Describe the difference between regular and diffuse reflection.

14.9 Direct and Indirect Lighting

Regular reflection occurs at smooth surfaces like mirrors; diffuse reflection occurs at rough surfaces such as wrinkled aluminum foil, concrete, and plaster. When a parallel beam of light is shone on a rough surface, the reflected light is **scattered** or **diffused**. This occurs because the incident rays do not meet the reflecting surface with equal angles of incidence. Therefore the reflected rays leave with different angles of reflection. When the diffused light meets a surface like a wall, it forms a spot that is larger and less bright than the original source.

Direct Lighting

When direct lighting is used in a room, light is allowed to travel directly from a lamp to all parts of the room. If the lamp is made of transparent glass, an unpleasant glare may be produced. This happens because parallel rays leave the source and undergo regular reflection at smooth surfaces. To lessen the glare, most incandescent lamps used in homes are made of translucent glass. Such glass diffuses the parallel rays and spreads the light more evenly over the room. A softer, more pleasant effect is produced. Still further softening of the light is obtained by placing the lamp in a shade. As the light passes through the shade, it is diffused even more.

You discovered in Investigation 14.8 that rough surfaces scatter light. Therefore glare in a room can be reduced still further by making the walls and ceiling rough. This may be done by using unglazed wallpaper, non-gloss paints, and rough plaster.

Indirect Lighting

Many homes and offices are illuminated by indirect lighting. Light is not allowed to travel directly from a lamp to objects in the room. Instead, it is first reflected off rough surfaces in order to scatter it. For example, a translucent shade may be mounted below the lamp to reflect most of the light off the ceiling and walls. If the ceiling and walls have rough, non-gloss surfaces, the light is scattered. A uniform and restful effect is produced. The shade is normally translucent to allow direct lighting of the space under the lamp, thereby producing a brighter area in that part of the room.

Discussion

1. **a)** What can be done at a light source to reduce glare?
   b) What can be done at a reflecting surface to reduce glare?
2. Explain the difference between direct and indirect lighting.
2. **a)** Describe an example of direct lighting in your home or school.
   b) Describe three things that were done in your home to reduce glare.
4. **a)** Why does waxing and polishing of furniture make it shiny?

b) A white surface reflects most of the light that falls on it. Why does it not act as a mirror?
c) Examine a fluorescent lamp. How is it equipped to reduce glare?

14.10 INVESTIGATION: Images in Plane Mirrors

In this investigation you discover the location and characteristics of the image in a plane mirror. In order to do this, you must be able to see through the mirror, yet still see a reflection in it. Therefore a piece of dusty glass is used as a mirror. It transmits some of the light and reflects some of it.

Before you begin this investigation, try to answer these questions:

a. What terms are used to describe the characteristics of an image? (See Investigation 14.4, page 387, if you have forgotten.)
b. If you stand 2 m in front of a plane mirror, how far behind the mirror does your image appear to be?
c. If you wave your right hand in front of a plane mirror, what hand does your image appear to wave back at you?
d. How tall does a plane mirror have to be in order that a person 1.8 m tall can see his/her entire image?
e. What are the position and characteristics of an image in a plane mirror?

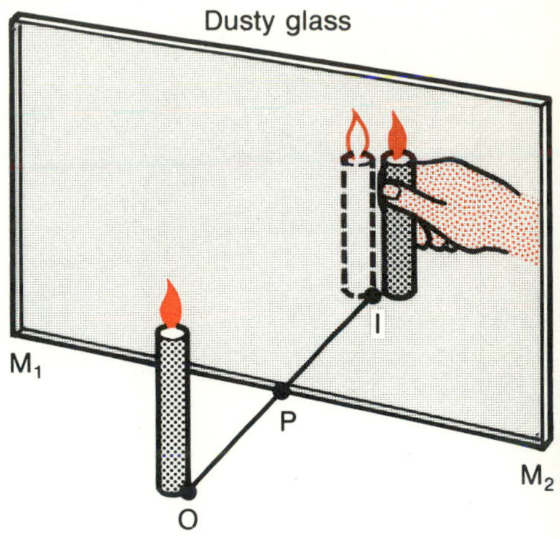

Figure 14-17 Locating the image of a candle in a plane mirror using the method of parallax.

Materials

piece of dusty glass
2 candles of the same size

Procedure

a. Draw a line M_1M_2 on a sheet of paper. (See Figure 14-17). Stand the piece of dusty glass on the line perpendicular to the paper.
b. Place a burning candle in front of the glass. This candle is the object. Label its position "O".
c. Light the second candle and place it behind the mirror so that it is exactly in the same place as the image of the first candle. Make sure that the second candle and the image are exactly in the same place by viewing them from several directions. This method of finding the position of the image is called the **method of no-parallax**. (**Parallax** *is the apparent motion of an object or an image with respect to another object as a*

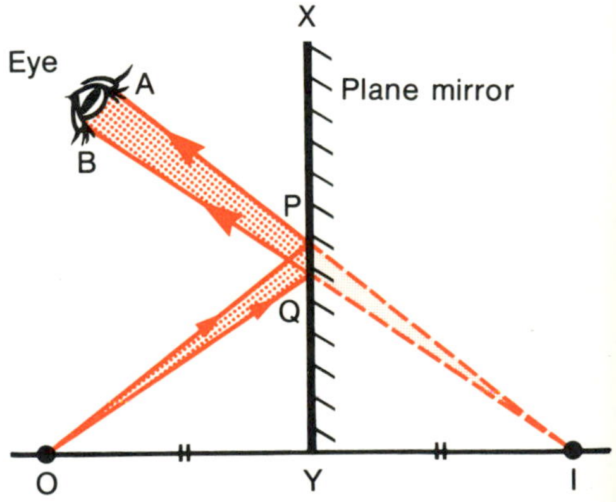

Figure 14-18 How the eye sees the image of a point in a plane mirror. Can you explain this optical illusion?

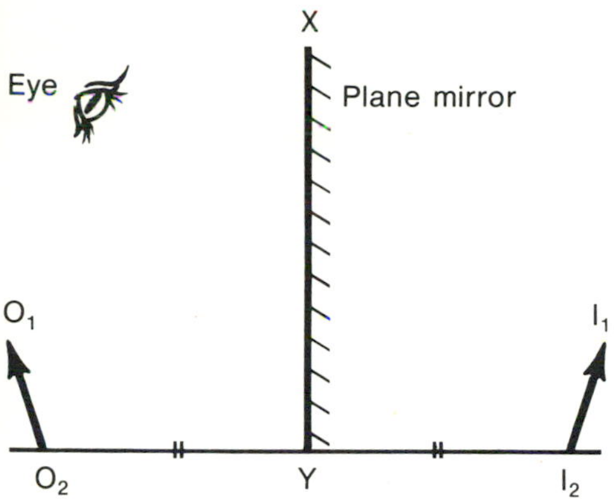

Figure 14-19 How does the eye see the image of an object in a plane mirror?

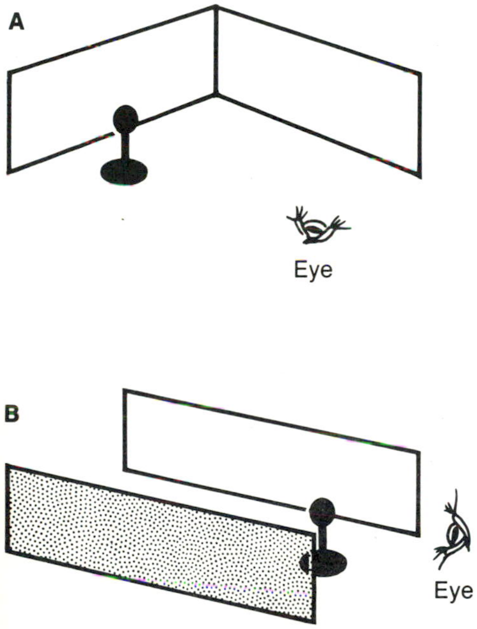

Figure 14-20 Reflections in two mirrors: A, two mirrors at right angles and B, two mirrors parallel to one another.

result of a change in the position of the observer.) Label the position of the image "I".

d. Replace the second candle with a sheet of paper. Try catching the image on the sheet of paper. Then try to get rid of it by inserting a piece of opaque material like a book between the image position and the mirror. Also, try to enclose the image in an opaque container like a can.

e. Remove the mirror. Join the points O and I with a straight line. Mark the point P.

Discussion

1. How long are the lines OP and IP? How are the lines OI and M_1M_2 related? Make a generalization that describes the **location** of the image in relation to the object and the mirror.
2. What is the **attitude** of the image?
3. What is the **size** of the image relative to the object?
4. What kind of image is formed in a plane mirror? Use your results from step d of the Procedure to explain your answer.
5. Figure 14-18 is a **ray diagram** that shows how the eye sees the image of a point source of light in a plane mirror. The object, O, sends out rays of light in all directions. Only those rays between OPA and OQB enter the pupil, AB, of the eye. These rays leave O and meet the mirror between P and Q. Then they are reflected to the eye. However, the eye is not able to detect the change in direction of the rays. Therefore it sees an image, I, behind the mirror which seems to send rays of light in straight lines to the eye. An **optical illusion** is formed; no light actually comes from behind the mirror.

   Now look at Figure 14-19. The point source, O, has been replaced by a true object, O_1O_2. Draw a ray diagram to show how the eye sees the image, I_1I_2, of this object. Hint: Draw ray diagrams for each of the points O_1 and O_2.
6. Use the ray diagram that you drew in question 5 to explain why you do not need a mirror as tall as yourself to see a full-height image.
7. Hold your right hand up in front of a plane mirror. Does the image of your hand appear to be a right hand or a left hand? Why is this so?
8. Stand 2 plane mirrors on their edges and at a right angle to one another as shown in Figure 14-20,A. Place an object between them and view its images from a position similar to that shown in the figure. How many images do you see? Draw a ray diagram to explain the formation of the images.
9. Stand 2 plane mirrors on their edges and parallel to one another as shown in Figure 14-20,B. Place an object between them, near one end. View its images from the position shown in the figure. How many images do you see? Draw a ray diagram to explain the formation of the images.

10. Stand 2 plane mirrors on their edges as you did in 8 and 9. This time use a variety of angles between the mirrors and try to discover the relationship between the angle and the number of images. Hint: Try angles of 30°, 60°, 90°, 120°, and 150° first.
11. State 2 uses of plane mirrors in the home and 2 uses of plane mirrors outside the home.

14.11 Curved Mirrors

You have probably seen several types of curved mirrors. Some common ones are the back and front of a metallic spoon, the side of a shiny kettle, and the security mirrors in supermarkets and department stores. Our studies will be restricted to **spherical mirrors**. These are curved mirrors which are simply portions of the surface of a polished sphere. If the inner surface is used as a mirror, it is called a **concave mirror**; if the outer surface is used as a mirror, it is called a **convex mirror** (Fig. 14-21).

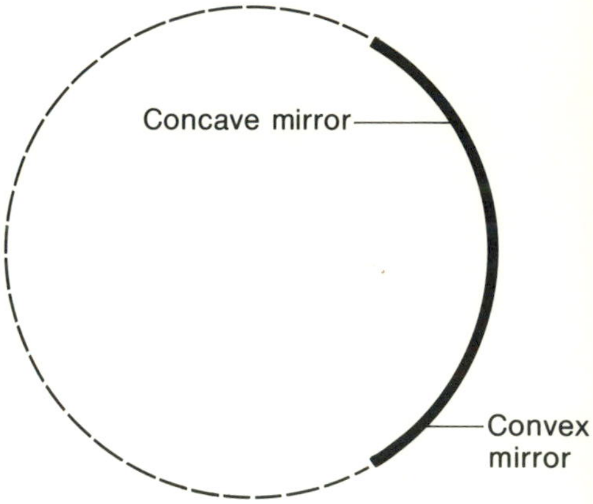

Figure 14-21 Two types of spherical mirrors.

Terminology

Before you can proceed with the investigation of the properties of curved mirrors, you need to know a few terms. Refer to Figure 14-22 as you read the following description.

The curve M_1M_2 represents a concave mirror. The point C is called the **centre of curvature** of the mirror since it is the centre of the imaginary sphere of which M_1M_2 is a part. Point V at the centre of the mirror is called the **vertex** or **optical centre**. The line that passes through C and V is called the **principal axis** of the mirror. Any other line that passes through C to the mirror is called a **secondary axis**. CA is an example of a secondary axis. The radius of a sphere always meets the surface of the sphere at right angles. Therefore the principal axis and all secondary axes are **normals** to the surface.

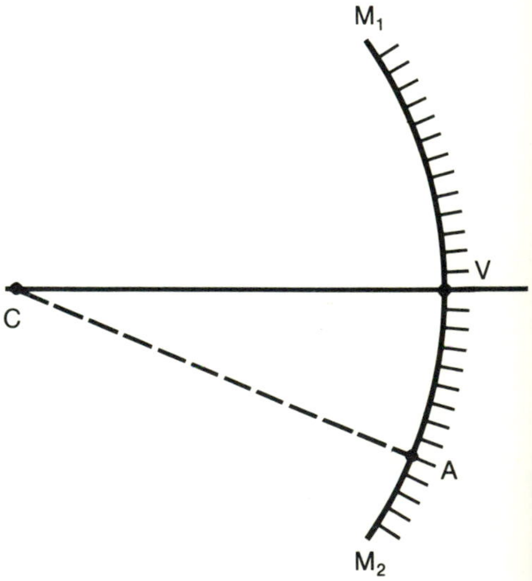

Figure 14-22 A section of a concave mirror.

Discussion

1. **a)** Distinguish between a concave and a convex mirror.
   b) Give 2 examples of concave mirrors and 2 examples of convex mirrors.
2. Explain these terms: centre of curvature, vertex, principal axis, secondary axis.
3. Do you think that the Laws of Reflection for plane mirrors apply to spherical mirrors? Why?

14.12 INVESTIGATION: Reflection at Spherical Surfaces

In this investigation you use a single ray of light to determine whether or not the Laws of Reflection for plane mirrors apply to the 2 types of spherical mirrors. Then you use several rays of light to study the effect of spherical mirrors on parallel incident rays.

Materials

ray box plus accessories kit
thin protractor
compasses
concave mirror
convex mirror

Procedure A The Laws of Reflection

a. Set up the ray box, protractor, and a concave mirror in a manner similar to that shown in Figure 14-15, page 394.
b. Direct a single ray of light from the ray box to the mirror so that the ray meets the mirror at the vertex. Draw the normal using a coloured pencil.
c. Record the angle of incidence and the angle of reflection in a table.
d. Move the ray box to get a different angle of incidence, but make sure the incident ray still meets the mirror at the vertex. Record the angle of incidence and the angle of reflection in your table.
e. Repeat step d for 3 more angles of incidence.
f. Direct a ray of light at a point on the mirror other than the vertex. Draw the normal using a coloured pencil. Measure the angle of incidence and the angle of reflection and record your values in the table.
g. Repeat step f three times using a different point and a different angle of incidence each time.
h. Repeat steps a to g using a convex mirror instead of the concave mirror.

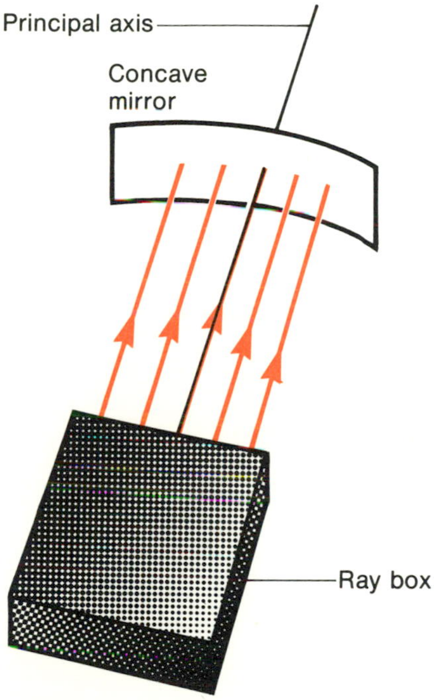

Figure 14-23 At what angle and at what point must the line and the mirror meet if the line is to represent the principal axis?

Procedure B Reflection from a Concave Mirror

a. Replace the single slit in the ray box with a multiple slit that produces 3 or 5 rays. Adjust the ray box until the rays are parallel.
b. Draw a straight line on a piece of plain paper. Stand a concave mirror on the paper so that the line represents the principal axis of the mirror (Fig. 14-23).

c. Direct the parallel rays at the mirror so that the centre ray coincides with the principal axis.
d. Use a pencil to trace the location of the mirror, the incident rays, and the reflected rays.
e. Label the point where the reflected rays come together F.
f. Use the compasses to find the centre of curvature of the mirror. Label this point C.
g. Move the ray box so that the reflected rays meet at a point other than F. Repeat this step 3 times.

Procedure C Reflection from a Convex Mirror

a. Repeat steps a to d of Procedure B using a convex mirror instead of the concave mirror.
b. Extend your tracings of the reflected rays backwards until they meet behind the mirror. Label the point where they meet F'.
c. Use the compasses to find the centre of curvature of the mirror. Label this point C.
d. Move the ray box so that the extensions of the reflected rays meet at a point other than F'. Repeat this step 3 times.

Discussion A

1. Do the 2 Laws of Reflection apply to concave mirrors? Explain how your data support your answer.

Discussion B

1. The points at which the reflected rays meet are called the **focal points** of the mirror. Why is this so? How many focal points can a concave mirror have?
2. The point F is called the **principal focus** of the mirror. Define the term ''principal focus'' in your own words.
3. The distance from the principal focus (F) to the mirror is called the **focal length** (f) of the mirror. The distance from the centre of curvature (C) to the mirror is called the **radius of curvature** (r) of the mirror. Compare f and r. Draw a diagram that shows F, C, f, and r.
4. A concave mirror is also called a **converging mirror**. Why?
5. Imagine that a small lamp is placed at the principal focus of a concave mirror. What would be the nature of the reflected beam? Give an example of a practical use of this fact.

Discussion C

1. How many focal points can a convex mirror have?
2. F' is usually called the virtual principal focus. Why is this so?

3. Compare f and r for a convex mirror.
4. A convex mirror is also called a **diverging mirror**. Why?

14.13 Geometrical Prediction of the Location and Characteristics of Images in Concave Mirrors

The purpose of this section is to use the information you discovered in Investigation 14.12 to draw ray diagrams that will help you predict the location and characteristics of images in a concave mirror. After you have made the predictions, you will have a chance to see how good they are by trying Investigation 14.14.

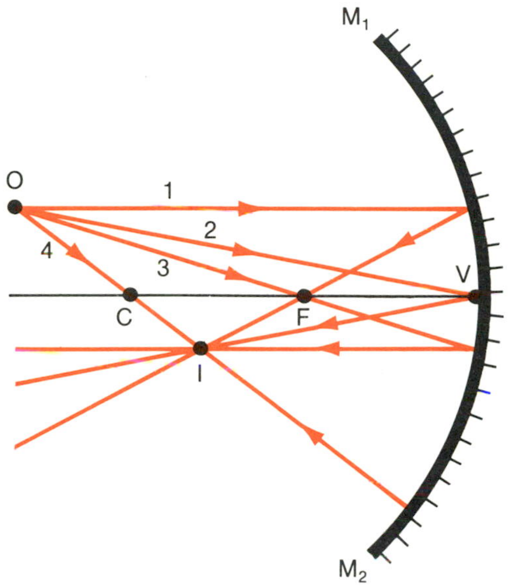

Figure 14-24 Some of the rays you can use to locate the image of a point in a concave mirror.

Rays to Use

In Figure 14-24, M_1M_2 represents a concave mirror. C is its centre of curvature, F its principal focus, and V its vertex. O is a point source object that is sending out rays of light in all directions. Some of those rays are shown in the figure. The following reasoning was used to draw them:

Ray 1 travels along a path that is parallel to the principal axis. You discovered in Investigation 14.12 that such a ray is reflected through F. Ray 2 meets the mirror at V. The First Law of Reflection says that the angle of reflection equals the angle of incidence. Ray 3 travels through F to the mirror and is reflected along a path that is parallel to the principal axis. Ray 4 travels through C to the mirror. It meets the mirror at a right angle. Therefore it is reflected back along its incident path.

Note that all of the reflected rays meet at point I. This is the location of the image of O. Clearly only 2 of the rays need to be used to locate an image of a point.

Geometrical Location of Images

Now that you know how to draw ray diagrams to locate the image of a point, let's move on to drawing ray diagrams to locate the image of an object. Examine Figure 14-25. To make the task easier, the object O_1O_2 has been placed with its base O_2 on the principal axis. This means that the base of the image, I_2, will also be on the principal axis. Therefore we can locate the image of O_1O_2 simply by locating the image of point O_1. This was done using rays 1 and 4 in Figure 14-24.

Our prediction is as follows: When the object is located beyond C, the image is located between C and F. It is smaller than the object, inverted, and real. (It can be caught on a screen since it is in front of the mirror.)

Discussion

The location and characteristics of the image in a concave mirror depend on the position of the object. Figure 14-25 helped us to predict the location and characteristics of the image when the object was beyond C. Draw 4 similiar ray diagrams to predict the location and characteristics of the image when the object is in each of these positions:

a. at C;
b. between C and F;
c. at F;
d. between F and V.

You must draw the diagrams carefully. Use compasses, ruler, and pencil. Summarize your predictions in a table similar to Table 92. The information obtained from Figure 14-25 has been placed in the table.

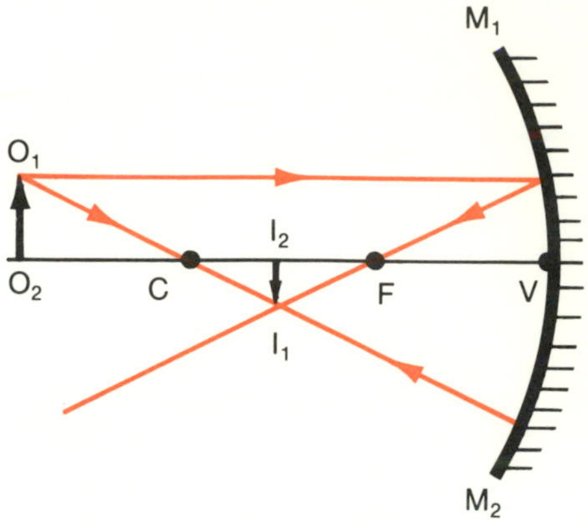

Figure 14-25 Two rays can be used to locate the image I_1 I_2 of the object O_1 O_2 when the base of the object is on the principal axis.

TABLE 92 Images in a Concave Mirror

Location of object	Location of image	Characteristics of image		
		Size	Attitude	Kind
Beyond C	Between C and F	Smaller than object	Inverted	Real
At C				
Between C and F				
At F				
Between F and V				

14.14 INVESTIGATION: Images in Concave Mirrors

In Section 14.13 you used ray diagrams to predict the location and characteristics of the image in a concave mirror when the object was in several different positions. In this investigation you test your predictions experimentally.

Materials

optical bench and accessories (metre stick, concave mirror and holder, screen)
candle
piece of chalk

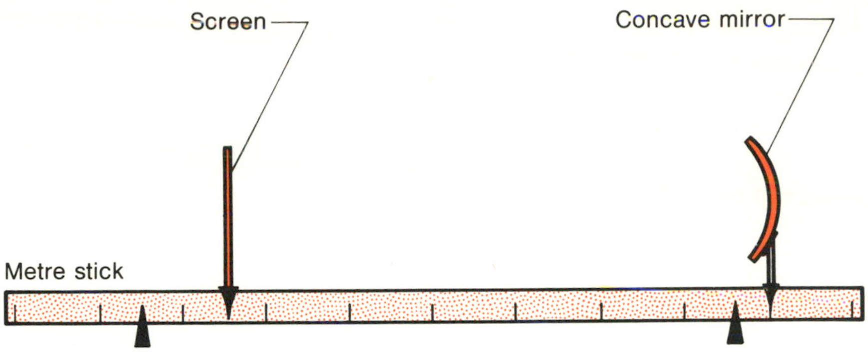

Figure 14-26 An optical bench can be used to determine the location and characteristics of the image in a concave mirror.

Procedure A Locating F and C

a. Set up the optical bench as shown in Figure 14-26.
b. Aim the mirror at a distant object such as a tree or house across the street from your school. Move the screen back and forth until a sharp image of the object can be seen on it. Use the chalk to mark the position of the screen as F.
c. Measure and record the focal length of the mirror, f.
d. Use the chalk to mark the centre of curvature as C. (Recall that the radius of curvature is twice the focal length.)

Procedure B Location and Characteristics of the Image

a. Prepare a data table similar to Table 92.
b. Place a burning candle at the end of the metre stick which is opposite to the mirror. Darken the room. Move the screen back and forth until you obtain a sharp image of the candle. Record the position of the object and image in your table. Record also the characteristics of the image.
c. Repeat step b with the candle at C, between C and F, at F, and between F and V.

Discussion A

1. Explain why the image of a distant object is at F.
2. Describe the size of the image of the distant object.
3. What kind of image was formed? How do you know?

Discussion B

1. Compare the predictions you made in Section 14.13 with the results you obtained in Investigation 14.14. If any of your predictions and results do not agree, try to determine why.
2. Where must the object be for a virtual image to be formed by a concave mirror?

3. If a man is using a concave shaving mirror, where must he place his face in relation to F in order that he can see an enlarged, erect image of his face?

14.15 Images in Convex Mirrors

In this section you are to determine the location and characteristics of the image in a convex mirror using procedures somewhat similar to those you used for the concave mirror. Proceed as follows:

a. Go back to Section 14.12 and review what you learned about convex mirrors in that section.

b. Study Figure 14-27 as you read this description of how the image is formed. Ray 1 is parallel to the principal axis. If the mirror were concave, the ray would be reflected through the principal focus. However, in a convex mirror the principal focus (F′) is behind the mirror. Therefore light cannot be reflected through it. Instead, it is reflected from the mirror as though it passed through F′. Ray 2 is heading for C. Therefore it meets the mirror at a right angle and is reflected back along the incident path as though it passed through C. The rays never cross but their virtual extensions (the broken lines) do cross behind the mirror. The image I_1I_2 is between F′ and C, smaller than the object, erect, and virtual.

c. Use a ray diagram to predict the location and characteristics of the image when the object is much further from the mirror than is the object in Figure 14-27.

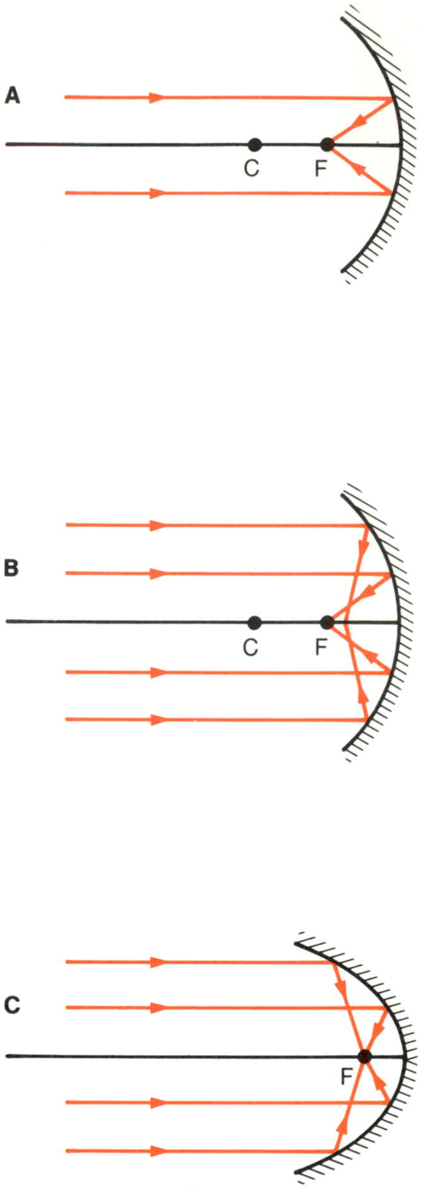

Figure 14-28 A parabolic mirror eliminates the spherical aberration of a concave mirror.

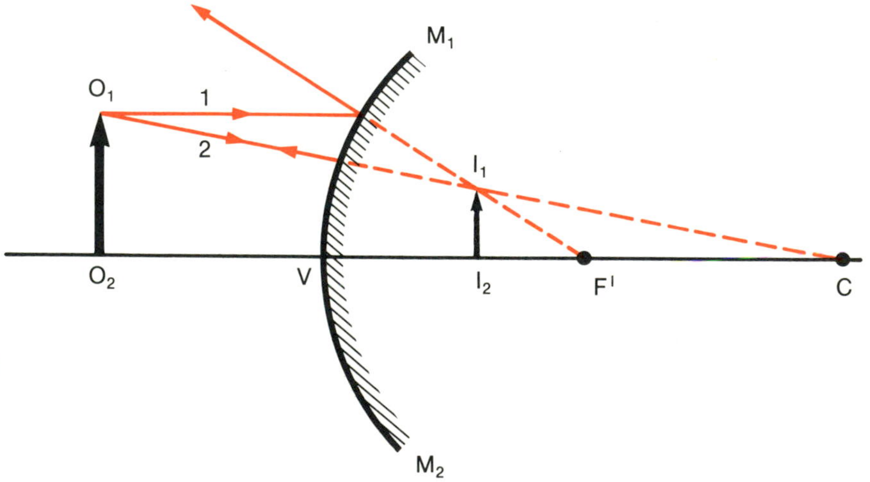

Figure 14-27 Geometrical location of the image in a convex mirror. What are the characteristics of this image?

d. Repeat step c with the object much closer to the mirror than is the object in Figure 14-27.
e. Use the large demonstration convex mirror provided by your teacher to check your predictions.
f. Make a generalization regarding the nature of the image in a convex mirror.

14.16 Uses of Curved Mirrors

During Investigation 14.12 you saw that incident rays of light that are parallel to the principal axis converge on the principal focus after reflection from a concave mirror. This convergence is shown in Figure 14-28,A. (The reflected rays are not drawn past F in order to avoid having a confusing maze of lines.) If any of the incident rays met the mirror near its edges, you may have noticed that the reflected rays did not pass exactly through F (Fig. 14-28,B). This scattering of the outer rays is called **spherical aberration** and it results in an image that is fuzzy around the edges. To eliminate this problem, parabolic mirrors are often used instead of concave mirrors. Notice in Figure 14-28,C how the curvature of the parabolic mirror focuses the outer rays through F.

The ability of parabolic mirrors to focus parallel rays has led to many uses for such mirrors. Astronomers use large parabolic mirrors in some of their telescopes to concentrate the faint light from distant stars. The Hale telescope on Mt. Palomar in California contains a parabolic mirror that is 5 m in diameter. Some people in India cook their meals with mirror ovens that use parabolic mirrors to collect the sun's energy. Scientists have used parabolic mirrors to concentrate enough sunlight to melt steel. In our energy-short world, such uses will undoubtedly increase.

Parabolic reflectors will focus rays other than those of light. For example, a radiotelescope uses a parabolic reflector to focus radio waves from outer space at F where a sensitive pick-up device is located. A parabolic reflector is also used to focus faint sounds for the purpose of recording them (Fig. 14-29). A microphone is located at F.

Figure 14-29 This parabolic reflector is being used to focus bird songs on a microphone that is connected to a tape recorder.

Many uses of parabolic reflectors depend on the fact that, if a source of energy is placed at F, it will send out parallel rays from the parabolic reflector. The rays simply follow paths that are the reverse of those shown in Figure 14-28,C. Parabolic reflectors are used in automobile headlights and flashlights. In both of these, the light source is placed at F. Rays from F that strike the mirror are reflected parallel to the principal axis. The parallel beam that is produced is very intense and can penetrate to great distances. The same principle is used in microwave transmission reflectors and some sound speakers. In each case the energy source is placed at F.

Discussion

1. **a)** What is spherical aberration?
   b) Describe two ways that spherical aberration can be eliminated.
2. Draw a ray diagram that shows how a mirror oven works.
3. Find a photograph of a radiotelescope. Make a sketch diagram that shows how it works.
4. Draw a ray diagram that shows how a parallel beam of light is produced in a flashlight.

C. Refraction of Light

A drinking straw appears to bend at the surface of the drink. A stick that is plunged into a pond appears to bend at the surface of the pond. A canoe paddle appears to bend at the surface of the water. The water in a lake or pond appears to be less deep than it really is. How can we account for these observations? One can only assume that the light rays must be bending as they pass from the water into the air. In the next few sections you study this bending or **refraction** of light rays. After you have gained an understanding of the nature of refraction, you will use your knowledge to study the operation of optical devices such as eyeglasses and binoculars. In addition you will study interesting natural phenomena such as mirages and the twinkling of stars.

14.17 INVESTIGATION: Refraction at a Glass-Air Interface

Refraction *is the bending of light as it passes across the interface (boundary) between two transparent media of different optical densities*. The optical density of a substance is a measure of the ease with which the substance transmits light. Water transmits light at about 0.7 of the speed of light in air. Therefore water is more optically dense than air. Glass is more optically dense than water or air since it transmits light at only 0.6 of the speed of light in air.

In this investigation you study the refraction of light as it crosses a glass-air interface. From this study you should be able to develop the Laws of Refraction.

Materials

ray box plus accessories kit
semi-circular glass slab
protector
coloured pencils

Procedure A Refraction from Air to Glass

a. Place the ray box and glass slab on a piece of plain paper. Direct a single ray of light from the ray box to meet the centre of the flat side of the glass slab at a 90° angle (see Figure 14-30,A). There should be no refraction of the ray. If there is, the angle is not 90°.

b. Mark the position of the glass slab and the ray path. The ray path you have marked is the **normal**.

c. Move the ray box so that the ray strikes the same central point on the flat side of the glass slab but at an angle of 10° to the normal (see Figure 14-30,B). This angle is the **angle of incidence**. Mark the new ray path using a coloured pencil. Measure the angle between the **refracted ray** and the normal. This is the **angle of refraction**.

d. Repeat step c using angles of incidence of 20°, 30°, 40°, 50°, and 60°. Use a different colour to draw each ray path, if possible.

e. Record the angles of incidence and refraction in a table.

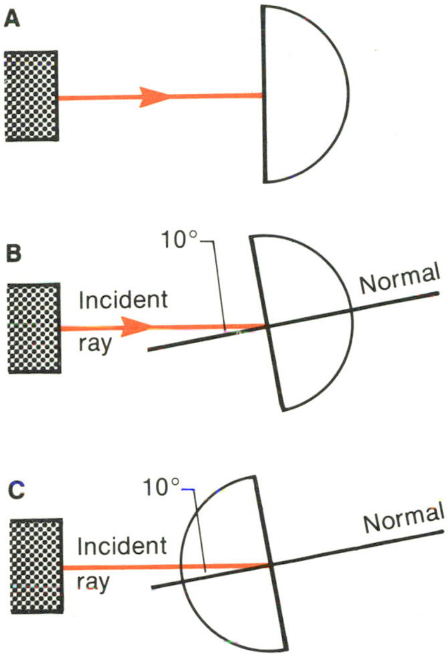

Figure 14-30 Refraction at a glass-air interface.

Procedure B Refraction from Glass to Air

In Procedure A you studied refraction as the light moved from air to glass. To study refraction from glass to air, simply turn the glass slab around as shown in Figure 14-30,C. You have already discovered that no refraction occurs at the curved surface provided the ray passes through the mid-point of the flat side. (A radius of a circle meets the circumference at 90°.) Therefore no refraction will occur until the ray emerges from the glass at the flat side.

Conduct studies similar to those of Procedure A. Record the angles of incidence and refraction in a table. Note carefully any evidence of *reflection*.

Discussion A

1. Define these terms: interface, optical density, refraction, refracted ray, angle of refraction.
2. Make a generalization that describes the behaviour of a light ray as it passes from one transparent medium to another of great optical density (air to glass).

Discussion B

1. Make a generalization that describes the behaviour of a light ray as it passes from one transparent medium to another of lower optical density (glass to air).
2. What unusual behaviour occurred when the angle of incidence was increased from 40° to 50°? This behaviour will be studied in Investigation 14.20.
3. Describe the behaviour of the reflected ray as the angle of incidence was increased.

14.18 The Laws of Refraction

Your studies of refraction were conducted at a glass-air interface. Similar studies have been conducted at the interface between many pairs of transparent media, including the water-air interface. In all cases the generalizations are similar to yours. Therefore the behaviour of light as it crosses the interface between transparent media of different optical densities can be summarized in the following generalizations that are called the **Laws of Refraction**.

1. *When a ray of light passes obliquely (at an angle other than 90°) from one transparent medium into another of greater optical density, the ray is refracted toward the normal.*
2. *When a ray of light passes obliquely from one transparent medium into another of lesser optical density, the ray is refracted away from the normal.*
3. *The incident ray, refracted ray, and normal are all in the same plane.*

Explanation of Refraction

The wave theory (see Section 14.6, page 392) can be used to explain the refraction of light as it passes from one medium into another. The wave theory says that light radiates from the source in the form of concentric spheres or wave fronts. Far from the source a small portion of a wave front is almost a straight line.

Figure 14-31 represents a parallel beam of light moving through air into glass at an oblique angle. AB is a portion of a wave front in that beam. The B end of the wave front meets the glass first (at D). Glass has a greater optical density than air. Therefore light travels more slowly in glass than in air. As a result, the B end of the wave front slows down at D. The rest of the wave front is still in air. Therefore it continues to move ahead at the original speed. As the B end of the wave front moves from

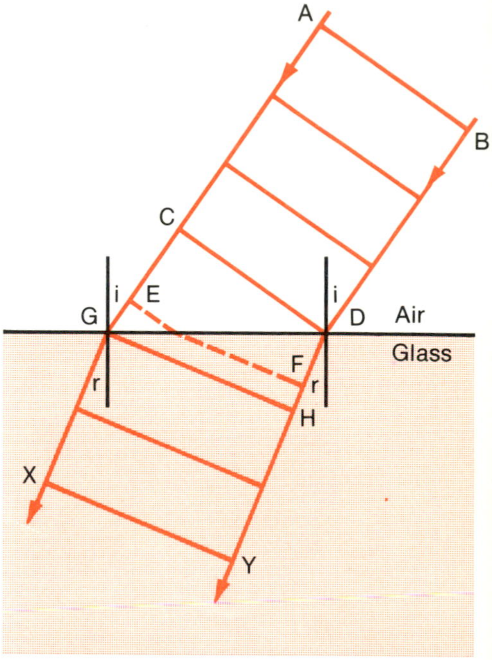

Figure 14-31 The wave theory can be used to explain refraction. The angle of incidence is i and the angle of refraction is r.

D to F in the glass, more and more of the wave front enters the glass and is slowed down. This slowing down is what causes the bending of the wave front toward the normal. When the A end of the wave front reaches the glass at G, it slows down to the same speed as the B end. Therefore it is refracted the same amount as the B end. The beam of light now continues as a parallel beam in the glass.

Factors Affecting the Angle of Refraction

Your studies of refraction at a glass-air interface suggest that two main factors determine the amount of refraction that occurs when light passes from one medium into another. These are the angle of incidence and the optical densities of the two media. You have observed the effect of angle of incidence on the angle of refraction. As you changed the angle of incidence in your studies, the angle of refraction changed. You have not observed the effect of optical density, but this explanation should help you understand it.

Light travels 1.33 times faster in a vacuum than in water. Thus

$$\frac{\text{speed of light in a vacuum}}{\text{speed of light in water}} = 1.33.$$

Refraction is caused by the change in speed when light moves from one medium into another of different optical density. Therefore the numeral 1.33 gives us an indication of the amount of refraction that occurs when light passes from a vacuum into water. Light travels 1.52 times faster in a vacuum than in crown glass. Thus

$$\frac{\text{speed of light in a vacuum}}{\text{speed of light in crown glass}} = 1.52.$$

Clearly, refraction is greater when light passes from a vacuum into crown glass than when it passes from a vacuum into water.

Scientists long ago recognized the value of ratios such as the 1.33 and 1.52 described above. As a result, they gave such a ratio a special name, the **index of refraction** (**n**). The index of refraction of a medium is defined as follows:

$$\mathbf{n} = \frac{\textbf{speed of light in a vacuum}}{\textbf{speed of light in the medium}}$$

Scientists use an instrument called a refractometer to measure indices of refraction of substances. Since the index of refraction is a characteristic physical property, it can be used to identify substances. Table 93 lists the indices of refraction of some common media. The index of refraction depends on the colour of the light (we will explore this aspect in Section 14.27). The values in Table 93 are for yellow light.

TABLE 93 Some Indices of Refraction

Medium	Index of refraction
Air	1.000 29
Water	1.33
Quartz	1.46
Crown glass	1.52
Zircon	1.90
Diamond	2.47

Discussion

1. Explain how your results from Investigation 14.17 support the Laws of Refraction.
2. Read the explanation of refraction from air to glass carefully. Then design a similar explanation using the wave theory to account for the refraction that occurs as light passes from glass into air. Hint: Use Figure 14-31 but start your discussion with the wave front XY in the glass.
3. **a)** Describe how the angle of incidence is related to the angle of refraction.
   b) Explain why the optical density of a medium affects the angle of refraction.
4. **a)** Define index of refraction.
   b) Index of refraction is a characteristic physical property. What does this mean?
   c) The index of refraction of quartz is 1.46. What does this mean?
   d) Substance A has an index of refraction of 1.50 and substance B has an index of refraction of 1.72. Compare the degree of refraction of a beam of light that passes from water into each of these substances.
5. Zircon, a relatively cheap mineral, is a fairly good substitute for diamond for use in jewellery. Why is this so? (See Table 93.)
6. Try to explain refraction of light as it passes from air into glass using the particle theory. You might begin with Figure 14-31 and replace the wave fronts with particles.
7. Try this simple experiment to see if you can develop an explanation for the fact that the water in a stream or lake generally appears less deep than it really is: Fasten a coin to the bottom of a can using clear cellulose tape. Close one eye and adjust your position so that the coin is just hidden from view for your open eye as shown in Figure 14-32. Without moving your head, slowly pour water into the can until it is full.

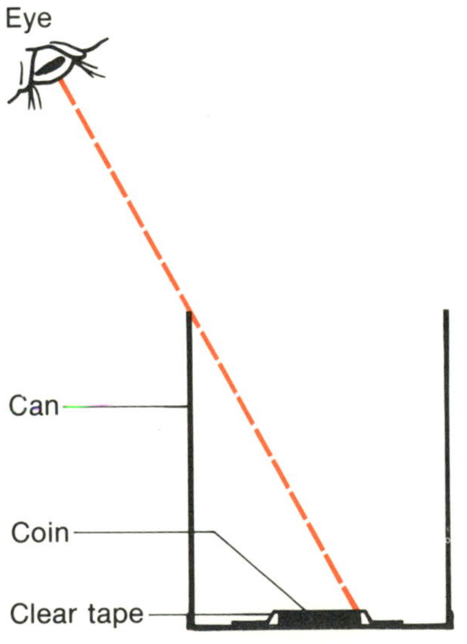

Figure 14-32 A simple set-up for investigating apparent depth.

a) Describe the results of this experiment.
b) Use your knowledge of refraction to explain the results.
c) Try to explain why the water in a stream or lake appears less deep than it really is.

14.19 INVESTIGATION: Refraction by Rectangular and Triangular Prisms

In Investigation 14.17 you used a semi-circular glass slab so that refraction would occur only at one side of the glass. (No refraction occurred at the curved side.) In this investigation you study the path of a ray of light that is refracted both when it enters and when it leaves a piece of glass. Two shapes of glass prisms are used, rectangular and triangular.

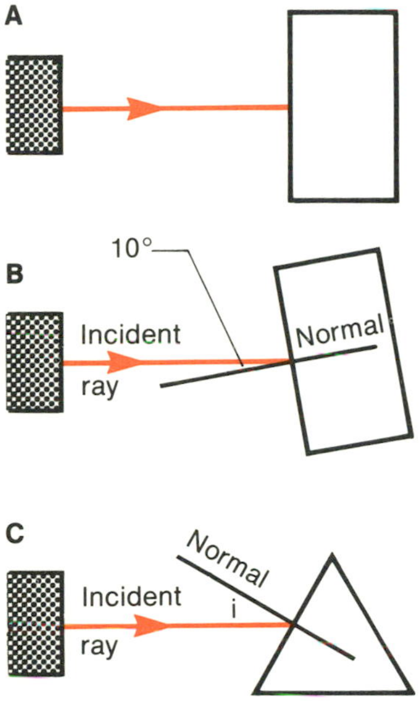

Figure 14-33 Studying the path of light through rectangular and triangular prisms.

Materials

- ray box plus accessories kit
- rectangular glass prism
- Triangular glass prism (with 60° angles)
- protector
- coloured pencils

Procedure A Refraction by a Rectangular Prism

a. Place the ray box and rectangular glass prism on a piece of plain paper. Direct a single ray of light from the ray box to meet the long side of the prism at a 90° angle (see Figure 14-33,A).
b. Mark the position of the prism and the ray path.
c. Move the ray box until the ray strikes the same spot on the long side of the prism but at an angle of incidence of 10° (see Figure 14-33,B). Mark the new ray path using a coloured pencil.
d. Draw in the normals to the interface where the ray enters and leaves the prism.
e. Measure and record the angles of incidence and refraction at both interfaces.

Procedure B Refraction by a Triangular Prism

a. Place the ray box and triangular prism on a piece of plain paper. Direct a single ray of light from the ray box so that it meets one face of the prism near its centre. The ray should be parallel to one of the other sides of the prism and it should emerge from the third side, having been refracted twice (see Figure 14-33,C). Note: The prism must be an equilateral triangular prism. That is, all of its three angles must be 60°.

b. Mark the position of the prism and the ray path.
c. Draw in the normals to the interface where the ray enters and leaves the prism.
d. Measure and record the angles of incidence and refraction at both interfaces.
e. Measure the total amount of refraction of the ray as it passes through the prism. Hint: Extend the incident ray and the emerging ray until they meet. Then measure the angle between them. This is called the **angle of deviation**.

Discussion A

1. Describe and account for what happens when a ray of light meets a rectangular prism at a 90° angle.
2. Describe the path followed by a ray of light when it meets a rectangular prism at an oblique angle.
3. Compare the refraction at the first surface with the refraction at the second surface.
4. Compare the direction of travel of the ray before it enters the prism with its direction after it emerges from the prism.
5. What is the overall effect of the rectangular prism on the light ray?

Discussion B

1. Describe the path followed by a ray of light when it meets one face of an equilateral triangular prism at an oblique angle.
2. Compare the refraction at the first surface with the refraction at the second surface.
3. Compare the direction of travel of the ray before it enters the prism with its direction after it emerges from the prism.
4. Design and try an experiment to check this hypothesis: The angle of deviation depends on the angle of incidence at which the ray enters the prism.
5. Refine your experiment to check this hypothesis: The smallest angle of deviation occurs when the refracted ray travels through the prism parallel to one side of the prism.

14.20 INVESTIGATION: Total Internal Reflection

During your studies of refraction from glass to air in Investigation 14.17, an interesting phenomenon occurred. When the angle of incidence was changed from 40° to 50°, the refracted ray disappeared and the incident ray was reflected from the inner surface of the flat side of the semi-circular glass slab as shown in Figure

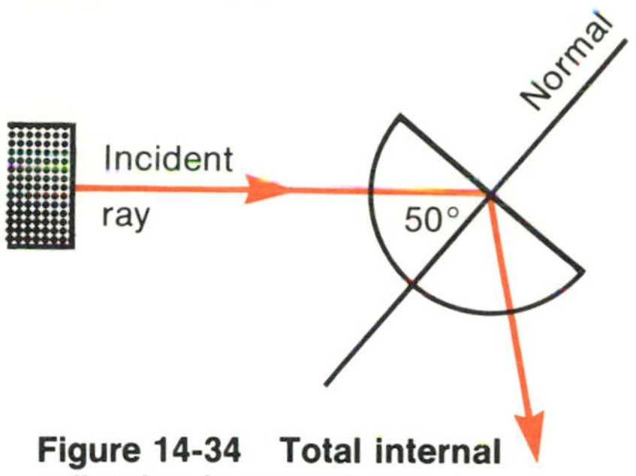

Figure 14-34 Total internal reflection in a semi-circular glass slab.

14-34. In this investigation you study this phenomenon, called **total internal reflection**, more closely. You also study total internal reflection in a triangular prism.

Materials

ray box plus accessories kit
semi-circular glass slab
90° triangular glass prism
protractor
coloured pencils

Procedure A Total Internal Reflection in a Semi-Circular Glass Slab

a. Set up the apparatus as shown in Figure 14-30, A (page 409). Locate the normal using the procedure outlined in step a of Procedure A of Investigation 14-17 (page 409).
b. Direct a single ray of light from the ray box to meet the centre of the flat side of the glass slab at an angle of incidence of 10°. Mark the ray path with a coloured pencil. Record the angle of refraction.
c. Repeat step b for angles of 20°, 30°, and 40°. Make careful notes on the intensity of the refracted ray and the reflected ray for each angle of incidence.
d. If you can, increase the angle of incidence in 1° steps, beginning at 40°. Record the angle of incidence when the angle of refraction is 90°. (At this angle, the refracted ray skims along the surface of the flat side.) This angle of incidence is called the **critical angle** for the glass.
e. Repeat step b for angles of 50° and 60°. Make careful notes on the phenomenon of total internal reflection.

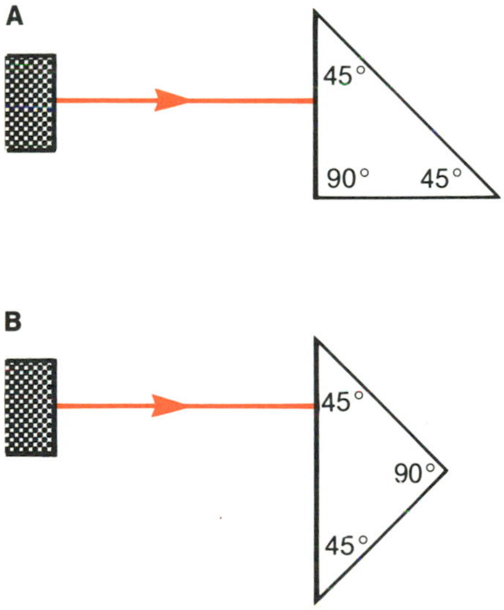

Figure 14-35 Studying total internal reflection in a triangular prism.

Procedure B Total Internal Reflection in a Triangular Prism

a. Place the ray box and 90° triangular glass prism on a piece of plain paper. Direct a single ray of light so that it meets one of the short sides at an angle of 90° as shown in Figure 14-35, A.
b. Mark the position of the prism and the ray path.
c. Direct a single ray of light so that it meets the hypotenuse (longest side) at an angle of 90° and toward one end as shown in Figure 14-35, B.
d. Mark the position of the prism and the ray path.

Discussion A

1. Define the terms "critical angle" and "total internal reflection."
2. What is the critical angle for your glass slab?

3. Account for the changes in the appearance of the refracted ray and the reflected ray as the angle of incidence was increased.

Discussion B

1. Use your knowledge of total internal reflection to account for the path of the ray in each of the studies of Procedure B.
2. Copy Figures 14-36,A and 14-36,B into your notebook. Complete the ray diagrams. Describe for each case what the object would look like to you if your eye were in the position shown.

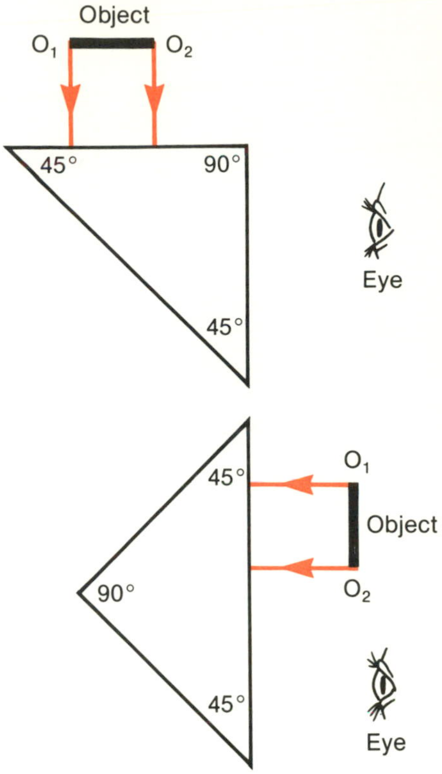

Figure 14-36 Applying your knowledge of total internal reflection in a triangular prism.

14.21 Applying Knowledge of Refraction and Total Internal Reflection

The knowledge you now have of refraction and total internal reflection makes it possible to understand many interesting natural phenomena and the operation of several optical instruments. This section gives some examples from both areas.

Atmospheric Refraction

Several interesting natural phenomena involve refraction of light by the atmosphere. **Refraction of sunlight** is one of these. Because of refraction, you can see the sun for several minutes before it actually rises above the horizon in the morning. Also, you can see it for several minutes after it has actually set below the horizon in the evening.

Figure 14-37 represents an observer standing in Greenland who is looking in the direction of the sun when the sun is actually in position A. Although the sun is really below the horizon, it can still be seen by the observer. The optical density of air is greater than the optical density of the near vacuum of space. Therefore the light slows down as it enters the atmosphere. This decrease in speed causes the refraction shown in the illustration. Since the atmosphere and space have no definite interface, no abrupt refraction occurs as you observed at the glass-air interface. Instead, a gradual bending occurs since the air gradually becomes more dense as the earth is approached.

Atmospheric refraction is also responsible for a phenomenon known as the **moon illusion**. You have likely noticed how large the moon appears to be when it is near the horizon. It also appears to have an elliptical (football-like) shape. This illusion is caused by the fact that rays of light from the lower part of the moon are refracted more than rays from the upper part.

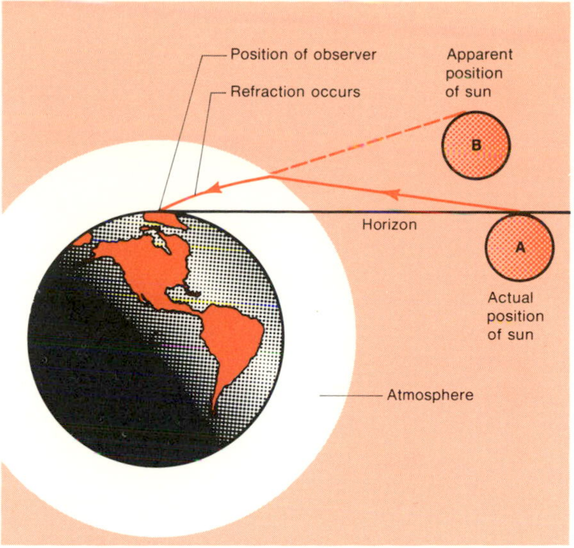

Figure 14-37 Refraction of sunlight causes an apparent change in the position of the sun. The diagram is not to scale; the bending is exaggerated. The true angle between the actual position and apparent position is not quite 2°.

Atmospheric refraction causes the **twinkling of stars**. The atmosphere is made up of many moving layers of air. These layers are of different temperatures and, therefore, of different densities. As a result, light from a star is refracted from side to side as it passes through these layers. Since the layers are moving, the stars appear to twinkle. The twinkling of stars hinders careful observation by astronomers. Therefore they generally make their observations on cold, calm winter nights when there is usually less movement in the atmosphere. Astronomers often locate their observatories high in mountains to get above some of the atmosphere. For many years some astronomers have avoided the atmosphere completely by sending telescopes and other equipment above the atmosphere in rockets and balloons.

You have likely seen **heat waves**. When an object is viewed through the air above a fire, hot pavement, or other heat sources, the object has a distorted, wriggling appearance. Many people refer to this phenomenon as heat waves. However, heat waves cannot be seen. You are simply seeing the effects of atmospheric refraction. The air just above a heat source is warmer and, therefore, less dense than the air further away. As a result, the light from the object you are viewing is refracted as it enters the warm air. Since the warm air is moving, the object appears to be distorted and wriggling.

A **mirage** is caused by total internal reflection. The most common mirage is the apparent layer of water that we see on paved roads in hot weather. The air next to the pavement becomes warm and, therefore, less dense than the air above it. As a result, light from the sky passes from a more dense to a less dense medium. If the light meets the less dense air at an angle of incidence that is greater than the critical angle for this interface, total internal reflection occurs and the observer sees an image of

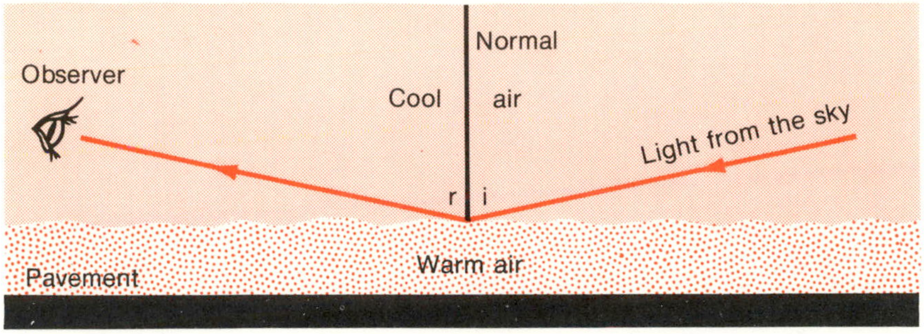

Figure 14-38 A mirage is caused by total internal reflection.

the sky as shown in Figure 14-38. The human eye cannot detect the change in direction of the light on reflection. Therefore the image of the sky appears to be on top of the pavement, causing the wet appearance.

Mirages of this type are very common in hot desert regions. Many stories are told of thirsty desert travelers who thought they saw pools of water in the sand. It would be incorrect to tell such people that their imaginations were fooling them. Mirages are real. They are formed by real rays of light and have often been photographed.

A less common type of mirage occurs when objects that are beyond the horizon appear to be lifted into view. Desert travelers have reported seeing a ship in the air above the sand or an oasis in the midst of a flat dry plain. This type of mirage is not easily explained but it is generally thought to be caused by a complex series of reflections and refractions at the surface of a warm layer of air.

Optical devices

The **periscope** is one optical device that uses total internal reflection (Fig. 14-39). The two prisms act as plane mirrors to change the direction of the light. The prisms are arranged so that total internal reflection occurs. Prisms are used instead of mirrors for two main reasons. First, the reflecting surface of a mirror is easily damaged, whereas the reflecting surface of the prism is on the inside where it cannot be easily damaged. Second, the prism reflects more of the incident light than does a mirror. Thus the image will be brighter.

You discovered in Investigation 14.20 that total internal reflection will occur twice in a 90° triangular prism if the incident light meets the hypotenuse at a 90° angle and near one end. This knowledge has been used in the designing of **binoculars** and other optical instruments that use "folded optics". One factor that determines the magnifying power of an optical instrument is the distance the light travels in the instrument. In order to avoid very long instruments, the light is bent back and forth ("folded") using

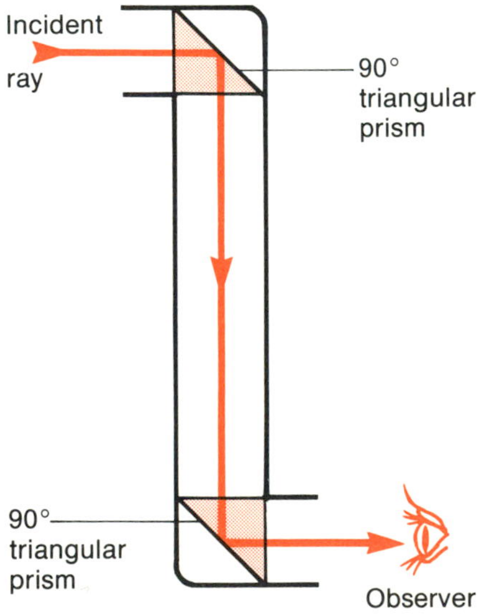

Figure 14-39 A periscope uses 90° triangular prisms to reflect light.

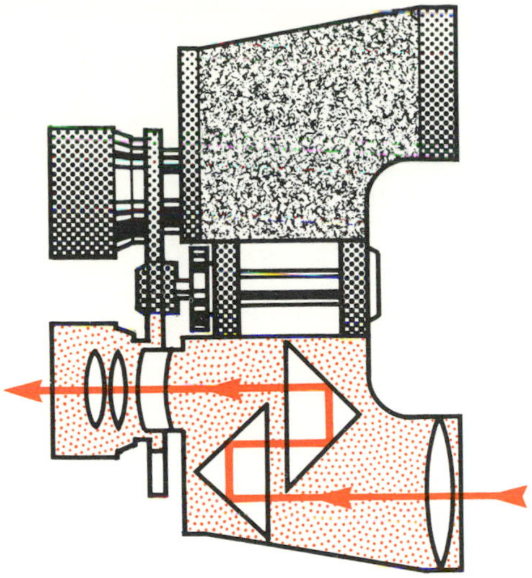

Figure 14-40 Binoculars contain prisms to increase the length of the path that light travels in the instrument.

prisms. Figure 14-40 shows how this is done in binoculars. The same principle is used in some telescopes and telephoto lenses for cameras.

Discussion

1. **a)** Explain how atmospheric refraction makes it possible for us to see the sun even though it may actually be below the horizon.
   b) The sun's rays are refracted whenever they enter the atmosphere at an oblique angle. However, the refraction is greatest when the sun is at the horizon. Why is this so? (There are two reasons. You should be able to discover one of them by studying Figure 14-37 and the other by comparing the angles of incidence and refraction you obtained in Procedure A of Investigation 14.17.)
2. **a)** What is the moon illusion? What causes it?
   b) What causes the twinkling of stars?
3. **a)** What causes heat waves?
   b) Explain in your own words why a pavement often appears wet on a hot day.
4. **a)** Draw a sketch of a periscope that uses plane mirrors instead of prisms. Show the light path.
   b) What advantages do prisms have over mirrors?
   c) Copy Figure 14-39 into your notebook, but without the light ray. Draw rays to show why the image of an object viewed through the periscope is erect.
5. **a)** What is "folded optics"?
   b) What advantage does folded optics offer?
   c) Name three optical devices that use folded optics.

14.22 INVESTIGATION: Properties of Convex and Concave Lenses

A **lens** is a transparent object with curved surfaces. Lenses refract light and, as a result, have many useful applications in optical devices such as cameras, telescopes, and microscopes. In this section you use your knowledge of refraction to predict the properties of two common types of lenses. You then check the predictions experimentally.

Prediction and Terminology

Examine Figure 14-41. Part A is an arrangement of prisms which, if put together and polished, could form the lens shown in Part C. Since this lens has two convex surfaces, it is called a **biconvex**

lens. Part B is an arrangement of prisms which, if put together and polished, could form the lens shown in Part D. Since this lens has two concave surfaces, it is called a **biconcave lens**.

Use your knowledge of refraction of light through triangular and rectangular prisms to predict the paths that will be followed by the incident rays shown in the diagram. Then predict whether a biconvex lens will converge or diverge light and whether a biconcave lens will converge or diverge light. After you have written your predictions in your notebook, proceed with the investigation.

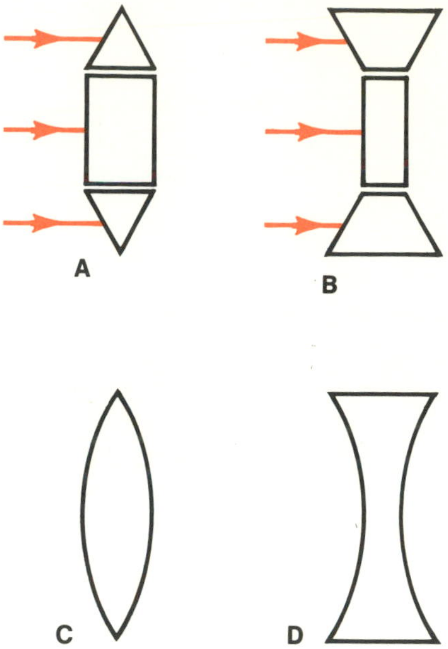

Figure 14-41 Theoretically lenses C and D can be produced by combining the prisms shown in A and B respectively.

Materials

ray box plus accessories kit
biconvex lens
biconcave lens
coloured pencils

Procedure A Refraction by a Biconvex Lens

a. Draw a straight line on a piece of plain paper. Stand a biconvex lens on the paper so the line represents the principal axis of the lens (Fig. 14-42, A).

b. Direct 3 or 5 parallel rays at the lens so that the centre ray coincides with the principal axis.

c. Use a pencil to trace the location of the lens, the incident rays, and the refracted rays.

d. Label the point where the rays come together F.

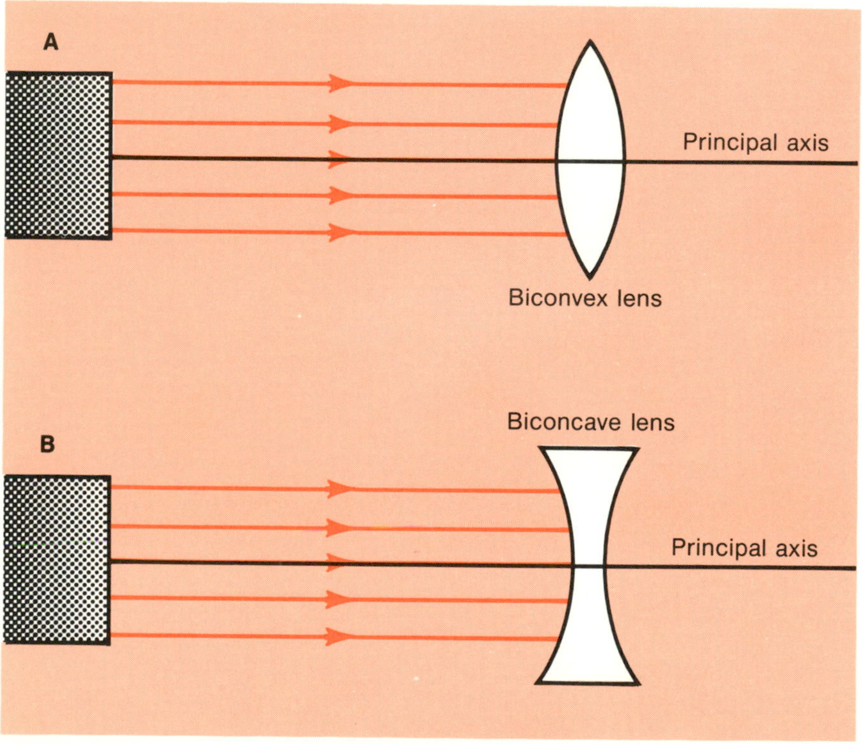

Figure 14-42 Refraction by two types of lenses.

Procedure B Refraction by a Biconcave Lens

a. Repeat steps a to c of Procedure A using a biconcave lens instead of the biconvex lens. (Fig. 14-42, B).

b. Extend your tracings of the refracted rays until they meet on the ray box side of the lens. Label the point where they meet F′.

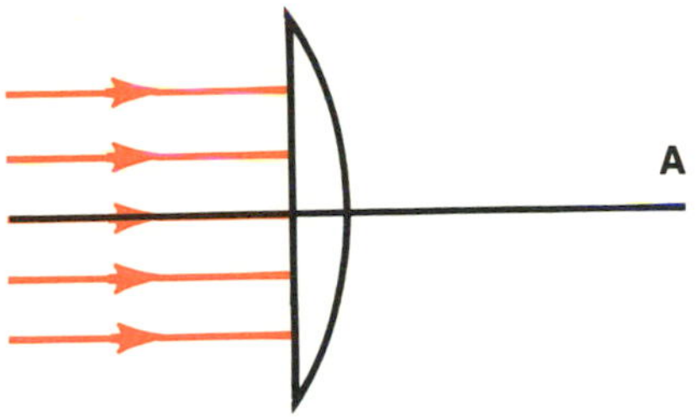

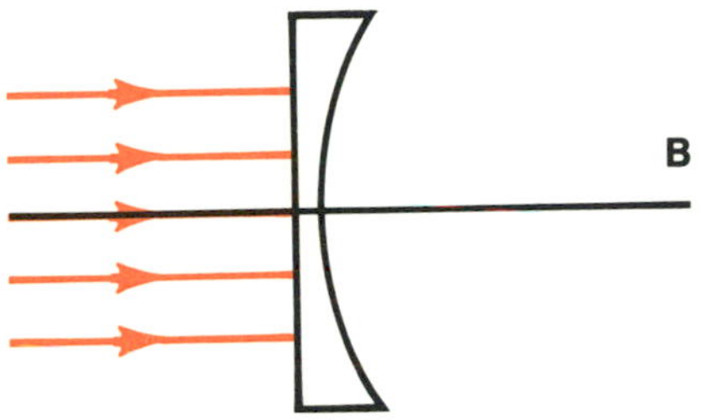

Figure 14-43 Which one of these lenses is a diverging lens and which one is a converging lens?

Discussion A

1. The point F is called the **principal focus** of the biconvex lens. Define this term in your own words.
2. A biconvex lens is also called a **converging lens**. Why?
3. Figure 14-43, A shows a **plano-convex lens**. Why is it given this name? Complete the ray diagram.

Discussion B

1. The point F′ is usually called the **virtual principal focus**: Why?
2. A biconcave lens is also called a **diverging lens**. Why?
3. Figure 14-43, B shows a **plano-concave lens**. Why is it given this name? Complete the ray diagram.

14.23 Geometrical Prediction of the Location and Characteristics of Images Formed by Convex Lenses

The purpose of this section is to draw ray diagrams that will help you predict the location and characteristics of images formed by a convex lens. After you have made the predictions, you will have a chance to see how good they are by trying Investigation 14.24.

Figure 14-44 shows two rays of light striking a biconvex lens. O is the **optical centre** of the lens, F is the **principal focus**, and

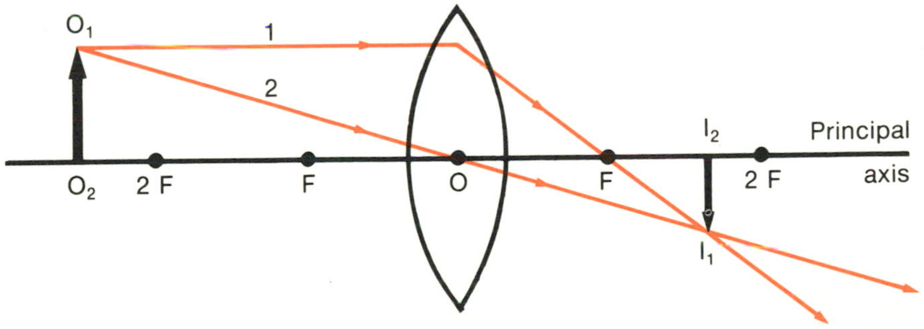

Figure 14-44 Two rays can be used to locate the image $I_1 I_2$ of the object O_1O_2 when the base of the object is on the principal axis.

2F is a point twice as far from O as is F. Note that a biconvex lens has two F's and two 2F's. The distance from F to O is the **focal length** (f) of the lens.

Ray 1 travels along a path that is parallel to the principal axis. You discovered in Investigation 14.22 that such a ray is refracted through F as shown. Ray 2 heads directly at O. Because of the symmetry of the lens, any refraction that occurs at the first surface is offset by an equal and opposite refraction at the second surface. These two rays are sufficient for locating and drawing the image.

Our prediction is as follows: When the object is located beyond 2F, the image is located between F and 2F on the other side of the lens. It is smaller than the object, inverted, and real.

Discussion

The location and characteristics of the image formed by a convex lens depend on the position of the object. Figure 14-44 helped us to predict the location and characteristics of the image when the object was beyond 2F. Draw 4 similar ray diagrams to predict the location and characteristics of the image when the object is in each of these positions:

a. at 2F;
b. between 2F and F;
c. at F;
d. between F and the lens.

Draw the diagrams carefully using compasses, ruler, and pencil. Summarize your predictions in a table similar to Table 94. The information obtained from Figure 14-44 has been placed in the table.

TABLE 94 Images Formed by a Convex Lens

Location of object	Location of image	Characteristics of image		
		Size	Attitude	Kind
Beyond 2F	Between F and 2F on opposite side	Smaller than object	Inverted	Real
At 2F				
Between F and 2F				
At F				
Between F and lens				

14.24 INVESTIGATION: Images Formed by a Convex Lens

In this investigation you test the predictions that you made in the preceding section.

Materials

candle
piece of chalk
optical bench and accessories (metre stick, convex lens and holder, screen)

Procedure A Locating F and 2F

a. Set up the optical bench as shown in Figure 14-45.
b. Aim the lens at a distant object such as a tree or house across the road from your school. Move the screen back and forth until a sharp image of the object can be seen on it. Use the chalk to mark the position of the screen F.
c. Measure and record the focal length of the lens, f.
d. Mark off a distance equal to twice f from the lens. Label this point 2F.
e. Label the F and 2F on the other side of the lens.

Procedure B Location and Characteristics of Image

a. Prepare a data table similar to Table 94.
b. Place a burning candle at one end of the metre stick. Darken the room. Move the screen back and forth until you obtain a sharp image of the candle. Record the position of the object and image in your table. Record also the characteristics of the image.
c. Repeat step b with the candle at 2F, between 2F and F, at F, and between F and the lens.

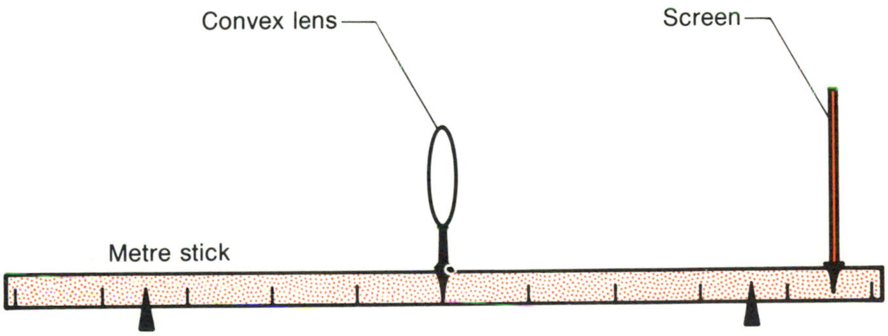

Figure 14-45 An optical bench is used to determine the location and characteristics of the image formed by a convex lens.

Discussion A

1. Why is the image of a distant object located at F?
2. Describe the size, attitude, and kind of image formed when a convex lens is directed at a distant object.

Discussion B

1. Compare the predictions you made in Section 14.23 with the results you obtained in this investigation. If any of your predictions and results do not agree, try to determine why.

14.25 Images Formed by a Concave Lens

In this section you are to determine the location and characteristics of the image formed by a concave lens using procedures somewhat similar to those you used for the convex lens. This time we are giving you very little guidance. You are to predict the location and characteristics of the image for various object positions. Then you are to design and try the experiment. Here are some suggestions:

a. Review Sections 14.14 and 14.15 on concave and convex mirrors.
b. Give some thought to the similarities between the two types of mirrors and the two types of lenses.
c. Draw ray diagrams to predict the location and characteristics of the image for chosen object positions.
d. Test your predictions experimentally.

14.26 Uses of Lenses

In this section one common use of lenses is discussed. Several others are listed, of which you are to select one to research as a project.

Correction of Vision Defects

Among the countless optical devices that use lenses are the eyeglasses that people wear to correct vision defects. Part A of Figure 14-46 shows that a normal relaxed eye focuses parallel rays of light on the retina. A far-sighted eye (B) tends to focus parallel rays behind the retina. A convex (converging) lens cor-

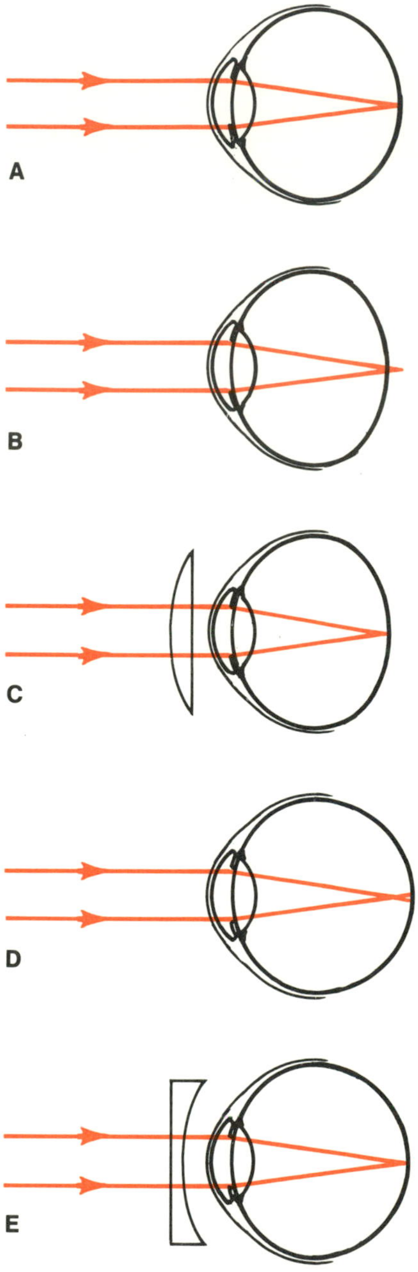

Figure 14-46 Correction of vision defects with lenses.

rects the problem as shown in C. A near-sighted eye (D) tends to focus parallel rays in front of the retina. A concave (diverging) lens corrects the problem as shown in E.

Other Uses

A number of applications of lenses are listed here. Select one that interests you, then identify the types of lenses used and state the reasons why those types are used. You should examine one of the devices closely. Then visit your school resource centre or community library and consult physics books and encyclopedias for ray diagrams and other information. Prepare a report that lists the types of lenses used and the reasons they are used. Include a ray diagram, if possible, and any other interesting information that you uncover.

a. magnifier (hand lens)
b. projector (slide or motion picture)
c. compound microscope
d. astronomical telescope (refracting type)
e. terrestrial telescope
f. binoculars
g. camera
h. photographic enlarger
i. human eye (How does the lens accommodate for various distances? What causes vision defects? What vision defects are there in addition to the two discussed here? How are they corrected?)

D. Colour

The purpose of the remaining sections of this unit is to provide you with sufficient knowledge of colour theory so that you can understand many interesting things that you may meet day to day. By the end of the unit you should be able to answer questions such as these: How does colour television work? Why do some laundry detergents contain a blue substance? Why do photographers use colour filters? Why is a red rose red?

The study of colour is intriguing but it is not simple, since colour perception by a person involves both physical and psychological factors. For example, two people may have the same physical reaction to a certain colour; that is, their eyes may have the same sensitivity to the colour. However, if their mental states are different, they may have entirely different psychological reactions to the colour. One person may feel warm and relaxed; the other may feel cold and tense.

Our studies deal almost entirely with the physical aspects and are based on this definition of colour: **Colour** *is that property of light energy that makes it possible for an organism to distinguish* *by sight two objects that are identical in shape, size, and structure*. According to this definition, grays are colours just as much as the reds, blues, and other hues. The reds, blues, and other hues are called **chromatic colours**; the grays are called **achromatic colours**. The white of new-fallen snow is an achromatic colour as are the black of a piece of coal and all shades of gray between these two extremes.

14.27 INVESTIGATION: Composition of White Light

For thousands of years people have been aware of the fact that diamonds, chunks of glass, and some other transparent media produce chromatic colours when white light is shone on them. Many early philosophers and scientists believed that the media added chromatic colours to the white light. Sir Isaac Newton felt that the chromatic colours were actually present in the white light. The media simply separated them. In 1666 Newton did experiments that proved who was correct. You repeat two of them in this investigation.

Materials

ray box plus accessories kit
equilateral triangular prism
white card (screen)
white cardboard
coloured pencils
protractor

Procedure A Dispersion of White Light

a. Arrange a ray box, equilateral triangular prism, and white card as shown in Figure 14-47, A.
b. Direct a narrow beam of white light at the prism so that it meets one side of the prism at an oblique angle as shown.
c. Adjust the position of the prism until the maximum refraction of the light is obtained.
d. Record your results.

Procedure B Recombination of the Spectral Hues

a. From a piece of white cardboard cut a circular piece that is 10-12 cm in diameter. Draw a diameter through the circle. Then use the protractor to divide the circle into 18 segments, all having a 20° angle.

A

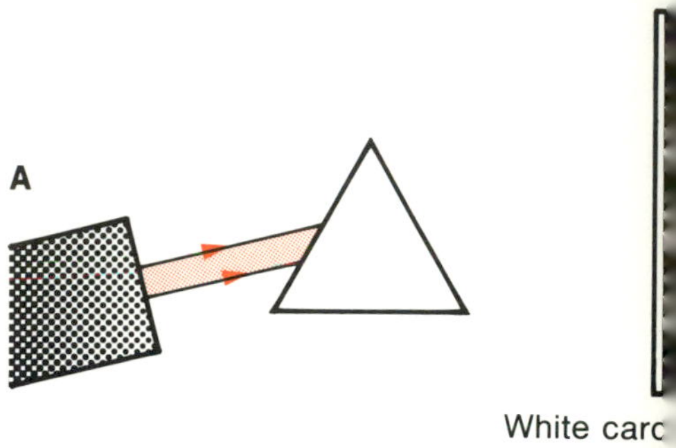

B

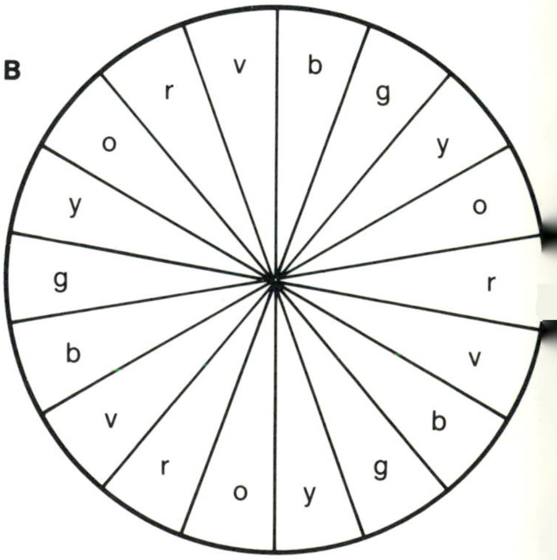

Figure 14-47 Materials for studying the composition of white light.

b. Colour the segments in a regular sequence with these 6 colours: red, orange, yellow, green, blue, violet. Repeat this sequence 3 times (See Figure 14-47, B.)
c. Push a common pin through the centre of the circle and into the eraser of a pencil.
d. Spin the disc rapidly and observe the effect. If possible, perform this step in bright light, preferably daylight.

Discussion A

The band of colours formed in this investigation is called a **spectrum**. Each of the colours is called a **spectral hue**. The separation of white light into its spectral hues is called **dispersion**.

1. Which edge of the spectrum is red and which edge is violet?
2. Name, in order, 6 spectral hues that you can see in the spectrum, beginning with red.
3. Refraction is caused by a change in speed of the light as it passes from one medium into another of different optical density. Which spectral hue was refracted the least? the most? Which spectral hue decreases in speed the least when it enters the prism and increases in speed the least when it leaves? Which spectral hue decreases in speed the most when it enters the prism and increases in speed the most when it leaves?

Discussion B

1. Describe what happens to the chromatic colours as the disc is spun rapidly.
2. How do the 6 colours used in Procedure B compare to the spectral hues obtained in Procedure A?
3. Make a generalization regarding the relationship between white light and the spectral hues. Does it agree with Newton's belief?

14.28 Light and the Electromagnetic Spectrum

You know that the refraction of light is due, in part, to the wave nature of light (see Section 14.18, page 410). You discovered in Investigation 14.27 that each of the spectral hues is refracted a different amount as it passes through a prism. Therefore it seems reasonable to assume that the spectral hues differ somehow in their wave properties. Scientists have discovered that they differ in wave length; red light has the longest wave length and violet light the shortest wave length of the spectral hues. What is "wave length"? The wave length of light is difficult to visualize but it may help if you try to visualize the wave length of a water wave. The wave length of a water wave in a lake is the distance from the peak of one crest to the peak of an adjacent crest.

Table 95 lists the wave lengths for the 6 common spectral hues. The wave lengths are given in nanometres (nm). Recall that 1 nm is 10^{-9}m, or 0.000 000 001 m.

TABLE 95 Wave Lengths of Spectral Hues

Spectral hue	Wave length (nm)
Red	620-770
Orange	590-620
Yellow	570-590
Green	500-570
Blue	450-500
Violet	390-450

Most people's eyes cannot detect radiation that has a longer wave length than 770 nm (red light) nor can they detect radiation that has a wave length shorter than 390 nm (violet light). However, the spectrum extends far beyond these wave lengths. In fact, the **visible band** makes up only a small portion of the **electromagnetic spectrum**, as it is called (Fig. 14-48). Next longest in wave length to red in the electromagnetic spectrum is the infrared band. These invisible rays are largely responsible for the propagation of heat through space. Next shortest in wave length to violet is the ultraviolet band. Certain wave lengths of these invisible rays cause tanning of human skin while other wave lengths cause sunburn.

Examine Figure 14-48 closely to see where other familiar bands of radiation are located. Keep in mind that all of these rays are similar to one another in basic nature and speed; but they differ in wave length.

Figure 14-48 The electromagnetic spectrum.

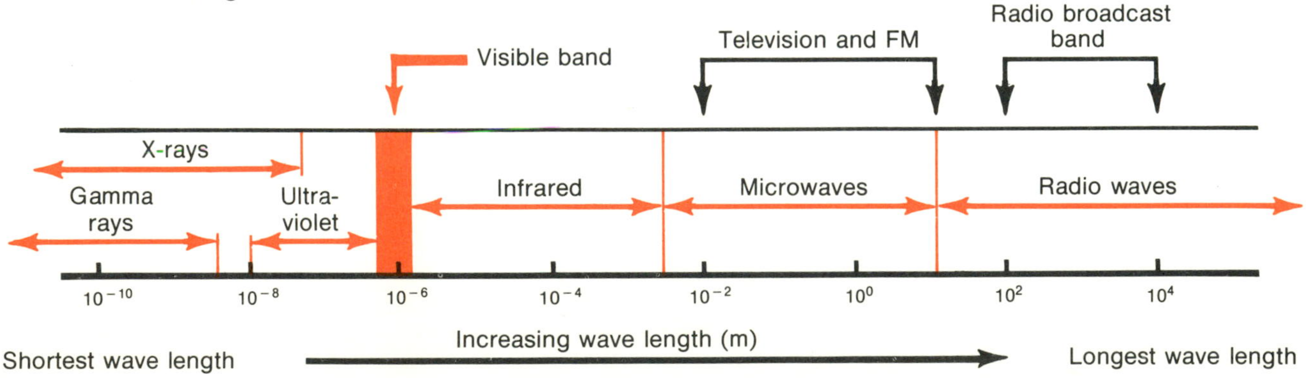

Discussion

1. **a)** What is the narrowest band of radiation in the electromagnetic spectrum?
   b) Which band of radiation has the shorter wave length, the infrared band or the ultraviolet band?
   c) What is the basic difference among the various types of electromagnetic radiation?
2. State 2 ways in which the visible band of radiation differs from the other forms of electromagnetic radiation.

14.29 INVESTIGATION: Introduction to Colour Theory

This investigation introduces colour theory to you by making a hypothesis that you are to check.

Hypothesis: The colour of an illuminated object depends upon

a) *the colour of the light that illuminates it;*

b) *the effect that the illuminated object has on the light that illuminates it.*

Think about part a of the hypothesis as you try to answer these questions: What colour will a white piece of paper be if only a red light is shone on it? What colour will the same piece of paper be if only a blue light is shone on it?

Think about part b of the hypothesis as you try to answer these questions: What colour will a piece of red paper be if it is illuminated by red light only? What colour will a blue piece of paper be if it is illuminated by red light?

Now try the investigation.

Materials

pieces of paper of the same size and shape but different colours
coloured light sources

Procedure

a. Spread the pieces of paper over a portion of a table.
b. Shine a red light on the pieces of paper.
c. Turn out the room lights and, without looking, change the positions of the pieces of paper on the table.
d. Try to decide what colour each piece of paper was when the room lights were on.
e. Turn on the room lights and check your results.
f. Repeat the experiment using 2 or 3 other colours of light source, if possible.

Discussion

Is the hypothesis correct? Defend your answer.
Note: Part a of the hypothesis is dealt with in Section 14.30 under the heading of **additive colour theory**. The concerns of this theory are the nature of coloured lights and the results obtained when 2 or more coloured lights are mixed. Part b of the hypothesis is dealt with in Section 14.31 under the heading of **subtractive colour theory**. The concerns of this theory are the effects of pigments in substances on the incident light and the results obtained when two or more of these pigments are mixed.

14.30 Additive Colour Theory

Additive Primaries and Additive Colour Mixing

Figure 14-49 shows 3 projectors, each of which is projecting a circle of light onto a white screen so that the 3 circles overlap as shown. Each of A, B, and C is a different colour; in colour theory they are called **components**. Note that regions A, B, and C are illuminated by only one component. Regions A + B, A + C, and B + C are illuminated by two components, and region A + B + C in the centre is illuminated by three components. When a region is illuminated by more than one component, the reflected light is called an **additive colour mixture**. It is called additive because the reflected light is made up of the part of A that is reflected plus the part of B that is reflected plus the part of C that is reflected. The colour of an additive mixture depends on two factors: the colours of the components and their relative intensities.

If the 3 component colours are **red**, **green**, and **blue** and they are of equal intensities, one obtains the result shown in Figure 14-50,B. Examine the figure closely. Note that the red, green, and blue add to produce what appears to be **white** light. This is a surprising result, since we learned in Section 14.27 that white light contains more spectral hues than just those three. Strictly speaking the white produced here is not a true white. However, the eye is unable to distinguish between it and the true white that is made up of all the spectral hues. Note, too, that the red and green lights add to produce what appears to be **yellow** light. This, too, is surprising since neither red nor green contains the wave lengths that produce the yellow hue in the spectrum. However, Figure 14-50,A shows that yellow light lies between red and green light in the spectrum. Apparently an averaging effect occurs in the eye. Note, further, that the blue and green lights add to produce **cyan**, a blue-green colour. Like yellow, cyan is

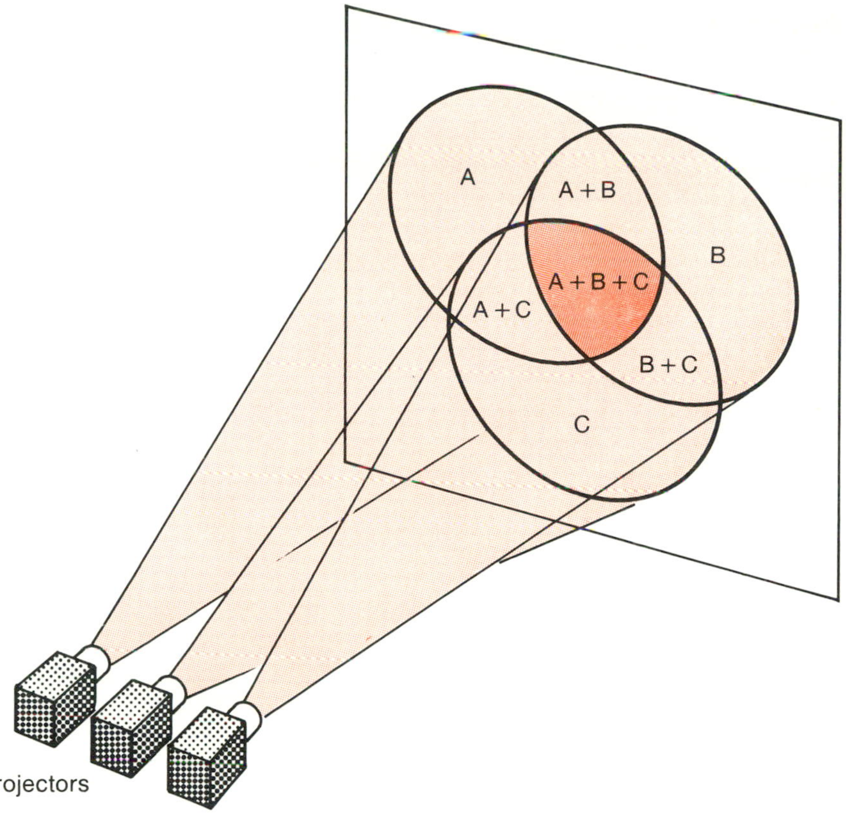

Figure 14-49 Additive colour mixing occurs when lights of different colours are shone together on a white screen.

between its component colours in the spectrum. Finally, the red and blue lights add to produce **magenta**, which is not in the spectrum.

Figure 14-50,B was obtained using red, green, and blue lights of equal intensities. The overlap produced yellow, cyan, magenta, and white. By varying the intensities, almost all of the spectral hues and many non-spectral hues can be produced. No other combination of coloured lights can produce as many colours as red, green, and blue. Further, no combination of coloured lights can produce red, green, and blue. Since these 3 colours are somewhat special, they are called the **primary colours of light** or the **additive primaries**. Yellow, cyan, and magenta are called the **secondary colours of light**.

Remember that additive colour mixing applies only to the mixing of coloured lights. It does not apply to the mixing of dyes and paints. The subtractive theory (Section 14.31) deals with that aspect of colour theory.

An Application of Additive Colour Theory

The additive colour theory has been used to explain **colour vision**. The retina of the human eye has two kinds of light-sensitive organs in it: rods and cones. The rods are sensitive to light and dark, but not to colour. The cones are sensitive to colour but only when the light is reasonably bright. For example, you usually cannot see colours outdoors at night since the light is too weak to affect the cones. Only the rods are affected; therefore you see a black-and-white image.

Some scientists feel that the human eye contains 3 kinds of cones. One kind responds to red light, one to blue light, and one to green light. Therefore, various degrees of stimulation of the 3 kinds of cones can produce the sensation of a wide range of colours. Other scientists feel that the human eye contains more than 3 kinds of cones. The famous scientist, Edwin Land, took a photograph of a scene using only yellow light. He then took a second photograph of the same scene using a yellowish-orange light. He projected the two pictures together on a white screen and asked several people to describe what they saw. Some people said they saw red, blue, brown, and yellow in the scene! This result casts doubt on the hypothesis that the human eye has only 3 kinds of cones. So far no one has actually determined experimentally the number of kinds of cones.

Complementary Colours

You have seen that the sensation of white light can be produced by adding the primary colours of light, red, green, and blue, in equal intensities. However, the same white light can be produced by adding only two coloured lights, provided you pick the right pair of colours. You need only add a primary colour and its complementary secondary colour. (**Complementary colours** are any two colours of light that add to produce white light.) Here is an example: The secondary colour yellow is the sum of the primary colours red and green. Therefore blue and yellow are complementary colours since they will produce the same result as blue plus red plus green, namely white. Similarly red and cyan are complementary colours as are green and magenta. Figure 14-50,D illustrates the concept of complementary colours.

An interesting example of the application of the concept of complementary colours occurs in laundry detergents that contain blueing agents. Many types of white cloth gradually turn yellow with age. The yellow cannot be removed by washing. Therefore some detergents contain a blue dye that remains in the cloth after washing. The dye reflects a blue light to your eye and the cloth reflects a yellow light to your eye. Since blue and yellow are complementary colours, a white sensation is produced and the cloth appears to be whiter than it really is.

Discussion

1. **a)** Name the additive primaries. Explain why the terms additive and primary are used to describe these colours.
   b) Name the secondary colours of light and explain how they can be formed.
   c) Explain how a stage crew could produce a wide variety of colours on the stage of a theatre, using only three coloured projection lanterns.
2. **a)** Outline an explanation of colour vision that is based on the hypothesis that the human eye has 3 kinds of cones in the retina.
   b) Based on this hypothesis, give what you feel is an explanation for the fact that some people are colour-blind to certain colours.
3. Examine the screen of a colour television set with a hand lens while the set is not operating. Record what you see. Turn on the set and allow it to warm up for 2 or 3 min. Now examine the screen again with the hand lens and record what you see. On the basis of your observations, explain how the set is able to produce a wide variety of colours, including black and white.
4. **a)** What are complementary colours?
   b) State the complementary colour for each of the 3 additive primaries.
   c) Explain how a blueing agent in a detergent makes white clothes which have yellowed appear to be whiter than they actually are.
5. If you stare at a red object too long, the red cones in your retina become "tired" and will eventually refuse to see red when you shift your gaze to another object. This effect, called retina fatigue, often results in an interesting phenomenon, the formation of after-images.
   a) Make a solid red circle about 1 cm in diameter on a sheet of white paper. Shine a bright light on the circle and stare at the circle intently for about 30 s. Then, without taking your eyes from the page, shift your gaze to the white paper nearby. Note the colour of the after-image. Use your knowledge of complementary colours and colour vision to explain this phenomenon.
   b) Predict the colour of the after-image you will get if you use a green circle instead of the red one. Check your prediction experimentally.
   c) Repeat part b using circles of these colours: blue, yellow, and, if you can find suitable coloured pencils, cyan and magenta.

14.31 Subtractive Colour Theory

The Colour of Opaque Objects

You discovered in Section 14.29 that an object cannot reflect a colour that is not present in the incident light. However, you also discovered that an object often reflects a colour that is different from that of the incident light. The **subtractive colour theory** explains these findings by stating that the object absorbs or **subtracts** one or more components of the incident light.

An object that looks white when it is illuminated with white light is reflecting most of the incident light (Fig. 14-51, A). On the other hand, an object that appears black when it is illuminated with white light is absorbing most of the incident light (Fig. 14-51, B). A red object such as a ripe tomato contains a pigment that absorbs all of the incident light except the red component which it reflects. Therefore the tomato appears red when illuminated with white light. It will also appear red when illuminated with red light. However, it will appear black if illuminated by blue or green light since the pigment absorbs all colours except red.

Usually a coloured object reflects, in addition to its main colour, portions of the colours that are next to it in the spectrum. Therefore a red object reflects some orange and a green object reflects some blue and yellow (see Figure 14-51, C and D).

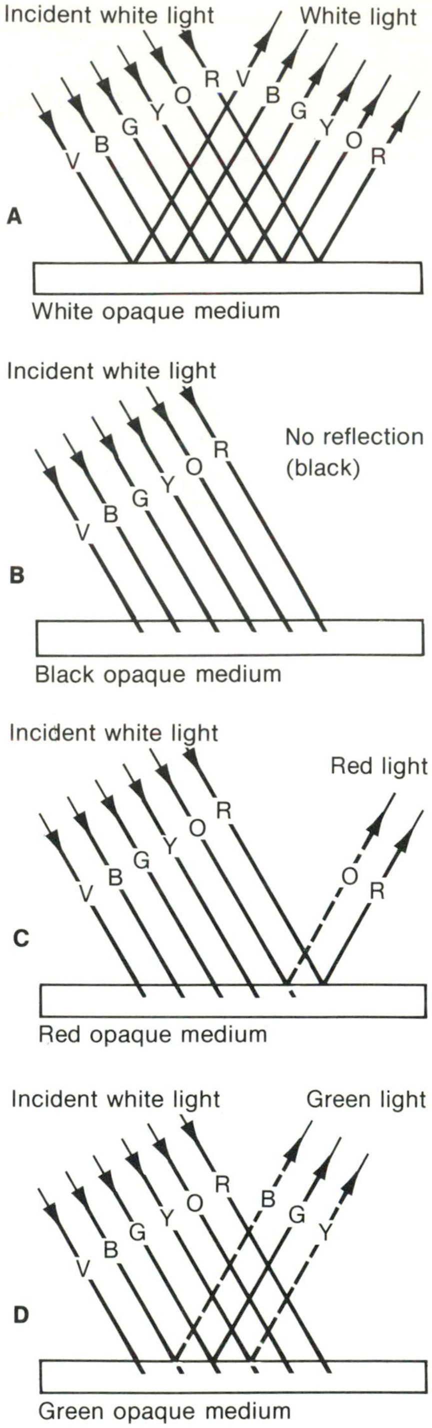

Figure 14-51 The subtractive colour theory can be used to account for the colour of opaque objects when they are illuminated with various colours of light.

The Colour of Transparent Objects

A transparent object that allows only one colour to pass through it is called a **colour filter**. The subtractive theory explains how colour filters work.

Colourless glass tramsmits all of the components of white light (Fig. 14-52, A). A red filter absorbs (subtracts) from white light all of the components except red, which it transmits (Fig. 14-52, B). However, unless the filter is a very expensive one, it will also transmit small amounts of the hues next to it in the spectrum. Thus the usual red filter transmits some orange light. A green filter absorbs from white light all of the components except green and portions of the yellow and blue components.

Subtractive Primaries and Subtractive Colour Mixing

In Section 14.30 you learned that, for *additive* colour mixing, the largest number of colours can be produced by *adding* red, green, and blue lights of various intensities. Much the same thing is true for *subtractive* colour mixing. The largest number of colours can be produced by controlling the same 3 colours, red, green, and

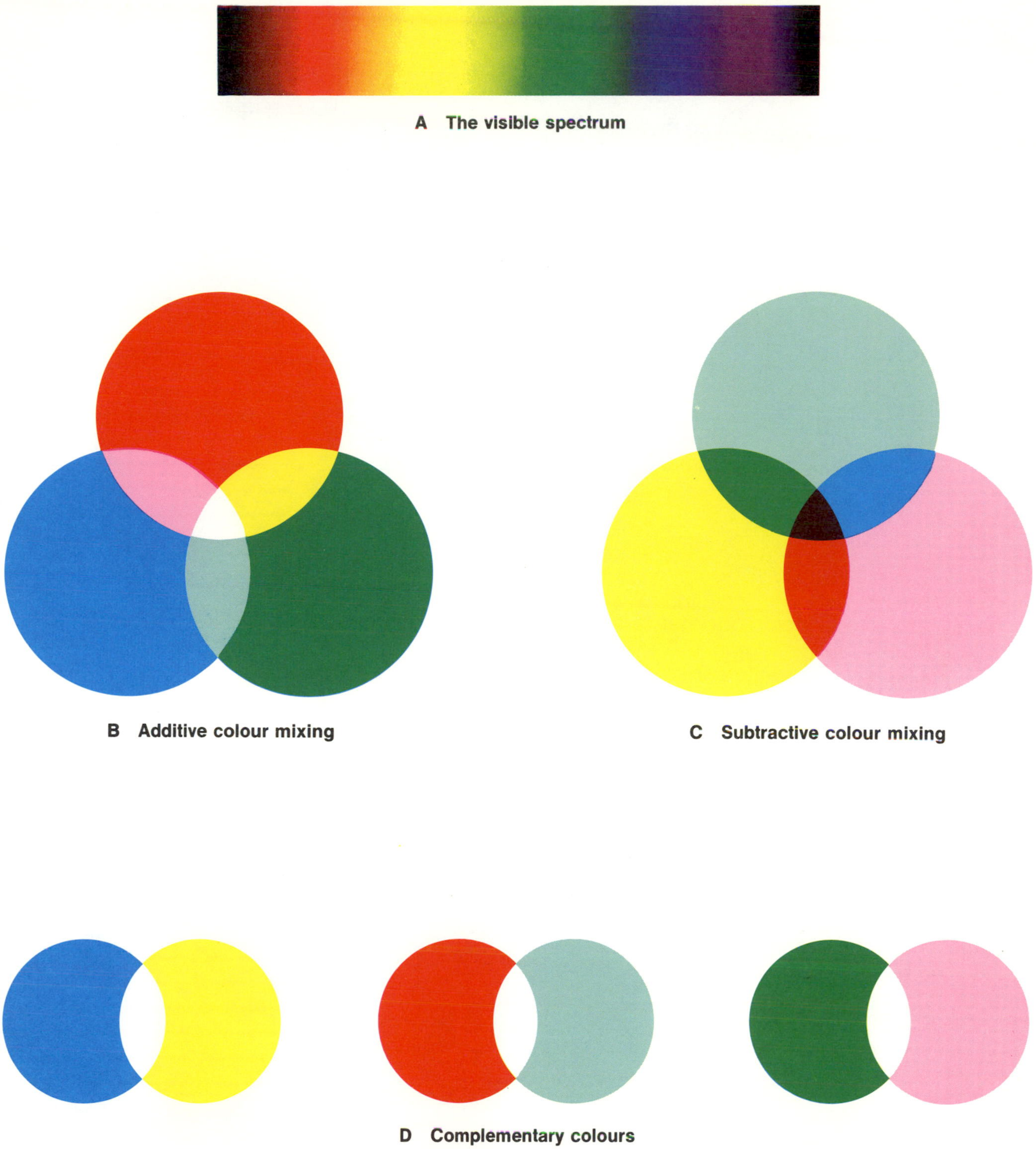

Figure 14-50 **Some aspects of colour theory.**

blue. However, red, green, and blue pigment colours are not the subtractive primaries. Rather, the pigment colours that control red, green, and blue are the subtractive primaries. These pigment colours control the red, green, and blue by absorbing or *subtracting* them from incident light. **Cyan** absorbs red and reflects blue and green; **magenta** absorbs green and reflects red and blue; **yellow** absorbs blue and reflects red and green. Therefore cyan, magenta, and yellow control red, green, and blue. As a result, they are called the **subtractive primaries** or the **primary pigment colours**. Most colours of dyes and paints can be made by mixing yellow, magenta, and cyan. However, no mixture of other pigments can produce yellow, magenta, and cyan.

Subtractive colour mixing is illustrated in Figure 14-50,C. Note that equal portions of yellow and magenta produce **red**, since red is the only colour not completely absorbed by either of these pigments. Similarly, equal portions of cyan and magenta produce **blue** and equal portions of cyan and yellow produce **green**. Red, green, and blue are the **secondary pigment colours**. Equal portions of the 3 subtractive primaries absorb all colours and produce **black**.

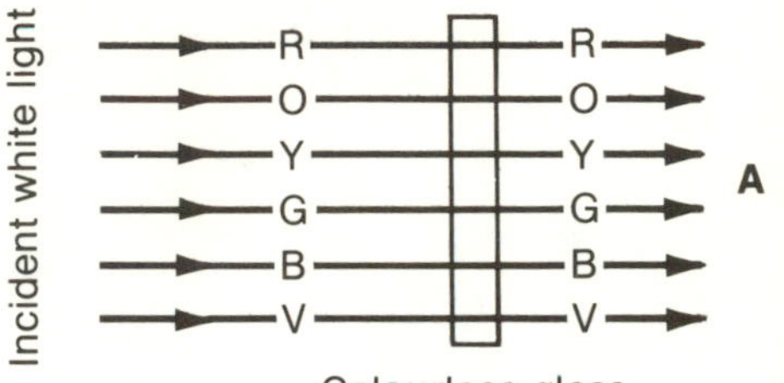

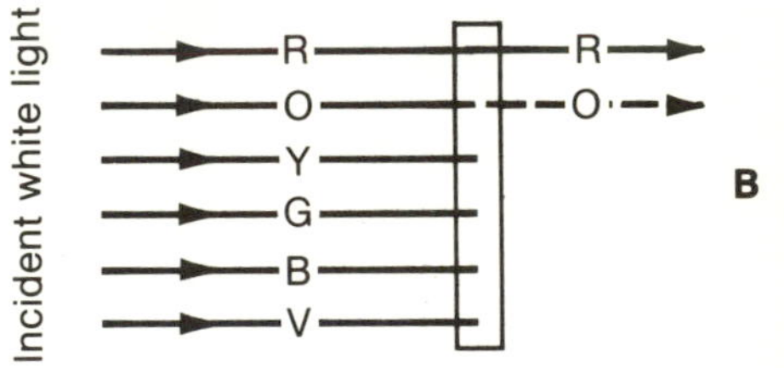

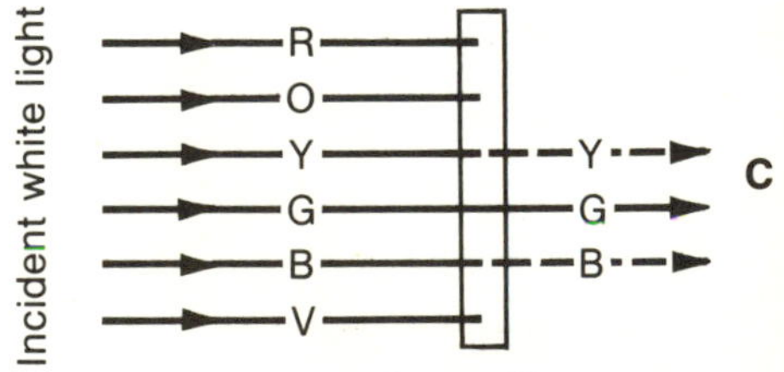

Figure 14-52 The subtractive theory can be used to account for the colours that are transmitted by colour filters.

Discussion

1. **a)** Use the Law of Conservation of Energy to explain what happens to the incident light that is absorbed by a coloured object.
   b) A black car and a white car are sitting in a parking lot on a sunny day. Which one do you think will have the hotter exterior? Why?
2. **a)** Why does a red rose appear red when it is illuminated with white light?
   b) What colour will the same rose appear to be if it is illuminated with only green light? Why?
3. Green plants contain a pigment called chlorophyll.
   a) What colour(s) of light does chlorophyll reflect if white light is shone on a green plant?
   b) What colour(s) of light does chlorophyll absorb if white light is shone on a green plant? What do you think happens to this absorbed energy?
   c) Is a green plant always green? Discuss.
4. **a)** What is a colour filter?
   b) Outline how the subtractive colour theory explains the operation of a colour filter.
5. **a)** Draw a diagram similar to those in Figure 14-52 to show how a blue filter affects incident white light.
   b) Use a diagram to explain how a yellow and a blue filter can be used together to allow only the green component of white light to be transmitted.
6. **a)** Serious photographers often use colour filters for black-

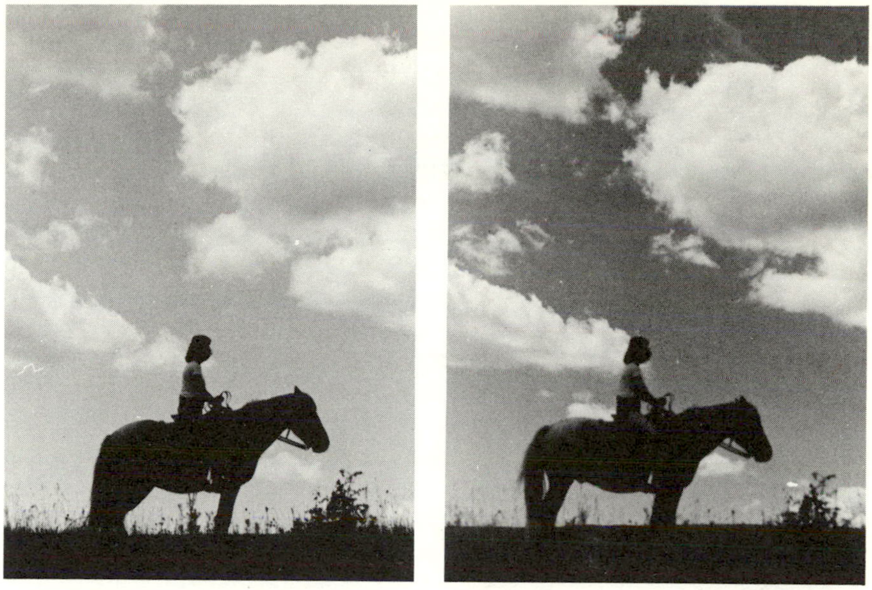

Figure 14-53 This scene was photographed without a filter (left) and, a few seconds later, with a filter (right).

and-white photography. Examine Figure 14-53. The photograph on the left was taken with no filter in front of the camera lens. The photograph on the right was taken a few seconds later with a red filter in front of the camera lens. Account for the apparent darkening of the sky.

b) Colour filters are also used by photographers for colour photography. For example, colour pictures taken at sunrise and sunset often have an undesirable orange appearance. To avoid this, a pale blue filter is placed over the lens of the camera. Explain how the filter eliminates the unwanted orange colour.

c) On overcast days colour pictures tend to have an unpleasant bluish appearance. What colour of filter could be used in front of the camera lens to eliminate this problem?

7. a) Name the 3 subtractive primaries. Explain why the words "subtractive" and "primary" are used to describe these pigments.

b) Name the 3 secondary pigment colours and explain how they are formed.

c) A person wants to colour the icing of a cake green. This person has no green food colouring but does have several other colours—cyan, magenta, blue, yellow, and red. What could the person do?

8. a) Figure 14-54 is called an additive colour triangle. It summarizes the additive colour theory by placing the additive primaries at the corners of the triangle and the secondary colours of light on the proper sides. Draw a subtractive colour triangle that summarizes the subtractive colour theory.

b) Compare the additive and subtractive colour theories in less than 150 words.

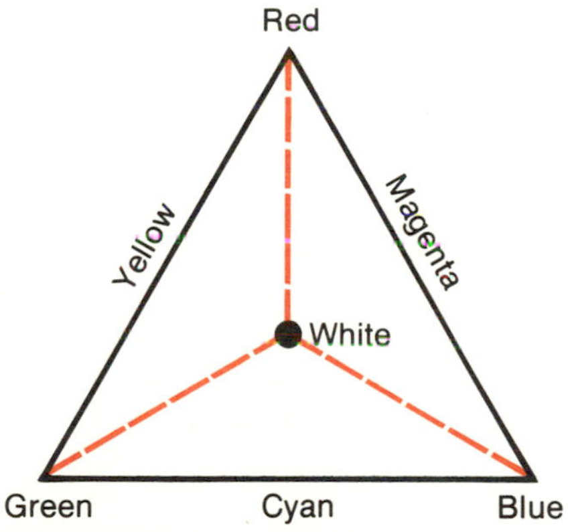

Figure 14-54 An additive colour triangle summarizes the additive colour theory. Draw a subtractive colour triangle that summarizes the subtractive colour theory.

Highlights

Light is that form of energy which the eye can detect. It can be formed from many other forms of energy in accordance with the Law of Conservation of Energy. Light travels in straight lines and will pass through a vacuum and transparent media. Its speed in a vacuum is 3×10^5 km/s. Its rectilinear propagation explains many phenomena, including eclipses and the operation of a pinhole camera. Three theories have been developed to explain the behaviour of light. Of these, the quantum theory is most widely accepted.

Light is reflected from both plane and curved surfaces in accordance with the two Laws of Reflection. The image in a plane mirror is erect, the same size as the object, virtual, and as far behind the mirror as the object is in front of it. The location and characteristics of the image in a curved mirror depend on the position of the object. Ray diagrams can be used to predict the location and characteristics.

Refraction can occur as light passes from one transparent medium into another of different optical density. The three Laws of Refraction summarize this behaviour. Many optical instruments depend on refraction and/or a related phenomenon called total internal reflection. Lenses are the most useful application of refraction. The ray diagrams for convex lenses bear many similarities to those for concave mirrors; likewise the ray diagrams for concave lenses bear many similarities to those for convex mirrors.

White light is composed of several spectral hues; most people can identify at least 6 hues in the spectrum. The Additive Colour Theory deals with the mixing of coloured lights. The additive primaries are red, green, and blue. The Subtractive Colour Theory deals with the mixing of pigments. The subtractive primaries are cyan, magenta, and yellow.

Index